Engineering Electromagnetics

Engineering Electromagnetics

Editor:

Harinirina Randrianarisoa, Ph.D.

ArclerPress

www.arclerpress.com

Engineering Electromagnetics

Editor: Harinirina Randrianarisoa, Ph.D.

© 2017 Arcler Press LLC
708 3rd Avenue, 6th Floor
New York
NY 10017
United States of America
www.arclerpress.com

ISBN: 978-1-68094-408-2

Library of Congress Control Number: 2016951053

About the Editor

Harinirina Randrianarisoa, Ph.D.

Hari Randrianarisoa obtained is Ph.D from Institute of Electronics, Microelectronics and Nanotechnology (IEMN) of the University of Science and Technology – Lille (France). His thesis focused for designing and manufacturing of thermal micro sensors. His interests span from on renewable energy especially solar energy. He has a strong knowledge in technology transfer field. He is currently working as a Business advisor for economic development organization in Québec City and as a consultant Scientist advisor.

Contents

List of Contributors

Qingyang Meng
Laboratory of Applied Research on Electromagnetics (ARE), Zhejiang University, Hangzhou 310027, China

Kuiwen Xu
Laboratory of Applied Research on Electromagnetics (ARE), Zhejiang University, Hangzhou 310027, China

Fazhong Shen
Laboratory of Applied Research on Electromagnetics (ARE), Zhejiang University, Hangzhou 310027, China

Bin Zhang
Laboratory of Applied Research on Electromagnetics (ARE), Zhejiang University, Hangzhou 310027, China

Dexin Ye
Laboratory of Applied Research on Electromagnetics (ARE), Zhejiang University, Hangzhou 310027, China

Jiangtao Huangfu
Laboratory of Applied Research on Electromagnetics (ARE), Zhejiang University, Hangzhou 310027, China

Changzhi Li
Department of Electrical and Computer Engineering, Texas Tech University, Lubbock, TX 79424, USA

Lixin Ran
Laboratory of Applied Research on Electromagnetics (ARE), Zhejiang University, Hangzhou 310027, China

Łukasz Januszkiewicz
Institute of Electronics, Lodz University of Technology, ul. Wólczańska 211/215, 90-924 Łódź, Poland

Paolo Di Barba
Department of Electrical, Computer and Biomedical Engineering, University of Pavia, via Ferrata 5, 27100 Pavia, Italy

Sławomir Hausman
Institute of Electronics, Lodz University of Technology, ul. Wólczańska 211/215, 90-924 Łódź, Poland

Zhenlong Xu
School of Mechatronics Engineering, Harbin Institute of Technology, Harbin 150001, China

Xiaobiao Shan
School of Mechatronics Engineering, Harbin Institute of Technology, Harbin 150001, China

Danpeng Chen
Zhejiang Guochen Robot Technology Co., Ltd., Hangzhou 311215, China

Tao Xie
School of Mechatronics Engineering, Harbin Institute of Technology, Harbin 150001, China

Chang-Woo Song
Department of Mechanical Engineering, Sogang University, 1 Shinsu-dong, Mapo-gu, Seoul 121-742, Korea

Seung-Yop Lee
Department of Mechanical Engineering, Sogang University, 1 Shinsu-dong, Mapo-gu, Seoul 121-742, Korea

Gabriele Gerardi
Department of Animal Medicine, Production and Health, University of Padova, Legnaro (PD), Italy

Antonella De Ninno
ENEA, C.R. Frascati, Department Fusion and Nuclear Technologies, Technologies for Safety and Health Division, Frascati (Roma), Italy

Vanni Ferrari
Department of Animal Medicine, Production and Health, University of Padova, Legnaro (PD), Italy

Sandro Mazzariol
Department of Comparative Biomedicine and Food Science, University of Padova, Legnaro (PD), Italy

Daniele Bernardini
Department of Animal Medicine, Production and Health, University of Padova, Legnaro (PD), Italy

Severino Segato
Department of Animal Medicine, Production and Health, University of Padova, Legnaro (PD), Italy

Yongxin Liu
Fujian Provincial Key Laboratory of Light Propagation and Transformation, Huaqiao University, Xiamen, China

Songjie Luo
Fujian Provincial Key Laboratory of Light Propagation and Transformation, Huaqiao University, Xiamen, China

Jixiong Pu
Fujian Provincial Key Laboratory of Light Propagation and Transformation, Huaqiao University, Xiamen, China

Zenghui Gao
Key Laboratory of Computational Physics, Yibin University, Yibin, China

Tran Vu La
Lab-STICC, UMR CNRS 6285, ENSTA Bretagne, Brest, France

Ali Khenchaf
Lab-STICC, UMR CNRS 6285, ENSTA Bretagne, Brest, France

Fabrice Comblet
Lab-STICC, UMR CNRS 6285, ENSTA Bretagne, Brest, France

Carole Nahum
French General Directorate for Armament (DGA), Paris, France

Helmi Ghanmi
Lab-STICC, UMR CNRS 6285, ENSTA Bretagne, Brest, France

Prerna Saxena
Department of Electronics and Communication Engineering, Visvesvaraya National Institute of Technology, Nagpur 440010, India

Ashwin Kothari
Department of Electronics and Communication Engineering, Visvesvaraya National Institute of Technology, Nagpur 440010, India

Di Tan
School of Transportation and Vehicle Engineering, Shandong University of Technology, Zibo 255049, China

Haitao Wang
School of Transportation and Vehicle Engineering, Shandong University of Technology, Zibo 255049, China

Qiang Wang
School of Transportation and Vehicle Engineering, Shandong University of Technology, Zibo 255049, China

Alain Vian
Université d'Angers, Campus du Végétal, UMR 1345 IRHS, CS 60057, SFR 4207 QUASAV, 49071 Beaucouzé Cedex, France

Eric Davies
Department of Plant and Microbial Biology, North Carolina State University, Raleigh, NC 27695, USA

Michel Gendraud
Université Blaise Pascal, 24 avenue des Landais, 63177 Aubière Cedex, France

Pierre Bonnet
Institut Pascal, Université Blaise Pascal, BP 10448, 63000 Clermont-Ferrand, France
CNRS, UMR 6602, 63171 Aubière, France

Ramón Díaz de León-Zapata
Instituto Tecnológico de San Luis Potosí, Avenida Tecnológico s/n, 78376 Soledad de Graciano Sánchez, SLP, Mexico
Universidad Autónoma de San Luis Potosí, Álvaro Obregón 64, 78000 San Luis Potosí, SLP, Mexico

Gabriel González
Universidad Autónoma de San Luis Potosí, Álvaro Obregón 64, 78000 San Luis Potosí, SLP, Mexico

Efrén Flores-García
Instituto Tecnológico de San Luis Potosí, Avenida Tecnológico s/n, 78376 Soledad de Graciano Sánchez, SLP, Mexico

Ángel Gabriel Rodríguez
Universidad Autónoma de San Luis Potosí, Álvaro Obregón 64, 78000 San Luis Potosí, SLP, Mexico

Francisco Javier González
Universidad Autónoma de San Luis Potosí, Álvaro Obregón 64, 78000 San Luis Potosí, SLP, Mexico

Charalampos Stergiou
Laboratory of Materials for Electrotechnics, Department of Electrical and Computer Engineering, Aristotle University, 54124 Thessaloniki, Greece
Laboratory of Inorganic Materials, Centre for Research and Technology Hellas, Thermi, 57001 Thessaloniki, Greece

Navid Mousavi
Young Researchers and Elite Club, Islamic Azad University, Najafabad Branch, Najafabad, Isfahan, Iran

Tohid Rahimi
Electrical and Computer Engineering Faculty, University of Tabriz, Tabriz, Iran
Electrical and Avionic Engineering Faculty, Graduate Malek Ashtar University, Iran

Homayoun Meshgin Kelk
Electrical Engineering Faculty, Tafresh University, Tafresh, Iran

Xiaowei Zuo
Key Laboratory of Electromagnetic Processing of Materials (Ministry of Education), Northeastern University, Shenyang 110004, China
School of Metallurgy, Northeastern University, Shenyang 110004, China

Congcong Zhao
Key Laboratory of Electromagnetic Processing of Materials (Ministry of Education), Northeastern University, Shenyang 110004, China
School of Materials Science and Engineering, Northeastern University, Shenyang 110004, China

Lin Zhang
Key Laboratory of Electromagnetic Processing of Materials (Ministry of Education), Northeastern University, Shenyang 110004, China
School of Metallurgy, Northeastern University, Shenyang 110004, China

Engang Wang
Key Laboratory of Electromagnetic Processing of Materials (Ministry of Education), Northeastern University, Shenyang 110004, China
School of Metallurgy, Northeastern University, Shenyang 110004, China

Yu Shao
Wireless Innovation Lab of Yunnan University, School of Information Science and Engineering, Kunming, Yunnan 650091, China

Jing Jing Yang
Wireless Innovation Lab of Yunnan University, School of Information Science and Engineering, Kunming, Yunnan 650091, China

Ming Huang
Wireless Innovation Lab of Yunnan University, School of Information Science and Engineering, Kunming, Yunnan 650091, China

Guo-hua Wang
Key Laboratory of Intelligent Computing & Signal Processing, Ministry of Education, Anhui University, Hefei 230039, China

Yu-fa Sun
Key Laboratory of Intelligent Computing & Signal Processing, Ministry of Education, Anhui University, Hefei 230039, China

Konstantin Vytovtov
Department of Physics, Electronics and Computing Systems, Oles Honchar Dnipropetrovsk National University, Gagarina 72, Dnipropetrovsk 49010, Ukraine

Said Zouhdi
Laboratoire de Génie Electrique de Paris, SUPELEC, 91192 Gif-sur-Yvette, France

Rostislav Dubrovka
Queen Mary University of London, UK

Volodymyr Hnatushenko
Department of Physics, Electronics and Computing Systems, Oles Honchar Dnipropetrovsk National University, Gagarina 72, Dnipropetrovsk 49010, Ukraine

Yi Cheng Zhong
Fundamental Science on Extreme High Frequency Laboratory, School of Electronic Engineering, University of Electronic Science and Technology of China, Chengdu 611731, China

Yu Jian Cheng
Fundamental Science on Extreme High Frequency Laboratory, School of Electronic Engineering, University of Electronic Science and Technology of China, Chengdu 611731, China

Yong Fan
Fundamental Science on Extreme High Frequency Laboratory, School of Electronic Engineering, University of Electronic Science and Technology of China, Chengdu 611731, China

Sree N. Koneru
Clinical Science and Engineering Research Laboratory, Binghamton University, Binghamton, NY, USA

Charles R. Westgate
Department of Electrical and Computer Engineering, Binghamton University, Binghamton, NY, USA

Kenneth J. McLeod
Clinical Science and Engineering Research Laboratory, Binghamton University, Binghamton, NY, USA

Weijie Xia
College of Electronic and Information Engineering, Nanjing University of Aeronautics and Astronautics, Nanjing 211100, China

Yuanyuan Qi
College of Electronic and Information Engineering, Nanjing University of Aeronautics and Astronautics, Nanjing 211100, China

Linlin Huang
College of Electronic and Information Engineering, Nanjing University of Aeronautics and Astronautics, Nanjing 211100, China

Xue Jin
College of Electronic and Information Engineering, Nanjing University of Aeronautics and Astronautics, Nanjing 211100, China

Maria Kogia
Brunel Innovation Centre (BIC), Brunel University, Cambridge CB21 2AL, UK

Liang Cheng
Brunel Innovation Centre (BIC), Brunel University, Cambridge CB21 2AL, UK

Abbas Mohimi
Brunel Innovation Centre (BIC), Brunel University, Cambridge CB21 2AL, UK
TWI Ltd., Granta Park, Great Abington, Cambridge CB21 6AL, UK

Vassilios Kappatos
Brunel Innovation Centre (BIC), Brunel University, Cambridge CB21 2AL, UK

Tat-Hean Gan
Brunel Innovation Centre (BIC), Brunel University, Cambridge CB21 2AL, UK
TWI Ltd., Granta Park, Great Abington, Cambridge CB21 6AL, UK

Wamadeva Balachandran
Brunel Innovation Centre (BIC), Brunel University, Cambridge CB21 2AL, UK

Cem Selcuk
Brunel Innovation Centre (BIC), Brunel University, Cambridge CB21 2AL, UK
TWI Ltd., Granta Park, Great Abington, Cambridge CB21 6AL, UK

Zhaoliang Chen
Research Center of Applied Electromagnetics, Nanjing University of Information Science and Technology, Nanjing 210044, China

Wen Geyi
Research Center of Applied Electromagnetics, Nanjing University of Information Science and Technology, Nanjing 210044, China

Ming Zhang
Huawei Technologies Co., Ltd., No. 360, Jiangsu Road, Binjiang District, Hangzhou 310052, China

Jun Wang
Huawei Technologies Co., Ltd., No. 360, Jiangsu Road, Binjiang District, Hangzhou 310052, China

R. Paramatti
INFN - Roma

X.G. Wang
ARC Centre of Excellence for Particle Physics at the Terascale and CSSM, Department of Physics, University of Adelaide, Adelaide SA 5005, Australia

A.W. Thomas
ARC Centre of Excellence for Particle Physics at the Terascale and CSSM, Department of Physics, University of Adelaide, Adelaide SA 5005, Australia

R.D. Young
ARC Centre of Excellence for Particle Physics at the Terascale and CSSM, Department of Physics, University of Adelaide, Adelaide SA 5005, Australia

Preface

The design of a device has long been a matter of experiences from the engineers in charge of the work. They were quickly faced with problems related with constraints. That is due to the increase data rates, operational frequencies and the miniaturization of devices to reduce their size and weight. With regards to engineering electromagnetics, this book discusses method and design to control the electromagnetic field effect in the systems. This literature proposed models using numerical electromagnetic calculation to simulate the behavior of device and to validate the efficiency and accuracy of the method which turns out to be highly efficient and accurate for the solution to electromagnetic problems. Also, to design a device, optimization methods that achieve the desired performance are discussed in this book. First chapter presents an approach on microwave imaging under oblique illumination. In second chapter, we discuss optimization of a WBAN system which uses wireless communication technology to monitor physiological and environmental parameters of firefighters and their exposure to hazardous environmental conditions. Third chapter explores on piezoelectric and electromagnetic conversion mechanisms, and fourth chapter aims to design and develop a solenoid for miniaturized segment robots with linear and turning motions. Fifth chapter covers on the effect of electromagnetic fields with the mg2+ cyclotron frequency and sixth chapter presents an overview of radially polarized random electromagnetic beam. Exploitation of electromagnetic models is highlighted in seventh chapter. The main aim of eighth chapter is to introduce the grey wolf optimization (GWO) algorithm to the electromagnetics and antenna community. Ninth chapter presents a study on the rollover characteristic of in-wheel-motor-driven electric vehicles considering road and electromagnetic excitation. In tenth chapter, we describe exposure devices, SAR determination methods, and biological responses observed after plant exposure to EMF. Eleventh chapter explores on evolutionary algorithm geometry optimization of optical antennas. Twelfth chapter examines the strong relation between grain size, porosity, density, and magnetic properties of Ni-Zn ferrite at high frequencies, with a view of forming certain material design directions. In thirteenth chapter, a new process is presented in order to design filters based on the leakage current and the frequency components of the DC link current. Influence of growth rate and magnetic field on microstructure and properties of directionally solidified Ag-Cu eutectic alloy is presented in fourteenth chapter. In fifteenth chapter, the electromagnetic simulation of graphene is introduced and some widely used computational electromagnetic methods for graphene modeling are reviewed. The advantages and drawbacks of each method are discussed and numerical examples of these methods are given to illustrate their performance and application. Sixteenth chapter presents a hybrid approach for fast computation of multiple incident angles electromagnetic scattering problems with compressive sensing and adaptive cross approximation. A theoretical study of the reflection of infrared radiation from antiferromagnets and M-type barium hexagonal ferrite using an attenuated total reflection geometry has been presented in seventeenth chapter. Eighteenth chapter will focus on how to model contacted balls and regular polyhedrons to simulate the packing particles. It can also be employed in the modeling of packing particles with complex shapes. Rectification of RF fields in load dependent coupled systems has been presented in nineteenth chapter. In twentieth chapter, we will fully take the target Radar Cross Section (RCS) feature into account to propose a novel RFAT echo generation method for missile-borne SAR. Twenty first chapter presents an overview on electromagnetic acoustic transducers and twenty second chapter presents a study on electromagnetic acoustic transducers. Twenty third chapter presents an approach on design options for the upgrade of the CMS electromagnetic calorimeter and last chapter reviews on electromagnetic contribution to charge symmetry violation in parton distributions.

Editor
Harinirina Randrianarisoa, Ph.D.

INTRODUCTION

Electromagnetics is a branch of physics or electrical engineering which is used to study the electric and magnetic phenomena. The electric and magnetic fields are closely related to each other. The comprehensive study of characteristics of electric, magnetic and combined fields, is nothing but the Electromagnetic Engineering. Many practical procedures and devices, not only related to electrical and electronics but related to other branches such as medical, chemical, agricultural are based on the basics of the engineering electromagnetics.

Electromagnetic (EM) is a branch of physics or electrical engineering in which electric and magnetic phenomena are studied. Electromagnetics (EM) may be regarded as the study of the interactions between electric charges at rest and in motion. It entails the analysis, synthesis, physical interpretation, and application of electric and magnetic fields.

EM principles find applications in various allied disciplines such as microwaves, antennas, electric machines, satellite communications, bioelectromagnetics, plasmas, nuclear research, fiber optics, electromagnetic interference and compatibility, electromechanical energy conversion, radar meteorology, and remote sensing.

EM devices include transformers, electric relays, radio/TV, telephone, electric motors, transmission lines, waveguides, antennas, optical fibers, radars, and lasers. The design of these devices requires thorough knowledge of the laws and principles of EM

Electromagnetics involves the macroscopic behavior of electric charges in vacuum and matter. This behavior can be accurately characterized by the Lorentz force law and Maxwell's equations, which were derived from experiments showing how forces on charges depend on the relative locations and motions of other charges nearby. Additional relevant laws of physics include Newton's law, photon quantization, and the conservation relations for charge, energy, power, and momentum. Electromagnetic phenomena underlie most of the "electrical" in "electrical engineering" and are basic to a sound understanding of that discipline.

Electromagnetic field theory is a discipline concerned with the study of charges, at rest and in motion, that produce currents and electric-magnetic fields. It is, therefore, fundamental to the study of electrical engineering and physics, and indispensable to the understanding, design, and operation of many practical systems using antennas, scattering, microwave circuits and devices, radio-frequency and optical communications, broadcasting, geosciences and remote sensing, radar, radio astronomy, quantum electronics, solid-state circuits and devices, electromechanical energy conversion, and even computers. Circuit theory, a required area in the study of electrical engineering, is a special case of electromagnetic theory and it is valid when the physical dimensions of the circuit are small compared to the wavelength. Circuit concepts, which deal primarily with lumped elements, must be modified to include distributed elements and coupling phenomena in studies of advanced systems. For example, signal propagation, distortion, and coupling in micro strip lines used in the design of sophisticated systems (such as computers) can be accounted for properly only by understanding the electromagnetic field interactions associated with them.

The study of electromagnetics includes both theoretical and applied concepts. The theoretical concepts are described by a set of basic laws formulated primarily through experiments conducted during the nineteenth century by many scientists—Faraday, Ampere, Gauss, Len; Coulomb, Volta, and others. They were then combined into a consistent set of vector equations by Maxwell. These are the widely acclaimed *Maxwell's equations.* The applied concepts of electromagnetics are formulated by applying the theoretical concepts to the design and operation of practical systems.

Maxwell's Equations

In general, electric and magnetic fields are vector quantities that have both magnitude and direction. The relations and variations of the electric and magnetic fields, charges, and currents associated with electromagnetic waves are governed by physical laws, which are known as Maxwell's equations. These equations, as we have indicated, were arrived at mostly through various experiments carried out by different investigators, but they *were* put in their final form by James Clerk Maxwell, a Scottish physicist and mathematician. These equations can be written either in differential or in integral form.

Differential Form of Maxwell's Equations

The differential form of Maxwell's equations is the most widely used representation to solve boundary-value electromagnetic problems. It is used to describe and relate the field vectors, current densities, and charge densities at *any point in space at any time.* For these expressions to be valid, it is assumed that the field vectors are *single-valued, bounded, continuous* functions of position and time and exhibit *continuous derivatives.* Field vectors associated with electromagnetic waves possess these characteristics except where there exist abrupt changes in charge and current densities. Discontinuous distributions of charges and currents usually occur at interfaces between media where there are discrete changes in the electrical parameters across the interface. The variations of the field vectors across such boundaries (interfaces) are related to the discontinuous distributions of charges and currents by what are usually referred to as the *boundary conditions.* Thus a complete description of the field vectors at any point (including discontinuities) at any time requires not only Maxwell's equations in differential form but also the *associated boundary conditions.*

Eleetmmagnetics entails the study of electric and magnetic phenomena in free space and material media. Electromagnetics comprises three branches: electrostatics concerning static electric fields, magneto statics concerning static magnetic fields, and electrodynamics concerning time-varying electric and magnetic fields. Electromagnetic theories are based on electromagnetic models that consist of sources such as electric charges and currents, basic quantities such as electric and magnetic field intensities, rules of operations such as vector algebra and coordinate systems, and fundamental laws such as Coulomb's law and Maxwell's equations. The first two chapters of the text discuss the rules of operation, including vector algebra, coordinates systems, and vector calculus. The basic quantities of cicaromagnctics are either scalars or vectors. A scalar is completely specified by a magnitude, as with electric potentials, whereas a vector is specified by a magnitude and a direction, as with electric forces. In sinusoidal steady-mate conditions in which all electromagnetic quantities vary sinusoidally with time, a complex number can be conveniently used to represent the sinusoidal time-dependence, regardless of whether or not the quantity is a vector. In general, electromagnetic quantities are functions of space and time, whereas the quantities of electrostatics or magneto statics vary with space coordinates only. Therefore, we usually deal with scalar and vector fields in electromagnetics. Mathematical rules and techniques are essential for not only expressing electromagnetic concepts in compact forms, but also for constructing electromagnetic models of problems for solutions. Vector algebra, vector calculus, and coordinate systems are three basic mathematical tools for electromagnetics. Vector algebra concerns vector operations such as vector addition, scaling, and scalar and vector products. Vector calculus deals with derivatives and integrals of a scalar or a vector field, some of which are formulated into vector operators called the gradient, divergence, and curl. A coordinate system allows us to express geometric elements such as points, lines, surfaces, and volumes in terms of mathematical equations. Since physical quantities and laws are independent of the coordinate system, we choose a coordinate system that best fits the geometry under consideration and thus facilitates the electromagnetic model of a problem. Although the definition of a vector is straightforward, we should be careful when we deal with a spatial distribution of vectors, called a vector field, because different types of vectors involved in a vector field may easily draw us into confusion. The position vector always starts at the origin and ends at a point in space, defining the point located at the terminal point of the vector. A distance vector starts at a point and ends at another, defining the distance between the two points and the direction from the initial to the terminal point. The distance and the direction are the quantities belonging to the terminal point of the vector. A vector in a vector field is specific to a point in space. Thus it would be meaningless if the vector is displaced to another point, because it represents the physical quantity observed at the original point. Its magnitude and direction are the quantities belonging to the initial point of the vector. In this case, the space coordinates of the terminal point have no significance. Base vectors arc a set of three mutually orthogonal unit vectors whose directions generally vary with position.

What is Electromagnetic Simulation?

Using approximate computations to Maxwell's equations and Faraday's law, electromagnetic simulations are models of electromagnetics and their effects on the environment and physical structures about them. An electromagnetic simulation can be used to aim a satellite antenna in the right direction for maximum channels and clarity and judge its performance or to determine wave propagation when not in free space. These simulations can aid in the efficient design of computer chips and point to how to improve performance in major electronics by locating the incompatibilities of components within them. Electromagnetic radiation being picked up and scattered and then absorbed by small particles is used in simulations for science projects at the European Organization for Nuclear Research (CERN) laboratories for their particle accelerators projects. Electromagnetic simulation programs are also being used as tools in college physics labs to teach more effectively as students receive hands-on experience with solving problems using them. Solving Maxwell's equations at each point in an orthogonal or non-orthogonal grid is one of the avenues to use grids to discretize space by creating a topological survey of the space. Solving these equations in an electromagnetic simulation often reveals problems in computer memory and power as they usually can only be done on supercomputers by time-stepping for each instant of time throughout a whole domain, solving the Maxwell equations as they go or split-stepping using time iterations and fast Fourier transforms. In fluid mechanics, the boundary method or "method of moments" (MoM) can be applied to solve engineering problems, acoustics, and electromagnetics. This focuses calculations only on the border areas of a space rather than the volume values at every time step of the entire space.

A kitchen microwave is analogous to what is known as a Faraday cage, which illustrates how an electromagnetic simulation model might be useful in electromagnetic protection. Electrical currents can be blocked by metal walls or other such shielding devices whereas magnetic currents merely can be moved around the obstruction. In Faraday's cage, when the walls of the cage are grounded, an electric current's path is disturbed by electrons acting as electric charge carriers in a mesh pattern and compensate for the field; this cause the electric current to dissipate. Just as the mesh screen on the front of a microwave door blocks microwaves from escaping the device because microwaves are larger than the tiny holes in the mesh, an electromagnetic mesh simulation can design good protective shielding from electrical currents.

An electromagnetic simulation method that solves Maxwell equations by cycling through an electric field for one instant and then cycling through a magnetic field for the next instant and repeatedly alternating over and over again is known as the finite-difference time-domain (FDTD) method for producing simulations. EM wave interaction with material structures engineering problems have been solved by this method more than any other in the US since around 1990. It is used in solving for radar signature technologies, wireless technologies and biomedical imaging, just to name a few of its applicable uses.

Wave modeling for electromagnetic simulation and analysis of circuits can be performed using the partial element equivalent circuit (PEEC) three-dimensional (3-D) full-wave modeling method. Integral equations are interpreted as Kirchhoff's voltage law and, using PEEC, are applied to a PEEC cell which gives the 3-D geometries solution of a complete circuit, allowing additional circuits to be piggy-backed onto the direct current design. Using models such as this in electromagnetic simulation saves time and money costs in the manufacture of integrated circuits.

College physics departments are starting to make use of video games designed to give students lessons via electromagnetic simulation to visually depict to the students the phenomena of physics representations. This can help students gain a better grasp of concepts and afford their brains experiences that reveal to them weaknesses in their own understanding and the steps to take to strengthen these. Both students and instructors alike have found that both faster and more in-depth learning can be facilitated using real-world examples of physics concept solving by means of electromagnetic simulation software.

What is an Electromagnetic Motor?

An electromagnetic motor is a machine that uses the magnetic forces produced by a wire with flowing electricity to power the movement of the motor. All electric motors operate on electromagnetic principles. Types of electric motors include alternating current (AC) motors and direct current (DC) motors. AC motors use wall current and DC motors use a battery as a source of power and magnetic forces to spin the motor. The electromagnetic motor is common in nearly every household, frequently found in popular household items such as fans, pool pumps, air conditioners, washing machines, and electric toothbrushes.

The rotating motion of an electromagnetic motor is based on the forces observed in magnetic poles. A magnet hung from a string in the center will naturally rotate to face one end to the north and one end to the south. The magnetic poles located on each end of the magnet attract if the poles are opposite and repel if the poles are alike. When two North Pole magnets are brought within the fields of the magnets' force, the magnets will push away from each other. If one North Pole and one South Pole are brought together, they will attract and stick to each other.

Though electromagnets used for science experiments in class are typically wires wrapped around a small stick of iron, a properly coiled conductive wire can become an electromagnet when supplied with current. The force becomes amplified when the wire is coiled. Magnetic force from a wire coil with current becomes stronger yet when the coil is wrapped around an iron magnet.

Electric current running through a wire produces a magnetic field, so electromagnetic force is present in wires with current running through them. Consequently, a wire with electric current running through it has an inherent magnetic force. When that wire is coiled into multiple parallel loops, it takes on the properties of a magnet when current runs through it. Wires not intended for use as electromagnets are often shielded to dampen the magnetic force and reduce its interference with other nearby electronics.

Some types of electromagnetic motors can be propelled with no electrified iron-core magnet, using only the magnetic force from carefully arranged coiled wires. Usually, electromagnetic motors have one or more iron-core electromagnets powering the movement of the motor. Though an electromagnet created with a magnetic core is more powerful, it also draws more electricity, putting more demand on the power source and draining batteries more quickly. High-powered electromagnetic motor assemblies sometimes use more than one electromagnet in the motor to give the motor a controlled power boost.

CHAPTER 1

MICROWAVE IMAGING UNDER OBLIQUE ILLUMINATION

Qingyang Meng, Kuiwen Xu, Fazhong Shen, Bin Zhang, Dexin Ye, Jiangtao Huangfu, Changzhi Li and Lixin Ran

Keywords: Inverse scattering problem; microwave imaging; SOM; oblique incidence

INTRODUCTION

The inverse scattering problem has been a popular topic in the microwave society for many years. Solving inverse scattering problems leads to diverse microwave imaging applications, such as non-destructive detection [1,2,3,4,5], medical examination [6,7,8,9,10,11,12,13,14], and through-wall imaging [15,16,17,18,19].

So far, various algorithms for inverse scattering problems have been studied [20,21,22,23,24,25,26,27,28,29,30,31,32,33,34,35]. Examples include the distorted Born iterative method [20], contrast source inversion method [21], the Gauss-Newton-type method [22], and stochastic-type inversion methods [23,24], and the recently proposed subspace-based optimization method (SOM) [27]. Among these algorithms, the SOM has been attracting many interests due to its unique advantages of fast convergence and insensitivity to noise [27,28,29,30,31,32,33,34,35]. The SOM performs spectrum analysis on the scattering operator, mapping from the induced source to the scattered field. Using the spectral property of the scattering operator, the SOM first determines a part of the induced source and then obtains the rest through optimization. Such a procedure simplifies the nonlinear optimization problem, stabilizes the solution, and accelerates the convergence speed. The SOM has been proposed for solving two-dimensional (2D) inverse scattering problems in both the transverse magnetic (TM) [27] and the transverse electric (TE) [28] scenarios. Moreover, it was extended to solve inverse scattering problems for perfectly electric conducting scatterers [29,30] and three-dimensional (3D) inverse scattering problems [31]. Meanwhile, extended SOM-based methods, such as TSOM [32], FFT-TSOM [33], and MR-FFT-TSOM [34], have been proposed to reduce the computational complexity, regularize the nonlinear inverse problem and improve the quality of imaging. Both numerical simulation and experimental results have validated the effectiveness of the SOM [33,35]. In recent studies, there are also results concerning "virtual experiments frameworks" [36,37]. Using the linear relationship between the incident and the scattered field, virtual experiments can be designed and special convenient conditions can be enforced. Thus, advantages can be gained from such knowledge. The inverse scattering technique is also widely used in ground penetrating radars (GPRs). GPRs are non-destructive imaging systems that can provide images of the subsurface. An overview of the microwave imaging strategies can be found in [38].

Citation: Meng, Q.; Xu, K.; Shen, F.; Zhang, B.; Ye, D.; Huangfu, J.; Li, C.; Ran, L. Microwave Imaging under Oblique Illumination.Sensors 2016, 16, 1046. doi:10.3390/s16071046.

Apart from inversion algorithms, different experimental setups for measuring scattered fields have also been proposed [39,40,41,42,43], based on which experiments were conducted to test the performance of such algorithms. These systems can be categorized into two typical types. Systems in [39,40] consist of a transmitting antenna, a receiving antenna, and some mechanical turning device, as described in Figure 1a. Usually, the transmitter is fixed at an azimuth angle and the receiver is rotated in the horizontal plane to measure the scattered fields in different directions. The scatterers can also be turned by a turntable, which is equivalent to turning the transmitter in the horizontal plane. Using such a setup, researchers are able to obtain multiview scattered data of targets under test. When this setup is used to measure 2D targets under normal far-field incidence, the receiving antenna could interfere, or even block the incident field, producing errors in the incidence modeling. Being a mechanical turning system, it also suffers from a long measurement period.

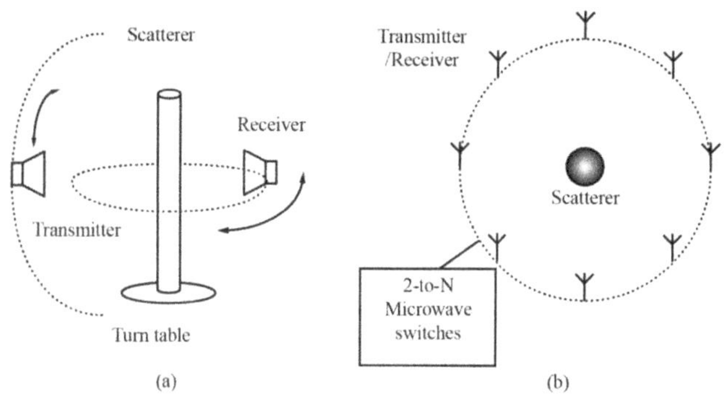

Figure 1. Typical setups for inverse scattering problem based microwave imaging. (**a**) Mechanical turning system; and (**b**) an electronically-controlled system with a 2-to-N microwave switching network.

Systems in [41,42,43] consist of a fixed antenna array, whose elements are uniformly placed on the perimeter of a circle and a complicated bi-directional microwave switch matrix, as described in Figure 1b. The switch matrix connects the elements to the measurement ports of microwave instruments, such that each antenna element can be switched to work in either transmitting or receiving mode. The advantage of such a setup is that the whole system does not have any mechanical turning device. Instead, it eliminates the complexity of physically turning the scatterers and antennas by electronic control and system calibration. So far, this kind of system is mainly used in near-field measurements.

In this paper, we experimentally investigate the performance of conventional SOM for 2D objects when it was applied to a setup designed for oblique incidence, as shown in Figure 2. This setup is able to combine the merits of the existing setups by using only one transmitting antenna and reducing the switching network to a 1-to-N one. In this case, the system calibration can be simplified, the transmitting antenna can be conveniently placed to ensure a plane wave incidence, and the detecting range can be independently optimized. Obviously, improved system calibration and reduced errors in incidence modeling and detected data will help to obtain better imaging results. However, the precondition to employ this setup is that the oblique incidence should not cause a notable loss of imaging quality.

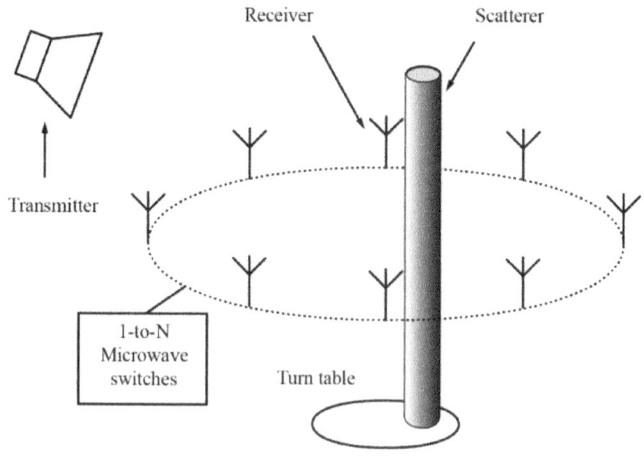

Figure 2. Proposed experimental setup with an oblique incidence and a 1-to-N microwave switching network.

In this paper, we investigate the application of SOM for this specific setup, focusing on the impact of the imaging quality due to the oblique incidence. In order to avoid the use of dual-polarized receiving antennas, which will introduce technical difficulty in system integration and bring additional calibration, the TM-polarized oblique incidence is used, and only the TM-polarized scattering is detected. Analytical, simulation and experimental results show that for 2D objects, neglecting the TE-polarized (cross-polarization) scattering will not cause a notable loss of information and, therefore, the quality of microwave imaging is not degraded compared with the case under normal illumination.

This paper is organized as follows. Section 2 presents the modeling of the forward scattering problem and the inversion algorithm for 2D objects under oblique microwave illumination. To validate the proposed algorithm, numerical simulation and experimental measurements are conducted in Section 3 and Section 4, respectively. Finally, a conclusion is drawn in Section 5.

ALGORITHMS

In this Section, we focus on the modeling of the forward and inverse problem for the scattering of a 2D dielectric object under oblique incidence, as described in Figure 3a. So far, most of the studies on the inverse scattering problems are focused on 2D or 3D scenarios under normal incidence [20,21,22,23 ,24,25,26,27,28,29,30,31,32,33,34,35,36,37,38,39]. Research on imaging 2D objects under oblique incidence are rarely studied. Inverse scattering problems under oblique incidence can be considered as reduced 3D problems. However, in engineering practice where the scatterers can be approximated as 2D objects [44], modeling the scatterers with 3D meshes is much more complicated and makes the computation more time and memory consuming. In this case, 2D approximation can potentially lead to some advantages in practical applications.

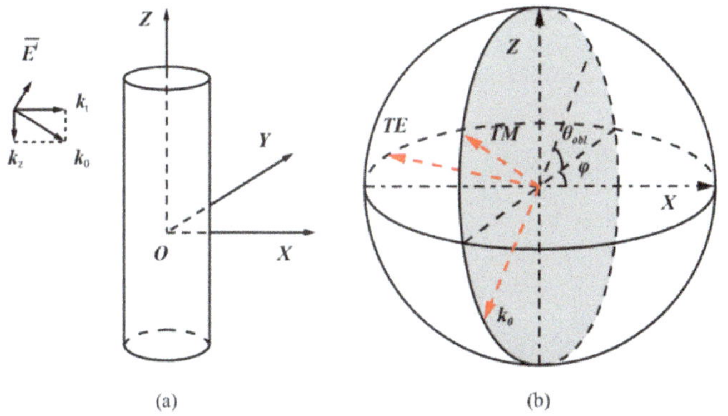

Figure 3. Problem description. (**a**) Analysis model of a two-dimensional object under oblique incidence; and (**b**) coordinate definition. It should be noted that the *TM* and *TE* components of the electric field and the wave vector k_0 are orthogonal to each other.

Forward Problem

There are many methods to obtain the solutions of forward problems [45,46,47,48,49]. Here we extend the method proposed in [45], using the inverse matrix, instead of the conjugate gradient method, to solve the equation we derived from the model of the forward scattering problem.

Consider the general equation that describes the scattering:

$$\overline{E}(\overline{r}) = \overline{E}^i(\overline{r}) - j\omega\overline{A} - \nabla\Phi \tag{1}$$

where $\overline{E}$ and $\overline{E}^i$ denote the total and incident electric fields. According to Equation (7) in [45], for an electric oblique incidence with wave number k_z in the axial direction, as shown in Figure 2, the solution of the electric field for a 2D object is:

$$\overline{E}(x,y) = \overline{E}^i(x,y) + k_0^2 \iint \overline{\overline{G}}(k_t\rho) \cdot \{\overline{\overline{\chi}}(x',y') \cdot \overline{E}(x',y')\}\, dx'dy' \tag{2}$$

where $\overline{\overline{G}}(k_t\rho)$ is known as the dyadic Green's function.

For a scatter made of an isotropic medium, when the incidence is normal and axially polarized, Equation (2) turns into a single, uncoupled integral equation since the total electric field $\overline{E}$ only has the axial component. However, in the oblique incidence case, $\overline{E}$ consists of three components and Equation (2) should be considered as three coupled integral equations.

To numerically solve the integral equations, a discretization procedure is needed. Based on [45], a set of algebraic equations can be obtained as:

$$\overline{E} = \overline{E}^i + \overline{\overline{G}}_D \cdot \overline{\overline{\chi}} \cdot \overline{E} \tag{3}$$

The domain of interest is discretized into $N = M \times M$ square cells. In Equation (3), $\overline{E}$ and $\overline{E}^i$ are both $3N \times 1$ vectors, denoting the total and incident electric fields, respectively. $\overline{\overline{G}}_D$ is a $3N \times 3N$ matrix. The elements of the discrete Green's function, $\overline{\overline{G}}_D$, can be estimated according to [45]. It should be noted that in [45], a numerical integration procedure is recommended for the estimation of $\overline{\overline{G}}_D$. Nevertheless, the computational cost of our problem does not increase much since $\overline{\overline{G}}_D$ does not change when solving the inverse problem. $\overline{\overline{\chi}}$ is a $3N \times 3N$ diagonal matrix. The first N diagonal elements of $\overline{\overline{\chi}}$ are given by $\chi_m = \varepsilon_m - 1$, where ε_m denotes the relative permittivity in cell m. The other diagonal elements of $\overline{\overline{\chi}}$ are given by $\chi_{m+2N} = \chi_{m+N} = \chi_m$.

Once $\overline{E}^i$, $\overline{\overline{G}}_D$, and $\overline{\overline{\chi}}$ are successfully constructed, $\overline{E}$ could be solved through a variety of methods. After the total electric field $\overline{E}$ is obtained, the scattered field is given by:

$$\overline{E}^{sca} = \overline{\overline{G}}_S \cdot \overline{\overline{\chi}} \cdot \overline{E} \tag{4}$$

where $\overline{E}^{sca}$ is a $3N_s \times 1$ vector that denotes the scattered field at N_s observation points. $\overline{\overline{G}}_s$ is a $3N_s \times 3N$ matrix. The estimation of the elements of $\overline{\overline{G}}_s$ is similar to that of $\overline{\overline{G}}_D$, but the Green's function is evaluated between N cells and the observation points.

Until now, the electric fields are described in a Cartesian coordinate system. It is more straightforward to represent the scattered electric field by the TE and TM components. Details of the polarization definition are described in Figure 3b. It should be noted that the TM and TE components of the electric field, as well as the wave vector k_0, are orthogonal to each other. The TE-TM frame is used because TE and TM components are orthogonal. This makes it clearer in the simulation part when comparing among the results inversed with different components of scattered fields. A similar definition is also used in [46]. The transformation of coordinates is thus:

$$\begin{bmatrix} E_{TE}^{sca} \\ E_{TM}^{sca} \end{bmatrix} = \begin{bmatrix} \sin(\phi) & -\cos(\phi) & 0 \\ 0 & 0 & 1/\cos(\theta_{obl}) \end{bmatrix} \cdot \begin{bmatrix} E_x^{sca} \\ E_y^{sca} \\ E_z^{sca} \end{bmatrix} \tag{5}$$

Here E_{TE}^{sca} and E_{TM}^{sca} denote the TE and TM components of the scattered electric fields, respectively. To combine this transformation with the derived algorithm, both sides of Equation (4) are left multiplied by the transformation matrix, which is equivalent to making a transformation to $\overline{\overline{G}}_s$. In the discussion below, we consider the scattered fields described by the TM and TE components without changing the form of the equations.

Inverse Problem

In this section, we discuss the inversion algorithm suitable for the imaging of 2D objects under oblique incidence based on the SOM.

The main idea of SOM is to determine a part of induced current by analyzing the spectral properties of the mapping from the induced current to scattered fields, and then obtain the rest part through optimization. Usually, the optimization procedure is to construct a cost function, and use a proper method to minimize it [27].

In our work, we construct the cost function following the method proposed in [27], i.e., using the singular value decomposition technique and choosing a truncation value L. It can be given by:

$$f(\overline{\alpha}, \overline{\overline{\chi}}) = \sum_{p=1}^{N_i} \left(\| \overline{\overline{G}}_s \cdot \left(\overline{I}_p^s + \overline{\overline{V}}^n \cdot \overline{\alpha}_p \right) - \overline{E}_p^{sca} \|^2 / \| \overline{E}_p^{sca} \|^2 + \| \overline{\overline{A}} \cdot \overline{\alpha}_p - \overline{B}_p \|^2 / \| \overline{I}_p^s \|^2 \right) \tag{6}$$

where N_i is the number of incidences, $\overline{\overline{V}}^n$ is composed of the last $3N - L$ right singular vectors, and $\overline{\alpha}_p$ is a $3N - L$ dimensional vector. $\overline{I}_p^s$ and $\overline{E}_p^{sca}$ denote the deterministic current and scattered field due to the p-th incidence, $\overline{\overline{A}} = \left(\overline{\overline{V}}^n - \overline{\overline{\chi}} \cdot \overline{\overline{G}}_D \cdot \overline{\overline{V}}^n\right)$ and $\overline{B}_p = \overline{\overline{\chi}} \cdot (\overline{E}_p^i + \overline{\overline{G}}_D \cdot \overline{I}_p^s) - \overline{I}_p^s$. Various methods can be used to minimize this cost function. In our work, the method proposed in [27] is used, i.e., updating the coefficients $\overline{\alpha}_p$ through the conjugate gradient method and then updating the contrast $\overline{\overline{\chi}}$ by solving a quadratic minimization problem.

In the 2D inverse scattering problem under normal TM incidence, $\overline{\overline{G}}_s$ is a $N_s \times N$ matrix; while under oblique incidence, $\overline{\overline{G}}_s$ is a 2Ns×3N matrix if the scattered field is decomposed into TM and TE components. $\overline{\overline{G}}_s$ can be written as $\overline{\overline{G}}_s = \left[\overline{\overline{G}}_{s,TM}; \overline{\overline{G}}_{s,TE}\right]$, where $\overline{\overline{G}}_{s,TM}$ is a Ns×3N matrix which denotes the mapping from the induced current to the TM component of the scattered field, and $\overline{\overline{G}}_{s,TE}$ is a Ns×3N matrix which denotes the mapping from the induced current to the TE component of the scattered field. In the following discussion about the influence of neglecting some components of the scattered fields, when full polarization data is used, $\overline{\overline{G}}_s$ is considered; when only TM polarization data is used, only $\overline{\overline{G}}_{s,TM}$ is considered; when only TE polarization data is used, only $\overline{\overline{G}}_{s,TE}$ is considered.

It should be noted that an oblique incidence may affect the imaging quality due to the fact that the main variable in the Green function is ktρ, which is different from k0·ρ in the normal incidence scenario [45]. Unlike k_0, k_t changes with the angle of oblique incidence. The change of the Green function and the incident field may affect the imaging performance.

NUMERICAL SIMULATIONS

In this Section, we present numerical simulations to test the performance of the SOM when it is applied to the proposed setup with oblique incidence.

The setup for the simulation is shown in Figure 4. Six TM-polarized plane waves whose propagation directions are oblique and equally spaced on a circle are used as incidences, and 24 linearly-polarized antennas placed on a horizontal circle are used to detect the scattered field. The computation domain is set as a square area of 25×25 cm^2, which is discretized into a 40×40 mesh grid. In the simulations, the frequency of the incident wave is 2.4 GHz, whose corresponding wavelength λ in free space is 12.5 cm. The background is set as air. The scatterer is a dielectric cylinder whose relative permittivity is 3. The radius of the cylinder is 0.2λ, i.e., 2.5 cm. The cylinder is firstly positioned at the coordinate (0, 0) and later at (5cos30°, −5sin30°). The units are both cm. In the inversion procedure, the simulated scattered field is introduced with an additive white Gaussian noise, such that the signal to noise ratio (SNR) of the scattered field is 10 dB. The singular value truncation number L is chosen as 25. It is chosen so that singular values can notably change the slope in the spectrum. Note that the spectrum varies with the angle of oblique incidence, so the smallest "knee" is chosen from scenarios with different incident angles. In our numerical simulations, the only prior information we used is that the scatterer in the computational domain is dielectric and, thus, has a positive contrast, i.e., the relative permittivity of the scatterer is greater than or equal to 1.

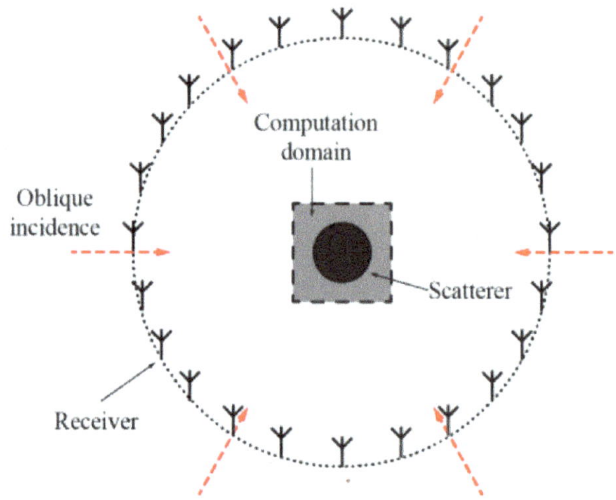

Figure 4. Simulation setup on the x-y plane.

Figure 5 and Figure 6 show the results of the numerical simulations obtained by the SOM algorithm. The original cross-section profiles of the target are illustrated in Figure 5a and Figure 6a, in which the 2D cylinder is placed at the center and off-center locations, respectively. Note that the profile has been meshed and the outline is, therefore, not a circle. It is seen that for all of the different positions of the target and the incidence angles of 0° (normal incidence), 15°, and 30°, the retrieved results of the relative permittivity, the location, and the cross-section profile of the target comply well with each other. To compare the difference between the inversed results under normal and oblique incidences, the relative error for each point in the image is calculated by:

$$error_1 = \left|\varepsilon_{oblique} - \varepsilon_{normal}\right| / \left|\varepsilon_{normal}\right| \times 100\% \quad (7)$$

and the error figures are given in Figure 5e,f and Figure 6e,f, respectively. It is seen that although ripples exist, the error remains very small in both the background and the scatterer regions, showing that the oblique incidence does not notably reduce information. It is worth noting that, for differently located targets, no notable differences exist between the results retrieved from the scattered fields under normal and oblique incidences. As a comparison, retrieved results using the first order Born approximation are shown in Figure 7. The truncated singular value decomposition (TSVD) method is used to solve the linearized inverse problem. It can be seen that in the off-center cylinder scenario, the scatterer can be identified. However, the permittivity is much lower due to the approximation. This indicates that the application of SOM is reasonable.

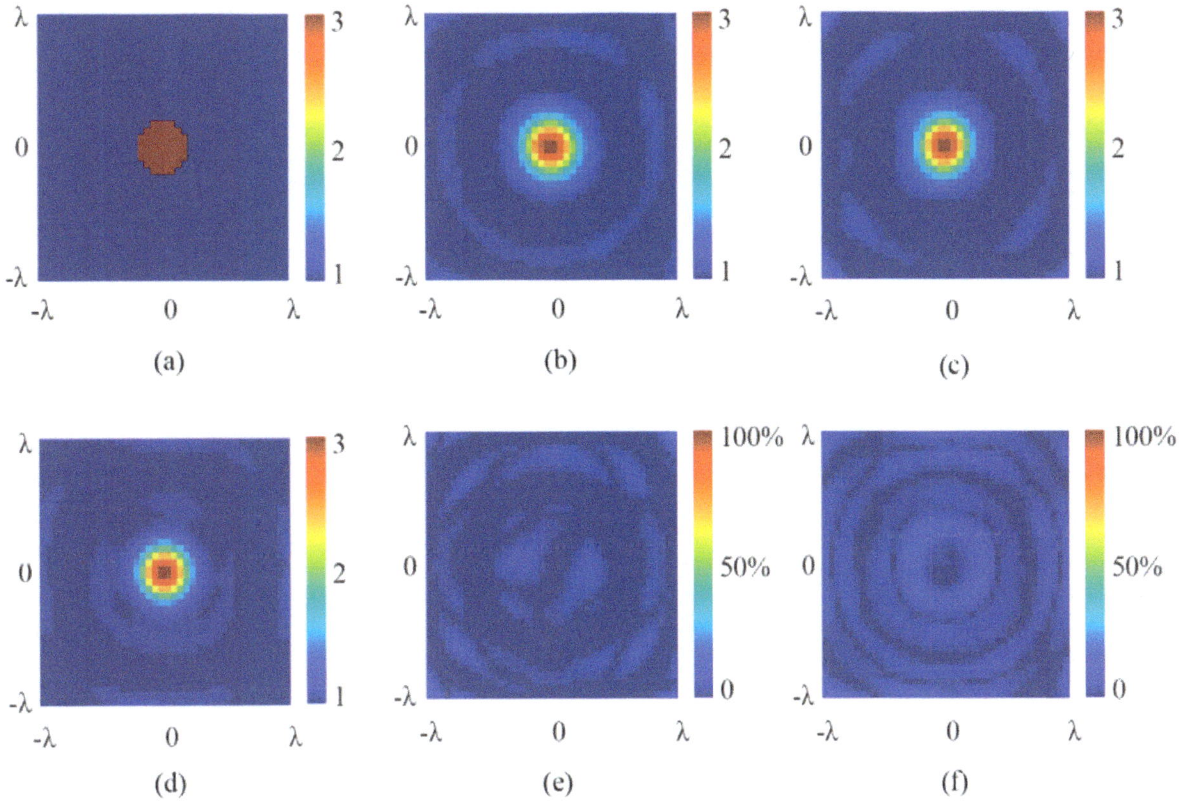

Figure 5. The retrieved cross-section profile of the centered 2D cylinder. (**a**) Original profile; (**b**) retrieved result at normal incidence; (**c**) retrieved result at 15° angle of incidence; (**d**) retrieved result at 30° angle of incidence; (**e**) relative error between 5c and 5b; and (**f**) relative error between 5d and 5b.

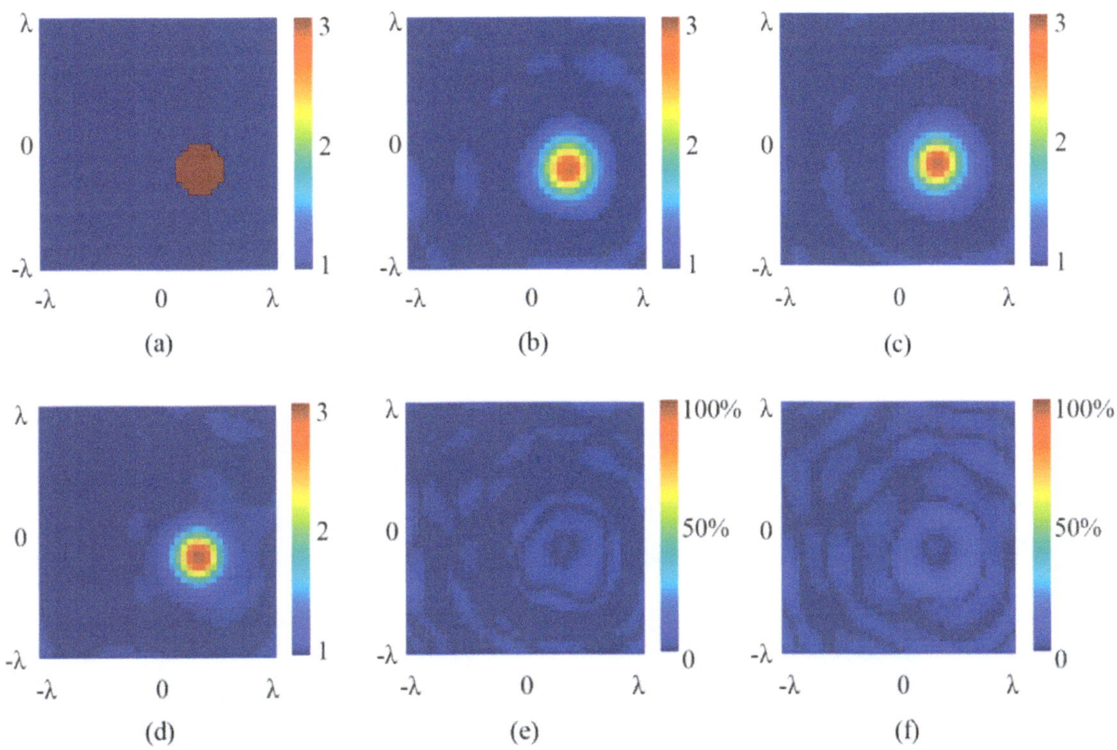

Figure 6. The retrieved cross-section profile of the off-centered 2D cylinder. (**a**) Original profile; (**b**) retrieved result at normal incidence; (**c**) retrieved result at 15° angle of incidence; (**d**) retrieved result at 30° angle of incidence; (**e**) relative error between 6**c** and 6b; and (**f**) relative error between 6d and 6b.

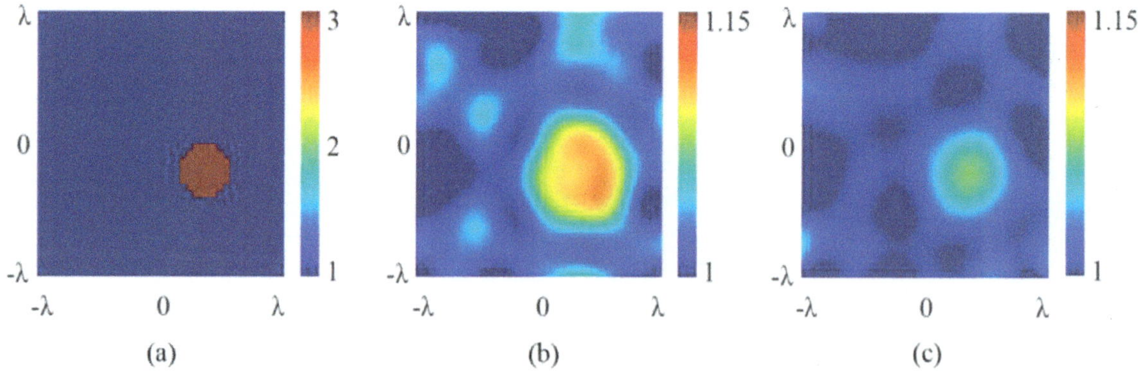

Figure 7. The retrieved cross-section profile of the off-centered 2D cylinder using first order Born approximation. (**a**) Original profile; (**b**) retrieved result at 15° angle of incidence; and (**c**) retrieved result at 30° angle of incidence.

The simulation results shown in Figure 5 and Figure 6 clearly validate the proposed algorithm. However, the above simulations used both the TE and the TM polarization information of the scattered field. In many practical cases, measuring the full polarization data is difficult, it may make the measurement setup complex and result in being much more time consuming in the inversion. To simplify the measurement in practice, we further test the proposed algorithm with part of the polarization of the scattered field. Since the incident wave is TM polarized in our simulation setup, and in the experimental setup later, we will use the TM component of the scattered field to retrieve the spatial distribution of the scatterer's relative permittivity.

Unlike the normal incidence scenario, the scattered field under oblique TM incidence also contains a cross-polarized TE component [46]. Therefore, the inversion with the TM component of the scattered field only means that the algorithm does not make full use of the scattered field data. In order to quantify the difference between inversion results with and without the full use of the polarization information, we further conduct more numerical simulations. The first simulation is the same as previous simulation based on the centered dielectric 2D cylinder. In the second simulation, a 2D scatterer with the so-called "Austria" cross-section profile is used [27], whose relative permittivity is set as 2. For both simulations, the inversion results solely rely on the TM and the TE polarization data, as well as the results using the full polarization information, are shown in Figure 8 for three different angles of incidence, i.e., 15°, 30°, and 45°. Since there are different unknowns in the TM and TE cases, in the optimization we used the same mesh and termination criterion for both cases, to make sure the comparison between them is reasonable. It is seen that in all cases, the profiles retrieved with the TM components are very similar to those retrieved with full polarization components. In the meantime, the retrieved profiles with the TE components have significant differences with the ones with both components.

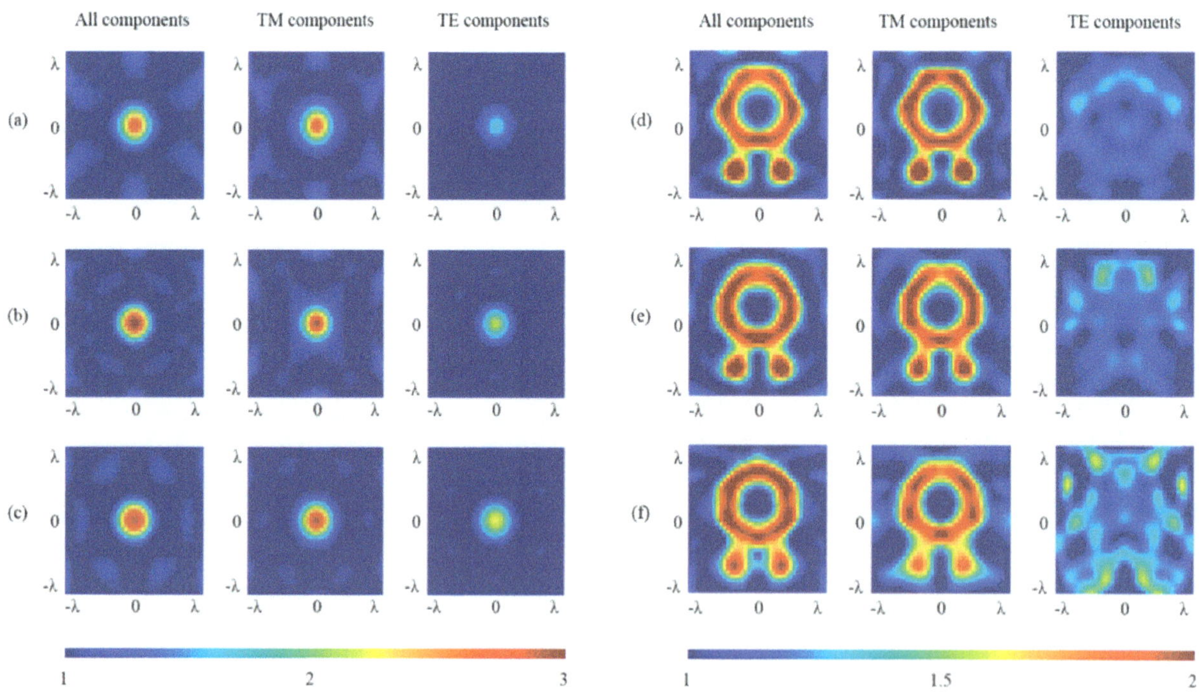

Figure 8. The retrieved cross section profiles of the 2D cylinder using different polarization components of the scattered field. (**a**) 15° angle of incidence; (**b**) 30° angle of incidence; and (**c**) 45° angle of incidence. The retrieved cross-section profiles of the 2D "Austria" profile using different polarization components of the scattered field; (**d**) 15° angle of incidence; (**e**) 30° angle of incidence; and (**f**) 45° angle of incidence.

To describe the difference between these cases, we define an error function as:

$$error_2 = \sum |\varepsilon_{inv} - \varepsilon_{all}|^2 / \sum |\varepsilon_{all}|^2 \times 100\% \quad (8)$$

where ε_{inv} and ε_{all} denote the inversed profile with only the TM or the TE polarization data and with all polarization information, respectively, and the summation is performed in the computational domain. The computed errors are listed in Table 1. It is seen that within 45° angle of incidence, the inversion results using only the TM component are satisfactory, keeping the error less than 1% in all incidences. On the other hand, the inversion results using only the TE component are with much larger error, especially for the target with the "Austria" profile. The large difference between these two cases implies that neglecting TE component does not lead to a notable loss of imaging quality.

Table 1. Error analysis for different 2D targets

	Cylinder Profile		"Austria" Profile	
Incident Angle	**TM Component**	**TE Component**	**TM Component**	**TE Component**
15°	0.16%	2.95%	0.22%	9.31%
30°	0.56%	2.14%	0.28%	9.03%
45°	0.24%	1.49%	0.81%	9.34%

These simulation results also comply with existing theory [45]. The scattering Green function matrix $\overline{\overline{G}}_s$ in Equation (4) can be written in the form $\overline{\overline{G}}_s$=[gxx, gxy, gxz; gyx, gyy, gyz; gzx, gzy, gzz], where gxy denotes the mapping to the *x*component of the scattered field from the *y* component of the induced source, etc., according to the Equations (10)–(12) in [45], $g_{xz} = C \cdot \frac{(x-x')}{\rho}$ and $g_{yz} = C \cdot \frac{(y-y')}{\rho}$, where C is a constant. Since the receivers are in the far field, $\frac{(x-x')}{\rho} \approx \cos\phi$ and $\frac{(y-y')}{\rho} \approx \sin\phi$. Consider the coordinate transformation described by (12) in our paper, the elements corresponding to the mapping from *z* component of induced source to TE component of the scattered field can be written as $g_{xz}\cdot\sin\phi - g_{yz}\cdot\cos\phi \approx C\cdot(\sin\phi\cdot\cos\phi - \sin\phi\cdot\cos\phi) = 0$. This indicates that the influence on the TE component of the scattered field from the *z* component of the induced source is weak under the precondition of that the incident field is TM polarized, Therefore, ignoring the TE component of the scattered field will not cause a notable degradation in the quality of the retrieved image.

The above simulation results and analysis strongly imply that in the case we discussed, i.e., scattering of 2D objects under oblique incidence, the TM component of the scattered field contains sufficient information for the retrieval of the scatterer, at least when the angle of incidence is within a limited range. In this case, imaging using the TM component only would not notably lose its quality. This conclusion can be used to simplify the measurement setup, as discussed in the following.

For even larger incidence angles, there is not much practical meaning since it is difficult for the measurement to be conducted in accordance with the approximation [50].

EXPERIMENTS

In this Section, we perform experiments to verify the proposed algorithm.

Experimental Setups

Figure 9a shows the implemented experimental setup, which is very similar to the simulation model depicted in Figure 4. The system is designed to work at 2.4 GHz, whose corresponding wavelength λ in free space is 12.5 cm. The main part of the detecting device is an array of 24 linearly-polarized patch antennas. These antennas are evenly spaced on a circle with a diameter of 113 cm, meaning that the distance between the center of the computation domain and the antenna is around 4.5λ. All antennas can be switched through commercial integrated microwave switches (Hittite HMC321LP4, Hittite Microwave, Chelmsford, MA, USA) mounted on three printed circuit boards, as shown in Figure 9b, and can be controlled by a micro-controller unit (Atmel's ATmega88, Atmel, San Jose, CA, USA). These patch antennas are used as receivers, and a standard horn antenna (Yinglian Microwave's LB34015CSF, Yinglian Microwave, Chengdu, China) is used as the transmitter to provide the oblique illumination.

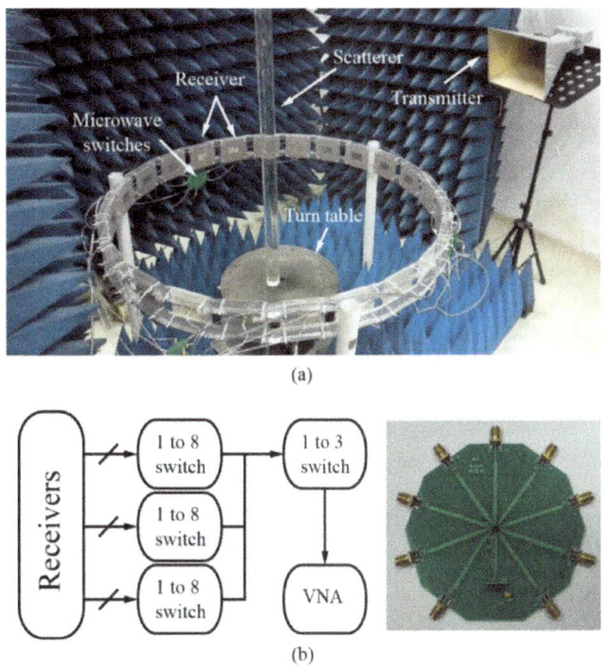

Figure 9. Experimental setup. Inset shows one of the three simple-structured unidirectional microwave switches. (**a**) Experimental setup; and (**b**) simple structured microwave switch.

A vector network analyzer, Agilent's 8722ES (Agilent, Santa Clara, CA, USA), is used to conduct the measurements.

A wideband amplifier, Agilent's 8449B, is used before the horn antenna to enhance the power level of the transmitted signal. Through switching between the receiving antennas, we can measure the scattered field at different angles. A turntable is placed in the center of the antenna array. By turning the object, we can also measure the scattered field at different incidences by using only one fixed transmitting antenna. In our experiment, the number of incidences is 6.

In the measurement, the electric field at each receiver's position in the absence and in the presence of the target should be measured separately. Then, the turn table rotates the target by 60°, and the electric fields in the presence of the target can be measured. Such rotating and measurement should be repeated six times. Through calculating the difference between the measured fields in the presence and in the absence of the target, we can obtain a 6 × 24 matrix of measured data of scattered fields. Since 2.4 GHz is located in an open ISM band, before experiments it is necessary to make sure that no nearby wireless device is using this frequency. It is seen that the above setup combines the advantages of the conventional setups shown in Figure 1a,b. While all antennas are fixed, the bi-directional microwave switch matrix can be replaced with simple-structured, unidirectional switches, as shown in the inset of Figure 9. These will bring significant convenience to the implementation and calibration of the measurement system.

Calibration of the Measurement System

Similar to all the other experimental setups, a precision calibration to the measurement setup is crucial to obtain satisfactory inversion results. In our work, we used a calibration method similar to that reported in [51]. The scattered field is calculated by subtracting the incident field from the total field. Measuring with the Vector Network Analyzer (VNA), we can obtain the S21 parameter of the measurement system. The S_{21} parameter of the entire system can be divided into four parts, i.e., $S_{21,i} = S_{21}^1 \cdot S_{21,i}^2 \cdot S_{21,i}^3 \cdot S_{21}^4, i = 1, \ldots, 24$. As shown in Figure 10, $S_{21}^1, S_{21,i}^2, S_{21,i}^3$ and S^4_{21} denote the S_{21} parameters from the VNA's output port to the transmitting antenna, the transmitting antenna to the *i*-th receiving antenna, the *i*-th receiving antenna to the switch, and the switch to the VNA input port, respectively. For different paths, S^1_{21} and $S^4_{21}$are the same, $S^3_{21,i}$ can be measured, $S^2_{21,i}$ has a definite relationship with the electric field. Therefore, we can measure the relative value of the electric field by measuring the S_{21} parameters of the measurement system.

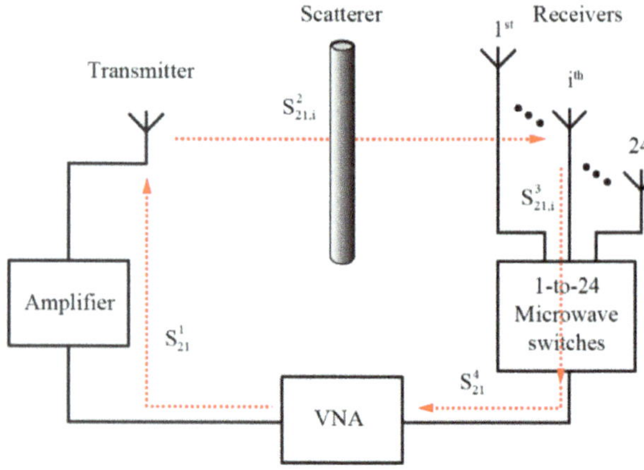

Figure 10. Calibration of the experimental setup.

In the calibration, the S_{21} parameter for each transmitter-receiver pair in the absence and in the presence of the scatterer, denoted as $\overline{E}^i$ and $\overline{E}^t$ respectively, should be measured first. The obtained $\overline{E}^i$ and $\overline{E}^t$ are both a 24 × 6 matrix. Each column of $\overline{E}^t$, written as $\overline{E}_p^t$, $p = 1, 2, \ldots, 6$, denotes the scattered field detected by 24 receivers due to the pth incidence. Finally, the scattered field at different incidences can be given by $\overline{E}_p^{sca} = \overline{E}_p^t - \overline{E}^i$.

It should be noted that the data we measured until now cannot be directly used to retrieve the scatterer. Firstly, there are differences in amplitude attenuation and phase shift between different transmitter-receiver pairs since the signals go through different paths. To solve this problem, a measurement of the switch matrix is also conducted. The measurement that needs to be performed is the S321,i of each path, including the switch boards and cables. Through this measurement, we can construct a series of complex coefficients that can be denoted as $C_i = S_{21,i}^3 / S_{21,ref}^3$, $i = 1, 2, \ldots, 24$. Here, $S^3_{21,ref}$ corresponds to the reference path, in which the receiving antenna can be chosen as the one directly facing the transmitting antenna. Then the measured scattered data can be divided by C_i for amplitude and phase compensation, i.e., $E_{mea,p}^{sca} = (\overline{E}_p^t - \overline{E}^i)/C$, $p = 1, 2, \ldots, 6$, where $E_{mea,p}^{sca}$ denotes the measured data due to the *p*-th incidence after compensation, and C is a column vector whose elements are C_i.

Secondly, there is a difference between the fields that we measured and the field that we used in the numerical calculation. It is easy to calibrate this due to the fact that $E_{ref}^i / E_{mea}^i = E_{cal}^{sca} / E_{mea}^{sca}$, where E^i_{ref} denotes the simulated incident field at the receiving antenna directly facing the transmitting antenna, which can be considered as a reference. E_{mea}^i and E_{mea}^{sca} denote the measured incident fields by the receiving antenna directly facing the transmitting antenna and the scattered fields, respectively, and E_{cal}^{sca} denotes the calibrated scattered field for the inversion algorithm.

Alternatively, the calibration of the measurement system can also be performed by measuring the scattered fields of a reference target. According to [51], this method may lead to more accurate results. However, much more effort is needed to carry out such a calibration. In this work, we chose the former one because the results we obtained after the simple calibration are acceptable.

Results

In our experiment, the scatterer is a 100-cm-long cylinder made of organic glass, whose nominal relative permittivity is around 3. The diameter of the cylinder is 5 cm.

In the first measurement, the cylinder is placed at the center of the receiving antennas. In the second measurement, the same cylinder is placed 5 cm off the center of the receiving antennas. The exact position in a Cartesian coordinate is (5cos30°, −5sin30°), or (4.33, −2.50) cm. In both measurements, the oblique incident angles are set as 15° and 30°, respectively, and the scattered fields are measured for six evenly-separated incident directions by rotating the turntable supporting the

2D object. Figure 11 shows the original and the retrieved profiles for the target under test with different locations and angles of incidence. It is seen that in both cases, the scatters are successfully retrieved. Compared with the original profile, the retrieved results in Figure 11a for the center-located target are satisfactory for both angles of incidence. The location of the target, the relative permittivity in the central area and the overall cross-section profile satisfy the expectation. For the off-center located target, while the retrieved location and the relative permittivity of the target are satisfactory, the cross-section shapes are slightly distorted from the original profile.

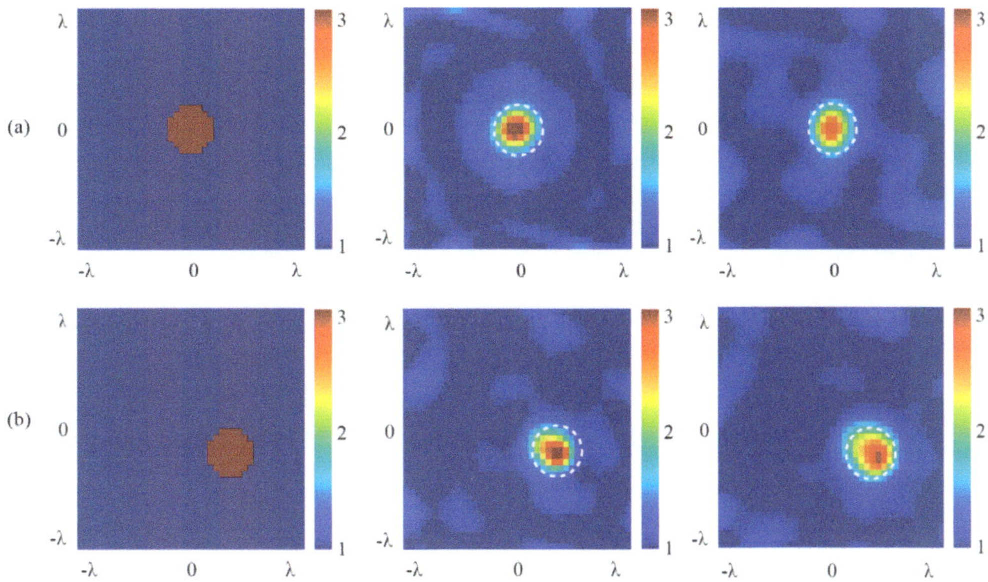

Figure 11. Retrieved results with experimental data for 15° and 30° angles of incidence. (**a**) Center-located target; and (**b**) off-center located target.

In order to investigate the source of errors that cause the differences between the original profiles and the retrieved results, we compare the magnitude and phase between the simulated and the measured data of the scattered fields. For demonstration, one group data for the off-centered cylinder are shown in Figure 12, where the dots represent the simulated and the measured data, and the dashed lines represent the corresponding fitting curves. It is seen that, the same as all experimental inversion problems, the measured data in our measurement also include errors and noise introduced by the environment and imperfections of the measurement system. Since the measured data, especially the measured amplitude data, notably deviate from their fitting curves, a degradation of the imaging quality would be inevitable.

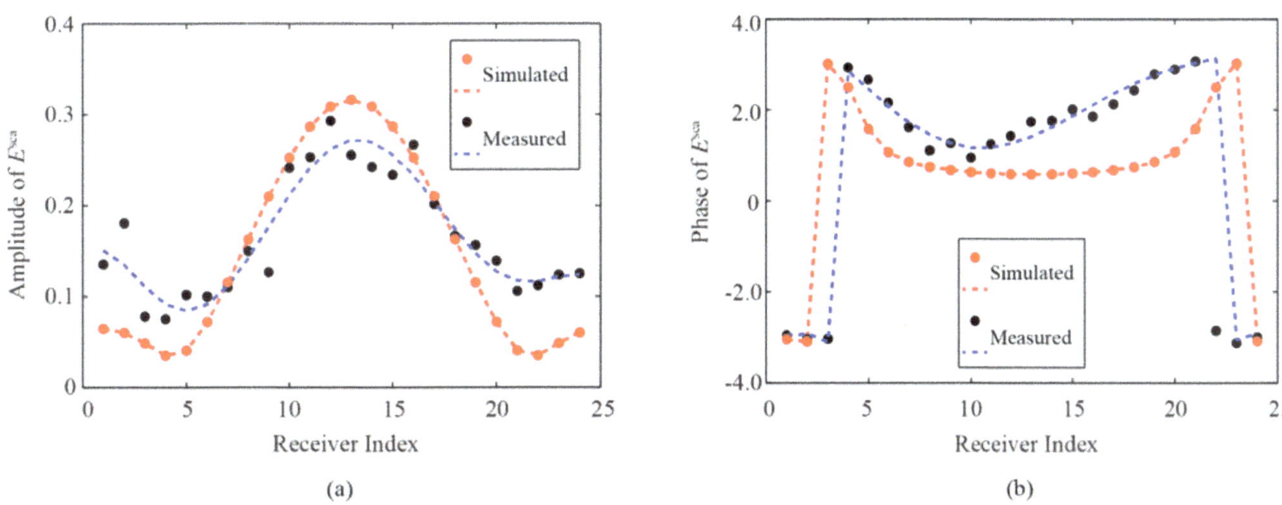

Figure 12. Comparison between simulated and measured scattered fields for a dielectric cylinder placed at the center of receiving antennas with 15° angle of incidence. (**a**) Amplitude; and (**b**) phase.

In our experiment, the home-made setup was not placed in an anechoic chamber and, therefore, the reflections due to the uncovered part of the ground, the edges and gaps between the absorber screens, the metal surface of the turntable, and the supporting structures of the antennas would easily result in such errors. Even so, compared with existing experimental results for similar microwave imaging, the results shown in Figure 11 are, overall, satisfactory [35,39,40]. The notable errors shown in Figure 12 imply that if the measurement setup can be further improved and calibrated, much better imaging quality can be obtained.

Discussion

As stated, the motivation inspired this work is to improve the engineering practicability of SOM-based imaging for scatters that can be approximated as 2D objects, other than to improve the SOM itself. Therefore, we only used the very basic SOM in the inversion. If performance-enhanced SOMs can be used, better inversion results can be expected. Our analytical, simulation, and experimental results showed that by introducing TM oblique incidence and only detecting TM scattering, the simplified setup is able to bring advantages, like less complexity, simpler calibration, and faster computation, to the inverse imaging without notably losing the imaging quality. This agrees with [52], where the depolarizing effects are taken into account in solving qualitative inverse scattering problems for targets showing electric and magnetic contrast.

It is worth noting that in the proposed method, the incidence is always TM polarized. Therefore, our results can be different from those obtained with differently-polarized incidences [22,53]. Since the wave vector of the plane wave incidence has a component along the axis direction, couplings exist between the TE and TM modes [46]. In this case, the method proposed in [54], i.e., decomposing the plane wave into separate TE and TM modes, cannot be used. In fact, TE incidence can also be used in our case. However, as indicated in [22], the obtained results between TE and TM inversions are similar when scattered fields are sampled in the far field. Unlike the TM data, TE data should be decomposed into *x-y* components for the inversion. Therefore, from a data processing point of view, TE incidence does not have obvious advantages for our setup.

It is also worth noting that in the proposed method dedicated for 2D objects, 3D-2D modelling errors always exists. In this case, measures must be taken to ensure the error due to the 3D-2D modelling is controllable. For the proposed setup with oblique incidence, the longitudinal separation between the receiver and transmitter planes should be sufficiently smaller than the longitudinal section of the scatter. For a specific 2D scatter, this condition can be satisfied by adjusting the oblique angle, the detecting distance, and the operating frequency. For a smaller oblique angle, a shorter detecting distance and a higher frequency, the requirement of the length of the 2D object can be reduced.

Finally, the main purpose of this work is to investigate the possibility of performing microwave imaging on 2D objects under oblique incidences, rather than to obtain the best results. However, since the inverse scattering problem is essentially ill-posed, it would always require a highly accurate measurement system for microwave imaging. In this sense, further improvement of the proposed measurement setup deserves continued research efforts.

CONCLUSIONS

In conclusion, an inversion algorithm based on the SOM to solve inverse scattering problems under oblique microwave illumination is mathematically derived. A corresponding experimental setup combining the advantages of two types of conventional setups is designed and fabricated to verify the effectiveness of the proposed approach. Our results imply that for 2D objects, oblique microwave illumination will not cause a notable loss of information and, therefore, the quality of the microwave imaging will not be degraded compared with the conventional case under normal illumination, and can potentially be used in microwave imaging applications for 2D objects.

ACKNOWLEDGMENTS

This work is supported by the NSFC under grant 61131002, 61471315 and 61528104.

AUTHOR CONTRIBUTIONS

Q.M., K.X., F.S., B.Z. and L.R. conceived and designed the experiments; Q.M. performed the experiments; Q.M. J.H. and D.Y. analyzed the data; Q.M., C.L. and L.R wrote the paper.

REFERENCES

1. Yu, C.; Yuan, M.; Zhang, Y.; Stang, J.; George, R.T.; Ybarra, G.A.; Joines, W.T.; Liu, Q.H. Microwave imaging in layered media: 3-D image reconstruction from experimental data. *IEEE Trans. Antennas Propag.* **2010**, *58*, 440–448.
2. Catapano, I.; Crocco, L.; Isernia, T. Improved sampling methods for shape reconstruction of 3-D buried targets. *IEEE Trans. Geosci. Remote Sens.* **2008**, *46*, 3265–3273.
3. Li, F.; Liu, Q.H.; Song, L.P. Three-dimensional reconstruction of objects buried in layered media using born and distorted born iterative methods. *IEEE Geosci. Remote Sens. Lett.* **2004**, *1*, 107–111.
4. Sheen, D.M.; Mcmakin, D.L.; Hall, T.E. Three-dimensional millimeter-wave imaging for concealed weapon detection.*IEEE Trans. Microw. Theory Tech.* **2001**, *49*, 1581–1592.
5. Caorsi, P.; Gragnani, G.L.; Pastorino, M. Numerical electromagnetic inverse-scattering solutions for two-dimensional infinite dielectric cylinders buried in a lossy half-space. *IEEE Trans. Microw. Theory Tech.* **1993**, *41*, 352–357.
6. Meaney, P.M.; Fanning, M.W.; Li, D.; Poplack, S.P. A clinical prototype for active microwave imaging of the breast.*IEEE Trans. Microw. Theory Tech.* **2000**, *48*, 1841–1853.
7. Shea, J.D.; Van Veen, B.D.; Hagness, S.C. A TSVD analysis of microwave inverse scattering for breast imaging. *IEEE Trans. Biomed. Eng.* **2012**, *59*, 936–945.
8. Meaney, P.M.; Goodwin, D.; Golnabi, A.H.; Zhou, T.; Pallone, M.; Geimer, S.D.; Burke, G.; Paulsen, K.D. Clinical microwave tomographic imaging of the calcaneus: A first-in-human case study of two subjects. *IEEE Trans. Biomed. Eng.* **2012**, *59*, 3304–3313.
9. Rubaek, T.; Kim, O.S.; Meincke, P. Computational validation of a 3-D microwave imaging system for breast-cancer screening. *IEEE Trans. Antennas Propag.*

2009, *57*, 2105–2115.

10. Abubakar, A.; van den Berg, P.M.; Mallorqui, J.J. Imaging of biomedical data using a multiplicative regularized contrast source inversion method. *IEEE Trans. Microw. Theory Tech.* **2002**, *50*, 1761–1771.
11. Klemm, M.; Leendertz, J.A.; Gibbins, D.; Craddock, I.J.; Preece, A.; Benjamin, R. Microwave radar-based breast cancer detection: Imaging in inhomogeneous breast phantoms. *IEEE Antennas Wirel. Propag. Lett.* **2009**, *8*, 1349–1352.
12. Sill, J.M.; Fear, E.C. Tissue sensing adaptive radar for breast cancer detection—Experimental investigation of simple tumor models. *IEEE Trans. Microw. Theory Tech.* **2005**, *53*, 3312–3319.
13. Fear, E.C.; Bourqui, J.; Curtis, C.; Mew, D. Microwave breast imaging with a monostatic radar-based system: A study of application to patients. *IEEE Trans. Microw. Theory Tech.* **2013**, *61*, 2119–2128.
14. Ostadrahimi, M.; Mojabi, P.; Zakaria, A.; Lovetri, J.; Shafai, L. Enhancement of Gauss-Newton inversion method for biological tissue imaging. *IEEE Trans. Microw. Theory Tech.* **2013**, *61*, 3424–3434.
15. Soldovieri, F.; Solimene, R.; Prisco, G. A multiarray tomographic approach for through-wall imaging. *IEEE Trans. Geosci. Remote Sens.* **2008**, *46*, 1192–1199.
16. Song, L.P.; Yu, C.; Liu, Q.H. Through-wall imaging (TWI) by radar: 2-D tomographic results and analyses. *IEEE Trans. Geosci. Remote Sens.* **2005**, *43*, 2793–2798.
17. Chen, P.H.; Shastry, M.C.; Lai, C.P.; Narayanan, R.M. A portable real-time digital noise radar system for through-the-wall imaging. *IEEE Trans. Geosci. Remote Sens.* **2012**, *50*, 4123–4134.
18. Ahmad, F.; Zhang, Y.; Amin, M.G. Three-dimensional wideband beamforming for imaging through a single wall.*IEEE Geosci. Remote Sens. Lett.* **2008**, *5*, 176–179.
19. Zhang, W.; Hoorfar, A. Three-dimensional real-time through-the-wall radar imaging with diffraction tomographic algorithm. *IEEE Trans. Geosci. Remote Sens.* **2013**, *51*, 4155–4163.
20. Chew, W.C.; Wang, Y.M. Reconstruction of two-dimensional permittivity distribution using the distorted born iterative method. *IEEE Trans. Med. Imaging* **1990**, *9*, 218–225.
21. Van den Berg, P.M.; Kleinman, R.E. A contrast source inversion method. *Inverse Probl.* **1997**, *13*, 1607–1620.
22. Mojabi, P.; LoVetri, J. Comparison of TE and TM inversions in the framework of the Gauss-Newton method. *IEEE Trans. Antennas Propag.* **2010**, *58*, 1336–1348.
23. Pastorino, M. Stochastic optimization methods applied to microwave imaging: A review. *IEEE Trans. Antennas Propag.***2007**, *55*, 538–548.
24. Huang, C.H.; Chiu, C.C.; Li, C.L.; Chen, K.C. Time domain inverse scattering of a two-dimensional homogenous dielectric object with arbitrary shape by particle swarm optimization. *Progr. Electromagn. Res.* **2008**, *82*, 381–400.
25. Cui, T.J.; Liang, C.H. Inverse scattering method for one-dimensional inhomogeneous lossy medium by using a microwave networking technique. *IEEE Trans. Microw. Theory Tech.* **1995**, *43*, 1773–1781.
26. Liu, Q.H.; Zhang, Z.Q.; Wang, T.T.; Bryan, J.A. Active microwave imaging. I. 2-D forward and inverse scattering methods. *IEEE Trans. Microw. Theory Tech.* **2002**, *50*, 123–133.
27. Chen, X. Subspace-based optimization method for solving inverse-scattering problems. *IEEE Trans. Geosci. Remote Sens.* **2010**, *48*, 42–49.
28. Pan, L.; Agarwal, K.; Zhong, Y.; Yeo, S.P.; Chen, X. Subspace-based optimization method for reconstructing extended scatterers: Transverse electric case. *J. Opt. Soc. Am. A* **2009**, *26*, 1932–1937.
29. Ye, X.; Chen, X.; Zhong, Y.; Agarwal, K. Subspace-based optimization method for reconstructing perfectly electric conductors. *Progr. Electromagn. Res.* **2010**, *100*, 119–128.
30. Shen, J.; Zhong, Y.; Chen, X.; Ran, L. Inverse scattering problems of reconstructing perfectly electric conductors with TE illumination. *IEEE Trans. Antennas Propag.* **2013**, *61*, 4713–4721.
31. Zhong, Y.; Chen, X.; Agarwal, K. An improved subspace-based optimization method and its implementation in solving three-dimensional inverse problems. *IEEE Trans. Geosci. Remote Sens.* **2010**, *48*, 3763–3768.
32. Zhong, Y.; Chen, X. Twofold subspace-based optimization method for solving inverse scattering problems. *Inverse Probl.* **2009**, *25*, 85003–85013.
33. Zhong, Y.; Chen, X. An FFT twofold subspace-based optimization method for solving electromagnetic inverse scattering problems. *IEEE Trans. Antennas Propag.* **2011**, *59*, 914–927.
34. Xu, K.; Zhong, Y.; Song, R.; Chen, X. Multiplicative-regularized FFT twofold subspace-based optimization method for inverse scattering problems. *IEEE Trans. Geosci. Remote Sens.* **2015**, *53*, 841–850.
35. Ye, X.; Chen, X.; Zhong, Y.; Song, R. Simultaneous reconstruction of dielectric and perfectly conducting scatterers via T-matrix method. *IEEE Antennas Propag. Mag.* **2013**, *61*, 3774–3781.
36. Donato, L.D.; Bevacqua, M.T.; Crocco, L.; Isernia, T. Inverse scattering via virtual experiments and contrast source regularization. *IEEE Trans. Antennas Propag.* **2015**, *63*, 1669–1677.
37. Bevacqua, M.; Crocco, L.; Donato, L.D.; Isernia, T. An algebraic solution method for non-linear inverse scattering.*IEEE Trans. Antennas Propag.* **2015**, *63*, 601–610.
38. Catapano, I.; Randazzo, A.; Slob, E.; Solimene, R. GPR imaging via qualitative and quantitative approaches. In *Civil Engineering Applications of Ground Penetrating Radar*; Benedetto, A., Pajewski, L., Eds.; Springer: Berlin, Germany, 2015; pp. 239–280.
39. Van den Berg, P.M.; Cote, M.G.; Kleinman, R.E. "Blind" shape reconstruction from experimental data. *IEEE Trans. Antennas Propag.* **1995**, *43*, 1389–1396.
40. Geffrin, J.M.; Sabouroux, P.; Eyraud, C. Free space

experimental scattering database continuation: Experimental set-up and measurement precision. *Inverse Probl.* **2005**, *21*, S117–S130.

41. Gilmore, C.; Zakaria, A.; Mojabi, P.; Ostadrahimi, M. The university of Manitoba microwave imaging repository: A two-dimensional microwave scattering database for testing inversion and calibration algorithms. *IEEE Antennas Propag. Mag.* **2011**, *53*, 126–133.
42. Ostadrahimi, M.; Zakaria, A.; Lovetri, J.; Shafai, L. A near-field dual polarized (TE–TM) microwave imaging system.*IEEE Trans. Microw. Theory Tech.* **2013**, *61*, 1376–1384.
43. Asefi, M.; Ostadrahimi, M.; Zakaria, A.; Lovetri, J. A 3-D dual-polarized near-field microwave imaging system. *IEEE Trans. Microw. Theory Tech.* **2014**, *62*, 1790–1797.
44. Meaney, P.M.; Paulsen, K.D.; Geimer, S.D.; Haider, S.A.; Fanning, M.W. Quantification of 3-D field effects during 2-D microwave imaging. *IEEE Trans. Biomed. Eng.* **2002**, *49*, 708–720.
45. Su, C.C. Calculation of electromagnetic scattering from a dielectric cylinder using the conjugate gradient method and FFT. *IEEE Trans. Antennas Propag.* **1987**, *35*, 1418–1425.
46. Richmond, J. Scattering by a dielectric cylinder of arbitrary cross section shape. *IEEE Trans. Antennas Propag.* **1965**, *13*, 334–341.
47. Rojas, R.G. Scattering by an inhomogeneous dielectric/ferrite cylinder of arbitrary cross-section shape-oblique incidence case. *IEEE Trans. Antennas Propag.* **1988**, *36*, 238–246.
48. Lucido, M.; Panariello, G.; Schettino, F. Scattering by polygonal cross-section dielectric cylinders at oblique incidence.*IEEE Trans. Antennas Propag.* **2010**, *58*, 540–551.
49. Wait, J.R. Scattering of a plane wave from a circular dielectric cylinder at oblique incidence. *Can. J. Phys.* **1955**, *33*, 189–195.
50. Bulcke, S.V.D.; Zhang, L.; Franchois, A.; Geffrin, J.M.; Stiens, J. Plane wave and Gaussian beam scattering by long dielectric cylinders: 2.5 D simulations versus measurements. *Int. J. Infrared Millim. Waves* **2008**, *29*, 1038–1047.
51. Litman, A.; Geffrin, J.M.; Tortel, H. On the calibration of a multistatic scattering matrix measured by a fixed circular array of antennas. *Progr. Electromagn. Res.* **2010**, *110*, 1–21.
52. Di Donato, L.; Catapano, I.; Soldovieri, F.; Crocco, L. Imaging of 3D magnetic targets from multiview multistatic GPR data. In Proceedings of the 13th International Conference on Ground Penetrating Radar (GPR), Lecce, Italy, 21–25 June 2010; pp. 1–6.
53. Otto, G.P.; Cho, C.W. Inverse scattering of Hz waves using local shape-function imaging: A T-matrix formulation. *Int. J. Imaging Syst. Technol.* **1994**, *5*, 22–27.
54. Chou, C.P.; Kiang, Y.W. Inverse scattering of dielectric cylinders by a cascaded TE-TM method. *IEEE Trans. Microw. Theory Tech.* **1999**, *47*, 1923–1930.

CHAPTER 2

FIELD-BASED OPTIMAL PLACEMENT OF ANTENNAS FOR BODY-WORN WIRELESS SENSORS

Łukasz Januszkiewicz, Paolo Di Barba and Sławomir Hausman

Keywords: Body area networks; wireless sensor network; computer optimization algorithms; computational electromagnetics

INTRODUCTION

Technological development, especially in the field of sensor technology, highly integrated electronics and telecommunications, allows for the introduction of new solutions in health monitoring of medical patients as well as monitoring of sportsmen, people working in hazardous environments or exposed to hazards at work, and other safety critical contexts [1]. Early solutions used wired sensors, involving wires and connectors, which adversely affected the user's comfort. A more recent approach is to use body-worn wireless sensor networks, known as Wireless Body Area Networks (WBAN) [2,3]. Due to the development of the integrated circuit technology it is possible to manufacture ever more complex and yet miniature electronic modules (including sensors), small enough to be integrated into clothing. A good example are Flora sensors made by the company Adafruit [4]. Their diameter is less than 45 mm and the weight is 4.4 g. Such clothing, fitted with miniature sensors, electronic devices and computing power of microcontrollers is known as smart clothing or e-textiles [5]. The functions performed by smart clothing may be diverse, including a mobile, remote measurement of human physiological parameters, vital signs monitoring, monitoring of physical activity or automatic detection of risks. A comprehensive introduction to WBAN and relevant design issues are provided in [1].

In this paper we discuss optimization of a WBAN system which uses wireless communication technology to monitor physiological and environmental parameters of firefighters and their exposure to hazardous environmental conditions. The system concept is presented in Figure 1. It consists of a head sensor, an arm sensor, and a chest sensor. The head sensor location is fixed in the helmet because it measures blood oxygen saturation (pulse oximetry) on the earlobe, which is one of typical locations used for this type of sensor [6]. The location of the arm node/sensors (accelerometers and a GPS receiver, temperature and humidity sensors) is fixed and determined by our experiments with a firefighter's comfort for a particular design of protective clothing in mind. The chest node is both an on-body sensor and an off-body node which is communicating with on-body components and with the remote monitoring system. The chest node measures the humidity, temperature, and respiratory activity of firefighters and therefore is located on the chest, in a position ensuring a low value of the path loss to the other two on-body sensors to achieve low required RF transmission power and consequently low power consumption

Citation: Januszkiewicz, Ł.; Di Barba, P.; Hausman, S. Field-Based Optimal Placement of Antennas for Body-Worn Wireless Sensors. Sensors 2016, 16, 713. doi:10.3390/s16050713.

leading to an increased network lifetime or less demanding energy harvesting. The objective of our optimization procedure is to minimize the path loss between the chest sensor and the other two sensors: it is achieved by changing the location of the chest sensor within a predefined region, the boundary of which represents a problem constraint. In that respect our optimization problem is significantly different from those where the number of WBAN nodes (sensors and sinks) is greater and network topology as well as routing optimization should be considered, e.g., [7,8,9]. We do not assume any empirical-statistical radio wave propagation model as in [9]. We use more accurate computational electromagnetics methods as explained in Section 2.1.

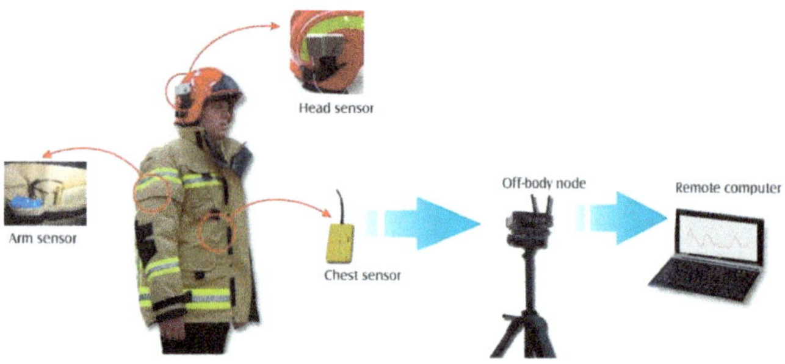

Figure 1. Structure of the wireless body-worn sensor system.

The path loss between the chest node and the other two nodes is defined by S_{21} and S_{31} which are parameters of the scattering matrix, widely used in the microwave technology. The parameters are defined as:

$$S_{21} = b_2/a_1 \text{ while } a_2 = 0 \text{ and } a_3 = 0$$
$$S_{31} = b_3/a_1 \text{ while } a_2 = 0 \text{ and } a_3 = 0 \quad (1)$$

where: a_i—input signal on port i and b_j—output signal on port j, which are explained in Figure 2. We assume that the ports for which the scattering parameters are defined are the feeding points of the antennas. In our investigation, half wave dipole antennas in a vertical position were considered.

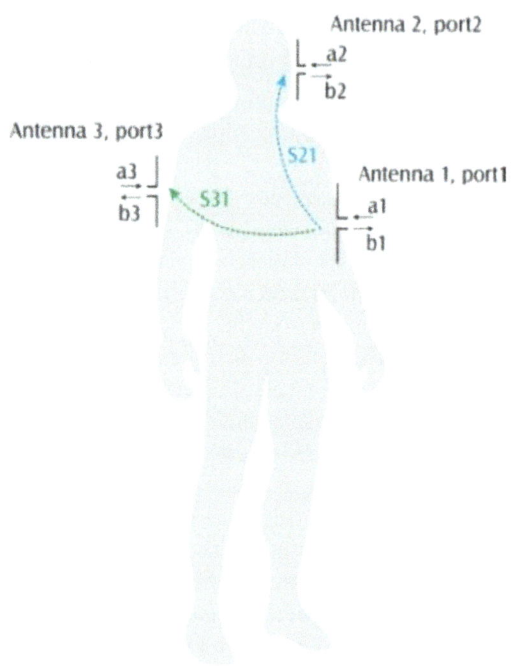

Figure 2. Scattering matrix parameters definition: $S_{21} = b_2/a_1|_{a2 = 0 \text{ and } a3 = 0}$, $S_{31} = b_3/a_1|_{a2 = 0 \text{ and } a3 = 0}$.

METHODS

The Numerical Human Body Model

An analysis of WBANs and sensor network operating in the proximity of a human body is often carried out with the use of full-wave electromagnetic simulation software which also implements a numerical model of the human body. The Finite Difference Time Domain (FDTD) method is widely used for this type of studies [10]. In our research we use the XFdtd®FDTD implementation from REMCOM, Inc. (State College, PA, USA) [11]. This package supports a scripting language which allows the user to execute customized functions, including real-time data exchange with external software.

A number of numerical human body models have been developed for the purpose of calculating the amount of electromagnetic power absorbed by the human body [12,13] (*i.e.*, Specific Absorption Rate) which can be also used for WBAN design. Currently, complex heterogeneous models which represent the inner structure of the human body reflecting electric properties of various tissue types are widely used. Those models are available in many contemporary electromagnetic CAD packages. A limitation of the multi-tissue, heterogeneous numerical models is the lack of their physical phantom equivalents because representations of the various internal organs would be very difficult to manufacture. Such models also require a lot of computer resources as the spatial discretization has to be dense to represent structural details of inner organs.

Experiments have been reported with the use of Specific Anthropomorphic Mannequin (SAM) which is a homogenous human body phantom [14,15]. The inner structure of the body is represented with homogenous tissue stimulant liquid with electromagnetic properties similar to the average value for the human body. Such models have a shape, proportions and general dimensions similar to an average human body. Those phantoms are expensive and difficult in use in the laboratory practice because they have to be filled with a liquid of precisely controlled electric properties which tend to vary with time and temperature. Simplified human body models have been also proposed for experimental studies e.g., [16] and for numerical simulation of WBANs [17,18]. They simplify a human body shape with cylinders filled with a homogenous tissue stimulant liquid but still are difficult to be re-created and require a careful control of dielectric properties of the liquid. This motivated us to propose an even more reduced complexity model which makes it very easy both for numerical implementation and manufacturing.

The simplified cylindrical model was also simulated with XFdtd®. Its geometry was optimized to approximate the proportion of the heterogeneous, anthropomorphic NMR Hershey model with six cylinders [19,20]. In Figure 3a, a reference NMR Hershey model is presented. The six cylinder model is presented in Figure 3b. Both the reference and the simplified models have the same height, H = 1800 mm. Other dimensions of the simplified model were fixed to fit the geometry of the reference model best, and to preserve the diameters of standard PVC pipes. The arm length is equal to A = 500 mm and the length of the legs is L = 790 mm, while the torso length is equal to T = 836 mm. The simplified model was designed in the way which makes it easy to manufacture. In order to be able to use the standard pipes available on the market, the following diameters of the cylinders were used in the model (see Figure 3): $\varphi_1 = 110$ mm, $\varphi_2 = 160$ mm, $\varphi_3 = 315$ mm. The pipe wall thickness varies from 3 to 6 mm depending on the pipe diameter. All pipes were filled with liquid (a water solution of sucrose and kitchen salt) of electrical properties similar to the human tissue. The electric constant and the conductivity of the liquid were examined with Di-Line measurement unit from the IndexSar (Newdigate, UK). It employs a slot transmission line which is filled with the liquid under test. Material parameters are then identified from network parameters measured with a vector network analyzer. The electric constant was equal to $\varepsilon \approx 54$ and the conductivity $\sigma \approx 1.5$ S/m. These values are close to the average electric parameters of human tissues ($\varepsilon = 52$, $\sigma = 1.8$ S/m [20]). Values of $\varepsilon = 54$ and $\sigma = 1.5$ S/m were also used in the simulations presented later in the paper.

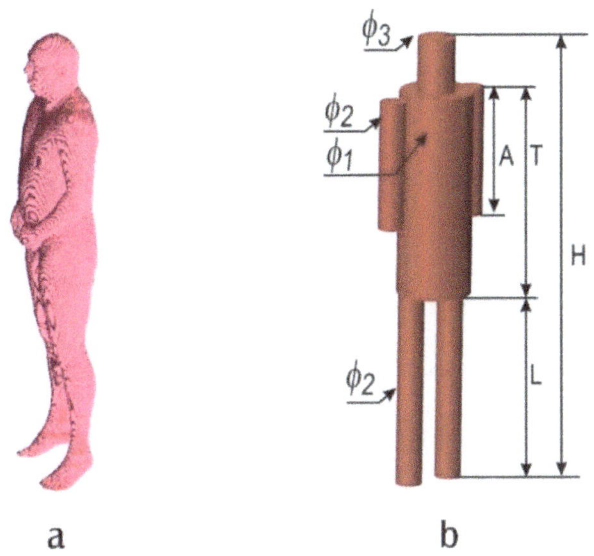

Figure 3. Human body model: (**a**) NMR Hershey; (**b**) simplified 6 cylinder model: H = 1800 mm, A = 500 mm, L = 790 mm, T = 836 mm, $\varphi_1 = 110$ mm, $\varphi_2 = 160$ mm, $\varphi_3 = 315$ mm.

A cylindrical human body model was used for further investigation to achieve independence of rotation angle phi and distance *d* of the antenna that was moved across the chest. For an irregular body shape that would be difficult (there is also a difficulty in defining and calculating distance *d*). For the simplified cylindrical phantom it is straightforward.

The motivation for the following work is the fact that the values of S_{21} and S_{31} strongly depend on the chest node antenna position, not just the antenna distance. This is evident in the example of simulation results obtained for two different locations of the chest antenna, which is presented in Figure 4.

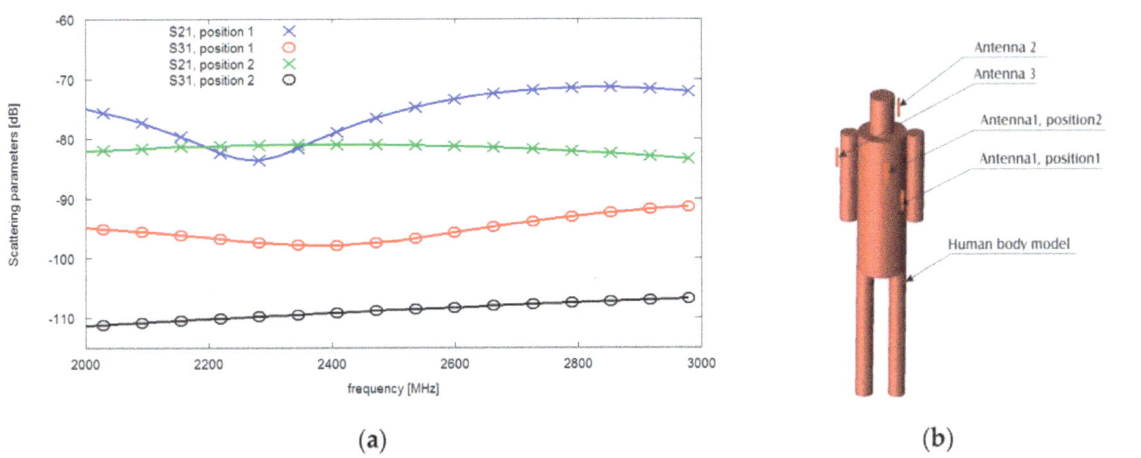

Figure 4. Scattering matrix parameter *vs.* frequency (**a**) and position of the chest node (**b**).

The antennas in this case were tuned to 2.4 GHz. The position 1 of the chest node corresponds to the initial location of the pocket in which the node was placed. In this case $S_{21} = -79$ dB and $S_{31} = -98$ dB. The position 2 of antenna No. 1 corresponds to the closest possible location of this antenna to the antenna No. 3 within the feasibility region (a possible location of the pocket). Even though the distance is smaller, the attenuation of the radio wave is higher due to propagation effects in the proximity of the body. For this case $S_{21} = -81$ dB and $S_{31} = -110$ dB. This is motivates us to carry out a thorough study of antenna placement with optimization algorithm and it makes the distance based approach inapplicable. The path loss also depends significantly on the surrounding multipath environment, but for clarity we neglect this issue in the paper and concentrate on a free-space case. A discussion of an indoor multipath propagation scenario can be found in [21].

Statement of the Optimal Design Problem

The optimal design problem can be posed as follows:

Having fixed the position of both a head sensor (antenna 2) and an arm sensor (antenna 3), find the position of the chest sensor (antenna 1) minimizing both S_{21} and S_{31} scattering parameters in the frequency band B between f_L and f_H (for 2.4 GHz ISM band we assumed that $f_L = 2.4$ GHz and $f_H = 2.5$ GHz while for 5GHz ISM band $f_L = 5.7$ GHz and $f_H = 5.9$ GHz).

The position of antenna 1 is defined by distance *d*, height z_p and azimuthal angle ϕ_p. Therefore, for a given frequency *f*, the scattering parameters depend on the triplet of design variables (d, z_p, ϕ_p); the formal statement follows:

Starting from a guess solution x_0, find

$$\inf_{x\in\Omega} \sup_{f\in B} [S_{21}(x,f), S_{31}(x,f)] \ , \ x = (d, z_p, \varphi_p) \ , \ \Omega \subset \Re^3 \ , \ B = [f_L, f_H] \tag{2}$$

subject to the solution of the field analysis problem; in Equation (2) Ω is the feasible region *i.e.*, the set of admissible positions for antenna 1 on the chest surface (Figure 5). Apparently, a min-max problem is originated: given a feasible solution x, the field analysis is performed for a discrete set of frequencies f_k, $k = 1, n_f$ (where n_f is equal to 227 for given FFT parameter of FDTD method) within the bandwidth B assuming the six-cylinder human body model. Subsequently, S_{21} and S_{31} scattering parameters are computed for each frequency f_k in the set. Eventually, the objective function value is the highest between S_{21} and S_{31} values within the bandwidth B.

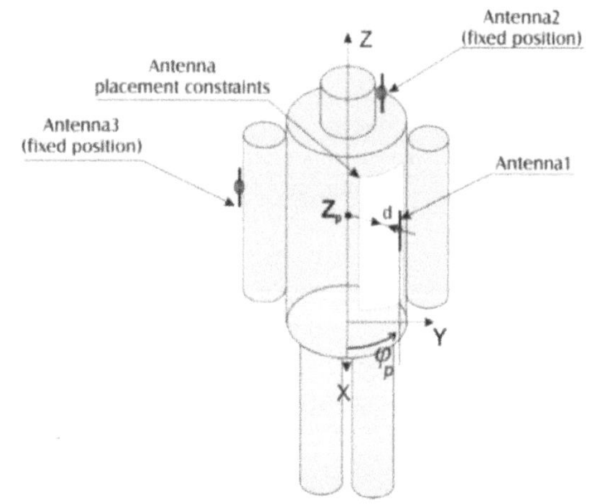

Figure 5. Six-cylinder model and feasible region (chest node region).

A numerical approximation to the solution of Equation (2) is found by means of the optimization algorithm presented in the following paragraph.

Optimization Algorithm

Over the last decade, evolutionary algorithms (EAs) have proven to be successful in solving optimization problems in the area of computational electromagnetism, see e.g., [22,23,24,25,26]. In fact, EAs do not require the objective function gradient to identify the search direction, and are global-optimum oriented in the case of non-convex objectives. Moreover, they are able to find a solution in a short time when the optimization problem is characterized by several constraints or exhibits discontinuities, while classical methods based on gradient information cannot be applied. Often, EAs are inspired by the models of natural phenomena: For instance, "island" paradigms mimic the phenomenon of populations evolving without exchange with the external environment [27]. In turn, the biogeography-based optimization (BBO) algorithm [28], which is inspired by the geographical distribution of species in an ecosystem, has been applied in the antenna design [29,30]. Particle swarm optimization, which imitates the behavior of a swarm of bees during their food-searching activity, was introduced in the antenna community [31] and stimulated numerous contributions, both under single-objective and under multi-objective context [32]. More recently, wind-driven optimization, inspired by the wind motion, which takes place in the high atmosphere to compensate the pressure imbalance, has been proposed and applied in the antenna array design [33]. Numerous techniques have been considered to reduce optimization time, such as more cost-effective electromagnetic codes (e.g., if applicable, the boundary element method instead of the finite difference time domain method), simplified simulation models (e.g., homogenous body phantoms instead of multi-tissue), massively parallel computing (e.g., using graphics card) or surrogate modelling which may require fewer evaluations of the objective function [34,35]. The approach taken in this paper, which adopts a simplified, homogenous model of the human body, turned out to sufficiently reduce the optimization time combined with an EA EStra [26] lowest order (*i.e.*, a single parent generates a single offspring) algorithm. With the direct approach involving Remcom XFdtd and a heterogeneous human body model the computation time turned out to be excessive, considering a great number of design variants taken into account.

In this paper EStra [27] was used as an evolutionary algorithm of the lowest order (*i.e.*, a single parent generates a single offspring).

In EStra each out of n_v design variables is represented by two pieces of information, *i.e.*, mean value m, and standard deviation d. The search in the n_v-dimensional design space begins in a region of radius d_0 (initial standard deviation) centered at the starting point m_0 (initial mean value); m_0 is externally provided, while d_0 is internally calculated based on the bounds boxing the variation of the design variables. After initialization, the optimization loop is based on three major steps as follows:

- Setting $\mathbf{m} = \mathbf{m}_0$ and $\mathbf{d} = \mathbf{d}_0$ at the initial iteration, the generation of the design vector $\mathbf{x} = \mathbf{m} + \mathbf{d}u$ takes place, resorting to a stochastic sample $u \in (0,1)$; generally, u is a normally distributed sample. Vector $\mathbf{x}$ is accepted if it fulfils bounds and constraints; otherwise, a new design vector is generated until the corresponding design point falls inside the feasible region.
- The associated objective function $f(\mathbf{x})$ is then evaluated, and the test $f(\mathbf{x}) < f(\mathbf{m})$ is performed; if the test is successful, $\mathbf{m}$ is replaced by $\mathbf{x}$ (*mutation* operator), otherwise $\mathbf{m}$ is conserved.
- The last step rules the size of the search region that will be used for the successive iteration: when a design point that is better than the current one is found, the radius of the search region is increased around the new point to search for further improvements; if no improvement is found, the radius of the search region is gradually decreased, up to convergence.

In contrast to gradient-based methods of numerical optimization, EStra is suitable for non-convex, non-smooth and discontinuous functions; in particular, it is able to approximate the global minimum of the objective function, regardless of the starting point. The implementation of lowest-order EStra originated a cost-effective algorithm of numerical optimization, with respect to a multi-member implementation, where at each iteration a set of parent individuals recombines to generate a set of offspring individuals; this feature is particularly important when the solution of the direct problem relies on a time-consuming field analysis.

In fact, successful applications of EStra to the design of low-frequency electromagnetic setups can be found in [26], while its combination with a FDTD solver for body-worn wireless sensor network optimization is an original concept presented in this paper. The subject is an emerging one. In fact, in [36] a flow optimization model for designing WBAN is provided; attention is focused on network installation cost and energy consumption of antennas, which are both design criteria to minimize. In [9] a robust optimization model and a new algorithm for designing WBAN in terms of topology and routing, taking into account data generation uncertainty, is presented. Also nature-mimicking algorithms are successfully exploited in WBAN design: for instance, [7] presents a fast evolutionary algorithm inspired by an ant colony; the case study is the optimal design of the body sensor network defined in [8].

As far as the implementation of the optimization procedure proposed in this paper is concerned, the following remarks can be put forward:

The XFdtd package provides users with a scripting language based on ECMAScript, in which one can prepare an antenna geometry model, as well as meshing and simulation procedures.

In contrast, EStra optimization algorithm was originally coded in Mathworks Matlab and for further work Octave [37] environment was applied because it is freeware and compatible with original Matlab scripts. The Octave and XFdtd scripts are combined and operate in turns to generate a new antenna position and calculate the objective function. Synchronization and data exchange between XFdtd and Octave have been solved by file exchange (Figure 6).

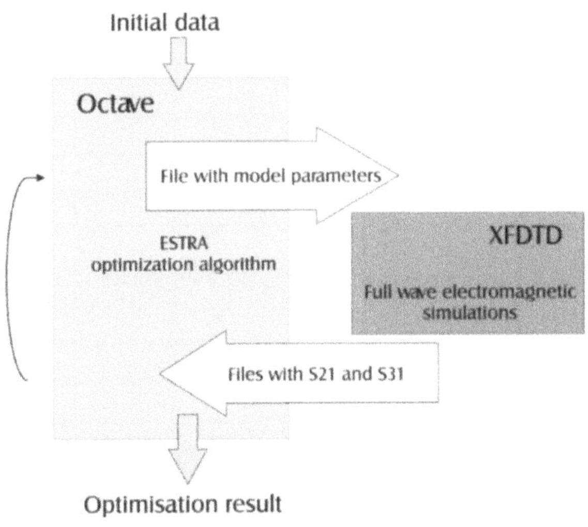

Figure 6. Optimization algorithm flow chart.

The position of the antenna towards the chest was controlled using a cylindrical coordinate system: Z_p—Antenna vertical position, ϕ_p—Antenna azimuthal position and d—Antenna distance to the body surface, presented in Figure 5. Due to the utilization of the simplified cylindrical body model it was easy to use three design variables for creating the antenna location constraints region which was equidistant to the cylindrical body surface.

The numerical model of the network under optimization consists of a human body model (1.8 m tall) and the dipole antennas of 1 mm diameter and the length equal to 53 mm for the 2.4 GHz case and 20 mm for the 5.8 GHz case. In order to decrease the computation time, we chose an adaptive meshing in XFdtd. For the simulations in 2.4 GHz band the voxel within the body model has an edge length equal to 10 mm, and 5 mm for 5.8 GHz. In the regions where antennas were located the voxel size was reduced to 0.5 mm. The model for 2.4 GHz consists of 8,276,136 voxels. With a computer equipped with two Nvidia Tesla C2075 processors the simulation time was ca 30 s. In the case of simulations for 5.8 GHz band, 13,306,865 voxels were used and the simulation time was ca 60 s.

As already clarified, we focus on the effect of the human body impact on the path loss assuming free-space environment in which the person is located. This simplification corresponds to a scenario for outdoor locations where multipath propagation is highly reduced. To simulate this effect an absorbing boundary condition was applied for simulations in Xfdtd.

SIMULATIONS AND RESULTS

In the optimization process the chest antenna location constraints were defined as a region by the three following ranges of design variable values: Rotation angle 0.2 rad < ϕ_p < 1.55 rad, antenna vertical position 0 m < z < 0.6 m and antenna distance to the body 0 < d < 0.01 m. These values follow from the application of our research to the firefighters system under development, in which the defined region corresponds to the possible predetermined localization of the jacket pocket for the node. Due to the presence of other firefighter equipment, it was only possible to place the pocket for the chest node in the left-front part of the jacket (two other sensors have a fixed location on the right arm and the left side of the head). We assumed that the distance between the antenna and the body could be adjusted slightly by placing the pocket on the inner or the outer side of the jacket or by using styrofoam spacers. We also assumed that the optimization process will start from the location of the chest antenna defined by the ϕ_p = 1 rad Z_p = 0.4 m and d = 0.001 m (antenna azimuthal position or rotation angle, antenna vertical position, antenna distance to the body surface according to Figure 5). It is the location of an already existing jacket pocket, which we intended to use for the chest module.

During the first stage of the optimization experiment a sub-problem of Equation (2) was considered, defined by Equation (3), where we are considering only two antennas operating in the ISM 2.4 GHz band: the head antenna with a fixed position and the chest antenna with a variable position.

$$\text{Find } \inf_{x\in\Omega} \sup_{f\in B} S_{21}(x,f) \ , \ x = (z_p, \varphi_p) \ , \ \Omega \subset \Re^3 \ , \ B = [f_L, f_H] \tag{3}$$

with f_L = 2.4 GHz and f_H = 2.5 GHz.

Therefore in Equation (3) only S_{21} is present. The objective function (to be maximized) was chosen as the minimum value of S_{21} in the 2.4–2.5 GHz frequency band. Consequently, a two-variable optimization was considered first. The third design variable—Antenna distance—Was fixed at d = 0.001 m. The results obtained after 51 iterations are presented in Figure 7 (blue line). The initial value of S_{21} objective function was −81 dB and the final was −69 dB. We may observe that the radio link quality has improved because the signal received in the optimized case has 12 dB (−81 dB *vs.* −69 dB) greater amplitude.

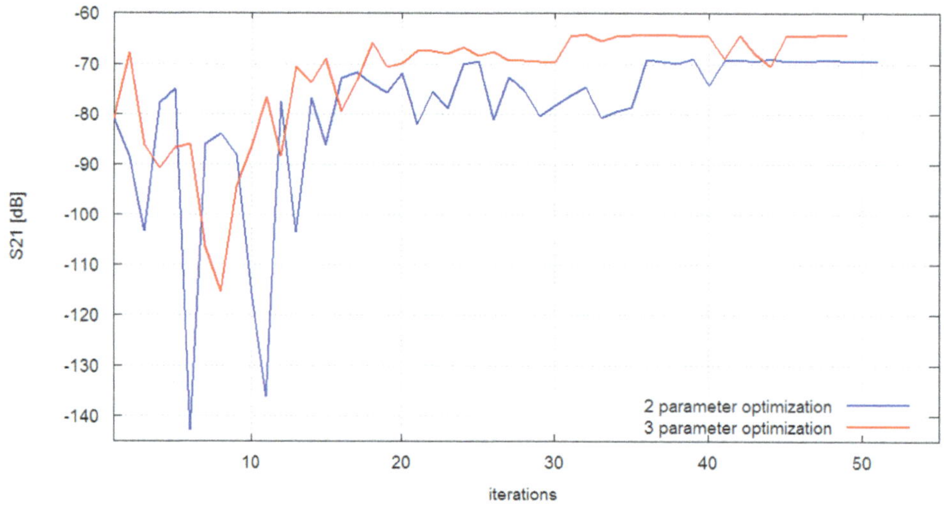

Figure 7. Optimization process for the two-antenna scenario; the objective function isS_{21}.

Subsequently, the full problem defined by Equation (2) was considered, *i.e.*, optimization experiments with two antennas operating at 2.4 GHz was repeated, but this time all three variables of the chest antenna position were identified (antenna rotation angle, antenna vertical position, antenna distance to the surface of the body). The results obtained after 49 iterations are presented in Figure 7 (red line). The initial value of S_{21} objective function was −81 dB, and the final was −64 dB. This means that also in this case the link quality has improved because the signal received in the optimized setup has 17 dB greater amplitude. It can be concluded that the three-variable optimization may provide a greater improvement of the objective function. This result may seem trivial considering that a better solution should exist in the three-parameter optimization space compared to the two-parameter optimization space. However, the problem is that this solution may not be found a particular optimization algorithm. The improvement achieved in our experiments suggests that EStra algorithm is suitable for the class of problems under consideration.

We used the three-variable approach to optimize the three-antenna scenario operating at 2.4 GHz. In this case the objective function was defined as the minimum value from among two parameters: S_{21} and S_{31} in the frequency range: 2.4–2.5 GHz. For the same starting point as previously, the initial value of objective function was −109 dB. After 61 iterations it was improved to −80 dB, which is a gain of 29 dB in the radio link budget. The improvement of the objective function in the optimization process *vs.* iteration number is presented in Figure 8.

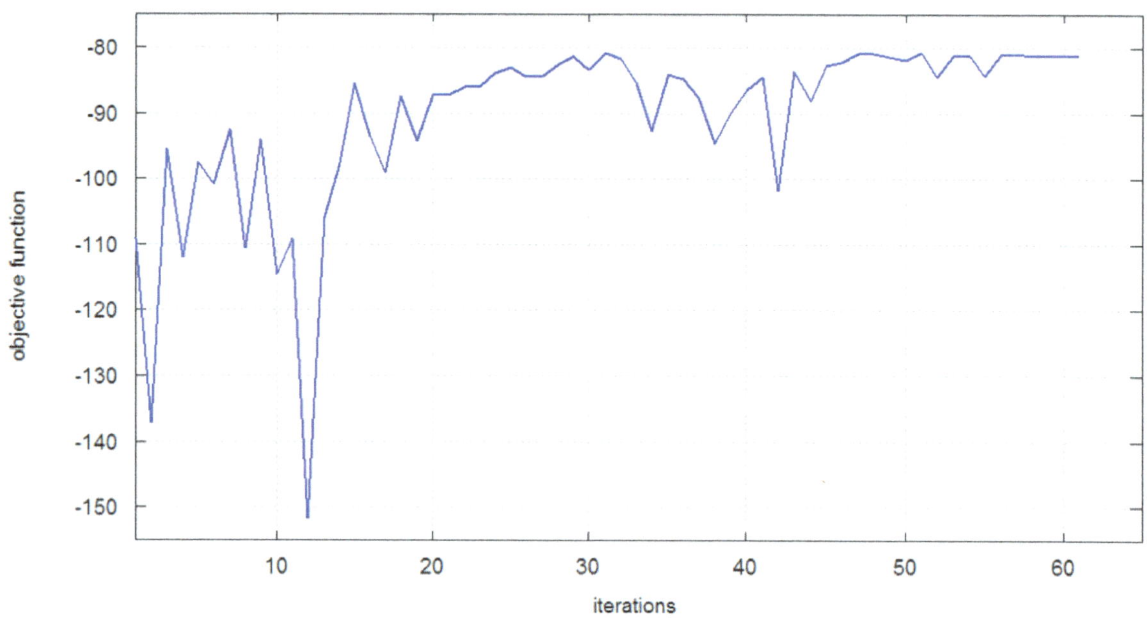

Figure 8. Optimization process with the three-antenna scenario; the objective function is the minimum of S_{21} and S_{31}.

In Figure 9 we present the scattering matrix parameter values (S_{21} and S_{31}) before and after optimization for 2.4 GHz. The improvement of S_{31} is visible while S_{21} changed only slightly.

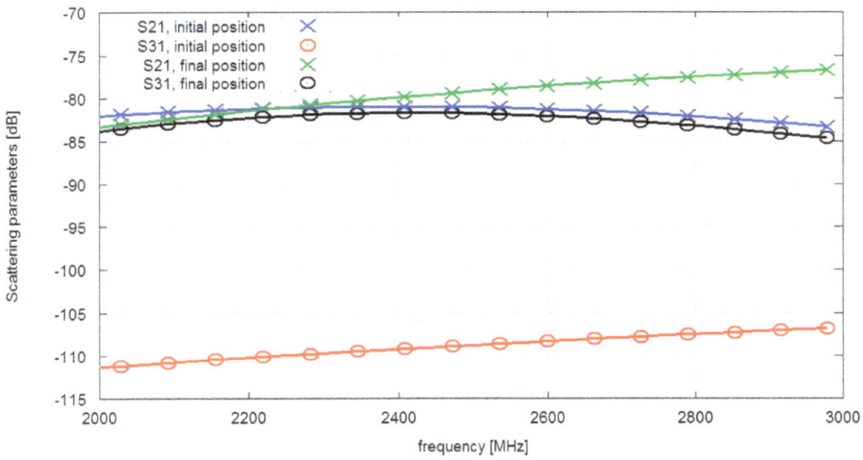

Figure 9. Scattering matrix parameter values *vs.* frequency before and after optimization for 2.4 GHz frequency band.

The good performance of the optimization procedure that we observed at 2.4 GHz was then verified in the 5.8 GHz ISM band. For this purpose we used shorter dipoles (20 mm long) which were tuned for the frequency equal to 5.8 GHz. The optimization starting point and the objective function definition were the same as for 2.4 GHz. The results were obtained after 67 iterations. The initial value of the objective function (the lowest value of S_{21} and S_{31} within the frequency range from 5.7 GHZ to 5.9 GHz) was −124 dB and it improved after 67 iterations to −96 dB. Figure 10 presents the scattering matrix parameter values before and after optimization. The results of all the described optimization experiments are gathered in Table 1.

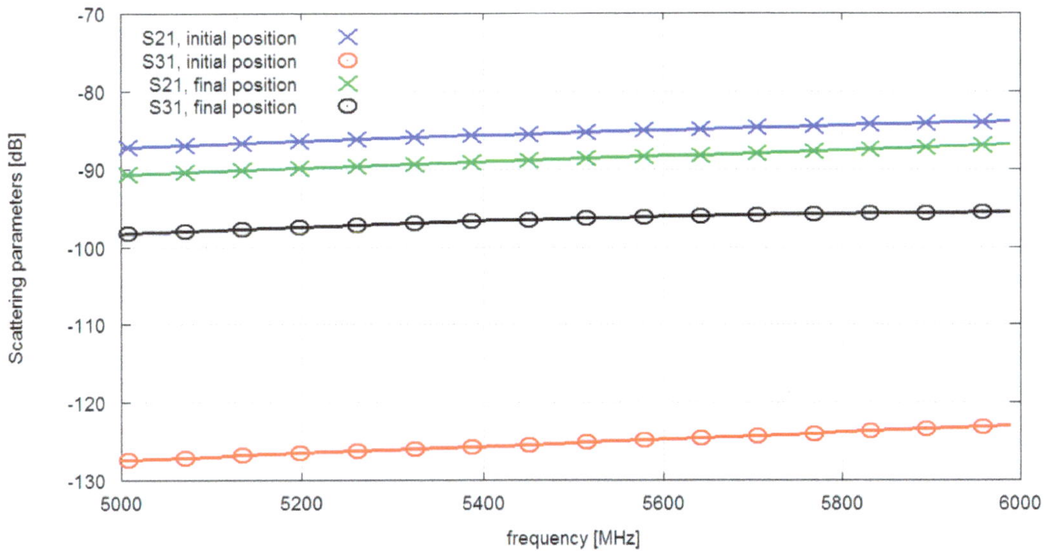

Figure 10. Scattering matrix parameter values *vs.* frequency before and after optimization for 5.8 GHz frequency band.

Table 1. Optimization with the evolution strategy algorithm

	Optimization Scenario			
	2 Antennas, 2.4 GHz Fixed Distance	**2 Antennas, 2.4 GHz Variable Distance**	**3 Antennas, 2.4 GHz Variable Distance**	**3 Antennas, 5 GHz Variable Distance**
Final Value of Angle [rad]	0.63	0.5	0.22	0.35
Final Value of Vertical Coordinate [m]	0.567	0.379	0.18	0.335
Final Value of Antenna Distance [m]	0.001	0.007	0.007	0.0045
Initial Value of Objective Function [dB]	−81	−81	−109	−124

Final Value of Objective Function [dB]	−69	−64	−80	−96
Improvement of Objective Function [dB]	12	17	29	28

We have assessed the validity of our simplified human body model using a much more detailed numerical anthropomorphic, heterogeneous body model (NMR Hershey) which is available in Remcom Xfdtd. We selected the model version comprising 5 mm voxels. The final antenna position obtained in the optimization process was applied to the model of the sensor network but this time with the heterogeneous numerical phantom. The simulations were performed at 2.4 GHz with three antennas/ nodes. The results are presented in Figure 11. A good agreement between the values obtained with the simplified model and the heterogeneous model can be observed, which advocates for the usage of the simplified one.

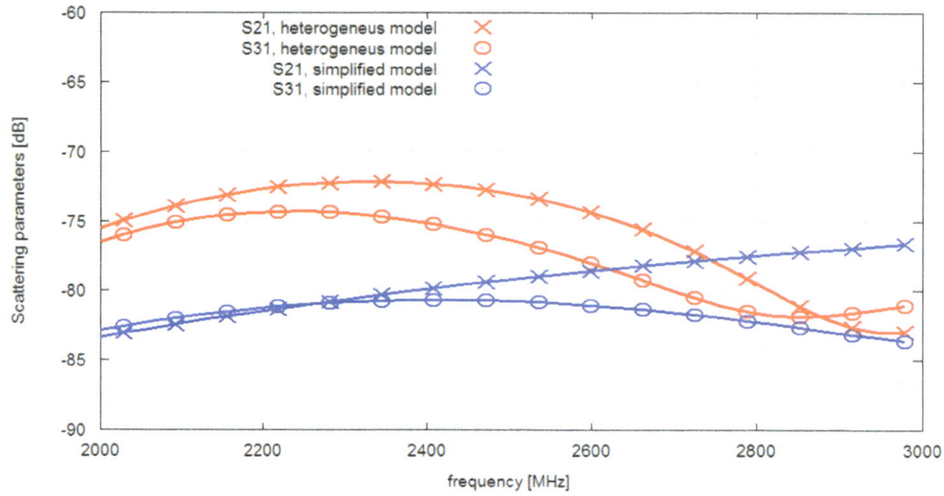

Figure 11. Comparison of results obtained with simplified model and heterogeneous model.

CONCLUSIONS

In this paper we present original results of our investigation of an evolutionary optimization of a simple body-worn wireless sensor network comprising three on-body nodes. This case was motivated by the practical need of improving performance of a WBAN designed to monitor firefighters' vital signs and environmental conditions. A number of design assumptions and constraints were imposed by our application, for example the fixed position of the head and arm nodes predetermined by the types of sensors they use. One degree of freedom available to us was the placement of the chest node in the front-left part of the jacket. In this paper we show that even with such strong design constraints it is possible to considerably decrease the path loss between nodes using an automated optimization procedure based on evolutionary computing (EStra) combined with a computational electromagnetics code (XFdtd) to evaluate the value of the objective function. Depending on the starting point, the improvement will differ but in our experiments it reached about 20 dB. This improvement would result in reduced power consumption of the WBAN node transmitters.

The three-variable optimization that allowed the change of both the antenna position and the antenna distance to the human body provided better results than the two-variable optimization with a fixed distance. However, this result may be difficult to apply in practice due to the difficulty of controlling the distance e.g., by adding dielectric spacers into the jacket. We show that the proposed methodology of sensor network optimization provides satisfactory results both in the 2.4 GHz band and in the 5.8 GHz band. Since the simplified human body model was positively verified for a broad range of frequencies [20], probably the same optimization routine could be successfully applied in other frequency bands.

We also show that the path loss values obtained at 2.4 GHz with the simplified homogeneous model are comparable to the results achieved with a much more detailed heterogeneous anthropomorphic multi-tissue model. In our study, one advantage of using the simplified cylindrical model is the ease of defining and controlling the antenna displacement with the rotation angle and height only, while keeping a constant antenna distance to the body. Another advantage is reduced computation time resulting from sparser gridding because the complexity of inner organs does not have to be represented. Our approach to the optimization problem is computationally effective. A single iteration lasted less than one minute and the overall simulation time was well below one hour.

Further investigation is planned for ultra wideband WBAN optimization.

AUTHOR CONTRIBUTIONS

This paper was prepared through a collective effort of all the authors. In particular:

Łukasz Januszkiewicz has contributed: Concept of the research, objective function definition, constraints definition, XFDTD script with a simplified human body model, Octave script for objective function calculation, performing the simulations, analysis of the results, preparation of diagrams and figures, preparation of the text;

Sławomir Hausman has contributed: XFdtd script for geometry generation, simulation setup and gridding control, research into the state-of-the-art in WBAN, compilation of references, interpretation of the simulation results, preparation of the text and its revision;

Paolo Di Barba has contributed: state-of-the-art and references, optimization algorithm concept, description of the optimization algorithm, Octave scripts encoding the optimization algorithm, formal definition of the optimization problem, statement of objective function and design variables.

REFERENCES

1. Alam, M.M.; Hamida, E.B. Surveying Wearable Human Assistive Technology for Life and Safety Critical Applications: Standards, Challenges and Opportunities. *Sensors* **2014**, *14*, 9153–9209.
2. Movassaghi, S.; Abolhasan, M.; Lipman, J.; Smith, D.; Jamalipour, A. Wireless Body Area Network: A Survey. *IEEE Commun.* **2014**, *16*, 1658–1686.
3. Lai, X.; Liu, Q.; Wei, X.; Wang, W.; Zhou, G.; Han, G. A Survey of Body Sensor Networks. *Sensors* **2013**, *13*, 5406–5447.
4. FLORA—Wearable Electronic Platform. Available online: https://www.adafruit.com/products/659 (accessed on 20 January 2016).
5. Cheng, J.; Lukowicz, P.; Henze, N.; Schmidt, A.; Amft, O.; Salvatore, G.A.; Troster, G. Smart Textiles: From Niche to Mainstream Publication. *IEEE Pervasive Comput. Mag.* **2013**, *12*, 81–84.
6. Buschmann, J.P.; Huang, J. Pulse Oximetry in the External Auditory Canal—A New Method of Mobile Vital Monitoring. *IEEE Sens. J.* **2012**, *12*, 671–676.
7. D'Andreagiovanni, F.; Nardin, A. Towards the fast and robust optimal design of wireless body area networks. *Appl. Soft Comput.* **2015**, *37*, 971–982.
8. D'Andreagiovanni, F.; Krolikowski, J.; Pulaj, J. A fast hybrid primal heuristic for multiband robust capacitated network design with multiple time periods. *Appl. Soft Comput.* **2015**, *26*, 497–507.
9. Elias, J. Optimal design of energy-efficient and cost-efficient wireless body area networks. *Ad Hoc Netw. Arch.* **2014**, *13*, 560–574.
10. Bringuier, J.N.; Mittra, R. Electromagnetic Wave Propagation in Body Area Networks Using the Finite-Difference-Time-Domain Method. *Sensors* **2012**, *12*, 9862–9883.
11. *XFDTD 7.3 Reference Manual*; Remcom Inc.: State College, PA, USA, 2012.
12. Xu, X.G.; Eckerman, K.F. *Handbook of Anatomical Models for Radiation Dosimetry*; Taylor & Francis: Boca Raton, FL, USA, 2010.
13. Robin, A. Electromagnetic Modelling of Human Tissues and its Application on the Interaction between Antenna and Human Body in the BAN Context. PhD Thesis, Université Paris-Est, Paris, France, 2009.
14. Panagamuwa, C.J.; Whittow, W.G.; Edwards, R.M.; Vardaxoglou, J.C. Experimental Verification of a Modified Specific Anthropomorphic Mannequin (SAM) Head used for SAR Measurements. In Proceedings of the Antennas and Propagation Conference, Loughborough, UK, 2–3 April 2007; pp. 261–264.
15. Arkko, A. Properties of numerical SAM phantom used in mobile phone antenna simulations. In Proceedings of the Antennas and Propagation Society International Symposium, Washington, DC, USA, 3–8 July 2005; pp. 450–453.
16. Uno, Y.; Saito, K.; Takahashi, M.; Ito, K. Structure of cylindrical tissue-equivalent phantom for medical applications. In Proceedings of the Electromagnetics in Advanced Applications (ICEAA), Sydney, Australia, 20–24 September 2010; pp. 406–409.
17. Lim, H.B.; Baumann, D.; Li, E.P. A Human Body Model for Efficient Numerical Characterization of UWB Signal Propagation in Wireless Body Area Networks. *IEEE Trans. Biomed. Eng.* **2011**, *58*, 689–697.
18. Iswandi; Aoyagi, T.; Kim, M.; Takada, J. The utilization of body skeleton model for modeling the dynamic BAN channels. In Proceedings of the 6th European Conference on Antennas and Propagation (EUCAP), Prague, Czech Republic, 26–30 March 2012; pp. 540–543.
19. Januszkiewicz, Ł.; Hausman, S. Simplified human phantoms for narrowband and ultra-wideband body area network modelling. *COMPEL Int. J. Comput. Math. Electr. Electron. Eng.* **2015**, *34*, 439–447.
20. Januszkiewicz, Ł. Simplified human body models for interference analysis in the cognitive radio for medical body area networks. In Proceedings of the 8th International Symposium on Medical Information and Communication Technology (ISMICT), Firenze, Italy, 2–4 April 2014; pp. 1–5.
21. Hausman, S.; Januszkiewicz, Ł. Impact of Indoor Environment on Path Loss in Body Area Networks. *Sensors* **2014**, *14*, 19551–19560.
22. Kim, H.K.; Chong, J.K.; Park, K.Y.; Lowther, D.A. Differential evolution strategy for constrained global optimization and application to practical engineering problems. *IEEE Trans. Magn.* **2007**, *43*, 1565–1568.
23. Wanner, E.F.; Guimarães, F.G.; Takahashi, R.H.C.; Lowther, D.A.; Ramírez, J.A. Multiobjective memetic algorithms with quadratic approximation-based local search for expensive optimization in electromagnetics. *IEEE Trans. Magn.***2008**, *44*, 1126–1129.
24. Crevecoeur, G.; Sergeant, P.; Dupré, L.; van de Walle, V. A two-level genetic algorithm for electromagnetic optimization. *IEEE Trans. Magn.* **2010**, *46*, 2585–2595.
25. Watanabe, K.; Campelo, F.; Iijima, Y.; Kawano, K.; Matsuo, T.; Mifune, T.; Igarashi, H. Optimization of inductors using evolutionary algorithms and its experimental validation. *IEEE Trans. Magn.* **2010**, *46*, 3393–3396.
26. Di Barba, P.; Mognaschi, M.E.; Palka, R.; Savini, A. Optimization of the MIT Field Exciter by a Multiobjective Design.*IEEE Trans. Magn.* **2009**, *45*, 1530–1533.

27. Coelho, L.S.; Alotto, P. Electromagnetic optimization using a cultural self-organizing migrating algorithm approach based on normative knowledge. *IEEE Trans. Magn.* **2009**, *45*, 1446–1449.

28. Simon, D. Biogeography-based optimization. *IEEE Trans. Evol. Comput.* **2008**, *12*, 702–713.

29. Singh, U.; Kumar, H.; Kamal, T.S. Design of Yagi-Uda Antenna Using Biogeography Based Optimization. *IEEE Trans. Antennas Propag.* **2010**, *58*, 3375–3379.

30. Singh, S.; Mittal, E.; Sachdeva, G. NSBBO for Gain-Impedance Optimization of Yagi-Uda Antenna Design. In Proceedings of the World Congress on Information and Communication Technologies, Trivandrum, India, 30 October–2 November 2012; pp. 856–860.

31. Robinson, J.; Rahmat-Samii, Y. Particle Swarm Optimization in Electromagnetics. *IEEE Trans. Antennas Propag.* **2004**,*52*, 397–407.

32. Jin, N.; Rahmat-Samii, Y. Advances in Particle Swarm Optimization for Antenna Designs: Real-Number, Binary, Single-Objective and Multiobjective Implementations. *IEEE Trans. Antennas Propag.* **2007**, *55*, 2745–2757.

33. Bayraktar, Z.; Komurcu, M.; Bossard, J.A.; Werner, D.H. The Wind Driven Optimization Technique and its Application in Electromagnetics. *IEEE Trans. Antennas Propag.* **2013**, *61*, 2745–2757.

34. Queipo, N.E.; Haftka, R.T.; Shyy, W.; Goel, T.; Vaidyanathan, R.; Tucker, P.K. Surrogate-based analysis and optimization. *Prog. Aerosp. Sci.* **2005**, *41*, 1–28.

35. Song, X.; Rotaru, M.; Sykulski, J.K. Exploration *versus* exploitation using kriging surrogate modelling in electromagnetic design. *COMPEL Int. J. Comput. Math. Electr. Electron. Eng.* **2012**, *31*, 1541–1551.

36. Chen, M.; Gonzalez, S.; Vasilakos, A.; Cao, H.; Leung, V.C. Body Area Networks: A Survey. *ACM/Springer Mob. Netw. Appl.* **2011**, *16*, 171–193.

37. Eaton, J.W.; Bateman, D.; Hauberg, S.; Wehbring, R. GNU Octave: Free Your Numbers. Available online: https://www.gnu.org/software/octave/octave.pdf (accessed on 18 October 2013).

CHAPTER 3 PIEZOELECTRIC AND ELECTROMAGNETIC CONVERSION MECHANISMS

Zhenlong Xu, Xiaobiao Shan, Danpeng Chen and Tao Xie

Keywords: Piezoelectric and electromagnetic; hybrid energy harvester; magnetic interaction; multi-frequency; low-frequency vibration

INTRODUCTION

Vibration energy harvesting is considered as a promising alternative approach to power wireless sensors nets and ultra low power microelectronics devices [1,2]. Piezoelectric and electromagnetic energy harvesting are the most reported transduction mechanisms. Traditional linear vibration energy harvester (VEH) has been usually considered as single degree-of-freedom (1DOF) model [3], which has a narrow operating frequency bandwidth. Once the excitation frequency shifts away from the resonant frequency, the performance of the VEH drops dramatically. In practice, the vibration source is random and time-varying. Some vibration sources especially exhibit multiple discrete frequency peaks in a large frequency range. For example, HVAC vents have three vibration frequencies of 21.8, 29 and 127.3 Hz [4]. Laptops exhibit two vibration frequencies of 43.2 and 90.2 Hz [5]. To effectively scavenge energy from vibration sources with multiple discrete frequencies is the first concerned research question. As we know, the resonant frequency of the piezoelectric energy harvester (PEH) increases sharply when the device size reduces to micro level. To date, most of the micro-scale PEHs operate at frequencies of more than 100 Hz [6,7], even at the level of 1 kHz [8]. This limits their applications to harvest some ambient vibration energy, which occurs in the frequency range below 30 Hz, such as those caused by wind or human motions. The micro electromagnetic energy harvester (EMEH) [9] exhibits lower power density due to the decreased flux density and smaller velocity of the magnet. Therefore, another concerning question is how to improve the performance of VEHs in the low frequency range.

To improve the performances of VEHs, a number of solutions have been proposed by researchers. These solutions include generator array [10,11,12], mechanical impact [13,14], nonlinear spring stiffness [15,16], magnetic coupling [17,18,19,20], hybrid conversion mechanism [21,22] and active/passive tuning techniques [23,24,25]. However, these generators are only effective in a continuous frequency range or at a single dominant resonant frequency. To harvest energy from vibration sources with discrete frequency peaks over a wide range, Chew and Li [26] reported a PEH composed by a series of piezoelectric beams, which are connected end-to-end. The resonant frequencies vary from 100 to 1000 Hz with different numbers of beams. Ferrari *et al.* [27] presented an array-type PEH made of three piezoelectric bimorphs. The

Citation: Xu, Z.; Shan, X.; Chen, D.; Xie, T. Piezoelectric and Electromagnetic Conversion Mechanisms. Appl. Sci. 2016, 6, 10. doi:10.3390/app6010010.

respective three fundamental resonant frequencies are about 113, 183 and 281 Hz. Yang *et al.* [28] designed a multi-frequency EMEH with three resonant modes of 369, 938 and 1184 Hz. Yang *et al.* [29] added a nonlinear spring oscillator to a PEH to scavenge the multi-frequency vibration energy. Liu *et al.* [30] developed a multi-mode EMEH for three-dimensional (3D) excitation at different frequencies. Tadesse *et al.* [31] reported a hybrid energy harvester (HEH) using piezoelectric and electromagnetic mechanisms. It consisted of a piezoelectric cantilever beam and a permanent magnet attached at the tip, which oscillated within a stationary coil. The first and the second resonances of such device were 20 Hz and 300 Hz. Wang *et al.* [32] and Shan*et al.* [33] analyzed the characteristics of the two degree-of-freedom (2DOF) HEH. Among these works, the resonant frequencies of VEHs are fixed when the prototypes are assembled. It is not convenient to tune the resonances according to the change of excitation frequency.

In this paper, we report a novel tunable multi-frequency HEH, using piezoelectric and electromagnetic conversion mechanisms. The design procedure is explained by considering two generating elements coupled with magnetic interaction. An electromechanical coupling model of the HEH is established. The energy harvesting characteristics are numerically simulated and verified by experiments. The proposed HEH can effectively scavenge the vibration energy of two distributed resonant frequencies. Compared with the single energy harvesting mechanism, the proposed HEH not only generates more electric power but also broadens the operation frequency bandwidth in the low-frequency range. The magnetic interaction can be used to alter the resonant frequencies of HEH.

DESIGN AND WORKING PRINCIPLE

Figure 1 illustrates the schematic diagram of the proposed tunable multi-frequency HEH. It consists of a cantilevered piezoelectric oscillator with a cylindrical magnet as a proof mass and a magnetic oscillator attached on the frame. Two oscillators are coupled through the magnetic interaction, which behaves as the repulsive force. An induction coil is attached around the magnetic oscillator. R_1 and R_2 are load resistances connected to piezoelectric patches and induction coil, respectively. When the HEH is subjected to the external excitation, the mechanical strain is converted into voltage in the piezoelectric layers. Meanwhile, the relative movement between the magnet and coil induces a current in the wire, based on Faraday's law. Note that the resonant frequencies of two oscillators are higher than their own natural frequencies due to the repulsive magnetic force. The natural frequency of the magnetic oscillator is designed to be lower than that of the piezoelectric oscillator, so that it can be used to scavenge the low-frequency vibration energy. Meanwhile, the piezoelectric oscillator can harvest the discrete high-frequency vibration energy.

The advantages of this design are as follows:

1. the output power is enhanced, because the kinetic energy can be simultaneously harnessed through piezoelectric effect and electromagnetic induction;
2. it is capable of simultaneously responding to excitation from multiple frequencies;
3. its resonant frequencies can be tuned by changing the separation distance between two magnets;
4. it has a wider bandwidth in the low-frequency range.

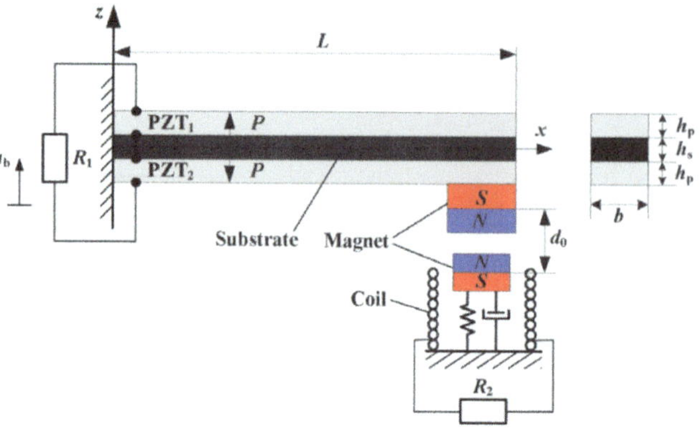

Figure 1. Schematic diagram of the proposed hybrid energy harvester.

THEORETICAL MODELING

Electromagnetic Coupling Coefficient and Magnetic Force

The electromagnetic transduction mechanism is based on Faraday's law of induction. In this paper, the magnetic oscillator is arranged to move inside the induction coil along the center axis, as depicted in Figure 2. The electromotive force U_{emf} is derived from the rate of change of the magnetic flux Φ_B as follows [34]

$$U_{\mathrm{emf}} = -\frac{\mathrm{d}\Phi_B}{\mathrm{d}t} = -\frac{\mathrm{d}\Phi_B}{\mathrm{d}z}\cdot\frac{\mathrm{d}z}{\mathrm{d}t} = -\frac{A\mathrm{d}B}{\mathrm{d}z}\cdot\dot{z} = \theta_{\mathrm{em}}\dot{z} \tag{1}$$

where B is the magnetic flux density. A is the area enclosed by the wire loop, which is considered as a constant value. θ_{em} is called the electromagnetic coupling coefficient.

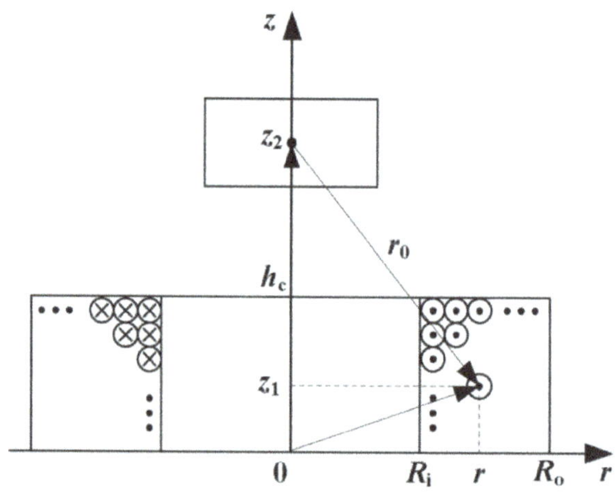

Figure 2. The relative position between magnet and coil.

The magnet is considered as a dipole with magnetic moment vector $\boldsymbol{m}_2$ [35], the magnetic field generated by the magnet can be expressed as

$$\boldsymbol{B} = -\frac{\mu_0}{4\pi}\left[\frac{\boldsymbol{m}_2}{r_0^3} - \frac{3\boldsymbol{r}_0(\boldsymbol{m}_2\cdot\boldsymbol{r}_0)}{r_0^5}\right] \tag{2}$$

where $\mu_0 = 4\pi \times 10^{-7}$ Hm^{-1} is the permeability of vacuum. $\boldsymbol{r}_0$ is the distance vector from the center of the magnet to an arbitrary single wire. Assumed that the position coordinates of an arbitrary wire and magnet core are (r, z_1) and $(0, z_2)$, respectively. The corresponding expression of $\boldsymbol{B}$ is

$$\boldsymbol{B}=\frac{B_rV_2}{4\pi}\cdot\frac{3(z_1-z_2)r\boldsymbol{e}_r-\left[r^2-2(z_1-z_2)^2\right]\boldsymbol{e}_z}{\left[r^2+(z_1-z_2)^2\right]^{5/2}} \tag{3}$$

where B_r and V_2 are the residual magnetic flux density and volume of the source magnet. $\boldsymbol{e}_r$ and $\boldsymbol{e}_z$ are the unit vectors alongr- and z-axis, respectively.

As can be seen from Equation (1), the magnetic flux gradient along the z-axis direction is related to the induced voltage. For this reason, the magnetic flux density along the z-axis is the focus of research. The magnetic flux through the area enclosed by the wire is

$$\Phi=\oint_s \boldsymbol{B}\mathrm{d}A=\frac{B_rV_2r^2}{2\left[r^2+(z_1-z_2)^2\right]^{3/2}} \tag{4}$$

The average magnetic flux over the cross-section of the coil can be given as

$$\overline{\Phi}=\frac{1}{A_c}\cdot\int_0^{h_c}\int_{R_i}^{R_o}\Phi\mathrm{d}r\mathrm{d}z_1 \tag{5}$$

where $A_c=(R_o-R_i)h_c$ is the coil cross-sectional area [36]. h_c, R_o, and R_i are the height, outer radius, and inner radius of the coil, respectively. Correspondingly, the total magnetic flux through the coil can be derived from $\Phi_B=f_cN\overline{\Phi}$. $f_c=NA_w/A_c$ is the coil fill factor [37]. N is the number of turns in the coil. A_w is the cross-sectional area of single wire. Consequently, the electromotive force U_{emf} can be expressed as

$$U_{\mathrm{emf}}=-\frac{\mathrm{d}\left(f_cN\overline{\Phi}\right)}{\mathrm{d}z}\dot{z}=\theta_{\mathrm{em}}\dot{z} \tag{6}$$

where θ_{em} is

$$\theta_{\mathrm{em}}=-\frac{B_rV_2f_cN}{2A_c}\left[\ln\frac{R_i+\sqrt{R_i^2+(z_2-h_c)^2}}{R_o+\sqrt{R_o^2+(z_2-h_c)^2}}+\ln\frac{R_o+\sqrt{R_o^2+z_2^2}}{R_i+\sqrt{R_i^2+z_2^2}}+\frac{R_o}{\sqrt{R_o^2+(z_2-h_c)^2}}-\frac{R_o}{\sqrt{R_o^2+z_2^2}}-\frac{R_i}{\sqrt{R_i^2+(z_2-h_c)^2}}+\frac{R_i}{\sqrt{R_i^2+z_2^2}}\right] \tag{7}$$

It is assumed that the centers of the two magnets are always vertically aligned during the vibration of the HEH. The magnetic force between the magnetic proof mass and magnetic oscillator can be calculated based on the dipole-dipole model [35], which is given as

$$F_{\mathrm{m}}=\frac{3\mu_0m_1m_2}{2\pi\left[u_1-u_2+d_0\right]^4} \tag{8}$$

where m_1 is the magnitude of the magnetic dipole moment of magnetic proof mass. u_1 and u_2 are the displacements of the magnetic proof mass and magnetic oscillator relative to the base, respectively. d_0 is the initial separation distance between the magnetic dipoles. The direction of the magnetic force depends on the magnetization directions of the magnets.

Under the small-amplitude excitation, the difference of the displacements (u_1-u_2) is small. The Taylor's expansion of magnetic force F_m around (u_1-u_2) can be expressed as

$$F_{\mathrm{m}}=\frac{3\mu_0m_1m_2}{2\pi}\left[d_0^{-4}+(-4)d_0^{-5}(u_1-u_2)+10d_0^{-6}(u_1-u_2)^2+\cdots+\frac{F_{\mathrm{m}}^{(n)}(0)}{n!}(u_1-u_2)^n\right] \tag{9}$$

If the high-order terms ($n \geq 2$) are neglected, the magnetic force will only induce the linear equivalent stiffness to the harvester, just as a linear spring.

Modeling of Linear 1DOF PEH

When both induction coil and magnetic oscillator are removed, the HEH becomes a conventional 1DOF linear PEH (LPEH). The LPEH consists of a bimorph piezoelectric cantilever beam and a proof mass. The electromechanical coupling model can be derived from the energy method [38,39], which is based on the modal analysis and small-signal piezoelectric material constitutive law. The proof mass is modeled as a mass point and rotation inertia is neglected. The beam is considered to be an Euler–Bernoulli beam, whose shear deformation and rotation inertia are neglected. The piezoelectric patches are assumed to be perfectly bonded with the substrate and the thicknesses of the electrodes are neglected. The electrical field is assumed to be constant across the thickness of the piezoelectric element. Geometric nonlinearity of the beam is ignored under the assumption of small deflection. Considering that the frequency of excitation source in the environment is below 200 Hz [40], the fundamental mode of the PEH is the research focus.

Assume that the displacement of the base excitation is $u_b(t)$. In the Rayleigh-Ritz approach, the displacement of the PEH relative to the base $u(x,t)$ and electrical potential for the electrode pair $\varphi(z,t)$ can be written as

$$u(x,t)=\psi_r(x)r(t) \tag{10}$$

$$\varphi(z,t)=\psi_v(z)v(t) \tag{11}$$

where x is the axial position coordinate of the beam. $\psi_r(x)$ is the mechanical mode shape function. $r(t)$ is the modal coordinate. $\psi_v(z)$ is the electrical potential distribution function. $v(t)$ is the electrical temporal coordinate.

The electromechanical coupling governing equations are simplified as

$$\begin{cases} M\ddot{r}+C\dot{r}+Kr-\theta_{\mathrm{p}}v_1=-B_{\mathrm{f}}\ddot{u}_{\mathrm{b}} \\ \theta_{\mathrm{p}}\dot{r}+C_p\dot{v}_1+v_1/R_1=0 \end{cases} \tag{12}$$

where the mass M, stiffness K, piezoelectric coupling coefficient θ_p, capacitive C_p, and forcing factor B_f are obtained as follows

$$M=\int_0^L m\psi_r^2\mathrm{d}x+M_t\psi_r^2(L) \tag{13}$$

$$K=\int_{V_s}c_s(-z\psi_r'')^2\mathrm{d}V_s+\int_{V_p}c_{11}^{\mathrm{E}}(-z\psi_r'')^2\mathrm{d}V_p \tag{14}$$

$$\theta_p = \int_{V_p} e_{31}(-z\psi_r'')(-\nabla\psi_v)dV_p \tag{15}$$

$$C_p = \int_{V_p} \varepsilon_{33}^S(-\nabla\psi_v)^2 dV_p \tag{16}$$

$$B_f = \int_0^L m\psi_r dx + M_t\psi_r(L) \tag{17}$$

where the subscripts p and s indicate the piezoelectric element and substrate, respectively. V_p and V_s denote the volumes. M_t is the proof mass. m is the mass per length in the axial direction. z indicates the distance from the neutral axis of the piezoelectric beam. v_1 is the voltage across external resistance R_1 for the piezoelectric patches. u¨b is the excitation acceleration. C denotes the mechanical damping, which can be measured by the logarithmic decrement method or bandwidth method [41]. The subscript 1 corresponds to the x direction and 3 corresponds to the z direction. c, e, and ε are the modulus of elasticity, piezoelectric constant, and dielectric constant, respectively. ∇ is the differential operator.

Modeling of the Proposed HEH

The electromagnetic energy harvesting part can be simplified as a 1DOF spring-mass-damper system, as shown in Figure 1. The magnetic force is considered as a concentrated force applied on the centers of both magnets along the z-axis direction. Assume that a distributed force $F(x,t)$ is applied on the beam. Based on the modal analysis, the corresponding modal force for the fundamental vibration mode, F_1, can be expressed as

$$F_1 = \int_0^L F(x,t)\psi_r(x)dx \tag{18}$$

Accordingly, the modal magnetic force applied on the piezoelectric beam is given as

$$F_1 = F_m(L-R,t)\psi_r(L-R) \tag{19}$$

where R is the radius of the magnetic proof mass.

Due to the low excitation frequency, the coil inductance is ignored [42]. Combing Equations (12) and (19), the electromechanically coupling governing equations of the HEH for the fundamental mode can be derived as

$$\begin{cases} M\ddot{r} + C\dot{r} + Kr - \theta_p v_1 - F_m\psi_r(L-R) = -B_f\ddot{u}_b \\ M_2\ddot{u}_2 + C_2\dot{u}_2 + K_2u_2 + \theta_{em}I_2 + F_m = -M_2\ddot{u}_b \\ \theta_p\dot{r} + C_p\dot{v}_1 + v_1/R_1 = 0 \\ \theta_{em}\dot{u}_2 - (R_c + R_2)I_2 = 0 \end{cases} \tag{20}$$

where M_2, C_2, and K_2 are the lumped mass, mechanical damping, and spring stiffness of the magnetic oscillator, respectively. R_c is the internal resistance of the coil.

Letting

$$\omega_1 = \sqrt{K/M}, \quad \omega_2 = \sqrt{K_2/M_2}, \quad \zeta_1 = \frac{C}{2\sqrt{KM}}, \quad \zeta_2 = \frac{C_2}{2\sqrt{K_2M_2}} \tag{21}$$

and defining a state vector

$$\mathbf{X} = \begin{bmatrix} X_1 & X_2 & X_3 & X_4 & X_5 \end{bmatrix}^t = \begin{bmatrix} r & \dot{r} & u_2 & \dot{u}_2 & v_1 \end{bmatrix}^t$$

(t denotes the transpose of the vector), the governing Equation (20) can be written in the state space form as

$$\dot{\mathbf{X}} = \begin{bmatrix} \dot{r} \\ \ddot{r} \\ \dot{u}_2 \\ \ddot{u}_2 \\ \dot{v}_1 \end{bmatrix} = \begin{bmatrix} X_2 \\ -\omega_1^2X_1 - 2\omega_1\zeta_1X_2 + \theta_pX_5/M + F_m\psi_r(L-R)/M - B_f\ddot{u}_b/M \\ X_4 \\ -\omega_2^2X_3 - (2\omega_2\zeta_2 + C_{em}/M_2)X_4 - F_m/M_2 - \ddot{u}_b \\ -\theta_pX_2/C_p - X_5/(R_1C_p) \end{bmatrix} \tag{22}$$

where $C_{em} = \theta_{em}^2/(R_2 + R_c)$ is the electromagnetic damping.

The output power delivered to the external loads R_1 and R_2 are respectively given as

$$P_p = \frac{1}{T}\int_0^T \frac{v_1^2}{R_1}dt \tag{23}$$

$$P_{em} = \frac{1}{T}\int_0^T \left(\frac{\theta_{em}\dot{u}_2}{R_2 + R_c}\right)^2 R_2 dt \tag{24}$$

where $T=2\pi/\Omega$ is the cycle of the base excitation. Ω is the angular velocity. The total output power of the HEH is

$$P = P_p + P_{em} \tag{25}$$

NUMERICAL SIMULATION

Upon Equations (22)–(24), the performances of the HEH depend on the magnetic force, mass, stiffness, and mechanical dampings of the system. In this section, we will discuss the effects of the above parameters on the harvesting performances of the HEH. The theoretical model is numerically simulated by using the ordinary differential equation solver "ode45" in MATLAB® (MathWorks, Natick, MA, USA). The piezoelectric material used is PZT-5H [43]. The geometric and physical properties of the proposed HEH are given in Table 1. Note that magnet A stands for the proof magnet, while magnet B stands for the magnetic oscillator. In the following parts, the electromechanical coupling coefficients θ_p and θ_{em} are determined based on these properties. The reference resistances connected to the HEH are assumed to be $R_1 = 1/C_p\omega_1$, $R_2 = Rc =$

240 Ω. The excitation acceleration is set 2 m/s² and the initial separation distance d_0 is 52.5 mm. The equilibrium position of the center of magnetic oscillator is set on the upper surface of the coil, *i.e.*, $z_2 = h_c$. The mechanical damping ratio of each oscillator is set 0.02.

Table 1. Geometric and physical properties of the hybrid energy harvester (HEH)

Parameter	Value
PZT length×width×thickness (mm³)	50 × 20 × 0.2
PZT density (kg/m³)	7500
Young's modulus of PZT layer (GPa)	60.6
Substrate length×width×thickness (mm³)	50 × 20 × 0.24
Substrate density (kg/m³)	8920
Young's modulus of substrate (GPa)	90
Magnet A radius×thickness (mm²)	10 × 10
Magnet B outer radius×inner radius×thickness (mm³)	9 × 4 × 7
Magnet density (kg/m³)	7800
Residual magnetic flux density (T)	1.25
Spring stiffness (N/m)	220

Electromagnetic Coupling Coefficient

Figure 3 shows the relationship between electromagnetic coupling coefficient θ_{em} and position of magnetic oscillator z_2. It seems that θ_{em} varies nonlinearly with the change of z_2 due to the change of magnetic field. θ_{em} reaches the peak values around the upper and lower surfaces of the coil, while the value is zero at the center of the coil. Because the maximum velocity of the magnetic oscillator occurs at the static balance position during the harmonic oscillation, the upper and lower surfaces of the coil are optimal equilibrium positions for the magnetic oscillator.

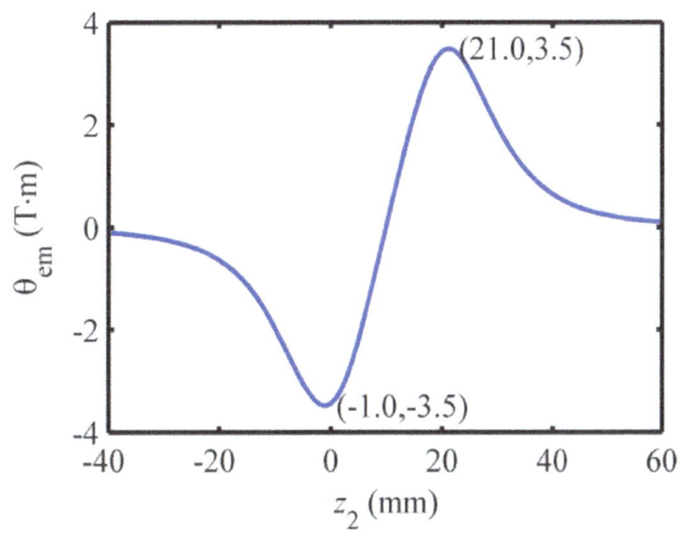

Figure 3. Electromagnetic coupling coefficient *versus* z_2.

Effect of Magnetic Force on the Output Power

Figure 4 shows the effect of magnetic force on the output power of the HEH. The first and second resonant frequencies are induced by magnetic and piezoelectric oscillators, respectively. It is the typical characteristic of the multi-frequency VEH. It can be seen that both of the two resonant frequencies increase with the decreasing of d_0, which is the same as the tendency of the difference of the two resonant frequencies. Therefore, we can tune the resonant frequencies of the system by changing the initial separation distance d_0. However, when d_0 is small enough and the repulsive force is large enough, the two resonances will merge into one resonance. In this situation, the magnetic interaction acts as a rigid connection between two oscillators, so that two oscillators have the same resonant frequency around the ω_1. With the decrease of d_0, the magnitude of the first peak power gradually increases, while the one of the second peak constantly decreases. When the second peak finally disappears, the first peak reaches the maximum value. Although the magnetic force is large enough when d_0 = 10.0 mm, there is still no obvious hardening or softening response, which is the typical phenomenon in magnetic coupling nonlinear VEH [20]. The main reason is that the difference of displacements between two oscillators (u_1-u_2) is very small under the small-amplitude excitation, so that the magnetic nonlinearity is invisible, which can be referred in Equation (9).

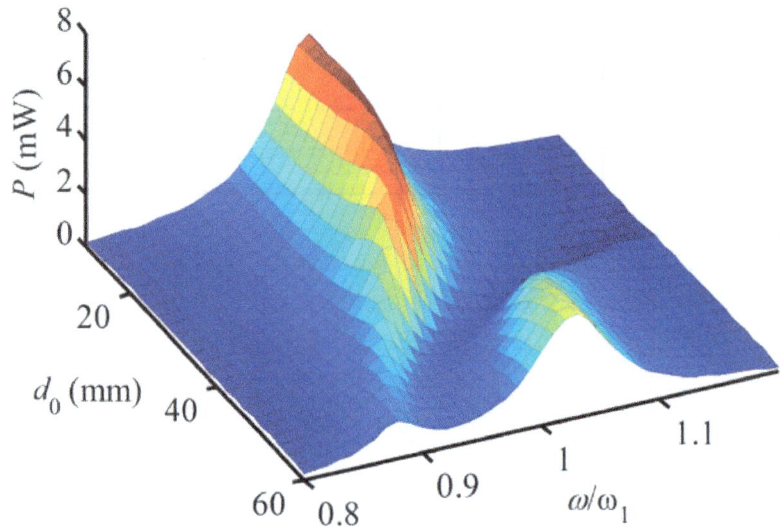

Figure 4. Output power *versus* excitation frequency for different d_0.

Effect of Mass Ratio on the Output Power

Define $M_1 = M/\psi_r^2(L)$, $K_1 = K/\psi_r^2(L)$ [17], $\mu = M_2/M_1$, and $\alpha = K_2/K_1$, where M_1 and K_1 are the lumped mass and stiffness of the piezoelectric oscillator, respectively. μ and α are mass and stiffness ratios, respectively. According to the properties in Table 1, $M_1 = 0.0257$ kg, $K_1 = 651.8$ N/m, $M_2 = 0.0118$ kg, $K_2 = 220.0$ N/m, $\mu = 0.46$, and $\alpha = 0.34$. The output power of the HEH *versus* excitation frequency for different mass ratio μ is illustrated in Figure 5. For a given α, the first resonance shifts to a higher frequency with the decrease of μ. It is difficult to distinguish the tendency of the second resonant frequency, due to the mixture of two peak power. Because the HEH is used in the low-frequency environment, the dynamic characteristic of the first peak power is our research focus. The magnitude of first peak initially decreases and then increases as μ decreases. Finally, it reduces to a smaller value when the first resonance is higher enough than the second one. In a word, the mass ratio affects the resonance distribution and output power of the HEH.

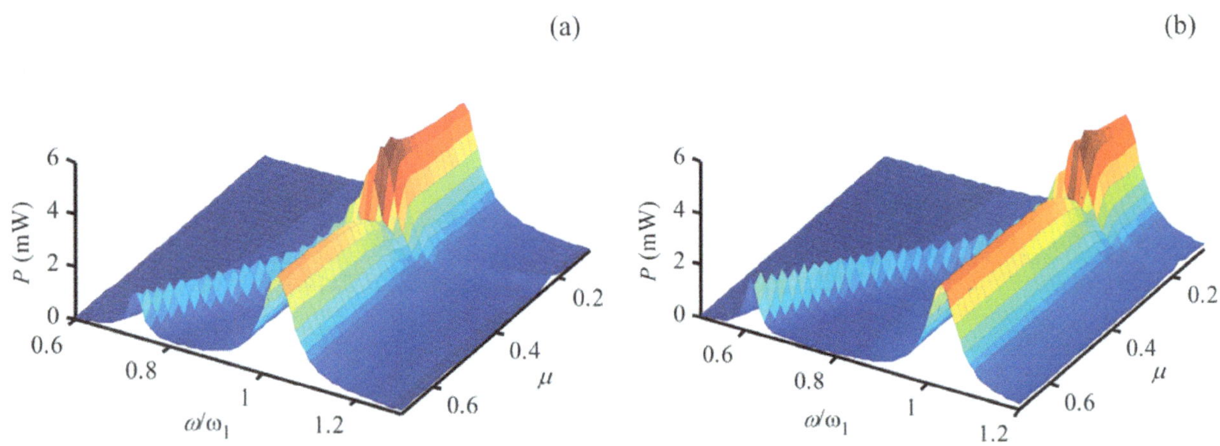

Figure 5. Output power *versus* excitation frequency for different μ: (**a**) $\alpha = 0.34$; (**b**) $\alpha = 0.23$.

For the given μ, the magnitude of the first peak power is enhanced with the smaller α ($\alpha = 0.23$), which is contrary to the change of the second peak power. However, the first resonance shifts to a lower value with the decreasing of α. Under the same excitation condition, the smaller α is, the more vibration energy can be harvested by magnetic oscillator. The second resonant frequency remains almost unchanged.

Effect of Stiffness Ratio on the Output Power

Figure 6 depicts the effect of stiffness ratio α on the output power of the HEH. For a given μ, the first resonant frequency increases as α increases, while the magnitude of the first peak initially decreases, then increases, and finally reduces to a smaller value. It also shows that α has almost no effect on the second resonance. When α remains unchanged, the larger μ is, the more the magnitude of the first peak increases. This conclusion is identical with that obtained before. Therefore, the stiffness ratio can also be used to alter the performance of the HEH.

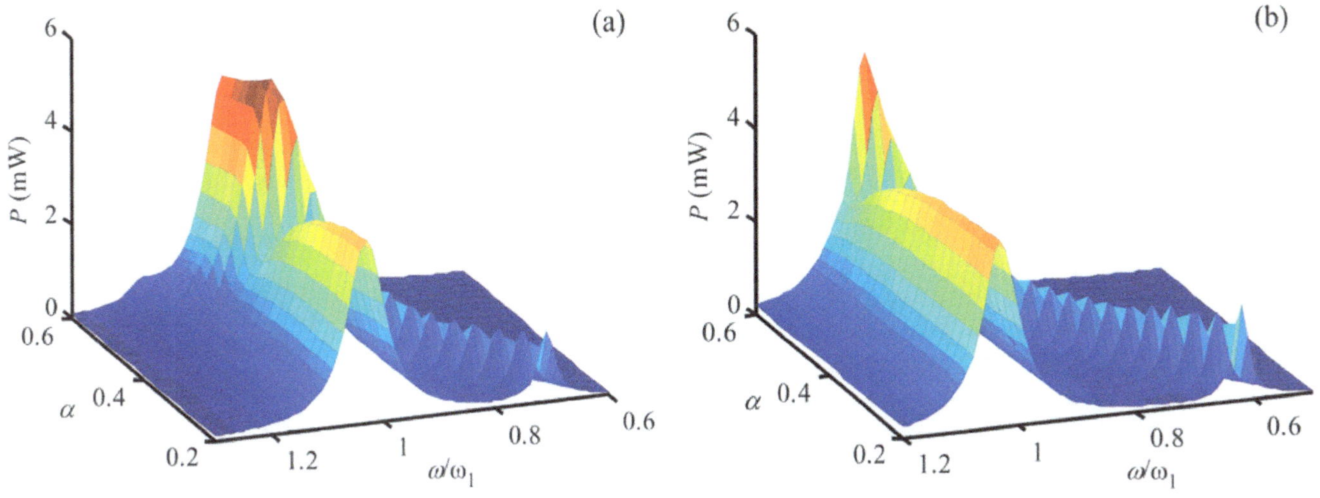

Figure 6. Output power *versus* excitation frequency for different α: (**a**) μ = 0.46; (**b**) μ = 0.60.

Effects of Mechanical Damping Ratios on the Output Power

Figure 7 shows the effects of mechanical damping ratios ζ_1 and ζ_2 on the output power of the HEH, respectively. It can be seen that ζ_1 mainly affects the performance of the second peak induced by the piezoelectric oscillator, while ζ_2 mainly affects that of the first peak induced by the magnetic oscillator. With the increase of ζ_1, the magnitude of the second peak output power constantly decreases, but the bandwidth increases. The magnitude of the first peak power slightly decreases. These effects are the same as that induced by ζ_2.

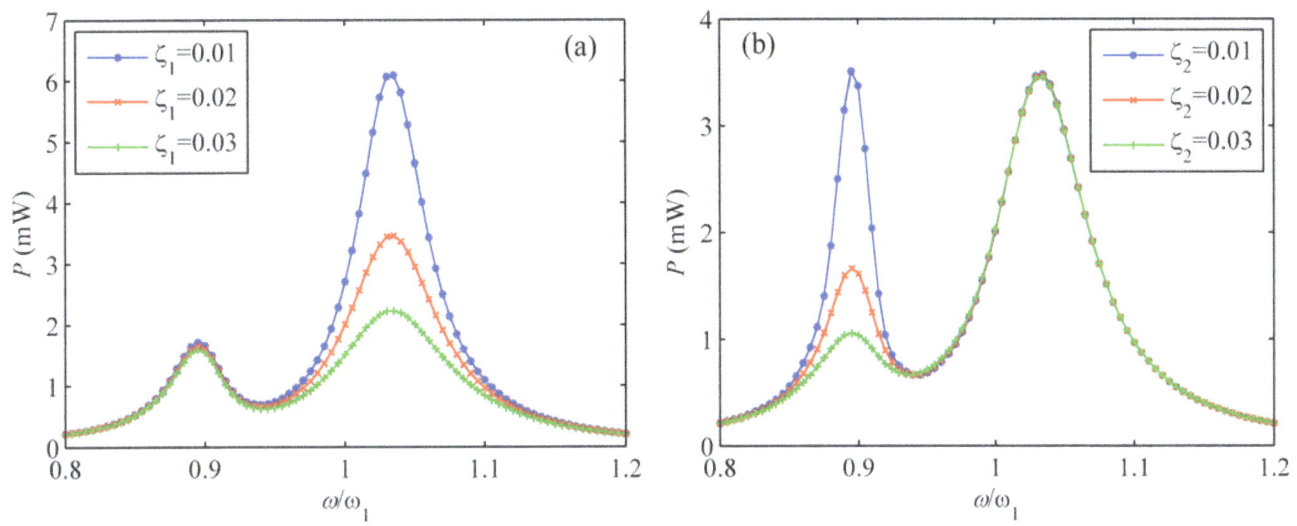

Figure 7. Output power *versus* excitation frequency for different mechanical damping ratios: (**a**) ζ_1; (**b**) ζ_2.

EXPERIMENTAL METHOD

Figure 8 shows the fabricated macro-scale prototype of the proposed HEH and the experimental system. The substrate of the cantilever beam is made of Phosphor Bronze (ALB Copper Alloys Co., Ltd., Xiamen, China). It is sandwiched between two PZT-5H patches (Baoding Hongsheng Acoustics Electron Apparatus Co., Ltd., Baoding, China), which are connected in series. Two permanent magnets are NdFeB (Ningbo Hony Technology Co.,Ltd., Ningbo, China). They have opposite directions of magnetization. The magnetic proof mass is cylindrical. The magnet of the magnetic oscillator is a magnetic ring. It is attached to an aluminum rod, which is bolted on the frame. The rod is used to keep the magnet moving on the vertical direction. The copper wire coil (Changzhou Wujin Enameled Wire Factory Co., Ltd., Changzhou, China) is placed along the motion direction of the magnetic oscillator. The geometric and physical properties of the prototype are according to Table 1. The internal resistance of the coil is 240 Ω. The experimental system consists of an electromagnetic shaker (JZK-50) (Sinocera Piezotronics Inc., Yangzhou, China), a signal generator (DG-1022) (Rigol Technologies Inc., Beijing, China), a power amplifier (YE5874A) (Sinocera Piezotronics Inc., Yangzhou, China). The base excitation acceleration was measured by an accelerometer (YD64-310) (Qinhuangdao Xinheng Electronic Technology Co. Ltd., Qinhuangdao, China). The output voltage was continuously monitored by the oscilloscopes DS-5102MA (Rigol Technologies Inc., Beijing, China) and TDS-1012C (Tektronix China Ltd., Shanghai, China).

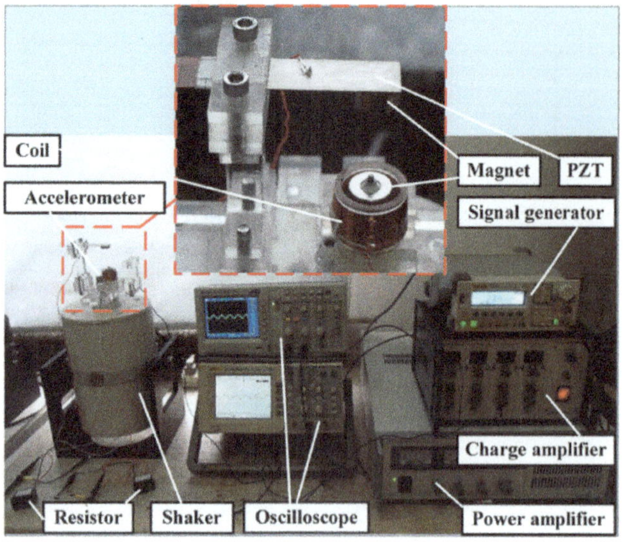

Figure 8. Prototype of the HEH and the experimental system.

Upon the numerical simulation, the theoretical resonances of the piezoelectric and magnetic oscillators are about 26 Hz and 22 Hz, respectively. Thus, we sweep the excitation frequency from 16 to 32 Hz in the following experiments. The base excitation generated by the electromagnetic shaker is harmonic and the acceleration is set 2 m/s². The initial separation distance between two magnets d_0 is 52.5 mm.

In order to evaluate the performance of the proposed HEH, four different energy harvesters with single conversion mechanism were also developed and experimentally verified for comparison under the same excitation condition. When the induction coil around the magnetic oscillator is open-circuit, there will be no electromagnetic damping induced in the energy harvesting system. Accordingly, the HEH changes into a conventional multi-frequency PEH (MPEH) with magnetic oscillator, as reported in reference [17]. When the piezoelectric layers are in the short-circuit condition, a conventional multi-frequency EMEH (MEMEH) with a magnetic oscillator can be developed. The magnetic oscillator and induction coil are the component units of the 1DOF linear EMEH (LEMEH). The combination of the LPEH and LEMEH is named LVEH. Its output power frequency response is compared with that of the HEH.

RESULTS AND DISCUSSION

To measure the optimal load resistance matched to the coil, the piezoelectric layers are short-circuit. There are two peak output powers for the configuration of two magnetic coupled oscillators. The first peak is induced by the magnetic oscillator, while the second one is induced by the piezoelectric oscillator. Figure 9 shows the first peak output power delivered to the load resistance R_2 for the MEMEH at resonant frequency. The maximum power is 1.2 mW at 22.8 Hz with the load resistance 370 Ω. The second peak output power delivered to the load resistance R_1 for the MPEH at resonant frequency is shown in Figure 10. The induction coil around the magnetic oscillator is open-circuit. With the increasing of R_1, the power reaches the maximum 2.57 mW at 57 kΩ. The corresponding resonant frequency is 25.8 Hz. By using the logarithmic decrement method, the measured mechanical damping ratios of piezoelectric beam and magnetic oscillator with magnetic interaction are 0.023 and 0.024, respectively.

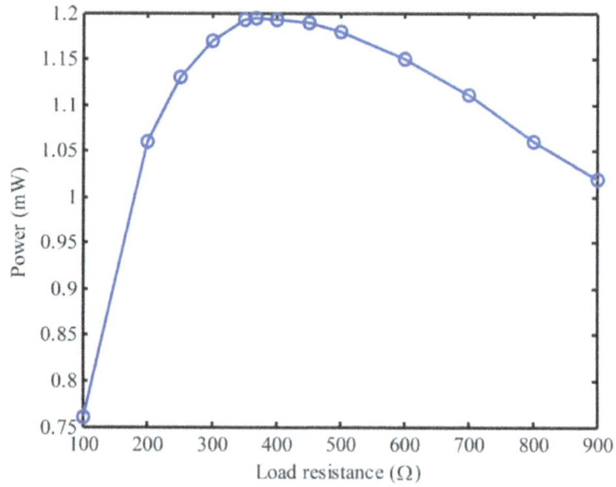

Figure 9. Output power with different load resistances for the MEMEH.

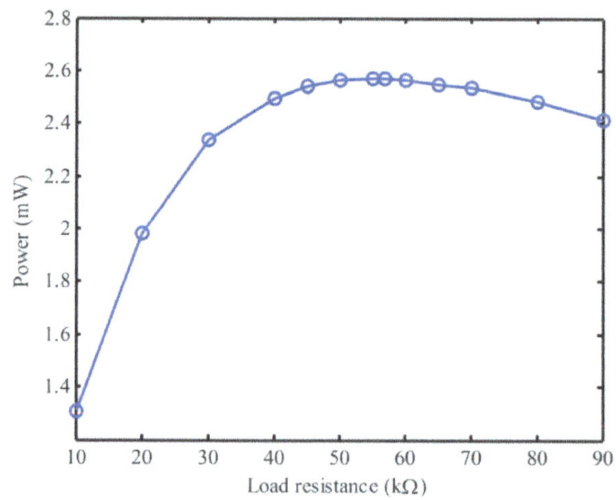

Figure 10. Output power with different load resistances for the MPEH.

Figure 11a,b shows the experimental and numerical comparisons of the frequency-response for the HEH, MPEH and MEMEH, respectively. The initial distance d_0 is still 52.5 mm. The loads connected to the HEH are the optimal values determined from the MPEH and MEMEH. It can be seen in Figure 11a that each harvester has two peak output powers. For HEH, the measured peak output powers are 1.40 mW at 22.6 Hz and 2.82 mW at 25.6 Hz. In view of the low-frequency applications, the first peak output power is the research focus. The first peak of the HEH increases by 16.7% and 833.3% compared with the first peaks of the MEMEH (1.2 mW) and the MPEH (0.15 mW), respectively. The frequency response of the HEH completely covers that of the MPEH and MEMEH. That is to say, the hybrid transduction mechanism is superior to the multi-frequency devices with single conversion mechanism in broadening the bandwidth and improving the generating performance. We can find that the simulation results of the HEH are in good agreement with the experimental results. The inaccuracies between theoretical and simulation results for the MPEH and MEMEH are significant. We think that these errors may be due to the damping change between the magnetic oscillator and the aluminum rod with varying vibration frequencies. Besides, the effective stiffness of the spring varies with the change of effective length during the oscillation, which may be another factor.

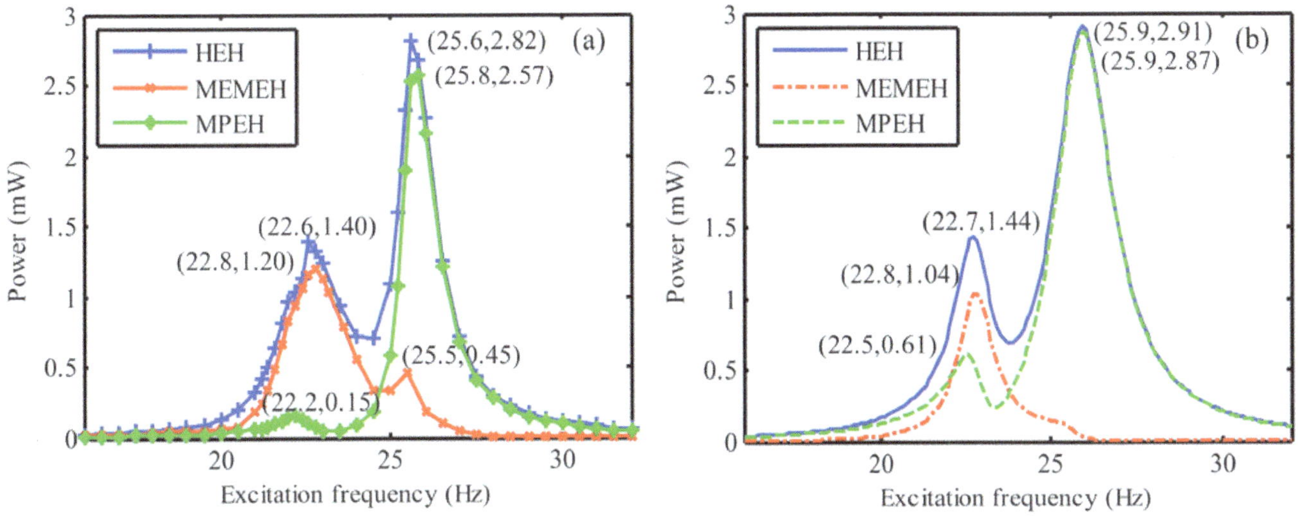

Figure 11. Frequency responses for the HEH, MEMEH, and MPEH: (**a**) experimental results; (**b**) numerical results.

Figure 12a,b illustrates the numerical and experimental comparisons of the output power for the HEH, LEMEH, LPEH and LVEHs, respectively. The connected load resistances keep still. Due to the repulsive magnetic force, the measured first resonance of the HEH (22.6 Hz) is higher than that of the LVEHs (21.6 Hz) and the second one (25.6 Hz) is higher than that of the LVEHs (25 Hz). Moreover, the first peak power of the HEH (1.4 mW) is 2.36 times more than that of the LVEHs (the sum of 0.08 mW from LPEH and 0.49 mW from LEMEH) at the same excitation frequency. Meanwhile, it increases by 45.8% compared to that of the LVEHs, although the second peak power of the HEH decreases by 13.8%. Obviously, the repulsive magnetic force can suppress the oscillation of the piezoelectric oscillator, but enhance the oscillation of the magnetic oscillator. The half-power bandwidth of the first peak power for the HEH is about 2.8 Hz, which is wider than that of the LEMEH (1.9 Hz). It is clear that the generating performances of the LPEH and LEMEH are greatly improved by the magnetic coupling multi-frequency structure and hybrid transduction mechanism in the low-frequency range. Consequently, magnetic interaction is an effective approach to improve the performance of the micro energy harvester in the low-frequency range. The numerical results show good agreement with the experimental results. Note that the measured mechanical damping ratios of piezoelectric and magnetic oscillators without magnetic interaction are 0.025 and 0.02, respectively. That is to say, magnetic interaction can add extra damping to the magnetic oscillator and cut down the damping of the piezoelectric oscillator. The simulation results for the LEMEH shows good agreement with the experimental results. Accordingly, the model of the electromagnetic coupling coefficient θ_{em} is reasonable and valid.

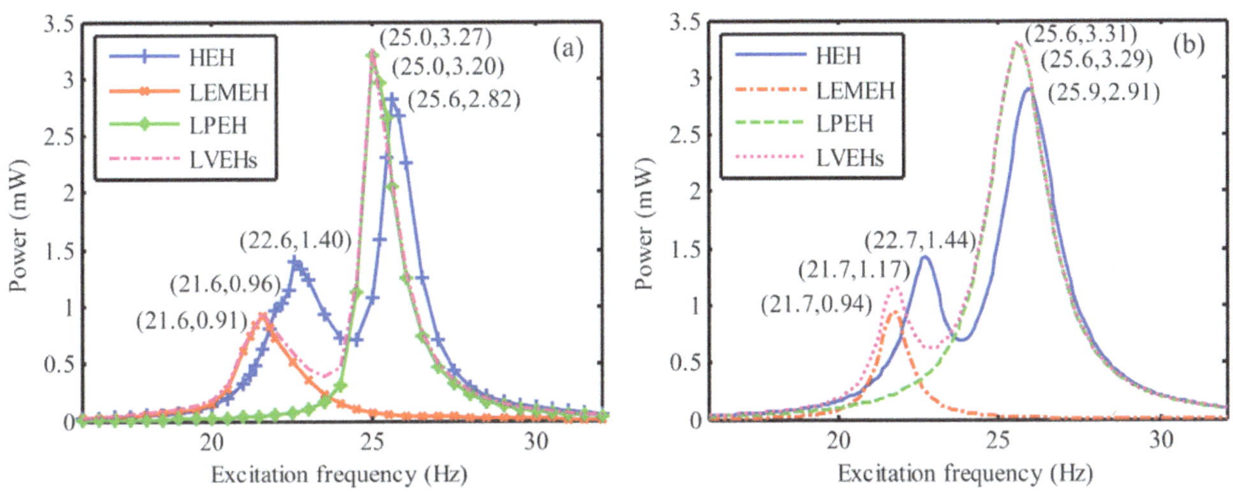

Figure 12. Frequency responses for the HEH, LEMEH, LPEH and LVEHs: (**a**) experimental results; (**b**) numerical results.

Figure 13 provides the investigation on the effects of the initial separation distance d_0 between two magnets on output power. The measurement was performed at three arbitrary different separation distances 42.5, 47.0, and 52.5 mm. For the sake of contrastive analysis, the excitation acceleration was controlled at 2 m/s^2. It shows that the resonant frequencies shift to the right with the decreasing of the initial separation distance, which is consistent with the simulation result. It can be attributed to the increasing repulsive magnetic force and the magnetic stiffness. Consequently, the resonant frequencies can be tuned by adjusting the separation distance according to the ambient excitation frequency. Furthermore, the amplitude of the first peak output power is another issue which deserves concern. Contrary to the resonant frequency, the peak output power for the first resonance goes up continuously with the decreasing of the separation distance. In the future designing, we need to make a trade-off between the resonance and peak output power. With the decrease of d_0, the error between experimental and numerical results of the first peak power increases, because the mechanical damping ratios were kept constant in the theoretical model, while the actual values are variable. Moreover, the damping ratio of magnetic oscillator increases more significantly than that of the piezoelectric oscillator.

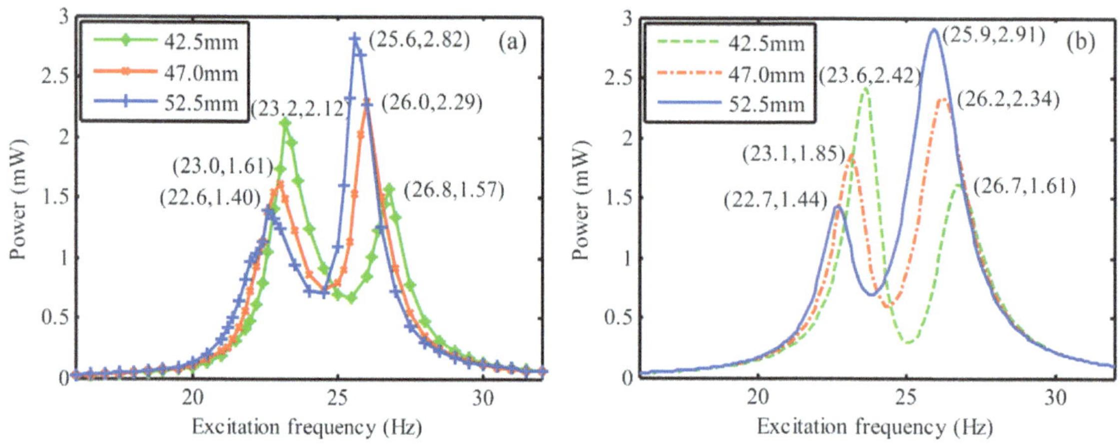

Figure 13. Frequency responses for the HEH with different separation distances d_0: (**a**) experimental results; (**b**) numerical results.

Figure 14 shows the output power for the HEH with different directions of magnetic force. Two magnets will attract each other with the same magnetization direction, while they will repel each other in opposite direction. Due to the attractive force, the initial distance reduces to 47 mm. As a result of the attractive force, the resonant frequencies shift to the left, while the first peak output power decreases sharply as compared to that of the HEH with repulsive force. On the contrary, the second peak output power is obviously enhanced. That is to say, the generating efficiency of piezoelectric element is much higher than that of the electromagnetic element with the attractive force. The theoretical value of the first peak power for the HEH is lower than the measured value. The reason is that attractive force induces more deformation of the piezoelectric beam. However, the mechanical damping of magnetic oscillator is also enhanced simultaneously. In this case, the HEH is more suitable for scavenging vibration energy in the high-frequency range. However, the attractive force is still a choice for energy harvesting at the lower excitation frequency. Similarly, the direction of magnetic force needs to be selected properly depending on the excitation frequency and electric power demands.

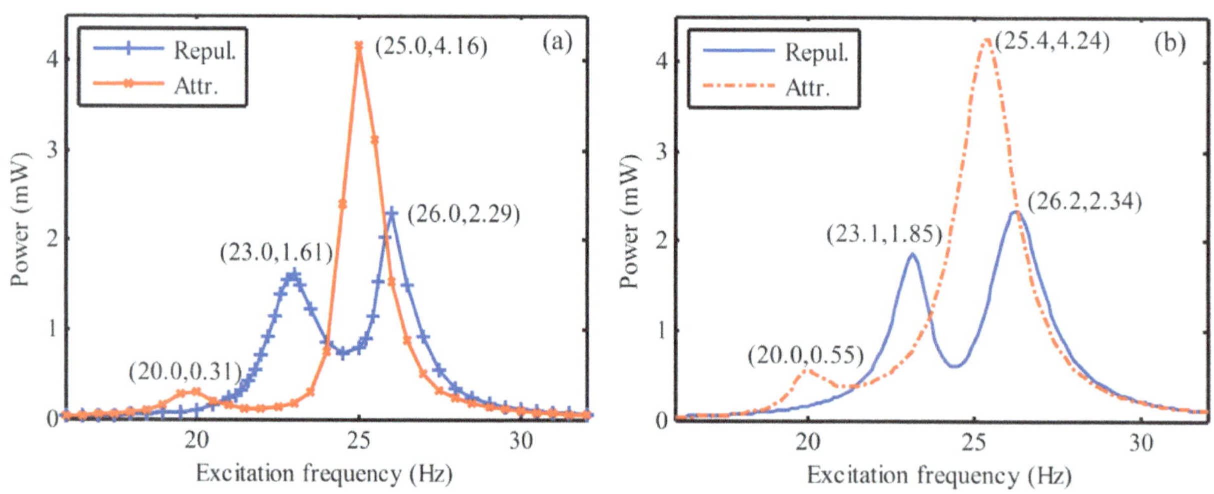

Figure 14. Frequency responses for the HEH with different directions of magnetic force: (**a**) experimental results; (**b**) numerical results.

CONCLUSIONS

In this paper, we proposed a novel tunable multi-frequency hybrid vibration energy harvester, by combining piezoelectric and electromagnetic transduction mechanisms. The electromechanical coupling model of the HEH was established and numerically simulated. The effects of magnetic fore, mass ratio, stiffness ratio, and mechanical damping ratios on the frequency response of the HEH were analyzed. A macro-scale prototype was fabricated and tested by experiments. It is demonstrated that magnetic interaction between piezoelectric and magnetic oscillators can tune the resonant frequencies of the HEH. It contributes to increasing energy harvesting efficiency of the micro energy harvesting devices in low-frequency range. The hybrid transduction mechanism is superior to the single mechanism in broadening the bandwidth and improving the output power. In addition, an initial separation distance between two magnets and the direction of magnetic force affect the frequency responses of the HEH significantly, which need to be carefully designed based on the excitation frequency and electric power demands. In a word, the proposed tunable multi-frequency hybrid energy harvesting mechanism is an effective approach to improve the performance of the micro-scale energy harvesting devices for the low-frequency environmental vibrations.

ACKNOWLEDGMENTS

The work described in this paper is supported by "The Fundamental Research Funds for the Central Universities" (Grant No. HIT. NSRIF. 2014059 and No. HIT. KISTP. 201412), China.

AUTHOR CONTRIBUTIONS

Zhenlong Xu established the theoretical model. All authors conceived and designed the experiments. Zhenlong Xu and Danpeng Chen performed the experiments. Zhenlong Xu, Xiaobiao Shan and Tao Xie analyzed the data. All authors contributed to the writing of the paper.

REFERENCES

1. Roundy, S.; Wright, P.K. A piezoelectric vibration based generator for wireless electronics. *Smart Mater. Struct.* **2004**,*13*, 1131–1142.
2. Beeby, S.P.; Tudor, M.J.; White, N.M. Energy harvesting vibration sources for microsystems applications. *Meas. Sci. Technol.* **2006**, *17*, R175–R195.
3. Williams, C.B.; Yates, R.B. Analysis of a micro-electric generator for microsystems. *Sens. Actuators A Phys.* **1996**, *52*, 8–11.
4. Reilly, E.; Miller, L.; Fain, R.; Wright, P.K. A study of ambient vibrations for piezoelectric energy conversion. In Proceedings of the PowerMEMS, Washington, DC, USA, 1–4 December 2009; pp. 312–315.
5. Swee, L.K.; Neil, M.W.; Harris, N.R. A novel piezoelectric thick-film free-standing cantilever energy harvester. In Proceedings of the Eurosensors XXII, Dresden, Germany, 7–11 September 2008.
6. Elfrink, R.; Kamel, T.M.; Goedbloed, M.; Matova, S.; Hohlfeld, D.; van Andel, Y.; van Schaijk, R. Vibration energy harvesting with aluminum nitride-based piezoelectric devices. *J. Micromech. Microeng.* **2009**, *19*.
7. Kim, M. Design, fabrication, and experimental demonstration of a piezoelectric cantilever for a low resonant frequency microelectromechanical system vibration energy harvester. *J. Micro Nanolithogr. MEMS MOEMS* **2012**, *11*.
8. Alamin Dow, A.B.; Schneider, M.; Koo, D.; Al-Rubaye, H.A.; Bittner, A.; Schmid, U.; Kherani, N. Modeling the performance of a micromachined piezoelectric energy harvester. *Microsyst. Technol.* **2012**, *18*, 1035–1043.
9. Beeby, S.P.; Tudor, M.J.; Torah, R.N.; Roberts, S.; O'Donnell, T.; Roy, S. Experimental comparison of macro and micro scale electromagnetic vibration powered generators. *Microsyst. Technol.* **2007**, *13*, 1647–1653.
10. Shan, X.B.; Yuan, J.B.; Xie, T.; Chen, W.S. Design and experiment of multiple piezoelectric bimorphs for scavenging vibration energy. *Int. J. Appl. Electromagn. Mech.* **2010**, *34*, 265–275.
11. Sari, I.; Balkan, T.; Kulah, H. An electromagnetic micro power generator for wideband environmental vibrations. *Sens. Actuators A Phys.* **2008**, *145–146*, 405–413.
12. Yang, Z.; Yang, J. Connected vibrating piezoelectric bimorph beams as a wide-band piezoelectric power harvester. *J. Intell. Mater. Syst. Struct.* **2009**, *20*, 569–574.
13. Liu, H.; Tay, C.J.; Quan, C.; Kobayashi, T.; Lee, C. A scrape-through piezoelectric mems energy harvester with frequency broadband and up-conversion behaviors. *Microsyst. Technol.* **2011**, *17*, 1747–1754.
14. Halim, M.A.; Park, J.Y. Theoretical modeling and analysis of mechanical impact driven and frequency up-converted piezoelectric energy harvester for low-frequency and wide-bandwidth operation. *Sens. Actuators A Phys.* **2014**, *208*, 56–65.
15. Hajati, A.; Kim, S.-G. Ultra-wide bandwidth piezoelectric energy harvesting. *Appl. Phys. Lett.* **2011**, *99*.
16. Cottone, F.; Basset, P.; Vocca, H.; Gammaitoni, L.; Bourouina, T. Bistable electromagnetic generator based on buckled beams for vibration energy harvesting. *J. Intell. Mater. Syst. Struct.* **2013**, *25*, 1484–1495.
17. Tang, L.; Yang, Y. A nonlinear piezoelectric energy harvester with magnetic oscillator. *Appl. Phys. Lett.* **2012**, *101*.
18. Mann, B.P.; Sims, N.D. Energy harvesting from the nonlinear oscillations of magnetic levitation. *J. Sound Vib.* **2009**,*319*, 515–530.
19. Stanton, S.C.; Owens, B.A.M.; Mann, B.P. Harmonic balance analysis of the bistable piezoelectric inertial generator. *J. Sound Vib.* **2012**, *331*, 3617–3627.
20. Stanton, S.C.; McGehee, C.C.; Mann, B.P. Reversible hysteresis for broadband magnetopiezoelastic energy harvesting.*Appl. Phys. Lett.* **2009**, *95*.
21. Mahmoudi, S.; Kacem, N.; Bouhaddi, N. Enhancement of the performance of a hybrid nonlinear vibration energy harvester based on piezoelectric and electromagnetic transductions. *Smart Mater. Struct.* **2014**, *23*.
22. Shan, X.; Xu, Z.; Song, R.; Xie, T. A new mathematical model for a piezoelectric-electromagnetic hybrid energy harvester. *Ferroelectrics* **2013**, *450*, 57–65.
23. Challa, V.R.; Prasad, M.G.; Shi, Y.; Fisher, F.T. A vibration energy harvesting device with bidirectional resonance frequency tunability. *Smart Mater. Struct.* **2008**, *17*.
24. Zhu, D.; Roberts, S.; Tudor, M.J.; Beeby, S.P. Design and experimental characterization of a tunable vibration-based electromagnetic micro-generator. *Sens. Actuators A Phys.* **2010**, *158*, 284–293.

25. Peters, C.; Maurath, D.; Schock, W.; Mezger, F.; Manoli, Y. A closed-loop wide-range tunable mechanical resonator for energy harvesting systems. *J. Micromech. Microeng.* **2009**, *19*.

26. Chew, Z.; Li, L. Design and characterisation of a piezoelectric scavenging device with multiple resonant frequencies.*Sens. Actuators A Phys.* **2010**, *162*, 82–92.

27. Ferrari, M.; Ferrari, V.; Guizzetti, M.; Marioli, D.; Taroni, A. piezoelectric multifrequency energy converter for power harvesting in autonomous microsystems. *Sens. Actuators A Phys.* **2008**, *142*, 329–335.

28. Yang, B.; Lee, C.; Xiang, W.; Xie, J.; He, J.H.; Kotlanka, R.K.; Low, S.P.; Feng, H. Electromagnetic energy harvesting from vibrations of multiple frequencies. *J. Micromech. Microeng.* **2009**, *19*.

29. Yang, B.; Liu, J.; Tang, G.; Luo, J.; Yang, C.; Li, Y. A generator with nonlinear spring oscillator to provide vibrations of multi-frequency. *Appl. Phys. Lett.* **2011**, *99*.

30. Liu, H.; Soon, B.W.; Wang, N.; Tay, C.J.; Quan, C.; Lee, C. Feasibility study of a 3D vibration-driven electromagnetic mems energy harvester with multiple vibration modes. *J. Micromech. Microeng.* **2012**, *22*.

31. Tadesse, Y.; Zhang, S.; Priya, S. Multimodal energy harvesting system: Piezoelectric and electromagnetic. *J. Intell. Mater. Syst. Struct.* **2009**, *20*, 625–632.

32. Wang, H.-Y.; Tang, L.-H.; Guo, Y.; Shan, X.-B.; Xie, T. A 2DOF hybrid energy harvester based on combined piezoelectric and electromagnetic conversion mechanisms. *J. Zhejiang Univ. Sci. A* **2014**, *15*, 711–722.

33. Shan, X.-B.; Guan, S.-W.; Liu, Z.-S.; Xu, Z.-L.; Xie, T. A new energy harvester using a piezoelectric and suspension electromagnetic mechanism. *J. Zhejiang Univ. Sci. A* **2013**, *14*, 890–897.

34. Spreemann, D.; Hoffmann, D.; Folkmer, B.; Manoli, Y. Numerical optimization approach for resonant electromagnetic vibration transducer designed for random vibration. *J. Micromech. Microeng.* **2008**, *18*.

35. Yung, K.W.; Landecker, P.B.; Villani, D.D. An analytic solution for the force between two magnetic dipoles. *Magn. Electr. Sep.* **1998**, *9*, 39–52.

36. Sneller, A.J.; Mann, B.P. On the nonlinear electromagnetic coupling between a coil and an oscillating magnet. *J. Phys. D Appl. Phys.* **2010**, *43*.

37. Priya, S.; Inman, D.J. *Energy Harvesting Technologies*; Springer: New York, NY, USA, 2009; p. 132.

38. DuToit, N.E.; Wardle, B.L. Experimental verification of models for microfabricated piezoelectric vibration energy harvesters. *AIAA J.* **2007**, *45*, 1126–1137.

39. Sodano, H.A.; Park, G.; Inman, D.J. Estimation of electric charge output for piezoelectric energy harvesting. *Strain***2004**, *40*, 49–58.

40. Roundy, S.; Wright, P.K.; Rabaey, J. A study of low level vibrations as a power source for wireless sensor nodes. *Comput. Commun.* **2003**, *26*, 1131–1144.

41. Inman, D.J.; Singh, R.C. *Engineering Vibration*, 3rd ed.; Pearson Education: Hackensack, NJ, USA, 2008; pp. 43–48.

42. Stephen, N.G. On energy harvesting from ambient vibration. *J. Sound Vib.* **2006**, *293*, 409–425.

43. Erturk, A.; Inman, D.J. *Piezoelectric Energy Harvesting*; John Wiley & Sons: West Sussex, UK, 2011; pp. 373–374.

CHAPTER 4

DESIGN OF A SOLENOID ACTUATOR WITH A MAGNETIC PLUNGER FOR MINIATURIZED SEGMENT ROBOTS

Chang-Woo Song and Seung-Yop Lee

Keywords: Solenoid; electromagnetic actuator; segmented robot; miniaturized robot; permanent magnet plunger

INTRODUCTION

Recently, there has been increasing research on segmented robots to mimic the peristaltic or serpentine motions and soft bodies of segmented animals. The developments of the miniaturized segment robots or ambulatory micro-robots have been applied to medical endoscopes, rescue robots, hazardous environment exploration, and industrial inspection systems [1,2,3].

The segmented or miniaturized robots can be classified according to actuation devices. In general, electromagnetic actuators have many advantages such as fast response, simple control law, and low manufacturing cost compared with the other actuators [4,5]. A solenoid has been widely used among electromagnetic actuators because it has a simple structure and both the direction and magnitude of the electromagnetic force can be easily controlled by changing the input current applied to the solenoid. The conventional solenoid uses a long loop of wire wrapped around a metallic core (plunger), and it produces a magnetic field to generate the linear motion of the plunge when an electric current is passed through the wire coil. Linear solenoids can be either unidirectional or bi-directional (push and pull) and they often have a spring return to bring the plunger back to the home position. Metallic plungers have often been replaced by a permanent magnet plunger to increase the electromagnetic force and response. Various solenoid actuators of bi-directional motions equipped with movable permanent magnets have been developed over a few decades [6,7]. Recently, Bammesberger *et al.* [8] and Kartmann*et al.* [9] have developed miniaturized solenoids with a ferromagnetic plunger for the application to disposal dispensing valves.

In a previous work [10], we designed a four-segmented robot using solenoids mimicking the earthworm's locomotion, where the segmented robot has a high-speed rectilinear motion. However, the miniaturized robot with turning locomotion has not been implemented. As the extension of the linear motion in the previous work [10], we design and develop a miniaturized solenoid with a ferromagnetic plunger to generate both rectilinear and turning motions of the segmented robot. We implement the theoretical analysis on the actuation force induced by the solenoid with a magnetic plunger based on the Biot-Savart law. Next, the design parameters of the solenoid for the miniaturized robot are determined using an optimization method. Experiments on the linear and turning motions are implemented and compared with the theoretical analysis using a prototype of the two-segmented robot equipped with a pair of solenoids.

Citation: Song, C.-W.; Lee, S.-Y. Design of a Solenoid Actuator with a Magnetic Plunger for Miniaturized Segment Robots. Appl. Sci. 2015,5, 595-607. doi:10.3390/app5030595.

SOLENOID DESIGN AND ANALYSIS

The objective of this research is to design and develop a solenoid for miniaturized segment robots with linear and turning motions. Figure 1a shows the design of the proposed robot with two segments. Each segment is actuated by a pair of solenoids to enable both linear and turning motions. The segmented robot has the rectilinear motion by actuating the solenoid pair in the same direction. Also, it can change direction by making the solenoid pair move in the opposite direction against each other. Moreover, to develop miniaturized segment robots, we have to reduce the size of a solenoid, on the one hand, and achieve an electromagnetic force strong enough to actuate each segment, on the other. In this chapter, we design and propose a solenoid based on the theoretical analysis.

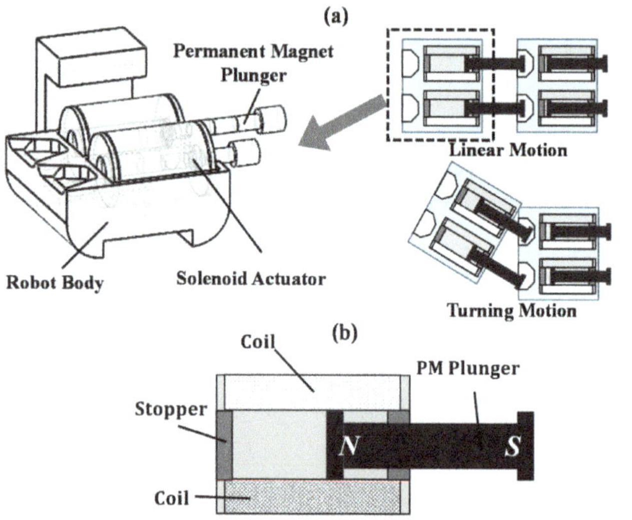

Figure 1. The design of a solenoid with a permanent magnet (PM) plunger: (**a**) the design concept of multi-segment robots with linear and turning motions using a pair of solenoids; (**b**) the cross-sectional view of the proposed solenoid actuator with PM plunger.

Solenoid Structure

The proposed solenoid consists of a coil and a movable plunger as shown in Figure 1b. The electromagnetically inductive coil is wound around the plunger. We use a permanent magnet (PM) as a plunger, rather than generally used metal rods, to generate a higher actuation force. Therefore, two kinds of forces are exerted in the solenoid. One is the electromagnetic force induced by the magnetic field inside the solenoid when a current flows into wound coils. The other is the magnetic force caused by a permanent magnet plunger.

Theoretical Analysis

The theoretical analysis on the electromagnetic force generated by the solenoid is implemented to design a solenoid actuator with a permanent magnet plunger. We develop a new type of solenoid actuator for bi-directional motion in the miniaturized robot. This solenoid actuator has a simple structure composed of coils and a permanent magnet plunger. The resultant force by the solenoid depends on the length, air-gap, number of coil turns, applied current, and the type of permanent magnet, *etc.* We induce the equation of electromagnetic force actuated by the solenoid as the functions of these design variables to develop a suitable solenoid for segmented robots.

The magnetic field along the axis of a circular current loop of radius r and steady current i can be expressed by the Biot-Savart law in Figure 1a:

$$\boldsymbol{B}_{sol} = \frac{\mu_0 i}{4\pi}\int \frac{\boldsymbol{dl} \times \boldsymbol{r}'}{|\boldsymbol{r}'|^3} \qquad (1)$$

Here, dl and r′ are the line segment and distance vector from the viewpoint to the source charge; $\mu_0=4\pi \times 10^{-7}$ Wb/A·m represents the permeability of vacuum. By the use of $dl = R\, d\psi\, \widehat{\boldsymbol{\psi}}$, $\boldsymbol{r}' = z\widehat{\boldsymbol{z}} - r\widehat{\boldsymbol{r}}$ and $|\boldsymbol{r}'| = \sqrt{z^2 + r^2}$, the magnetic field $B_{sol,z}$ at an arbitrary point P on the axis of the current loop becomes

$$\boldsymbol{B}_{sol,z} = \frac{\mu_0 i r^2}{4\pi}\int_0^{2\pi} \frac{d\psi}{(r^2+z^2)^{3/2}}\,\widehat{\boldsymbol{z}} = \frac{\mu_0}{2}\frac{ir^2}{(r^2+z^2)^{3/2}}\,\widehat{\boldsymbol{z}} \qquad (2)$$

Figure 2b shows the cross-sectional view of the solenoid actuator. In particular, for currents with special symmetry, Ampere's law in integral form offers an extraordinarily efficient means for finding the electromagnetic field. The axial electromagnetic field of the solenoid is obtained by Equation (2). Using the design parameters of the solenoid actuator such as inner diameter d, outer diameter D, number of coil turns N, and length L, Equation (2) becomes

$$\boldsymbol{B}_{sol,z} = \int_0^L \int_{d/2}^{D/2} \frac{\mu_0 n i r^2}{2\left[r^2 + \left(z_p + L/2 - z\right)^2\right]^{3/2}}\,\widehat{\boldsymbol{z}}\, dr\, dz \qquad (3)$$

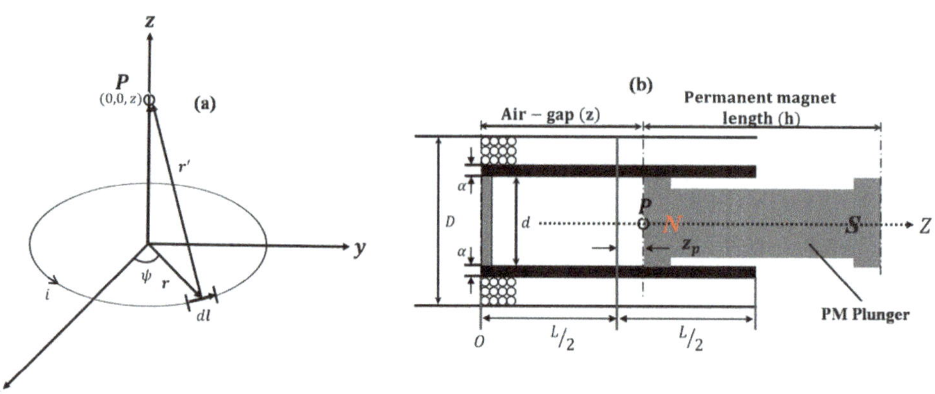

Figure 2. The solenoid dimensions for the electromagnetic analysis: (**a**) one turn of coil flowing with current *i*; (**b**) the cross-sectional view of a solenoid with PM plunger and design parameters.

Here, n= N/L and zp represent the wounded coils per unit length and the distance of point P from the center of the solenoid; z= L/2+zp is the moving distance of the PM plunger inside the solenoid or the length of the air-gap. By the integration of Equation (3), the electromagnetic field at point P becomes

$$\boldsymbol{B}_{sol,z} = \frac{\mu_0 Ni}{2L(D-d)}\left(L+2z_p\right) ln\left[\frac{D+\sqrt{D^2+\left(L+2z_p\right)^2}}{d+\sqrt{d^2+\left(L+2z_p\right)^2}}\right] + \frac{\mu_0 Ni}{2L(D-d)}\left(L-2z_p\right) ln\left[\frac{D+\sqrt{D^2+\left(L-2z_p\right)^2}}{d+\sqrt{d^2+\left(L-2z_p\right)^2}}\right]\hat{\boldsymbol{z}} \quad (4)$$

The electromagnetic force Fz(z, i) acting on the permanent magnet plunger by the solenoid actuator can be expressed as [11]

$$\boldsymbol{F}_z(\boldsymbol{z},\boldsymbol{i}) = \int(-\nabla\cdot\boldsymbol{M}_{mag})\boldsymbol{B}_{sol,z}\,dV + \int(\boldsymbol{M}_{mag}\cdot\hat{\boldsymbol{n}})\boldsymbol{B}_{sol,z}\,ds \quad (5)$$

Here, $\boldsymbol{B}_{sol,z} = B_{sol,z}\hat{\boldsymbol{z}}$ is the external magnetic field produced by the wounded coils. The direction of external magnetic field Bsol,z equals that of magnetic flux ϕ, and its magnitude is calculated by Equation (4). It is assumed that the magnet has a fixed and uniform magnetization (Mmag= $-M\hat{\boldsymbol{z}}$) in the direction of the longitudinal axis of the solenoid. Since the unit surface normal vector $\hat{\boldsymbol{n}} = -\hat{\boldsymbol{z}}$ at z = 0 and $\hat{\boldsymbol{n}} = \hat{\boldsymbol{z}}$ at z = h, the surface charge density σm is calculated as

$$\sigma_m = \boldsymbol{M}_{mag}\cdot\hat{\boldsymbol{n}} = \begin{cases} M, & z=0 \\ -M, & z=h \end{cases} \quad (6)$$

Only the north pole of the permanent magnet is located in the field of the solenoid. Since the uniform magnetization M_{mag} has no divergence, it has M = B_r/μ_0 where B_r represents the remanent flux density or remanence occurred by the permanent magnet plunger. Also, the magnetic volume charge density $\rho_m = -\nabla\cdot M_{mag} = 0$ [11], then, finally, the electromagnetic force is

$$\boldsymbol{F}_z(\boldsymbol{z},\boldsymbol{i}) = \int\sigma_m\boldsymbol{B}_{sol,z}\,ds = \frac{B_r}{\mu_0}\boldsymbol{B}_{sol,z}\int_0^{2\pi}\int_0^{d/2+\alpha} r\,dr\,d\psi = \frac{B_r B_{sol,z}A}{\mu_0}\hat{\boldsymbol{z}} \quad (7)$$

Here, A= π(d/2 + α)2 is the area by the inner diameter of the solenoid. The electromagnetic force is proportional to the magnetic field of the solenoid actuator ($B_{sol,z}$) and the magnetic plunger (B_r). Using Equation (7), the electromagnetic force is evaluated analytically or numerically. In this theoretical analysis, the magnetic circuit is regarded as ideal, where the magnetic fringe and leakage were ignored.

Solenoid Design

Equation (7) shows that the electromagnetic force from the solenoid actuator is proportional to the change in inductance of the coil with respect to the change in the position of the PM plunger as well as the current flowing through the coil. The solenoid force depends on various design parameters such as the inner and outer diameters of the solenoid, the number of coil turns, the air-gap, solenoid length, input current, and the remanence of the permanent magnet. We determine the optimal design parameters of the proposed solenoid for the segmented robots. Each segment is actuated by a pair of solenoids to enable both linear and turning motions. In order for the solenoid to actuate each segment, the pushing or retracting force of the solenoid is required to overcome the friction force of each segment.

To determine the design parameters of the solenoid, we fix the mass of each segment at *m* = 60 g for the light robot body. We use a ferromagnetic material (NdFeB) for the PM plunger. Then the total electromagnetic force of each solenoid is set to have F_z = 0.6 N, which is larger than the friction force ($F_z > F_{fric} = \mu_k mg$). Secondly, we have also fixed the following parameters: the current applied to the coil (*i* = 0.6 A), the air-gap from the center of solenoid (z_p = 0), the diameter of the wire (0.2 mm), and the remanence of the PM (4000 Gauss).

Then, we determine the optimal values of the following design parameters: the number of coil turns *N*, the solenoid length *L*, and the inner diameter *d* of the solenoid. The outer diameter of solenoid *D* is calculated by the relation among the solenoid length *L*, diameter of the wire (0.2 mm), the number of coil turns *N*, and the solenoid inner diameter *d*. Using the commercial software MATLAB (MathWorks Inc., Natick, MA, USA), we calculate the electromagnetic force of the solenoid given by Equation (7). Figure 3a,b show the solenoid forces for different inner diameters and numbers of coil turns, and the force distribution depending on the solenoid length and number of coil turns, respectively. Finally, the solenoid length (*L* = 20 mm), the number of coil turns (*N* = 1250 turns), and the inner diameter (*d* = 5 mm) are determined for the solenoid to actuate the segmented robot.

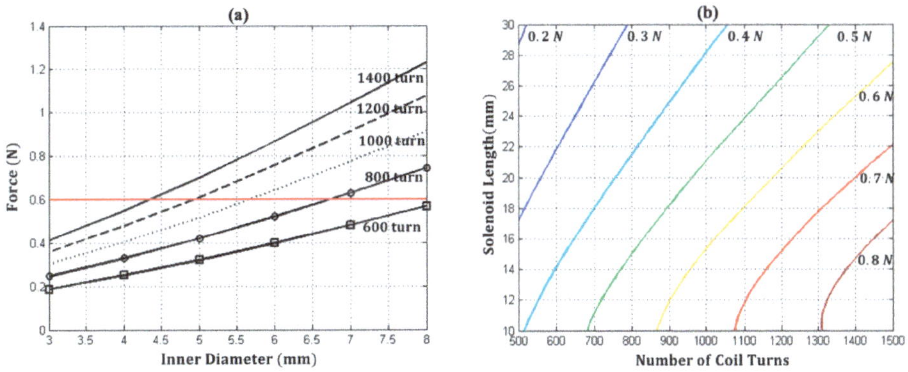

Figure 3. The theoretical analysis of the solenoid force: (**a**) the solenoid force for different inner diameters and numbers of coil turns at the solenoid length of 20 mm; (**b**) the distribution of the solenoid force depending on the solenoid length and number of coil turns at the inner diameter of 5 mm.

FABRICATION OF THE SOLENOID AND SEGMENTED ROBOT

Fabrication of Solenoid Actuator

We made a winding machine to manufacture the solenoid actuator to develop various solenoid actuators. Figure 4a shows that the winding machine consists of a direct current (DC) motor, encoder, controller, and body kits. The body kits are produced by using Acrylonitrile Butadiene Styrene (ABS). Figure 4b shows the manufactured solenoid actuator prototype with the solenoid length (L = 20 mm), with the number of coil turns (N = 1250 turns) based on the design optimization in Section 2.3. Inner and outer diameters of the solenoid are 5 mm and 15 mm, respectively. Neodymium iron boron (NdFeB) with the remanence of 4000 Gauss was chosen as the PM material for the plunger. The PM plunger is enveloped in a cylinder-type guide made of ABS. The plunger can move back and forth inside the guide by 15 mm. A stopper is located at the end of the inner solenoid to ensure that the end of the PM stops in this position. The air-gap is the distance between the inner bottom of the solenoid and the end of the PM plunger.

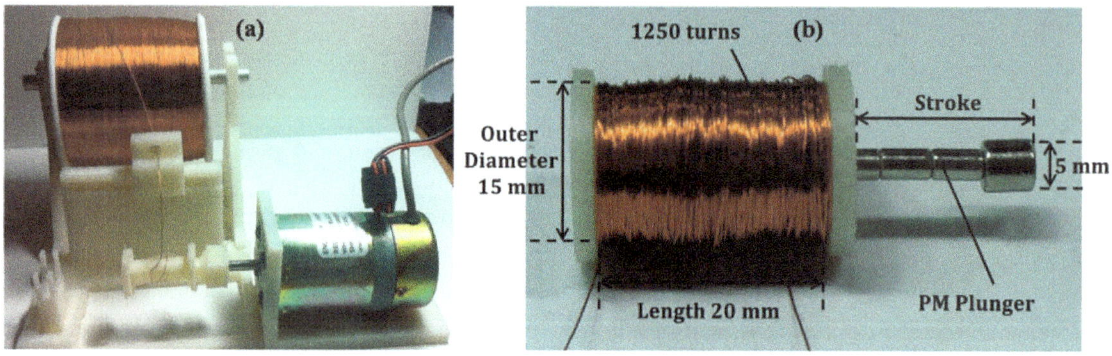

Figure 4. Design of a solenoid actuator with a PM plunger: (**a**) winding machine for manufacturing the solenoid actuators; (**b**) the photograph of a developed solenoid actuator prototype.

Segmented Robot with Solenoids

The prototype of the two-segmented robot with a pair of solenoids is shown in Figure 5a. The size of each module is 37 mm × 41 mm × 31 mm (length/width/height) and the mass is 50 g. If two segments adjoin each other with the end of the PM plunger attached to the stopper, the gap between the two segments is 5 mm (Figure 5b). The total maximum stroke becomes 15 mm if the PM plungers extrude from the solenoids.

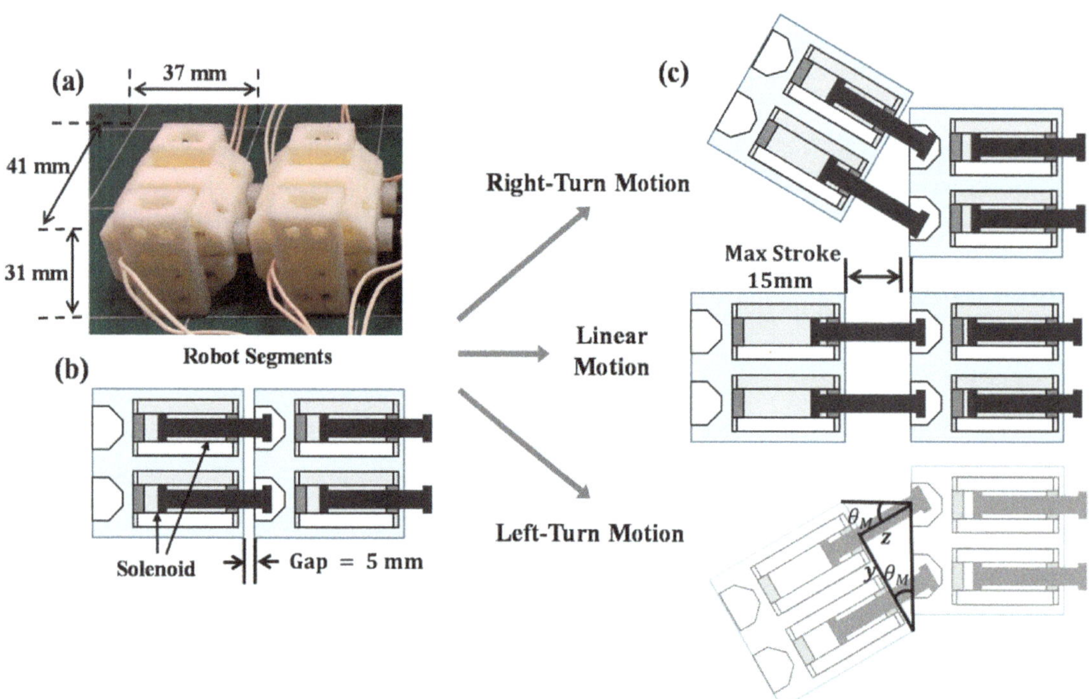

Figure 5. (**a**) Prototype of a two-segmented robot; (**b**) two segments adjoin each other; (**c**) the linear and turning motions of the segmented robot by symmetrical and asymmetrical actuations of the solenoid pair.

When the input current is applied to the coils, the plunger protrudes to push a neighboring segment in contact with it. For the linear motion, both solenoid plungers in each module are operated to move together and generate the synchronized force in the same direction, as shown in Figure 5c. However, for the change of direction or rotation of the segmented robot, the solenoid pair moves 180 degrees out of phase to generate an asymmetrical force and bend the two segmented body as shown in Figure 5. The asymmetrical force by the pair of solenoids easily generates the right- or left-turning motion of the segmented robot.

In the rectilinear motion, the moving distance and speed over the cycle are easily calculated to be proportional to the stroke of the plunger. However, the angular displacement by the turning motion depends on the module length and the distance between the two solenoids, as well as the stroke of the plunger. When one solenoid pushes and the other side retracts, the angular displacement θM in Figure 5 caused by the asymmetrical motion of the solenoid pair becomes

$$\theta_M = \tan^{-1}(z/y) \qquad (8)$$

Here, z is the stroke of the pushing plunger and *y* is the distance between the PM plunger and the side of module. Also, *y*depends on the constraint for the plunger rotation at the second segment, the stroke of the retracting plunger and the distance between the two solenoids. The theoretical angular displacement is calculated to have 8.82 degrees for the plunger stroke of 15 mm.

EXPERIMENTS AND DISCUSSIONS

We set up the experimental apparatus to measure the force and the displacement by the solenoid actuator to examine the performance of this solenoid actuator. As shown in Figure 6a, we construct an experimental set-up to measure the force generated by the solenoid using a load cell (CASTM PW4MC3, HBM, Darmstadt, Germany) the fixture to hold the solenoid, and the power supply. We use an infrared sensor (2Y0A02, Sharp Corporation, Osaka, Japan) as shown in Figure 6b to measure the displacement of the segmented robot. The distance measuring range of the infrared sensor system is from 20 to 150 cm. Additionally, the tolerance of this sensor system is approximately 0.3 mm.

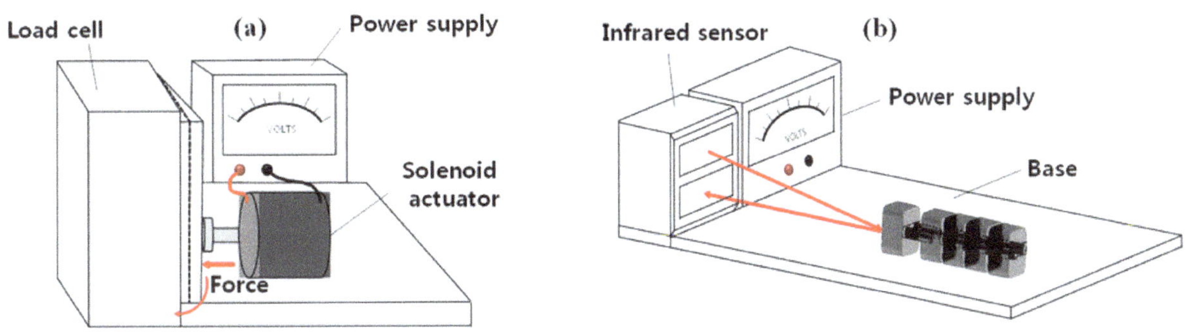

Figure 6. Experimental apparatus: (**a**) the measurement of the output force by the solenoid actuator; (**b**) the displacement measurement of the segmented robot using an infrared sensor.

Table 1 shows the theoretical and experimental results of electromagnetic forces acting on the PM plunger by the solenoid at the center of the solenoid (zp = 0), using the optimal design parameters: the solenoid length (*L* = 20 mm), the number of coil turns (*N* = 1250 turns), and the inner diameter (*d* = 5 mm). The experiments are performed 10 times and the averaged values are compared to the theoretical ones. The theoretical forces are calculated using Equation (7) for various numbers of coil turns, from 1100 to 1600 turns, at the constant input current (±0.6 A). The resistance of the proposed solenoid actuator is about 20 ohms. The experimental pushing and pulling forces are measured while changing the turns of wounded coils and the direction of the constant current. The electromagnetic force by the solenoid depends on the positions of the solenoid and the plunger. The theoretical results show that the pushing and pulling forces are the same at the constant input current. However, the experimental results indicate that the pushing and pulling forces are slightly different when the input current is reversed. The proportional dependence of the solenoid-induced force increasing on the coil turns is similar in both theory and experiments.

Table 1. The pulling and pushing electromagnetic forces from the solenoid actuator and the comparison of theoretical forces with experimental results for various coil turns

No. of Turns of Coils	Theoretical Force (N)		Experimental Force (N)	
	Pushing $(\hat{z})$	Pulling $(-\hat{z})$	Pushing $(\hat{z})$	Pulling $(-\hat{z})$
1100	0.5644	0.5664	0.5335	0.5077
1200	0.6104	0.6104	0.6036	0.5881
1300	0.6554	0.6554	0.6392	0.6580
1400	0.6996	0.6996	0.6557	0.6926
1500	0.7429	0.7429	0.7082	0.7906
1600	0.7854	0.7854	0.7659	0.8370

The theoretical electromagnetic forces from Equation (9) and the experimental data are plotted in Figure 7 for various coil turns when the plunger positions are changed from zp = −5 to 5 mm at the center of the solenoid. Figure 7a–f represent the pushing and pulling forces for six cases of coil turns: 1100, 1200, 1300, 1400, 1500, and 1600 turns. Other variables are fixed for the experiments such as the current applied to the coil (i = 0.6 A), the diameter of coil (0.2 mm), the solenoid length (L = 20 mm), and the remanence of PM (4000 Gauss). The electromagnetic force from the solenoid also depends on the positions of the solenoid and the plunger. The dashed circle and dashed star lines in the graphs indicate the experimental pushing and pulling electromagnetic forces, respectively. As mentioned in Table 1, the theoretical pushing and pulling forces are the same.

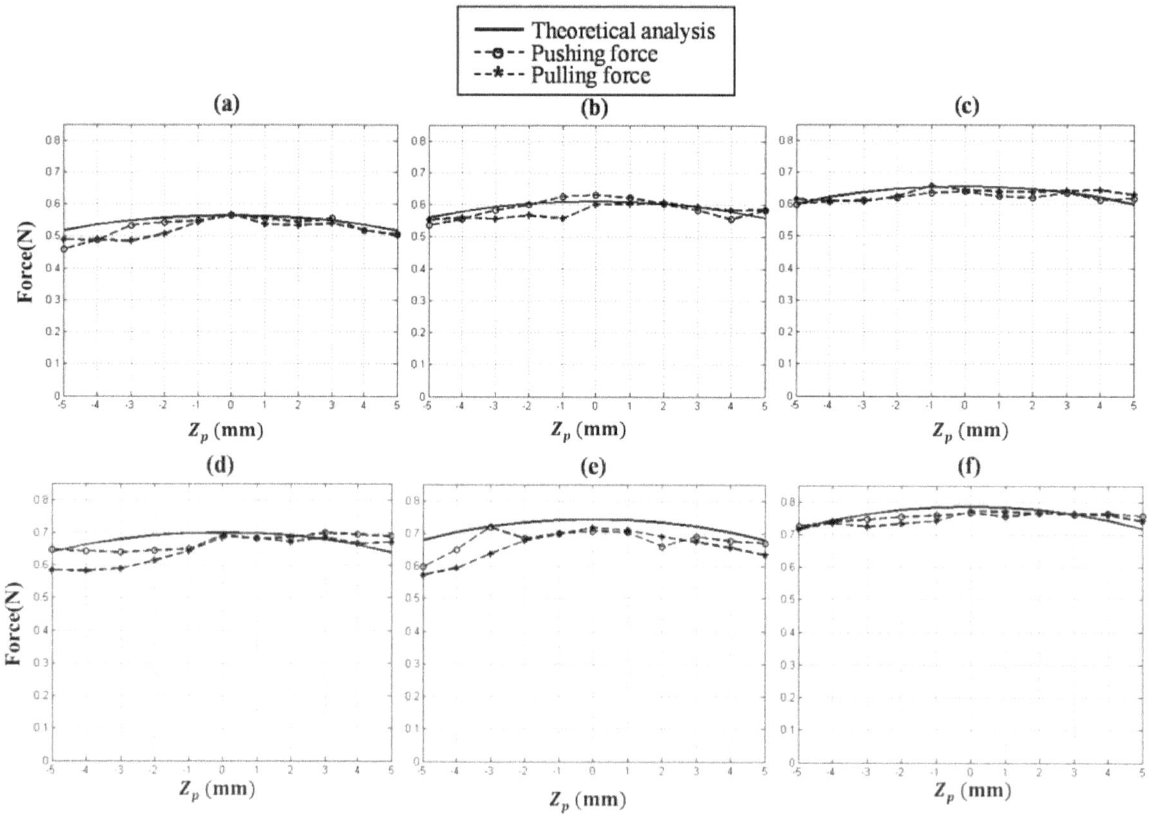

Figure 7. The resultant magnetic forces as a function of the plunger position z_p from the solenoid center for various coil turns: (**a**) 1100 turns; (**b**) 1200 turns; (**c**) 1300 turns; (**d**) 1400 turns; (**e**) 1500 turns; (**f**) 1600 turns.

Figure 8a,b show the time profiles of the input currents applied to the left and right solenoids of the segmented robot for the linear and right-turning motions, respectively. To generate the linear motion, the current signals of ±0.6 A are applied to a pair of solenoid actuators, and then these signals actuate the pair of PM plungers at the same time. For the turning motion, reverse input signals are applied to the pair of solenoids, indicating the solid and dashed lines of the graph. Figure 8c shows the photos of the experimental results when the solenoid pair is actuated for the linear and right/left-turning motions over 0.2 s. We measure the displacement of the front segment moving on the surface with kinetic frictional coefficient μ k = 0.2 using an infrared sensor (2Y0A02, Sharp Corporation, Osaka, Japan). The captured images over a cycle are shown in Figure 8c. It takes 0.1 s to move one segment and 0.2 s for two segments to complete the linear motion over one cycle. For the rectilinear motion, both plungers of the solenoid pair drive the first segment in the same direction. However, unwanted backward or slip motion occurs between two segments. It is because the back segment is not weighty enough to keep stationary during expansion ($0 < t < 0.1$ s) and the front segment has a backward motion during contraction ($0.1 < t < 0.2$ s). We measure the linear displacement over 10 cycles and the averaged displacement becomes 4.9 mm. The experimental displacement over the cycle is smaller than the theoretical one which is the plunger stroke of 15 mm. For the turning motion, the solenoid pair moves in the opposite direction from each other to generate an asymmetrical force. The experimental angular displacement (1.84 degrees) is lower than the theoretical one of θM = 8.82 degree from Equation (8) due to the backward or slip motion between the two segments.

To evaluate the effect of friction between the surface and the segmented robot, we implement experiments for the four cases of μk = 0.2, 0.35, 0.5, and a fixed condition holding the back segment stationary. We measure the linear and turning motions over 10 cycles for the four cases and the results are summarized in Table 2. The discrepancy between the theory and experiments decreases with the increasing friction coefficient. The preliminary experiments using two segments can be extended to multi-segmented robots by attaching additional segments with a pair of solenoids.

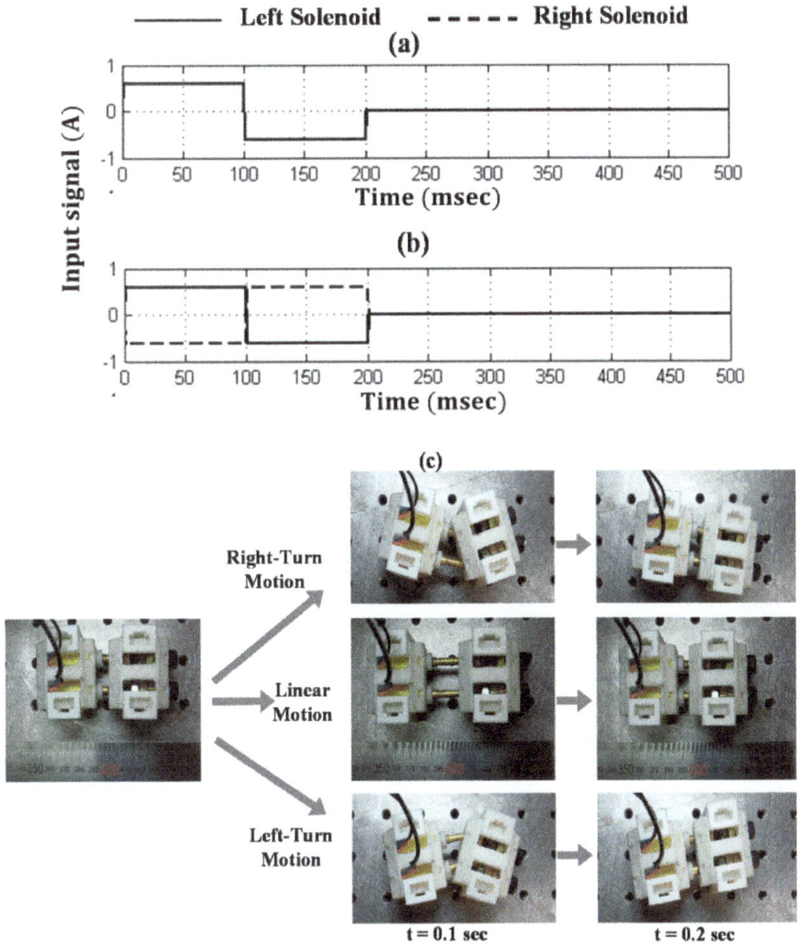

Figure 8. Profiles of input currents applied to the solenoid pair of the segmented robot (**a**) for the linear motion; (**b**) for the right-turning motion; (**c**) the captured images of the linear and right/left-turning motions of the two segments at 0.2 s.

Table 2. Linear and angular displacements of the two-segmented robot for various friction conditions: the comparison of theoretical values with experimental results

	Theory	Experiment			
		$\mu_k = 0.2$	$\mu_k = 0.35$	$\mu_k = 0.5$	fixed
Linear displacement (mm)	15	4.9	8.1	8.6	14.3
Angular displacement (deg)	8.82	1.84	2.66	2.81	7.97

CONCLUSIONS

We develop a solenoid with a ferromagnetic plunger to generate both rectilinear and turning motions of multi-segmented robots. The linear and turning motions of the segmented body are easily generated by the in-phase and out-of-phase actuation of the solenoid pair. The theoretical analysis on the actuation force imposed by the solenoid with the magnetic plunger is implemented using the Biot-Savart law, and then the optimal design parameters of the solenoid are determined to actuate the segmented robot. We manufacture prototypes of the miniaturized robot consisting of two segments and a pair of solenoids. Experiments show that slip or backward motion occurs between two segments. The linear and angular displacements increase when the segmented body moves on the surfaces with higher frictional coefficients. The novel mechanism to generate both linear and turning motions by the simple and low-cost solenoid pair could be applied to mimic the peristaltic and serpentine motions of various flexible or segmented insects by adding segment bodies.

ACKNOWLEDGMENTS

This research was supported by both National Nuclear R & D Program (2012M2A8A4055325) and Basic Science Research Program (2013R1A2A2A01068159) through the National Research Foundation of Korea funded by Ministry of Science ICT & Future Planning.

AUTHOR CONTRIBUTIONS

Chang-Woo Song designed the solenoid actuator and implemented experiments. Seung-Yop Lee contributed the theoretical analysis.

REFERENCES

1. Trivedi, D.; Rahn, C.; Kier, W.; Walker, D. Soft robotics: Biological inspiration, state of the art and future research. *Appl. Bionics Biomech.* **2008**, *5*, 99–117.
2. Hosokawa, D.; Ishikawa, T.; Morikawa, H.; Imai, Y.; Yamaguchi, T. Development of a biologically inspired locomotion system for a capsule endoscope. *Int. J. Med. Robotics Comput. Assist. Surg.* **2009**, *5*, 471–478. [PubMed]
3. Suzumori, K.; Miyagawa, T.; Kimura, M.; Hasegawa, Y. Micro inspection robot for 1-in pipes. *IEEE/AMSE Trans Mech.* **1999**, *4*, 286–292.
4. Shin, B.H.; Oh, D.; Lee, S.-Y. A two-dimensional laser scanning mirror using motion-decoupling electromagnetic actuators. *Sensors* **2013**, *13*, 4146–4156. [PubMed]
5. Shin, B.H.; Lee, K.-M.; Lee, S.-Y. A miniaturized tadpole robot using an electromagnetic oscillatory actuator. *J. Bionic Eng.* **2015**, *13*, 29–36.
6. Lequesne, B. Fast-acting long-stroke bistable solenoid with moving permanent magnets. *IEEE Trans. Ind. Appl.* **1990**,*26*, 401–407.
7. Ebihara, D.; Watada, M. Development of a Single-winding linear oscillatory actuator. *IEEE Trans. Magn.* **1992**, *28*, 3030–3032.
8. Bammesberger, S.B.; Kartmann, S.; Tanguy, L.; Liang, D.; Mutschler, K.; Ernst, A.; Zengerle, R.; Koltay, P. A low-cost, normally closed, solenoid valve for non-contact dispensing in the sub-μL range. *Micromachines* **2013**, *4*, 9–21.
9. Kartmann, S.; Koltay, P.; Zengerle, R.; Ernst, A. A disposable dispensing valve for non-contact microliter applications in a 96-well plate format. *Micromachines* **2015**, *6*, 423–436.
10. Shin, B.H.; Choi, S.W.; Bang, Y.B.; Lee, S.-Y. An earthworm-like actuator using segmented solenoids. *Smart Mater. Struct.* **2011**.
11. Furlani, E.P. *Permanent Magnet and Electromechanical Devices*, 1st ed.; Academic Press: Waltham, MA, USA, 2001; pp. 1–205.

CHAPTER 5

THE EFFECT OF ELECTROMAGNETIC FIELDS WITH THE MG2+ CYCLOTRON FREQUENCY

Gabriele Gerardi, Antonella De Ninno, Vanni Ferrari, Sandro Mazzariol, Daniele Bernardini, Severino Segato

Keywords: ELF-EMF, Cyclotron Frequency, Mouse, Birth-Rate, Postnatal Growth, Histological Findings

INTRODUCTION

The effects of exposure to external non-ionizing electromagnetic fields and their ability to induce carcinogenesis, teratogenesis and mutagenesis-related effects are of specific importance to human health. This is especially true in industrialized societies, in which individuals are continuously exposed to increasingly high levels of electromagnetic fields (EMFs) that are emitted by various electrical installations and telecommunication systems. Therefore, a wide body of literature has been produced regarding this subject. The 3 kHz - 300 GHz range of electromagnetic radiation has been focused on; this range is referred to, for convenience, as radiofrequency electromagnetic fields (RF EMFs) [1] - [5] . Even if the existence of a clear and strong effect on embryonic development has not been widely accepted by the scientific community for RF [2] or extremely low frequency (ELF) (<300 Hz) magnetic fields with non-thermal intensities [4] , several studies have shown that relatively low intensity EMFs are capable of interacting with molecular, cellular and systemic processes [6] - [9] . The primary issue that confronts this field is the lack of a conclusive understanding of the mechanism of interaction between EMFs and living organisms. This lack of data creates limits with respect to both the design and the critical evaluation of different experiments. Among the available evidence, two experiments [10] [11] point towards the existence of a resonance-based mechanism that relies on the coupling of the electric charge of biologically relevant ions and the applied magnetic field. It was observed that very weak magnetic fields, when applied to living organisms, produce variations in intracellular ionic concentrations when the frequency of the applied field matches the characteristic frequency of the involved ion species. This frequency is referred to as the ion cyclotron frequency (ICR), which can be defined as $f_c = (1/2p) \times (q/m) \times B_0$, where q and m are the electrical charge and the mass of the ion, respectively, and B_0 is a constant magnetic field. Another mechanism that is quoted in the literature is the ionic parametric resonance model [12] , for which the amplitude of the field is a critical parameter in explaining both the non-thermal effect and the so called "window effect", i.e. the amplitude range in which an effect is observed. The main results that were observed in such experiments include the modification of ionic flux through cell membranes. These models suggest a novel approach for experiments that are based on animal exposure to EMFs. Specifically, ions that are involved in cell division, such as Mg^{2+}, can be targeted using

Citation: Gabriele Gerardi, Antonella De Ninno, Vanni Ferrari, Sandro Mazzariol, Daniele Bernardini, Severino Segato (2016) The Effect of Electromagnetic Fields with the Mg2+ Cyclotron Frequency. Journal of Electromagnetic Analysis and Applications, 08,115-123. doi: 10.4236/jemaa.2016.87012

a resonant frequency that is chosen according to the q/m ratio and the static ambient magnetic field.

The aim of this study was to compare an experimental and a control group of mice, the first one exposed to a magnetic field with a frequency that was based on the cyclotron frequency of Mg^{2+}(derived using the above formula), during gestation, birth and the first 11 postnatal days. The overall experimental trial lasted 100 days.

ANIMALS

All of the animal handling, anaesthesia and euthanasia procedures were performed in accordance with the Italian and European codes of ethics. All of the animal procedures conformed to the Italian D.L. no. 116 of 27 January 1992 and the associated guidelines in the European Communities Council Directive of 24 November 1986 (86/609/ECC). Furthermore, this study was supported by a grant from the Ministero dell'Istruzione, dell'Università e della Ricerca, Italy (registration number 60A08-5987/10).

Twenty female and five male (8 weeks old) CD1 SPF (Specific Pathogen Free)/VAF (Virus and Antibody Free) mice from Charles River Laboratories Italia (Calco-LC) were used for this study. The animals were kept in the following controlled conditions over the duration of the experiment: environmental temperature of 21°C ± 1°C; humidity between 40% and 70%; air change per hour equal to 8 to 12 volumes; nyctohemeral cycle of 12/12 h; lighting intensity of 300 to 600 lux, with the light suspended 1 m over the floor; noise < 60 db. The bases of the cages were made of polypropylene, and the ceilings were made of a meshed stainless steel. A complete maintenance dry diet for mice (Standard Diet GLP - 4 RF 21; Charles River, Italy) was administered ad libitum. Water was also available ad libitum. Following 15 days of adaptation, 4 groups, each of which consisted of 1 male mouse and 4 female mice, were exposed to low intensity and low frequency electromagnetic fields (ELF-EMF groups). Another group, which consisted of 1 male mouse and 4 female mice, was used as a control (Control group). Each group was housed in a single cage.

MATERIALS AND METHODS

Exposure System

To create a uniform exposure zone, we used 6 coils with radii of 380 mm and which were spaced 380 mm apart in a Helmholtz configuration (Figure 1). Each of the coils was made of glass fibre, with copper wire windings wrapped within the carbon fibre. An alternating magnetic field was generated using a waveform generator mod. AFG310 by Tektronix (Beaverton, OR 97077, USA) with 2 Volt peak-to-peak with a 10 Ohm resistance, compensated using an external resistor. The environmental, static, magnetic field and the alternating magnetic field that was produced inside of the cage were measured over the entire exposure period using a F.W. Bell gauss meter (F.W. Bell, Orlando, FL, USA) Mod. 7010. The gauss meter was equipped with a low-field Hall sensor that had a resolution of 0.1 μT. The static magnetic field that was measured at the plane of the cages was $B_0 = 47.4$ μT. The amplitude of the alternating field that was produced by the coils was therefore set to a range of 38 - 47 μT. The measured amplitude of the static field was used to calculate the sinusoidal field frequency according to the cyclotron frequency of Mg^{2+}, namely $f_c = 1.26 \times B_0$. Both the intensity and frequency of the employed ELF-EMFs were monitored daily throughout the trial. During the exposure, the temperature inside of the cages was at room temperature (21°C ± 1°C).

Reproductive Performance and Blood Analysis

During the 100-day trial, the reproductive variables of the dams (gestations, births, offspring) were recorded and used to evaluate the effect of ELF-EMFs on fertility. Moreover, mice were weighed and data were used to calculate the average body weight (ABW) and average daily gain (ADG) of the offspring as indicators of the ELF-EMFs influence on growth and feed efficiency. The presence or absence of macroscopic alterations in the adults and the offspring, as well as the amount of food that was consumed by the adults, was also monitored.

As described above, on the 100th day of trial, blood samples were intracardially obtained from all 25 of the adult mice, which had not been subjected to any food restrictions. The blood samples were treated with K_2- EDTA and were used to examine the following haematological parameters: haematocrit percentage, haemoglo- bin concentration, erythrocyte count, mean corpuscular volume, mean corpuscular haemoglobin levels, mean corpuscular haemoglobin concentrations, platelet count, leukocyte count. A differential leukocyte count was also performed for neutrophils, lymphocytes, monocytes, eosinophils and basophils. All of these analyses were performed with the same commercial system (ADVIA 120 Haematology System-BAYER Corp. Diagnostic Division, Tarrytown, NY, USA). The morphological evaluation of the blood cells was microscopically performed on blood smears that were stained with Wright-Giemsa. Furthermore, the plasma that was obtained following the centrifugation of the blood samples was treated with K_2-EDTA and used to measure the following clinical chemistry parameters: total cholesterol, triglycerides, blood glucose, total protein, albumin, calcium, inorganic phosphorus, creatinine, blood urea nitrogen, alanine aminotransferase, aspartate aminotransferase, lactate dehydrogenase, creatine kinase, sodium, potassium, and chloride (automatic analyser ROCHE Hitachi 912 Plus-ROCHE Diagnostic Corp., Indianapolis, USA).

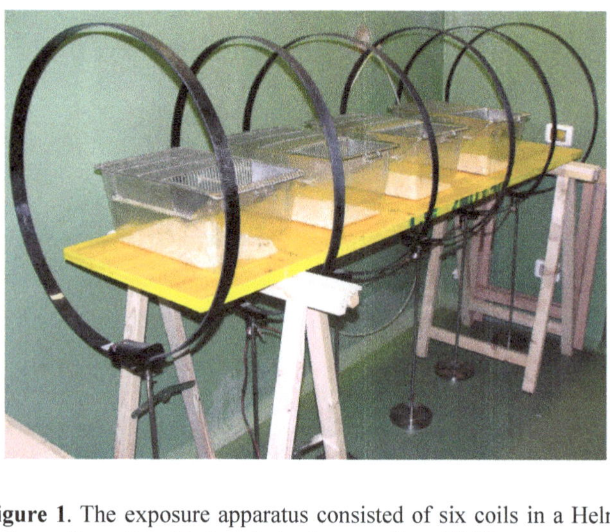

Figure 1. The exposure apparatus consisted of six coils in a Helmholtz configuration. Each coil, having a radius of 0.38 m, was made out of glass fibre, in which was embedded the copper wire. Coils are

spaced 0.38 m, the total length of the exposure space was 1.9 m. Magnetic fields are generated by currents induced in the coils by a waveform generator.

Post Mortem Examination

All of the newborn mice were euthanized 11 days after birth. According to a random selection method, one offspring per birth (72 exposed and 19 control, respectively), together with all of the dams and adult males, were submitted to a complete post mortem examination to assess any morphologic or topographic abnormalities. The entire body was fixed in 10% buffered formalin, and main organs and tissues (i.e., brain, thymus, spleen, lungs, heart, liver, pancreas, stomach, intestine, kidneys, skin, muscles and bone marrow) were sampled. All sampled organs and tissues were then formalin-fixed, paraffin-embedded, routinely processed and stained with haema- toxilin and eosin.

Microscopic examination were performed by the same pathologist who performed post mortem examination using a camera (Eclipse E100, Nikon) in a blinded fashion: three slide per each organs were observed reporting any pathological changes according to standard diagnosis (UNI EN ISO 10993-10, 2004). Specifically, the vertebrae of the mouse pups were counted to assess the effect of ELF-EMF exposure on bone formation. Changes in haematopoietic activity were evaluated by counting the splenic megakaryocytes in 10 random high-magnification fields (40×).

Statistical Analysis

The average daily gain (ADG) was calculated by performing linear regression analysis of the body weights that were measured on the 3rd, 7th and 11th postnatal days. The assumption of normality and variance homogeneity of the following variables was tested using the Shapiro-Wilk test (PROC UNIVARIATE): the number of gestations, the number of births, the number of offspring per dam, the average number of offspring per birth, the average body weight (ABW) per nest, the ADG per mouse and nest, the food consumption of each group per day and the haematological profile. Because these variables were normally distributed, the data were submitted to a one-way ANOVA (PROC GLM) to assess the effect of the ELF-EMF. The histological data (in percentages) were not normally distributed. These data were therefore analyzed by performing a multiple proportions test using the Marascuillo procedure. Bonferroni adjustments were performed to verify the effects of the ELF-EMFs on the offsprings and dams. For the histological variables that exhibited significantly (P £ 0.01) different k-proportions between the groups, the risk ratios and the 95% confidence intervals were calculated in terms of the ratio between the exposed and unexposed animals. The Marascuillo procedure was also used to analyse the histological findings of the treated dams and their offsprings.

All the statistical analysis were performed using SAS software (2010; release 9.3), except for the Marascuillo parametric test that was carried out by using XL-Stat software (Version 2015.2.02, Addinsoft, GMSL, Milano, Italy).

RESULTS

Reproductive Performance and Haematological Profile

The number of gestations, births and total offspring for each dam were not affected by ELF-EMF exposure (Table 1). However, the average number of offspring per birth was significantly (P = 0.006) lower in the exposed dams. As reported in Table 1, the ELF-EMFs produced a significantly lower ABW in the offspring at 7 (P = 0.044) and 11 (P = 0.048) postnatal days; this result was confirmed by the observed lower ADG, which was determined by performing a linear regression analysis of the BW of the pups over time. The amount of food that was consumed did not differ between the groups (4 exposed and 1 control) of adult mice during the trial (31.3 vs. 32.7 g per day; P = 0.589).

The haematological and clinical chemistry parameters in the adult animals were not affected by the exposure. However, a moderate decrease in the white blood cell count (4.0 vs. 5.7 × 10^9/L; P = 0.071) was observed in the exposed mice. Furthermore, a reduction in the number of both neutrophils (20.7 vs. 27.9 × 10^9/L; P = 0.034) and lymphocytes (2.7 vs. 4.0 × 10^9/L; P = 0.028) in the exposed mice were also detected.

Histological Findings

Neither the offspring, the dams nor the adult males exhibited any gross changes, although hydrocephalus was observed in one exposed pup. The post mortem examination of the offspring revealed no significant differences for any of the histological findings. However, it was observed a mild mononucleated inflammatory population in hepatic portal spaces, a condition for which there was a significantly (P = 0.001) higher k-proportion in the exposed mice and a relative risk ratio of 2.53 (Table 2). As reported in Table 3, the post mortem analyses of exposed adult females revealed that the ELF-EMF induced a significantly (P = 0.0023) higher incidence of myocardial hypertrophy. Moreover, splenic follicular hyperplasia was significantly higher in the control group (12 vs. 75%; P = 0.008), although the risk ratio was very low (0.18).

Table 1. The effect of ELF-EMFs on gestations, births, offspring number, average body weight (ABW) per nest at 3, 7 and 11 postnatal days, and average daily gain (ADG) per mouse and nest

	Treatment		ANOVA	
	ELF-EMF	Control	P-value	RMSE
Number of gestations, n°	4.44	4.75	0.288	0.51
Number of births, n°	4.44	4.75	0.288	0.51
Average number of offspring per birth, n°	11.0	11.6	0.006	0.35
Average number of deaths per birth, n°	0.20	0.19	0.942	0.18
Total number of offspring per dam, n°	48.9	55.2	0.103	6.6

ABW per nest at 3 days, g	112	123	0.107	12
ABW per nest at 7 days, g	257	294	0.044	30
ABW per nest at 11 days, g	404	463	0.048	51
ADG per mouse, g/d	0.74	0.77	0.002	0.01
ADG per nest, g/d	36.5	42.5	0.045	5.0

RMSE, root mean square error.

The post mortem analysis of the adult males revealed a higher incidence of mild hepatic degeneration, splenic follicular hyperplasia, mild multifocal acute splenitis and chronic interstitial pneumonia in the exposed groups, although these effects were not statistically significant given the very low number of examined animals. Lastly, no difference was observed in the adult animals with respect to the number of splenic megakaryocytes (data not tabulated for brevity).

DISCUSSION

Reproductive Performance and Haematological Profile

In the present study, extremely low frequency fields of a specific resonant frequency, namely the cyclotron frequency of Mg^{2+}, were applied to test their effects on mouse reproduction. The importance of Mg^{2+} in the duplication process of cells is well established [13] [14] . Thus, the effect of electromagnetic fields on Mg^{2+} may be enhanced during embryogenesis. With respect to the birth rate data, the ELF-EMF exposure elicited a mild effect, which was limited to a significant reduction of the offspring number per birth; although the total offspring during the entire period of the trial did not change. In agreement with the literature [15] - [18] , EMF exposure did not cause adverse effects on reproductive performance.

The ELF-EMFs significantly affected the growth of the offspring after 3 postnatal days as a consequence of altering lactation and/or their metabolism. These data are in agreement with other authors [19] , who reported that the mean BW of exposed offspring at postnatal day 7 days was significantly lower than the weight of the unexposed pups. Moreover, a negative influence on BW increase seems to persist into the second and third weeks of age [18] - [20] . Because it was recorded that food consumption was the same between the two groups of dams, a result that was also observed [16] , it could be hypothesized that the lower ADG of the exposed offspring was the consequence of a change in the efficiency with which dams converted dietary gross energy to net energy of lactation.

The results of the haematological and blood chemistry analyses in the dams revealed no reproducible alterations due to ELF-EMF exposure. Consistent with reports in the literature, the observed changes were sporadic and did not affect organogenesis in the offspring [17] - [22] .

Table 2. The effect of ELF-EMFs on the histological parameters that were examined in the offspring (n = 72 exposed; n = 19 control)

	Treatment		**P-value**
	ELF-EMF	**Control**	
Mild multifocal interstitial pneumonia, %	47[a]	22	0.133
Mild centro-lobular hepatosis, %	31[ab]	58	0.188
Mild chronic aspecific reactive hepatitis, %	28[abc]	0	0.001
Chronic enteritis, %	3[bc]	0	0.310
Splenic follicular hyperplasia, %	3[bc]	0	0.310
Mild chronic rhinitis, %	0[c]	11	0.289
Mild chronic pleuritis, %	0[c]	11	0.289

P-value indicates the significance of the k-proportion based on the Marascuillo test (Bonferroni adjustment). In the case of mild chronic aspecific reactive hepatitis, a risk ratio of 2.53 was estimated, and the 95% confidence interval was 1 - 17.4. [a,b,c]The column proportions differ significantly (P £ 0.01; $c^2 = 51.6$); the analysis was performed using treated offspring.

Table 3. The effect of ELF-EMFs on the histological parameters that were examined in the dams (n = 16 exposed; n = 4 control)

	Treatment		**P-value**
	ELF-EMF	**Control**	
Mild hepatic degeneration, %	81[a]	75	0.835
Mild multifocal acute splenitis, %	31[ab]	25	0.737
Myocardial hypertrophy, %	25[ab]	0	0.023
Mild multifocal interstial nephritis, %	12[b]	0	0.129
Renal tubules degeneration/ectasia, %	12[b]	25	0.618
Splenic erythrocatheresis, %	12[b]	0	0.129
Splenic follicular hyperplasia, %	12[b]	75	0.008
Chronic interstial pneumonia, %	12[b]	0	0.129
Apoptotic findings, %	12[b]	0	0.129
Splenic congestion, %	6[b]	25	0.417
Chronic enteritis, %	0[b]	25	0.248

P-value indicates the significance of the k-proportion comparison based on the Marascuillo test (Bonferroni adjustment). With respect to splenic follicular hyperplasia, a risk ratio of 0.18 was estimated, and the 95% confidence interval was 0.04 - 0.73. With respect to the observed myocardial hypertrophy, the risk ratio was not estimable. [a,b]The column proportions differ significantly (P £ 0.01; c^2 = 42.8); the analysis was performed using treated dams.

Histological Findings

In the present study, post mortem examinations of postnatal day 11 pups verified the absence of any changes in either organ or vertebrae genesis or development. These results confirm the results of previous studies that have been performed on rats and mice that were chronically exposed to electromagnetic fields [3] - [23] . The only noticeable histological finding among the offspring was an instance of hydrocephalus, which was observed in one exposed pup. Congenital hydrocephalus can be experimentally induced by numerous insults, such as radiation, infections, chemicals and nutritional deficiencies [24] - [27] ; however, in all these studies, the incidences of the resulting hydrocephalus was not sufficiently high to allow for an explanation of its primary causes.

The absence of gross changes in the exposed dams is in agreement with the studies of other authors [16] [17] , which reported that were no significant macroscopic alterations between EMF-exposed and control dams. A higher observation of mild hepatic chronic inflammatory population, without other associated changes, indicated a slight influence of ELF-EMF on offspring pathologies. Similar results were also observed in literature [21] , who reported that the pathological changes observed in 21-day-old mice that were exposed to fields of different intensities and frequencies were not specific to EMF exposure.

In the present study, the only statistically significant observations in the dams were myocardial hypertrophy in the exposed and splenic erythrocatheresis in the control ones. This result confirms that the majority of the observed microscopic findings are primarily due to spontaneous pathologic conditions. These changes are not related to specific pathogen but they could be spontaneous or secondary to opportunistic biological agents exposure not revealed by microscopic observation [28] . Furthermore, the observed mild hepatic degeneration was the most significant histological finding for both the exposed and control dams. This effect was a likely consequence of the repeated pregnancies and/or lactation rather than of the ELF-EMF exposure.

Summarizing the results of this study and of the current literature, it can be noticed that the epidemiological evidence of the effects of non-thermal electromagnetic fields on living organisms is still weak and inconclusive, primarily due to inaccurate assessment of the intensity and frequency of the exposure, which makes difficult to compare different experimental trials. Alternatively, there is a wide body of evidence that electromagnetic field exposure induces noteworthy effects in vitro [29] , including specific outcomes on processes that are related to embryogenesis [30] . With respect to in vivo studies, there are many reports on high frequency field exposure; however, fewer data are available regarding exposure to low frequency fields even though many positive reports are available [7] . Particularly, it was previously reported [31] that long-term exposure to the Ca^{2+}cyclotron frequency affects the growth and metabolism of rats.

CONCLUSION

The present experimental protocol was designed on the basis of the hypothesis that embryogenesis is the life stage ideal for studying the effects of EMF exposure. However, the effect of the fields on a very multifaceted system, such as a living organism, involves highly complex interactions between multiple parameters and it is not plain to relate an effect to a single cause or a sequence of external agents. The exposure of mice to a magnetic field with the cyclotron frequency of Mg^{2+} during pregnancy appears to weakly perturb the reproduction of the exposed animals in terms of the average number of offspring per birth. Such a fact together as the minor anomalies observed in adult mice during 100 days of exposure may be considered as a warning about the importance of environmental magnetic fields on the stability of individual species or even eco-systems. The present data confirm the need for a conceptual bio-physical model that allows for the clear identification of how electromagnetic fields exert their biological effects even at non-thermal intensities and very low frequencies.

ACKNOWLEDGEMENTS

We want to thank Marco Prosdocimi and Filippo Barbaro (PROMETEO S.r.l., Via Marostica n. 2, 35100 Padova, Italy) for their technical support.

REFERENCES

1. Ryan, B., Polen, M., Gauger, J.R., Mallett Jr., E., Kearns, M.B., Bryan, T.L. and McCormick, D.L. (2000) Evaluation of the Developmental Toxicity of 60 Hz Magnetic Fields and Harmonic Frequencies in Sprague-Dawley Rats. Radiation Research, 153, 637-641. http://dx.doi.org/10.1667/0033-7587(2000)153[0637:EOTDTO]2.0.CO;2
2. Chung, M.K., Kim, J.C. and Myung, S.H. (2004) Lack of Adverse Effects in Pregnant/Lactating Female Rats and Their Offspring Following Pre-and Postnatal Exposure to ELF Magnetic Fields. Bioelectromagnetics, 25, 236-244.http://dx.doi.org/10.1002/bem.10182
3. Nishimura, I., Oshima, A., Shibuya, K. and Negishi, T. (2011) Lack of Teratological Effects in Rats Exposed to 20 or 60 kHz Magnetic Fields. Birth Defects Research Part B, Developmental and Reproductive Toxicology, 92, 469-477.http://dx.doi.org/10.1002/bdrb.20316
4. Vallejo, D. and Hidalgo, M.A. (2012) Growth Variations in OF1 Mice Following Chronic Exposure of Parental and Filial Generations to a 15 μT, 50 Hz Magnetic Field. Electromagnetic Biology and Medicine, 31, 19-33.http://dx.doi.org/10.3109/15368378.2011.620203
5. Berman, E., Carter, H.B. and House, D. (1982) Reduced Weight in Mice Offspring after in Utero Exposure to 2450-MHz (CW) Microwaves. Bioelectromagnetics, 3, 285-291. http://dx.doi.org/10.1002/bem.2250030212
6. Sienkiewicz, Z.J., Robbins, L., Haylock, R.G. and Saunders, R.D. (1994) Effects of Prenatal Exposure to 50 Hz Magnetic Fields on Development in Mice: II. Postnatal Development and Behaviour. Bioelectromagnetics, 15, 363-375.http://dx.doi.org/10.1002/bem.2250150410

7. Gagnon, Z.E., Newkirk, C., Conetta, J.A., Sama, M.A. and Sisselman, S. (2003) Teratogenic Effect of Broad-Band Electromagnetic Field on Neonatal Mice (Mus musculus). Journal of Environmental Science and Health Part A, Toxic/Hazardous Substances & Environmental Engineering, 38, 2465-2481. http://dx.doi.org/10.1081/ESE-120024449

8. Kim, S.H., Song, J.E., Kim, S.R., Oh, H., Gimm, Y.M., Yoo, D.S., Pack, J.K. and Lee, Y.S. (2004) Teratological Studies of Prenatal Exposure of Mice to a 20 kHz Sawtooth Magnetic Field. Bioelectromagnetics, 25, 114-117.http://dx.doi.org/10.1002/bem.10164

9. Juutilainen, J. (2003) Developmental Effects of Extremely Low Frequency Electric and Magnetic Fields. Radiation Protection Dosimetry, 106, 385-390. http://dx.doi.org/10.1093/oxfordjournals.rpd.a006376

10. Inouye, M. and Kajiwara, Y. (1990) Strain Difference of the Mouse in Manifestation of Hydrocephalus Following Prenatal Methylmercury Exposure. Teratology, 41, 205-210. http://dx.doi.org/10.1002/tera.1420410212

11. Aolad, H.M., Inouye, M., Darmanto, W., Hayasaka, S. and Murata, Y. (2000) Hydrocephalus in Mice Following X-Irradiation at Early Gestational Stage: Possibly Due to Persistent Deceleration of Cell Proliferation. Journal of Radiation Research, 41, 213-226. http://dx.doi.org/10.1269/jrr.41.213

12. Crews, L., Wyss-Coray, T. and Masliah, E. (2004) Insights into the Pathogenesis of Hydrocephalus from Transgenic and Experimental Animal Models. Brain Pathology, 14, 312-316. http://dx.doi.org/10.1111/j.1750-3639.2004.tb00070.x

13. Harada, T., Takamoto, M., Jin, D.H., Tada, T. and Sugane, K. (2007) Young C3H Mice Infected with Toxoplasma gondii Are a Novel Experimental Model of Communicating Hydrocephalus. Neurological Research, 29, 615-621.http://dx.doi.org/10.1179/016164107X164201

14. Suttie, A.W. (2006) Histopathology of the Spleen. Toxicologic Pathology, 34, 466-503.http://dx.doi.org/10.1080/01926230600867750

15. Blank, M. (2008) Protein and DNA Reactions Stimulated by Electromagnetic Fields. Electromagnetic Biology and Medicine, 27, 3-23. http://dx.doi.org/10.1080/15368370701878820

16. Berman, E., Chacon, L., House, D., Koch, B.A., Koch, W.E., Leal, J., LØvtrup, S., Mantiply, E., Martin, A.H., Martucci, G.I., et al. (1990) Development of Chicken Embryos in a Pulsed Magnetic Field. Bioelectromagnetics, 11, 169-187.http://dx.doi.org/10.1002/bem.2250110208

17. Gerardi, G., De Ninno, A., Prosdocimi, M., Ferrari, V., Barbaro, F., Mazzariol, S., Bernardini, D. and Talpo, G. (2008) Effects of Electromagnetic Fields of Low Frequency and Low Intensity on Rats Metabolism. Biomagnetic Research and Technology, 6, 3. http://dx.doi.org/10.1186/1477-044X-6-3

18. Yang, L., Arora, K., Beard, W.A., Wilson, S.H. and Schlick, T. (2004) Critical Role of Magnesium Ions in DNA Polymerase Beta's Closing and Active Site Assembly. Journal of the American Chemical Society, 126, 8441-8453. http://dx.doi.org/10.1021/ja049412o

19. Lyons-Darden, T. and Topal, M.D. (1999) Effects of Temperature, Mg2+ Concentration and Mismatches on Triplet-Repeat Expansion during DNA Replication in Vitro. Nucleic Acids Research, 27, 2235-2240. http://dx.doi.org/10.1093/nar/27.11.2235

20. Vincze, G., Szasz, A. and Liboff, A.R. (2008) New Theoretical Treatment of Ion Resonance Phenomena. Bioelectromagnetics, 29, 380-386. http://dx.doi.org/10.1002/bem.20406

21. Liboff, A.R. (1985) Geomagnetic Cyclotron Resonance in Living Cells. Journal of Biological Physics, 13, 99-102. http://dx.doi.org/10.1007/BF01878387

22. Blackman, C.F., Benane, S.G., Rabinowitz, J.R., House, D.E. and Joines, W.T. (1985) A Role for the Magnetic Field in the Radiation-Induced Efflux of Calcium ions from Brain Tissue in Vitro. Bioelectromagnetics, 6, 327-337. http://dx.doi.org/10.1002/bem.2250060402

23. Lisi, A., Ledda, M., De Carlo, F., Foletti, A., Giuliani, L., D'Emilia, E. and Grimaldi, S. (2008) Calcium ion Cyclotron Resonance (ICR) Transfers Information to Living Systems: Effects on Human Epithelial Cell Differentiation. Electromagnetic Biology and Medicine, 27, 230-240. http://dx.doi.org/10.1080/15368370802269135

24. BioInitiative Working Group, Sage, C. and Carpenter, D.O. (Eds.) (2012) BioInitiative Report: A Rationale for a Biologically-Based Public Exposure Standard for Electromagnetic Radiation. www.bioinitiative.org

25. Liboff, A.R. (2005) The Charge-to-Mass ICR Signature in Weak ELF Bioelectromagnetic Effects. In: Lin, E.C., Ed., Advances in Electromagnetic Fields in Living Systems, Springer Verlag, Berlin, Volume 4, 189-218. http://dx.doi.org/10.1007/0-387-24024-1_6

26. Zwirska-Korczala, K., Adamczyk-Sowa, M., Polaniak, R., Sowa, P., Birkner, E., Drzazga, Z., Brzozowski, T. and Konturek, S.J. (2004) Influence of Extremely-Low-Frequency Magnetic Field on Antioxidative Melatonin Properties in AT478 Murine Squamous Cell Carcinoma Culture. Biological Trace Element Research, 102, 227-243. http://dx.doi.org/10.1385/BTER:102:1-3:227

27. Huuskonen, H., Lindbohm, M.L. and Juutilainen, J. (1998) Teratogenic and Reproductive Effects of Low-Frequency Magnetic Fields. Mutation Research, 410, 167-183. http://dx.doi.org/10.1016/S1383-5742(97)00038-0

28. Lee, H.J., Lee, J.S., Pack, J.K., Choi, H.D., Kim, N., Kim, S.H. and Lee, Y.S. (2009) Lack of Teratogenity after Combined Exposure of Pregnant Mice to CDMA and WCDMA Radiofrequency Electromagnetic Fields. Radiation Research, 172, 648-652. http://dx.doi.org/10.1667/RR1771.1

29. Juutilainen, J. (2005) Developmental Effects of Electromagnetic Fields. Bioelectromagnetics, 7, 107-115. http://dx.doi.org/10.1002/bem.20125

30. Heynick, L.N. and Merritt, J.H. (2003) Radiofrequency Fields and Teratogenesis. Bioelectromagnetics, 24, 174-186. http://dx.doi.org/10.1002/bem.10127

31. Klug, S., Hetscher, M., Giles, S., Kohlsmann, S. and Kramer, K. (1997) The Lack of Effects of Non Thermal RF Electromagnetic Fields on the Development of Rat Embryos Grown in Culture. Life Sciences, 61, 1789-1802. http://dx.doi.org/10.1016/S0024-3205(97)00803-5

CHAPTER 6

RADIALLY POLARIZED RANDOM ELECTROMAGNETIC BEAM

Yongxin Liu, Songjie Luo, Jixiong Pu, Zenghui Gao

Keywords: Generalized Stokes Parameters, Electric Cross-Spectral Density Matrix, Young's Double Slits, Mach-Zehnder Interferometer, Radially Polarized Stochastic Electromagnetic Beam

INTRODUCTION

Coherence and polarization of the electromagnetic beams are always considered as two unrelated subjects to investigate and discuss. However, both of them have the same physical phenomenon, that is, the correlation between the fluctuations of the electric field. Coherence is caused by the fluctuation of the electric field of two or more spatial points. Polarization is related to the fluctuation of the component of the electric field of the same spatial point.

In 2003, Emil Wolf developed the unified theory of coherence and polarization of a random electromagnetic beam, which illuminated an intimate relationship existing between the coherence and polarization of electromagnetic beams [1] . Then, how to use the cross spectral density matrix of the electric field to calculate the degree of polarization, the degree of coherence and the spectrum of a random electromagnetic beam in the propagation is analyzed theoretically [2] - [4] . And next, Emil Wolf presented a method of how the four elements of the cross spectral density matrix may be determined from measurements of spectra in the Young's interference experiment in which polarizers and rotators are placed in front of the pinholes [5] .

Recently, the generalized Stokes parameter, which is a new extension associated with two spatial points, also depends on the cross spectral density matrix of the electric field [6] . Therefore, the generalized Stokes parameter can be used to study the coherence and polarization of the electromagnetic beam.

In 2008, Kamdpal H. proposed an experimental method of measuring the cross spectral density matrix and the generalized Stokes parameters of the laser beam [7] , but it is still difficult to be realized in the ordinary laboratory, because of the demanding experimental structures. In 2009, C. Kamdpal H. also pointed out another method which can determine the generalized Stokes parameters of the beam by directly and experimentally measuring the ordinary Stokes parameters, but the use of the beam is completely coherent laser beam [8] .

Meanwhile, the radially polarized beam can be focused to generate a strong longitudinal and non-propagating electric field at the focal plane through a high numerical aperture objective [9] [10] . Due to this unique feature, the radially polarized beam

Citation: Yongxin Liu, Songjie Luo, Jixiong Pu, Zenghui Gao (2016) Radially Polarized Random Electromagnetic Beam. Journal of Electromagnetic Analysis and Applications,08,109-114. doi: 10.4236/jemaa.2016.86011

has attracted significant interest [11] [12] . In this paper, we will investigate the generalized Stokes parameters of a radially polarized random electromagnetic beam experimentally. We propose an experimental method utilizing a double-slit aperture and a Mach-Zehnder interferometer to measure the cross spectral density matrix and generalized Stokes parameters of a radially polarized random electromagnetic beam. A rotating ground glass and a radial polarization converter are placed in front of the double-slit aperture to produce a radially polarized partially coherent beam, and the polarizers and half wave plates in the Mach-Zehnder interferometer are used to take measurements of the four elements of the cross spectral density matrix.

FORMULATION

Let us consider a quasi monochromatic light propagating along the Z direction. The 2×2 cross-spectral density matrix can be given by:

$$W(r_1,r_2,\omega)=\begin{bmatrix} W_{xx}(r_1,r_2,\omega) & W_{xy}(r_1,r_2,\omega) \\ W_{yx}(r_1,r_2,\omega) & W_{yy}(r_1,r_2,\omega) \end{bmatrix} \quad (1)$$

where the r_1 and r_2 are the coordinate vectors on the plane Z, and

$$W_{ij}(r_1,r_2,\omega)=\langle E_i^*(r_1,\omega)E_j(r_2,\omega)\rangle, i=x,y, j=x,y, \quad (2)$$

where the asterisk denotes the complex conjugate and the angular brackets denote the ensemble average. E_i and E_j are components of the electric field.

The spectral degree of coherence of the electric field of the beam at the points r_1 and r_2 can be expressed as [5] :

$$\eta(r_1,r_2,\omega)=\frac{\mathrm{Tr}W(r_1,r_2,\omega)}{\sqrt{\mathrm{Tr}W(r_1,r_1,\omega)}\sqrt{\mathrm{Tr}W(r_2,r_2,\omega)}} \quad (3)$$

where, Tr is the matrix trace.

The spectral density can be given by:

$$S(r_1,\omega)=\mathrm{Tr}W(r_1,r_1,\omega) \quad (4a)$$

$$S(r_2,\omega)=\mathrm{Tr}W(r_2,r_2,\omega) \quad (4b)$$

Then, Equation (3) can be simplified as:

$$\mathrm{Tr}W(r_1,r_2,\omega)=\sqrt{S(r_1,\omega)}\sqrt{S(r_2,\omega)}\eta(r_1,r_2,\omega) \quad (5)$$

According to Ref. [5] , using the Young's double slits, all the four elements of the electric cross-spectral density matrix are determined as follows:

$$W_{xx}(r_1,r_2,\omega)=\sqrt{S_x(r_x,\omega)}\sqrt{S_x(r_x,\omega)}\eta_{xx}(r_1,r_2,\omega) \quad (6a)$$

$$W_{yy}(r_1,r_2,\omega)=\sqrt{S_y(r_1,\omega)}\sqrt{S_y(r_2,\omega)}\eta_{yy}(r_1,r_2,\omega) \quad (6b)$$

$$W_{xy}(r_1,r_2,\omega)=\sqrt{S_x(r_1,\omega)}\sqrt{S_y(r_2,\omega)}\eta_{xy}(r_1,r_2,\omega) \quad (6c)$$

$$W_{yx}(r_1,r_2,\omega)=\sqrt{S_y(r_1,\omega)}\sqrt{S_x(r_2,\omega)}\eta_{yx}(r_1,r_2,\omega) \quad (6d)$$

The generalized stokes parameters can be given by:

$$S_0(r_1,r_2,\omega)=W_{xx}(r_1,r_2,\omega)+W_{yy}(r_1,r_2,\omega) \quad (7a)$$

$$S_1(r_1,r_2,\omega)=W_{xx}(r_1,r_2,\omega)-W_{yy}(r_1,r_2,\omega) \quad (7b)$$

$$S_2(r_1,r_2,\omega)=W_{xy}(r_1,r_2,\omega)+W_{yx}(r_1,r_2,\omega) \quad (7c)$$

$$S_3(r_1,r_2,\omega)=i\left[W_{yx}(r_1,r_2,\omega)-W_{xy}(r_1,r_2,\omega)\right] \quad (7d)$$

In the coming section, we will present a method to measure the four elements of the electric cross-spectral density matrix and the generalized stokes parameters of a beam on the basis of Equation (6) and Equation (7).

RESULTS AND DISCUSSIONS

The schematic of the experimental setup is shown in Figure 1. A linear polarized helium-neon laser is incident over a radially polarization converter to produce the radially polarized beam. Then, the beam is expanded by an expanding system which containing a rotating ground glass in the focus to reduce the coherence of the light. The newly generated radially polarized partially coherent beam passes through a Young's double slit and a Mach- Zehnder interferometer consisting of apertures, quarter-wave plates and polarizers. The interference fringes can be obtained in the observation plane as shown in Figure 2.

To calculate the W_{xx} (r_1, r_2, w) component, it is necessary to keep the polarization of two polarizers in x axis direction. And the direction of optic axis of two half wave plates should be the same with x axis (to ensure the same optical loss of different condition). This allows only the x component of the electric field to pass through.

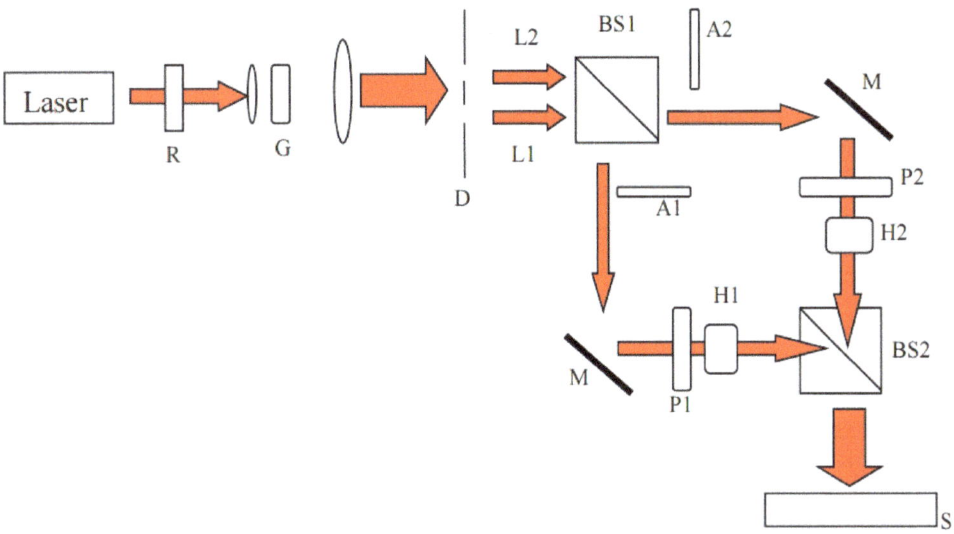

Figure 1. The schematics of the experimental setup. R is a radial polarization converter; G is a ground glass; D is a Young's double slits with slit width 0.1 mm and slit separation 1.6 mm; L_1 and L_2 are beams coming from the Young's double slits; BS is a beam splitter; A is an aperture; P is a polarizing film, H is a half wave plate; S is a spectrometer.

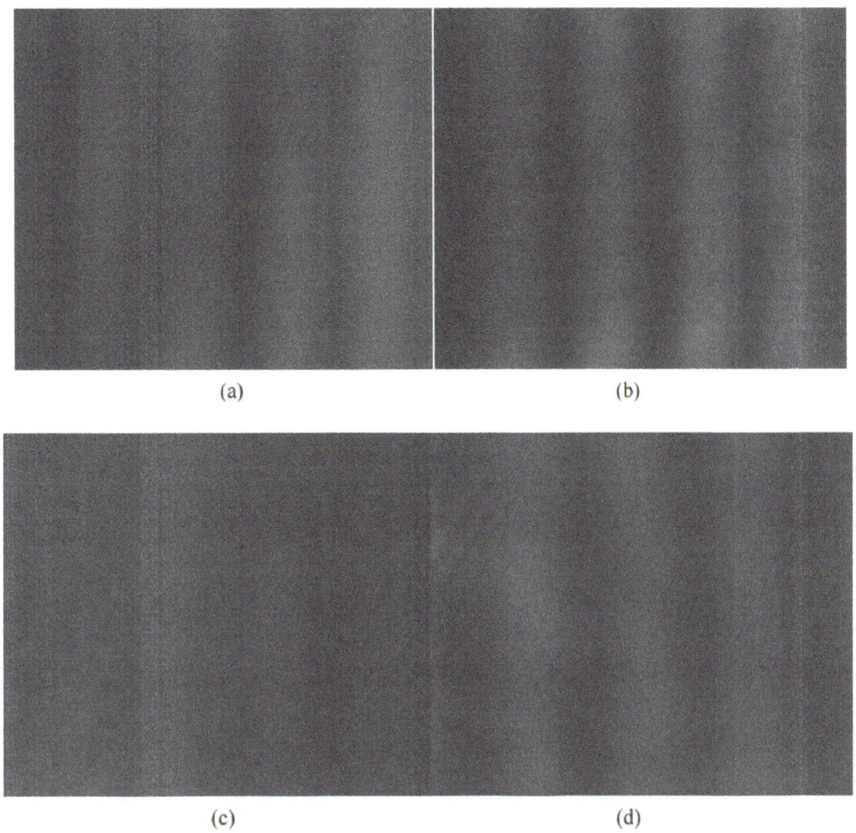

Figure 2. The figure of interference fringe at the plate S. (a) xx polarization; (b) yy polarization; (c) xy polarization; (d) yx polarization.

The aperture A_1 only allows L_2 to pass and A_2 allows L_1 to pass. Then, the interference fringes of the double- slits are obtained on the observation plane. The maximum and minimum values of the spectral density of the central fringe are recorded, and the spectral densities due to each individual slits are also recorded at the same point by a spectrometer. To get the W_{yy} (r_1, r_2, w) component, the polarizers and half wave plates should be rotated clockwise by 90° (this change the direction of polarization to y axis) and repeat the similar steps.

The W_{xy} (r_1, r_2, w) component can be obtained when the directions of polarization of two polarizers are perpendicular. The fast axis of the half wave plate (H_1) which is followed by the polarizer (P_1) with x direction of polarization should keep the 45° with the x axis. Therefore, the polarization of incident light will change from the x direction to y direction. At this time, the interference fringes appear when the polarization direction of the polarizer (P_2) and the half wave plate (H_2) fast axis are kept along y axis in another arm. The maximum and minimum values around the central fringe and spectral densities of each single slit in the same position should be recorded. The W_{yx} (r_1, r_2, w) component can be gotten when the direction of polarization of polarizer (P_1) and the fast axis of the half wave plate (H_1) are kept the same with y axis. Meanwhile, the polarizer (P_2) is along the x axis and the half wave plate (H_2), having optics axis at 45° with x axis, works as a 90° polarization rotator in another arm. Similar measurements for both the slits as well as the individual slits are taken.

As the Figure 2 shows, the visibility of interference fringe of radially polarized random electromagnetic beam due to different components is similar.

With the help of values from spectrometer, the spectral degree of coherence can be calculated by the equation:

$$\eta = \frac{S_{\max} - S_{\min}}{S_{\max} + S_{\min}} \qquad (8)$$

where the S_{max} and S_{min} are the maximum and minimum values of spectral intensity around the central fringe when both the slits are open.

The coherences of the four conditions are given by Table 1. Utilizing the spectral intensity of single slit, the electric cross-spectral density matrix can be calculated. After repeated measurement experiments four times, the average values and the deviation have been worked out. According to Equations (7a)-(7d), the generalized Stokes parameters can be obtained immediately as shown in Table 2.

According to the Table 1, the spectral coherence of the same polarized direction component (xx and yy) of the radially polarized light is similar with that for different polarized direction component (xy and yx). It is because that there is almost the same correlation between the two orthogonal components or the parallel components of the radially polarized partially coherent beam.

For comparison, the interference fringes of a randomly polarized partially coherent laser have been experimentally obtained as shown in Figure 3. From the Figure 3, we can see that the visibility of interference fringes for xy components and yx components is less than that for xx components and yy components. It is due to the fact that there is very little correlation between the two orthogonal components of a randomly polarized light.

Table 1. Components of the spectral degree of coherence and the cross-spectral density matrix obtained from the laser light

Polarization	Degree of coherence η (r_1, r_2, w)	Cross-spectral density matrix W (r_1, r_2, w)
xx	0.33 ± 0.02	52.24 ± 4.71
xy	0.26 ± 0.03	40.04 ± 4.62
yx	0.32 ± 0.03	51.79 ± 4.86
yy	0.37 ± 0.04	68.31 ± 7.38

Table 2. Generalized stokes parameters for a pair of points in the random electromagnetic beam

Generalized stokes parameters	Value
$S_0(r_1,r_2,w)$	120.55± 6.37
$S_1(r_1,r_2,w)$	16.07± 2.90
$S_2(r_1,r_2,w)$	91.83± 2.33
$S_3(r_1,r_2,w)$	(16.52± 1.41)i

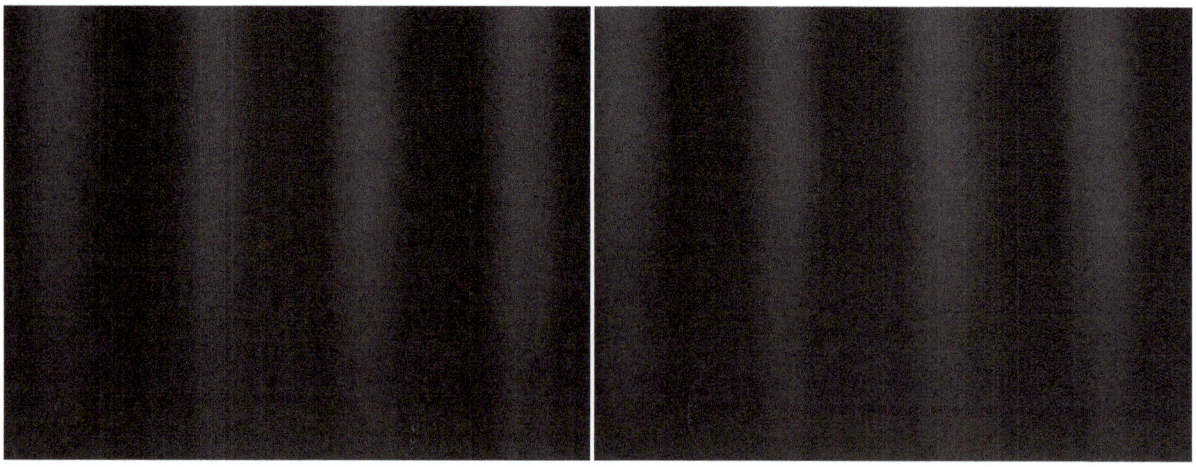

(a) (b)

(c) (d)

Figure 3. The interference distribution of a randomly polarized partially coherent laser beam at the plate S. (a) xx polarization; (b) yy polarization; (c) xy polarization; (d) yx polarization.

CONCLUSION

In conclusion, we have experimentally measured the four elements of the electric cross-spectral density matrix of a radially polarized partially coherent laser by the simple equipment. The generalized Stocks parameters can be computed by the cross-spectral density matrix. The method measures the coherence of two different points of the radially polarized laser. The result shows that there is almost the same correlation between the two orthogonal components or the parallel components of the radially polarized partially coherent beam, which is different from the randomly polarized partially coherent laser. The generalized Stocks parameters help us to study the properties of coherence and polarization of the radially polarized stochastic electromagnetic beams in propagation.

ACKNOWLEDGEMENTS

This research was supported by the National Natural Science Foundation of China (Nos. 61275203; 61505059; 61515070), the High level talent research of Huaqiao University (12BS231), and the Natural Science Foundation of Fujian Province of China (No. 2014J05007).

REFERENCES

1. Wolf, E. (2003) Unified Theory of Coherence and Polarization of Random Electromagnetic Beams. Physics Letters A, 312, 263-267. http://dx.doi.org/10.1016/S0375-9601(03)00684-4
2. Wolf, E. (2003) Correlation-Induced Changes in the Degree of Polarization, the Degree of Coherence, and the Spectrum of Random Electromagnetic Beams on Propagation. Optics Letters, 28, 1078-1080. http://dx.doi.org/10.1364/OL.28.001078
3. Agrawal, G.P. and Wolf, E. (2009) Propagation-Induced Polarization Changes Impartially Coherent Optical Beams. Optical Society of America A, 17, 2019. http://dx.doi.org/10.1364/JOSAA.17.002019
4. Wolf, E. (2007) Introduction to Theory of Coherence and Polarization of Light. Cambridge University Press, Cambridge.
5. Roychowdlhury, H. and Wolf, E. (2003) Determination of the Electric Cross-Spectral Density Matrix of a Random Electromagnetic Beam. Optics Communications, 226, 57-60. http://dx.doi.org/10.1016/j.optcom.2003.07.054 [
6. Korotkova, O. and Wolf, E. (2005) Generalized Stokes Parameters of Random Electromagnetic Beams. Optics Letters, 30, 198-200. http://dx.doi.org/10.1364/OL.30.000198
7. Kanseri, B. and Kamdpal, H.C. (2008) Experimental Determination of Electric Cross-Spectral Density Matrix and Generalized Stokes Parameters for a Laser Beam. Optics Letters, 33, 2410-2412. http://dx.doi.org/10.1364/OL.33.002410
8. Kanseri, B., Rath, S. and Kandpal, H.C. (2009) Direct Determination of the Generalized Stokes Parameters from the Usual Stokes Parameters. Optics Letters, 34,719-721. http://dx.doi.org/10.1364/OL.34.000719
9. Dorn, R., Quabis, S. and Leuchs, G. (2003) Sharper Focus for a Radially Polarized Light Beam. Physical Review Letters, 91, 233901. http://dx.doi.org/10.1103/PhysRevLett.91.233901
10. Lin, H., Jia, B.H. and Gu, M. (2011) Generation of an Axially Super-Resolved Quasi-Spherical Focal Spot Using an Amplitude-Modulated Radially Polarized Beam. Optics Letters, 36, 2471-2473. http://dx.doi.org/10.1364/OL.36.002471
11. Zhang, Y., Ding, B. and Taikei, S. (2010) Trapping Two Types of Particles Using a Double-Ring-Shaped Radially Polarized Beam. Physical Review A, 81, 023831. http://dx.doi.org/10.1103/PhysRevA.81.023831
12. Yan, S. and Yao, B. (2007) Radiation Forces of a Highly Focused Radially Polarized Beam on Spherical Particles. Physical Review A, 76, 053836. http://dx.doi.org/10.1103/PhysRevA.76.053836

CHAPTER 7 EXPLOITATION OF ELECTROMAGNETIC MODELS

Tran Vu La, Ali Khenchaf, Fabrice Comblet, Carole Nahum, Helmi Ghanmi

Keywords: C-Band Synthetic Aperture Radar (SAR), Electromagnetic (EM) Scattering Models, Radar Cross-Section (RCS), Remote Sensing, Small Perturbation Model (SPM), Sea Surface Wind

INTRODUCTION

The exploitations of Synthetic Aperture Radar (SAR) images to retrieve oceanic parameters have been widely studied in the literature, due to many advantages of SAR systems: stable operations in most meteorological conditions, revisit period, high resolution. Among the important oceanic parameters, sea surface wind plays a crucial role for the studies of the other parameters, i.e. waves, currents, marine meteorology, and the coupling of oceanic and atmospheric systems. As well as, surface wind speed is an important parameter in the studies of many oceanic applications, i.e. oil slick observation [1] , and ship detection [2] . The retrieval of wind fields has been carried out from different SAR data via different approaches [3] - [8] . In such studies, the resolutions of the estimated wind vectors are significantly dependent on the signal-to-noise ratio (SNR) of each SAR system. In practice, one can obtain a 1 km × 1 km wind cell in cases where the speckle noise level is quite low. In spite of the diverse approaches, the retrieval of wind vectors by using the physical models is particularly preferred, due to their generality. For this approach, one can find two principal methods: via empirical (EP) models, and electromagnetic (EM) models.

The common point of both EP and EM models is the use of the geophysical model functions (GMFs) to describe the dependency of normalized radar cross section (NRCS) on wind speed and the geometry of observations. Therefore, by knowing NRCS and geometric parameters, it is possible to invert the GMFs to estimate wind speed. For the EP models (also known as scatterometry-based approaches), the GMFs are constructed and validated by the series of satellite scattero meter missions. For instance, the C-band GMFs, known as CMOD.4 [9] , CMOD.IFR2 [10] , CMOD.5 [11] , CMOD.5N [12] , were derived by using the data from scattero meters onboard the European Remote Sensing 1 and 2 satellites (ERS-1 and ERS-2). The CMOD.5 and CMOD.5N are more widely used than the others, since they can offer good wind speed estimation for most wind regimes. The common point of the EP GMFs is that they are defined for VV-polarization (VV-pol). Thus, a polarization ratio (PR) [13] - [19] should be used when the SAR image is in HH-polarization (HH-pol). In contrast to the EP models, the EM ones are constructed and validated by the physical calculations for the reflections and diffractions

Citation: Tran Vu La, Ali Khenchaf, Fabrice Comblet, Carole Nahum, Helmi Ghanmi (2016) Exploitation of Electromagnetic Models. Journal of Electromagnetic Analysis and Applications,08,42-55. doi: 10.4236/jemaa.2016.83005

between radar scattering and sea surface. Depending on different approaches, we can find a variety of EM models based on the exact methods, i.e. Method of Moment (MoM) [20] , Forward?Backward Method (FBM) [21] , or the asymptotic methods, i.e. Small Slope Approximation (SSA) [22] , Two-Scale Model (TSM) [23] , Small Perturbation Method (SPM) [23] . The advantages and limits of these models have been analyzed and discussed in many references [20] - [23] . In fact, the TSM, SSA, MoM and FBM models can describe radar backscattering from sea surface with different scales of ocean waves, i.e. long-, intermediate-, and short-scale waves. Their descriptions are generally complicated due to a lot of variables, relations, and conditions. Meanwhile, for the SPM the description of the relation between radar backscattering and sea surface is quite simpler, since it concerns principally short-scale waves. Under the point of view of the inverse problem, the SPM seems to be more appropriate than the other EM models to estimate wind speed, due to its flexibility of inversion. This is also the initial objective of this study.

In both EM and EP models, wind direction is a crucial parameter for wind speed estimation. It can be obtained from different available wind sources, i.e. in situ measurements, numeric weather models, retrieval from SAR data, etc. The last approach is the most widely used in the literature, since it can give (almost) immediately the information of wind directions with high resolution. The methods to retrieve wind directions can be divided into two categories: one is in the spectral domain, and the other is in the spatial domain. The most well-known method in the spectral domain is the Fourier Fast Transform (FFT) [24] , while the Local Gradient (LG) [25] is the most widely used one in the spatial domain. Generally, wind directions retrieved by the LG have (much) higher resolution than the ones given by the FFT. In fact, the LG can offer a wind cell of 3 km × 3 km, even 1 km × 1 km in some conditions [25] , instead of a cell of 20 km × 20 km given by the FFT. The common issue of both the LG and FFT is the 180° directional ambiguity of the retrieved wind directions. It can be removed only if wind shadow is present and visible enough in the lee of coastlines, or if there are additional wind source data, i.e. in situ measurements, SCAT wind data, etc.

The outline of this paper is as follows. In Section 2, the descriptions of the SPM and CMOD.5 are presented in detail. It also includes the analyses of validity domain, and the impact of different variables on the relation between radar backscattering and sea surface. Section 3 presents the data used for the studied models to retrieve wind speed. Section 4mentions the retrieval of wind directions from SAR images. Section 5 presents the estimation of wind speed by using the SPM and CMOD.5. The obtained results are compared to in situ measurements to evaluate. Section 6 discusses the advantages and limits of the SPM for wind speed estimation. Section 7 summarizes the main points of this paper, and proposes the perspectives to improve the performance of the EM models for wind speed retrieval.

METHODOLOGY

Small Perturbation Method (SPM)

The SPM, also known as Bragg resonance, was first identified as an important mechanism of radar backscattering from the water surface a long time ago. It was then developed for radar backscattering from the short-scale waves in the ocean by Wright (1966, 1968) [26] , Valenzuela (1978) [27] , and Plant (1990) [28] . Depending on different approaches, the descriptions of the SPM in terms of equation, parameters, and validity domain are diverse. In this paper, for the monostatic case of radar scattering, we use the definition presented in [23] as shown by (1).

$$\sigma_{pp}^{s} = 8\left|k^2 \sigma \cos^2 \theta \alpha_{pp}\right|^2 W\left(k_x + k\sin\theta, k_y\right) \quad (1)$$

where σ_{pp}^{s} denotes the radar backscattering (or NRCS) from sea surface in VV- or HH-pol; k is the radar wave number; k_x, k_y are the x- and y-axis components of k, respectively; θ indicates the radar incident angle (and also backscattering angle in monostatic case); α_{pp} denotes the Bragg coefficients in VV-or HH-pol; σ and W represent the standard deviation and normalized roughness spectrum of sea surface waves, respectively.

The Bragg coefficients in VV- and HH-pol, α_{VV} and α_{HH}, are described as a function of sea water permittivity ε_r and incident angle θ, as given by (2) and (3).

$$\alpha_{VV} = \frac{(\varepsilon_r - 1)\left[\varepsilon_r\left(1+\sin^2\theta\right) - \sin^2\theta\right]}{\left[\varepsilon_r \cos\theta + \sqrt{\left(\varepsilon_r - \sin^2\theta\right)}\right]^2} \quad (2)$$

$$\alpha_{HH} = \frac{\varepsilon_r - 1}{\left[\cos\theta + \sqrt{\left(\varepsilon_r - \sin^2\theta\right)}\right]^2} \quad (3)$$

The calculations of sea water permittivity have been shown in many references [29] - [32] . The most widely used model is probably the study of Debye [31] , and then the double-Debye model [32] . In practice, the significant difference of these models is only noted at very high frequencies. In C-band (f = 5.4 GHz), the sea water permittivity calculated by the two Debye models is quite similar [32] . In this paper, for the simple calculations we use the Debye model to calculate ε_r, as given by (4).

$$\varepsilon_r = \varepsilon_\infty(T,S) + \frac{\varepsilon_s(T,S) - \varepsilon_\infty(T,S)}{1 + j2\pi v\tau(T,S)} - j\frac{\sigma}{2\pi v \varepsilon_0} \quad (4)$$

where $\varepsilon_s(T,S)$ and $\varepsilon_\infty(T,S)$ are the static and high-frequency dielectric constants, respectively; τ is the relaxation time; ε_0 is the permittivity of free space (ε_0 = 8.854 × 10-12 F/m); σ is the ionic conductivity. The temperature T and salinity S of the sea water of the Iroise Coast (France) are collected via www.previmer.org, from Oct. 2014 to Apr. 2015. The temperature T varies between 4°C (Feb. 2015) and 18°C (Oct. 2014), while S is quite stable around 32 - 35 ppt. By using (4), we note that the permittivity of sea water varies slightly from ε_r = 62.54 + j39.82 (Feb 2015) to ε_r = 66.81 + j34.93 (Oct 2014). It signifies that temperature and salinity do not have significant effects on the calculation of radar backscattering as in (1).

In contrast to temperature and salinity, the roughness spectrum of sea surface is an important parameter for the NRCS calculations. In fact, it concerns a lot of parameters, particularly wind vector which is the favorite one in this paper. For a simple description, σ and W in (1) are replaced by the directional surface wave spectrum S(K,), with wave propagation referenced to wind direction. The spectrum S(K,Φ) is generally described by the two parts as given by (5).

$$S(K,\Phi) = S(K,0) \times f(K,\Phi) \tag{5}$$

where S(K,0) is the omnidirectional spectrum, and f(K,Φ) is the angular spreading function. In general, S(K,0) is influenced by wind speed, while both wind speed and wind direction affect in the description of f(K, Φ). By knowing S(K,Φ) and Φ, it is possible to estimate wind speed. One should note that K in (5) is the wavenumber of ocean waves. For the SPM (or Bragg resonance), K relates to the radar wavenumber k in (1) as $K = 2k \times \sin\theta$, where θ is always the radar incident angle.

The theoretical modelling of sea surface roughness spectrum has a history as long as the investigation of radar backscattering from the ocean surface. The initial studies were proposed by Cox and Munk (1954) [33] , Pierson and Moskowitz (1964) [34] . It was then developed by Bjerkass and Riedel (1979) [35] , Donelan and Pierson (1987) [36] , and Apel's (1994) [37] . Based on the previous studies, Elfouhaily et al. (1997) [38] were proposed a unified form of sea surface spectrum. This is one of the most widely used models in the literature because it concerns both long- and short-scale waves of the ocean. Then, Kudryavtsev et al. (2003) [39] were proposed a new model of sea short-scale wave spectrum. It is preferred in some references, especially for the studies of rain effects on radar backscattering [40] ; since it was developed from a balance of wind input, parasitic wave generation, and wave dissipation, and therefore is convenient to modify to include rain effects. However, this model does not work well for wind speed below 5 m/s [39] . Recently, for the reason of rapid and simple calculations of radar backscattering, Hwang et al. [41] - [43] have developed an empirical form of sea surface spectrum. This study is based on the conclusion about the dependency of curvature wave spectrum on wind friction velocity and wave phase speed. From the analysis of the studies of sea surface spectrum, we select the Elfouhaily model due to its generality. The comparison of different sea spectrum models will be discussed in the other papers.

The validity condition of the SPM has been discussed in [23] [44] , but it seems to be difficult to give the obvious conclusions in terms of wind speed and incident angle, since this condition concerns the mean square slope (mss) of sea wave spectrum used in the SPM. Generally, as indicated in [44] the SPM works well for the root mean square slope (rmss) < 0.3, or mean square slope (mss) < 0.09. The relation of mss and wind speed has been studied and validated in many studies as discussed in [38] . Among them, the most successful measurements were proposed by Cox and Munk (1954) [33] . For a clean sea surface, the mss is defined as in (6).

$$mss_{clean} = 10^{-3}\left(3 + 5.12U_{10}\right) \pm 0.004 \tag{6}$$

where U_{10} is the wind speed at the 10 m height. Equation (6) is only validated under the wind condition: 1 m/s < U_{10} < 13 m/s. By combining (6), the conditions of mss < 0.09, and 1 m/s < U_{10} < 13 m/s, we realize that the validity domain of SPM limits for wind speed below 13 m/s. According to the Beaufort scale, for this level of wind speed, there are only wavelets, sometimes short-scale waves on the sea surface. It signifies that the SPM should be used under the condition of slightly rough sea surface, without the presence of intermediate- or/ and long- scale waves. In comparison to the other EM models, which can describe sea surface roughness more generally (i.e. TSM, SSA), the SPM works well for moderate incident angles from 32.5° to 45°.

Comparison of SPM and CMOD5

A general form of the GMFs described in the EP models is defined as in (7) [9] - [12] .

$$\sigma_{VV}^0 = A\left[1 + b_1 \times \cos\varnothing + b_2 \times \cos 2\varnothing\right]^B \tag{7}$$

where Φ is the wind direction relative to the radar look direction, A, B, b_1, b_2 are the functions of U_{10} and θ. Thus, by determining σ_{VV}^0 and θ from the SAR data, and Φ, wind speed U_{10} can be estimated by inverting (7). As indicated, we select the CMOD.5 for the estimation of wind speed, since it can work well for most wind regimes. As well as the SPM, the validity domain of the CMOD.5 is limited under some conditions. The CMOD.5 can work well for the incident angles of 20° - 50°, and for wind speed below 25 - 30 m/s [11] . One should know that the range of 20° - 50° is widely used to acquire images for most SAR systems, and wind speed is rarely beyond 25 - 30 m/s. This signifies that the CMOD.5 is generally a good empirical model to estimate wind speed from SAR data.

Since the EP GMFs are only defined for VV-pol, a polarization ratio, PR, should be used for the SAR images in HH-pol. The calculations of PR have been studied in many references [13] - [18] . They can be divided into two approaches: Thompson model [13] , and Mouche model [16] . In the Thompson model, the PR is defined as a function of radar incident angle, as given by (8).

$$PR = \frac{\left(1 + 2\tan^2\theta\right)^2}{\left(1 + \alpha\tan^2\theta\right)^2} \tag{8}$$

where α is an adjustable parameter. It is calibrated via the in situ measurements. In fact, one can find some values of α: α = 0.6 in [13] , α = 1.0 in [14] , and α = 1.2 in [15] . Contrary to the Thompson model, in 16], Mouche et al. concluded that the PR depends not only on radar incident angle, but also on wind direction. Hence, they proposed a new model for PR calculation as given by (9).

$$PR = A_\Phi \exp\left(B_\Phi\right) + C_\Phi \tag{9}$$

where A_Φ, B_Φ, and C_Φ are the parameters determined by the in situ measurements and SAR images, with regard to the impact of wind directions. Based on the study of Mouche, some studies [17] [18] have re-calculated A_Φ, B_Φ, and C_Φ to better agree with the different C-band SAR data and in situ measurements. In this paper, we use the model proposed by Liu et al. [18] for wind speed retrieval in HH-pol.

Figure 1 presents the comparison of the NRCS calculated by the SPM with the Elfouhaily wave model (SPM- Elf) and CMOD.5 in VV-poland HH-pol. The studied values of wind speed and incident angle limit at u_{10} = 13 m/s and θ = 30° - 45° which correspond to the validity domain of the SPM. At θ = 30°, for both up-wind (Figure 1(a)) and cross-wind directions (Figure 1(b)), the NRCS in VV-pol calculated by the SPM-Elf is (about) 2 - 4 dB lower than that offered by the CMOD.5. At θ = 45°, the two models give quite similar NRCS level for up-wind direction. Meanwhile, for cross-wind direction the level calculated by the SPM-Elf is (about) 2 - 3 dB higher than that of the CMOD.5. In HH-pol, in most cases of incident angle and wind direction, the SPM-Elf underestimates NRCS level in respect with wind speed (about 5 - 7 dB).

The results obtained in Figure 1 show that in VV-pol, the SPM-Elf can offer similar wind speed estimation to the CMOD.5for up-wind direction, and a little more for cross-wind direction. This conclusion is noted for moderate incident angles (32.5° - 45°). On the contrary, in HH-pol for most cases of incident angle and wind direction, wind speed obtained by the SPM-Elf is overestimated in comparison to the one given by the CMOD5. This is particularly noted for quite low incident angles (30° - 32.5°).

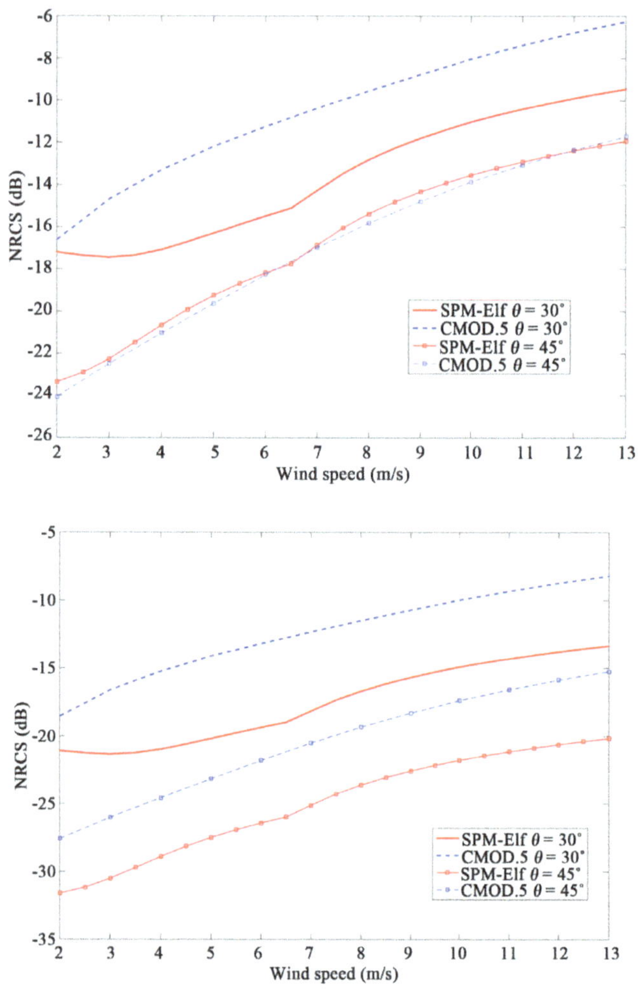

(a)

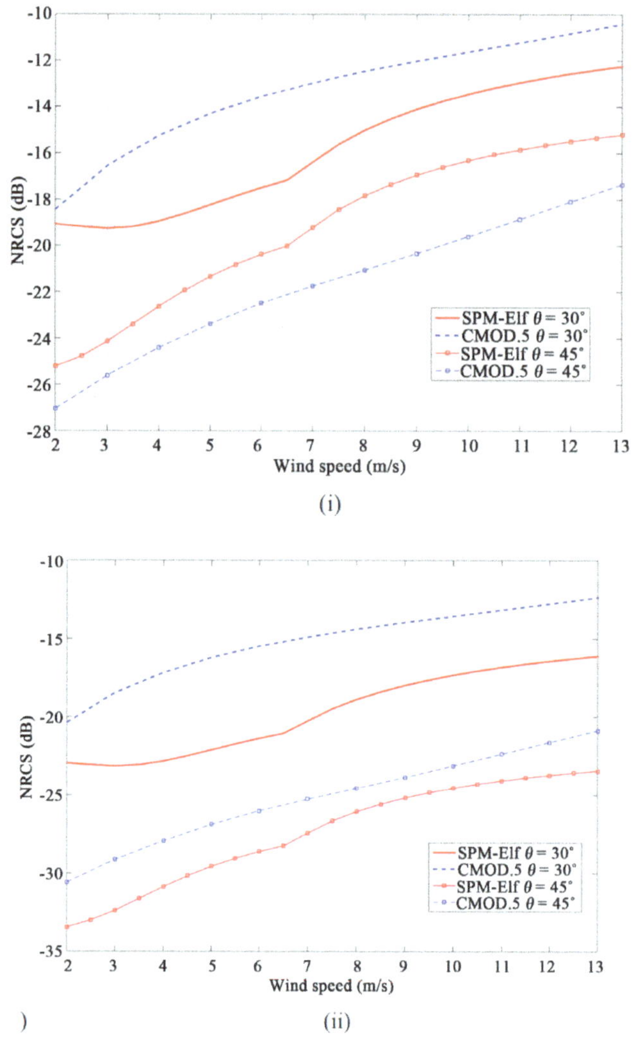

(b)

Figure 1. Comparisons of NRCS calculated by SPM with Elfouhaily spectrum (SPM-Elf) and CMOD.5 for θ = 30° and θ = 45°. (i) VV-pol. (ii) HH-pol. (a) Up-wind direction. (b) Cross-wind direction.

DATA PREPARATION

Sentinel-1 Images

SAR images studied in this paper are acquired by the C-band Sentinel-1 satellite. It was launched in April 2014 by the European Space Agency (ESA), with the aim of providing an independent operational capability for continuous radar mapping of the Earth. The acquisition modes of the Sentinel-1 include: Stripmap (SM), Interferometric Wide Swath (IWS), Extra Wide Swath (EWS), and Wave Model (WM). The images can be acquired in single polarization (VV/HH), or in dual-polarization (VV+VH / HH+HV). In this study, we use the Level-1 images acquired with SM mode in HH-pol, and with IWS mode in VV-pol. They are downloaded via https://scihub.esa.int. The acquired date are the type of GRD (Ground Range, Multi-Look, Detected), with high resolution (HR). This type allows reducing speckle noise, but it also decreases the spatial resolution of the image. For instance, the spatial resolution of the studied image in SM mode is reduced to 23 m × 23 m, instead of 3.6 m × 4.9 m. As well, the image in IWS mode has a spatial resolution of 20 m × 22 m, instead of 3.5 m × 22 m.

In Situ Measurements

The in situ measurements used in this paper are collected at the meteorological stations of Météo France along the Iroise Coast (Figure 2), since there are not any fixed meteorological buoys in this area. These stations (red points) are located very close to the open sea. This allows reducing as possible the difference between the measured data along the coast and on the open sea. Since wind speed is measured at different heights, it needs to be corrected to the reference level of 10 m height for comparisons with estimated results. The correction methods of wind speed can be found in [45] [46] . However, the obtained results with the studied models are not much different. In this study, we use a simple power-law varying wind profile as given by (10) [45] .

$$\frac{u_2}{u_1} = \left(\frac{z_2}{z_1}\right)^{0.1} \qquad (10)$$

where u_2 stands for the wind speed at height z_2 and u_1 and z_1 are the known wind speed and height, respectively.

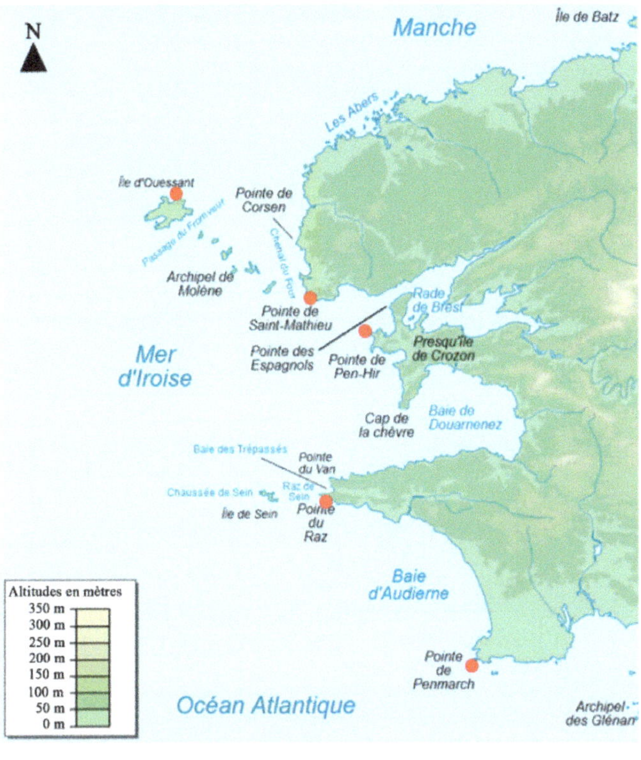

Figure 2. Meteorological stations (red points) of Météo France along the Iroise Coast (France).

WIND DIRECTION EXTRACTION

For both SPM and CMOD.5, wind direction is a crucial parameter for wind speed estimation, in particular for high wind speed. However, its impact is not the same for the two models. In fact, for the SPM the effect of wind direction is noted in the angular spreading function of sea roughness spectrum (as in (5)), while it is represented by the $\cos\Phi$ and $\cos 2\Phi$ parameters in the CMOD.5 (as in (7)). As shown in Figure 1, the NRCS for up-wind direction is higher than the one in cross-wind direction, and this result is noted for both polarizations. Meanwhile, as shown in Figure 3, the difference of NRCS in respect to wind direction calculated by the CMOD.5 is more significant than the one calculated by the SPM, in particular between up- and cross-wind directions, and for high wind speed.

The wind direction data can be obtained via many ways, i.e. from: a) in situ measurements, b) numeric weather models, or c) extraction from SAR images. Among them, the retrieval of wind directions from SAR data is the most widely used, since it can give high wind resolution. In addition, it can be one of the rare choices to have wind direction information in the areas where the fixed meteorological buoys are not installed. As indicated, the LG method is preferred to the FFT one since it can give wind directions with smaller resolution cells (even 1 km × 1 km in some cases [25]). Since the fundamentals of the LG method have been discussed in many references [25] [47] , we only summarize here the main steps and present some important remarks to obtain good extractions.

- First, the filtered images should be reduced dimension sizes to enhance wind streaks, and divided into the sub-images according to expected spatial resolution.
- Then, for each defined sub-image, the histogram of LG directions must be weighted and smoothed.
- After that, the main LG direction of each sub-image is estimated, and the orthogonal of the most frequent gradient direction is assigned as the most available wind direction.
- Finally, the 180° ambiguity of wind direction can be solved either by in situ measurements, or via the Continuous Wavelet Transform (CWT) method.

The first step is a crucial task to obtain good extractions in respect to spatial resolution. The spatial resolution of extracted wind directions depends on the acquisition model of SAR images. It can be 1 km × 1 km if the quality of SAR image is good enough. For the images used in this paper (SM and IWS modes), a wind direction cell of 3 km × 3 km seems to be enough for the SPM and CMOD.5 to have good spatial resolution of wind speed . The importance of the second step can be verified in Figure 4. Due to being weighted and smoothed, the extracted wind directions are very close to the measured data (red arrows). As shown in the right, the difference between the not-smoothed wind direction and measured data is up to 35°, while it is reduced to 5° for the smoothed wind direction.

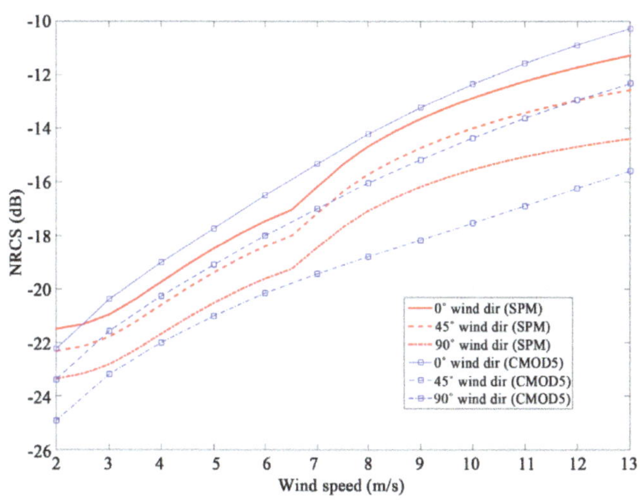

Figure 3. NRCS in respect to wind direction and wind speed studied by the SPM and CMOD.5.

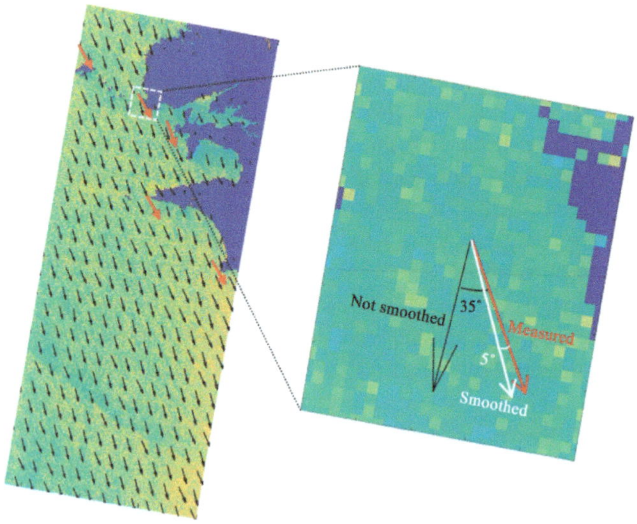

Figure 4. Extraction of wind direction by LG method on a Sentinel-1 image (SM mode), acquired on 27th Dec. 2014 from 06:23:46 UTC to 06:24:15 UTC. (Left) Smoothed vs. measured (red arrows) wind directions. (Right) Main LG direction of a sub-image before and after smoothing, compared to measured wind direction.

One of the most important constraints of the LG method is the 180° ambiguity of the extracted wind directions. It can be solved by in situ measurements or numerical weather models. However, the measured data are not always available for all cases. Recently, the CWT method has been developed to solve this problem. Nevertheless, it seems to be complicated to apply. There is another way to solve the 180° ambiguity of extracted wind direction, if wind streaks on SAR images are visible enough to observe. As shown in Figure 5, after reducing the dimension sizes of the SAR image presented in Figure 4, the wind streaks are much more visible to be noted. Based on this observation, the 180° ambiguity of extracted wind directions can be easily solved. The efficiency of this method depends on the visibility of wind streaks which depends on wind regimes and speckle noise level. For low wind regimes (below 5 m/s), it is generally difficult to determine wind streaks. This is the same in the case where speckle noise is significant.

WIND SPEED ESTIMATION

Figure 6 describes the indispensable steps to retrieve wind speed from SAR images. The Level-1 Sentinel-1 images are processed by the toolbox for data calibration, land masking, and sub-image division. Thanks to this step, the NRCS of the image is determined. Then, by using the LG method described above, wind directions with different spatial resolutions can be determined. They are compared to the measured data to verify. Finally, the NRCS and wind directions extracted from SAR images are applied in the SPM and CMOD.5 to retrieve wind speed with different spatial resolutions. To evaluate, the estimated wind speed will be compared to the measured data collected at the stations in Figure 2.

VV-Pol Image

Wind speed estimation on a VV-pol Sentinel-1 image by using the SPM and CMOD.5 is presented in Figure 7. This image is acquired in IWS mode with radar incident angles from 32.5° to 45°. The spatial resolution of wind direction and wind speed in this case is 3 km × 3 km. As discussed in Section 4, for this level of spatial resolution, the different wind speed areas may be well visualized. As expected from Figure 1, for the range of θ = 32.5° - 35°, the wind speed estimated by the SPM is 2 - 3 m/s higher than that given by the CMOD.5. This is also noted for θ = 35° - 40°, but the deviation is less significant (only about 1 - 2 m/s). For θ = 40° - 45°, the SPM and CMOD.5 give (quite) similar wind speed estimation. They are also very close to the measured data (the same range of 5 - 7 m/s). Compared to the SPM, the CMOD.5 seems to give higher spatial resolution of wind speed, especially for the range of 1 - 5 m/s. Indeed, the noted minimum value of the SPM wind speed is about 4 - 5 m/s, while it is 0 - 1 m/s for the CMOD.5 wind speed.

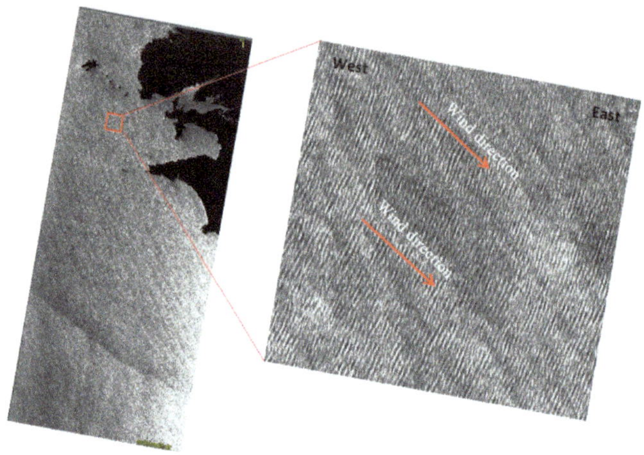

Figure 5. Enhancement of wind streaks on the SAR image presented in Figure 4 to obtain exact wind direction.

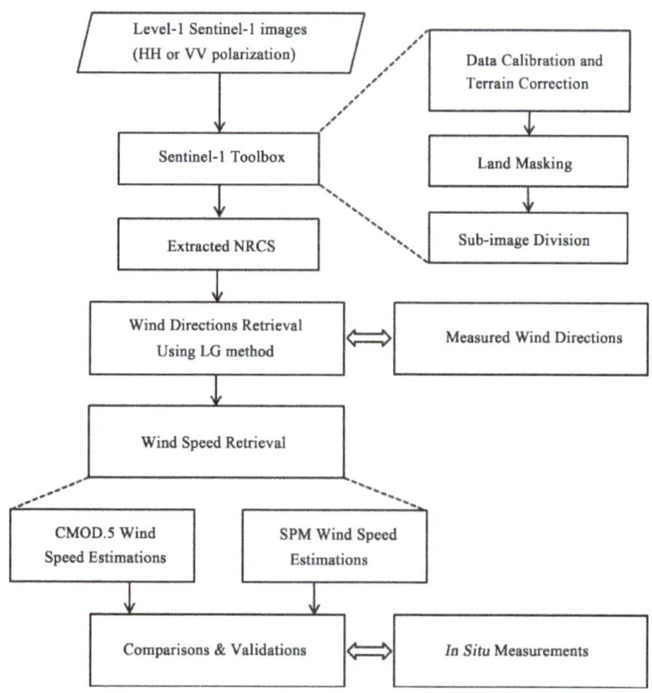

Figure 6. Diagram of wind speed retrieval carried out by the SPM and CMOD.5, from Level-1 Sentinel-1 images in VV- and HH-pol.

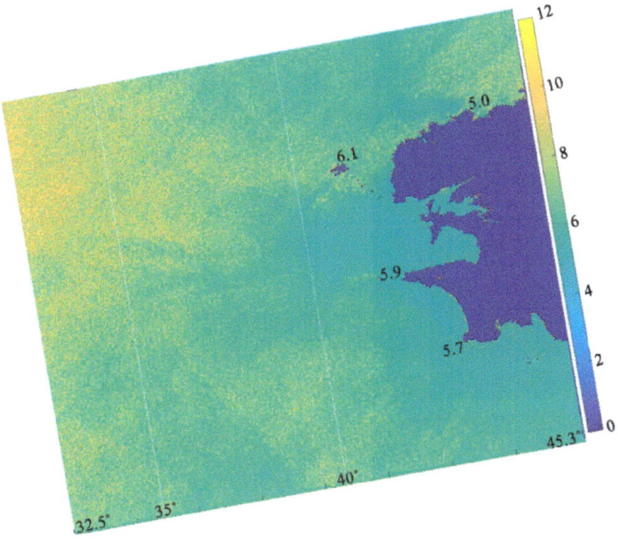

(a)

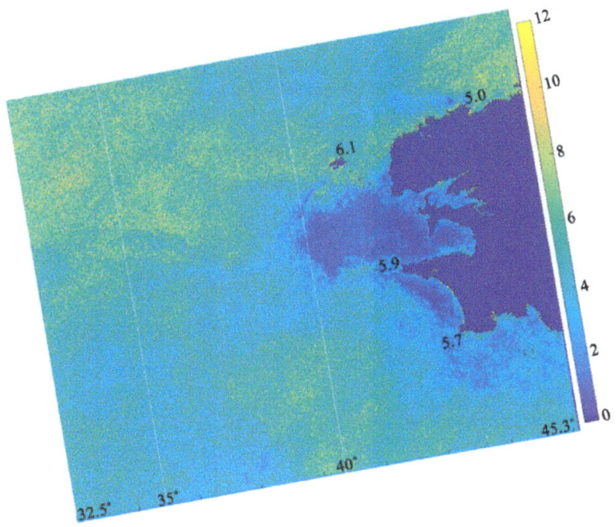

(b)

Figure 7. Map of wind speed estimated by the SPM (a) and CMOD.5 (b) from a VV-pol Sentinel-1 image in IWS mode acquired on 15th Apr 2015, from 18:12:46 UTC to 18:13:15 UTC.

HH-Pol Image

Figure 8 shows the wind speed estimations on a HH-pol Sentinel-1 image by using the SPM and CMOD.5. This image is acquired in SM mode with radar incident angles from 42° to 46°. As well as in the previous case, the spatial resolution of wind direction and wind speed in this case is 3 km × 3 km. As expected from Figure 1, SPM overestimates wind speed in most cases of incident angle and wind direction. That can be verified in this case. While the wind speed estimated by the CMOD.5 is only in the range of 10 m/s - 15 m/s, that of the SPM is 10 - 15 m/s higher. This is confirmed by the comparisons with the measured data which are only in the range of 10 - 16 m/s. As well as in Figure 7, the 3 km × 3 km spatial resolution gives the good distinction of wind speed areas. In particular, in contrast to the previous case, the wind streaks can be well noted in this case. This is an important sign to solve the 180° ambiguity of extracted wind direction, as discussed in Figure 4and Figure 5.

DISCUSSION AND VALIDATION

For the Sentinel-1 data in VV-pol, wind speed estimated by the SPM is quite similar to the one offered by the CMOD.5 for the range of $\theta = 32.5°$ to 45°. The deviation of 2 - 3 m/s is tolerable in some applications (e.g. weather forecasting). Certainly, this conclusion is only validated in the wind speed range of 1 - 12 m/s, corresponding to the validity domain of the SPM. Nevertheless, from our study, the SPM still gives (quite) similar wind speed estimation to the CMOD.5 for wind speed above 12 m/s. This result is only acceptable under the point of view of mathematics. It has not any physical significance. However, if the SPM is corrected to widen the validity domain of wind speed and incident angle, it can be used to estimate wind speed from SAR images as the EP models. Indeed, the TSM is a good successive model of the SPM to study radar backscattering with higher wind speed and different ranges of incident angles [23] . In Figure 7, the wind speed estimated by the SPM has lower spatial resolution than that of the CMOD.5. This is probably from the sea surface roughness spectrum studied in this paper, or from the SPM. Therefore, to well understand, the other wave spectrum models should be taken into account in the next paper.

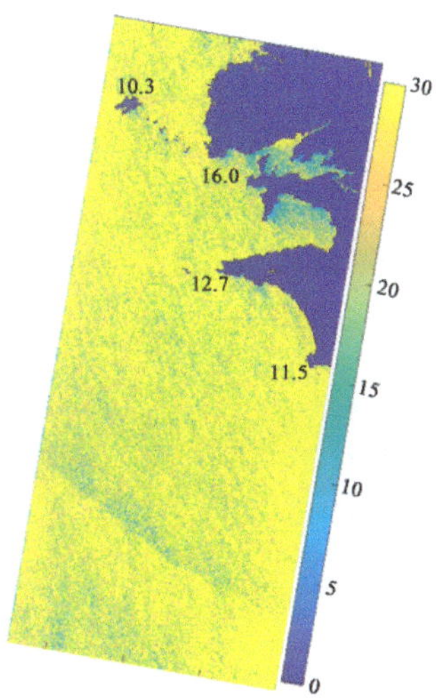

(a)

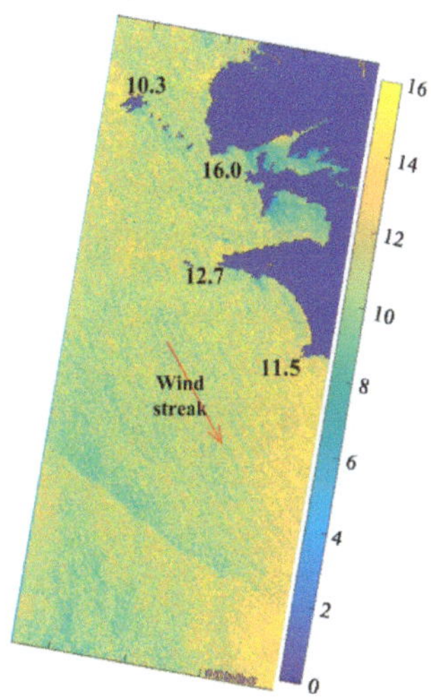

(b)

Figure 8. Map of wind speed estimated by the SPM (a) and CMOD.5 (b) from a HH-pol Sentinel-1 image in SM mode acquired on 27th Dec 2014, from 18:12:46 UTC to 18:13:15 UTC.

For the images in HH-pol, the SPM overestimates wind speed in most cases of incident angle and wind direction. This result can be explained by some facts. First, the polarizations of radar backscattering respond to sea surface roughness and wave breaking differently. In fact, as shown in many studies [26] - [28] , VV-pol is dominated by Bragg resonance scattering, while non-Bragg scattering mechanisms is more appropriate for HH-pol (maybe VH-pol). In other words, the SPM can describe well radar backscattering in VV-pol, while for HH-pol the other EM models (i.e. TSM, SSA) should be used. Second, long wind waves are not aligned with local wind, and there may be waves, i.e. swell, not related to the local wind. Furthermore, the sea can be partially developed, and this makes the wave spectrum different from the standard reference ones (i.e. Elfouhaily). In other words, while the standard models assume that the roughness statistics are homogenous, in practice, they vary nonlinearly. This constraint for HH-pol is more sensitive than for VV-pol because NRCS in HH-pol is normally several dBs lower than NRCS in VV-pol. Therefore, a non-linear spectral description of sea surface roughness should be used to improve wind speed estimation.

CONCLUSIONS AND PERSPECTIVES

This paper has discussed about the application of the EM models to estimate wind speed from SAR data. Under the point of view of the inverse problem, the SPM has been selected due to its flexibility of inversion. Nevertheless, it can only work well for wind speed below 13 m/s, and for moderate incident angles of 32.5° - 45°. The wind speed estimated by the SPM is specifically compared to the one offered by the CMOD.5. The CMOD.5 is a widely used empirical model in the literature to estimate wind speed from SAR data. In both studied models, wind direction plays an important role to estimate accurately wind speed. In the indicated validity domain (wind speed below 13 m/s and incident angles of 32.5° - 45°), the SPM gives quite similar wind speed estimation to the CMOD.5 in VV-pol. However, the resolution of wind speed obtained by the SPM is lower than that of the CMOD.5. In HH-pol, the SPM overestimates wind speed in most cases of incident angle and wind direction. This result can be explained by some facts which concern the different behavior of radar backscattering to polarization, and the difference between the standard wave spectrum used in the SPM and the practical one at the moment of SAR image acquisition.

In the next steps, in order to improve the validity domain of the SPM, the other EM models like TSM or SSA should be used. They are expected to possibly estimate wind speed above 13 m/s, and for different ranges of incident angles, notably for $\theta = 20° - 30°$. In particular, to improve wind speed estimation in HH-pol, together with the other EM models, a study of sea surface roughness spectrum should be done.

ACKNOWLEDGEMENTS

This work is supported by the French General Directorate for Armament (DGA) in the frame of Project SIMUSO. TheSentine-1 images are provided by the European Space Agency (ESA) viahttps://scihub.esa.int/. The in situ measurements are collected via http://www.infoclimat.fr/.

REFERENCES

1. Ghanmi, H., Khenchaf, A. and Comblet, F. (2014) Numerical Modeling of Electromagnetic Scattering from Sea Surface Covered by Oil. Journal of Electromagnetic Analysis & Applications, 6, 15-24.http://dx.doi.org/10.4236/jemaa.2014.61003
2. Pino, M.R., Landesa, L., Rodriguez, J.L., Obelleiro, F. and Burkholder, R.J. (1999) The Generalized Forward-Backward Method for Analyzing the Scattering from Targets on Ocean-Like Rough Surfaces. IEEE Transactions on Antennas and Propagation, 47, 961-969.
3. Shimada, T., Kawamura, H. and Shimada, M. (2003) An L-Band Geophysical Model Function for SAR Wind Retrieval Using JERS-1 SAR. IEEE Transactions on Geoscience and Remote Sensing, 41, 518-531.
4. Lehner, S., Schulz-Stellenfleth, J., Schättler, B., Breit, H. and Horstmann, J. (2000) Wind and Wave Measurements Using Complex ERS-2 SAR Wave Mode Data. IEEE Transactions on Geoscience and Remote Sensing, 38, 2246-2257.
5. Carvajal, G.K., Eriksson, L.E.B. and Ulander L.M.H. (2014) Retrieval and Quality Assessment of Wind Velocity Vectors on the Ocean with C-Band SAR. IEEE Transactions on Geoscience and Remote Sensing, 52, 2519-2537.
6. Horstmann, J., Koch, W., Lehner, S. and Tonboe, R. (2000) Wind Retrieval over the Ocean Using Synthetic Aperture Radar with C-Band HH Polarization. IEEE Transactions on Geoscience and Remote Sensing, 38, 2122-2131.
7. Li, X.-M. and Lehner, S. (2014) Algorithm for Sea Surface Wind Retrieval From TerraSAR-X and TanDEM-X Data. IEEE Transactions on Geoscience and Remote Sensing, 52, 2928-2939.
8. Montuori, A., de Ruggiero, P., Migliaccio, M., Pierini, S. and Spezie, G. (2013) X-Band COSMO-SkyMed wind Field Retrieval, with Application to Coastal Circulation Modeling. Ocean Science, 9, 121-132.http://dx.doi.org/10.5194/os-9-121-2013
9. Stoffelen, A. and Anderson, D. (1997) Scatterometer Data Interpretation: Estimation and Validation of the Transfer Function CMOD4. Journal of Geophysical Research, 102, 5767-5780. http://dx.doi.org/10.1029/96JC02860
10. Quilfen, Y., Chapron, B., Elfouhaily, T., Katsaros, K. and Tournadre, J. (1998) Observation of Tropical Cyclones by High-Resolutions Scatterometry. Journal of Geophysical Research, 103, 7767-7786. http://dx.doi.org/10.1029/97JC01911
11. Hersbach, H., Stoffelen, A. and de Haan, S. (2007) An Improved C-Band Scatterometer Ocean Geophysical Model Function: CMOD5. Journal of Geophysical Research, 112, C03006. http://dx.doi.org/10.1029/2006JC003743
12. Verspeek, J., Stoffelen, A., Portabella, M., Bonekamp, H., Anderson, C. and Saldaña, J.F. (2010) Validation and Calibration of ASCAT Using CMOD5.n. IEEE Transactions on Geoscience and Remote Sensing, 48, 386-395.

13. Thompson, D.R., Elfouhaily, T.M. and Chapron, B. (1998) Polarization Ratio for Microwave Backscattering from the Ocean Surface at Low to Moderate Incidence Angles. Geoscience and Remote Sensing Symposium Proceedings (IGARSS'98), Seattle, 6-10 July 1998, 1671-1673. http://dx.doi.org/10.1109/igarss.1998.692411

14. Horstmann, J., Koch, W., Lehner, S. and Tonboe, R. (2002) Ocean Winds from RADARSAT-1 ScanSAR. Canadian Journal of Remote Sensing, 28, 524-533. http://dx.doi.org/10.5589/m02-043

15. Vachon, P.W. and Dobson, F.W. (2000) Wind Retrieval from RADARSAT SAR Images: Selection of a Suitable C-Band Polarization Wind Retrieval Model. Canadian Journal of Remote Sensing, 26, 306-313. http://dx.doi.org/10.1080/07038992.2000.10874781

16. Mouche, A.A., Hauser, D. and Kudryavtsev, V. (2006) Radar Scattering of the Ocean Surface and Sea-Roughness Properties: A Combined Analysis from Dual-Polarizations Airborne Radar Observations and Models in C Band. Journal of Geophysical Research: Oceans, 111, C09039. http://dx.doi.org/10.1029/2005JC003166

17. Zhang, B.A., Perrie, W. and He, Y.J. (2011) Wind Speed Retrieval from RADARSAT-2 Quad-Polarization Images Using a New Polarization Ratio Model. Journal of Geophysical Research: Oceans, 116, C00D11.

18. Liu, G., Yang, X., Li, X., Zhang, B., Pichel, W., Li, Z. and Zhou, X. (2013) A Systematic Comparison of the Effect of Polarization Ratio Models on Sea Surface Wind Retrieval from C-Band Synthetic Aperture Radar. IEEE Journal of Selected Topics in Applied Earth Observations and Remote Sensing, 6, 1100-1108.

19. Ren, Y., Lehner, S., Brusch, S., Li, X.-M. and He, M. (2012) An Algorithm for the Retrieval of Sea Surface Wind Fields Using X-Band TerraSAR-X Data. International Journal of Remote Sensing, 33, 7310-7336. http://dx.doi.org/10.1080/01431161.2012.685977

20. Colliander, A. and Ylä-Oijala, P. (2007) Electromagnetic Scattering from Rough Surface Using Single Integral Equation and Adaptive Integral Method. IEEE Transactions on Antennas and Propagation, 55, 3639-3646.

21. Holliday, D., DeRaad, L.L. and St-Cyr, G.J. (1996) Forward-Backward: A New Method for Computing Low-Grazing Angle Scattering. IEEE Transactions on Antennas and Propagation, 44, 722-729.

22. Awada, A., Ayari, Y., Khenchaf, A. and Coatanhay, A. (2007) Bistatic Scattering from an Anisotropic Sea Surface: Numerical Comparison between the First-Order SSA and the TSM Models. Waves in Random and Complex Media, 16, 383-394. http://dx.doi.org/10.1080/17455030600844089

23. Khenchaf, A. (2001) Bistatic Scattering and Depolarization by Randomly Rough Surface: Application to Natural Rough Surface in X-Band. Waves in Random and Complex Media, 11, 61-89. http://dx.doi.org/10.1088/0959-7174/11/2/301 [Paper reference:6]

24. Fetterer, F., Gineris, D. and Wackerman, C.C. (1998) Validating a Scatterometer Wind Algorithm for ERS-1 SAR. IEEE Transactions on Geoscience and Remote Sensing, 36, 479-492.

25. Koch, W. (2004) Directional Analysis of SAR Images Aiming at Wind Direction. IEEE Transactions on Geoscience and Remote Sensing, 42, 702-710.

26. Wright, J.W. (1966) Backscattering from Capillary Waves with Application to Sea Clutter. IEEE Transactions on Antennas and Propagation, 14, 749-754.

27. Valenzuela, G.R. (1978) Theories for the Interaction of Electromagnetic and Oceanic Waves—A Review. Boundary-Layer Meteorology, 13, 61-85. http://dx.doi.org/10.1007/BF00913863

28. Plant, W.J. (1990) Bragg Scattering of Electromagnetic Waves from the Air/Sea Interface. In: Geernaert, G.L. and Plant, W.L., Eds., Environmental Fluid Mechanics. Vol. 8. Surface Waves and Fluxes, Kluwer Academic Publishers, Dordrecht, 41-108.

29. Klein, L.A. and Swift, C.T. (1977) An Improved Model for the Dielectric Constant of Sea Water at Microwave Frequencies. IEEE Journal of Oceanic Engineering, 2, 104-111. http://dx.doi.org/10.1109/JOE.1977.1145319

30. Stogryn, A. (1971) Equations for Calculating the Dielectric Constant of Saline Water. IEEE Transactions on Microwave Theory and Techniques, 19, 733-736.

31. Guillou, C., Ellison, W., Eymard, L., Lamkaouchi, K., Prigent, C., Delbos, G., Balana, G. and Boukabara, S.A. (1998) Impact of New Permittivity Measurements on Sea Surface Emissivity Modeling in Microwaves. Radio Science, 33, 649-667. http://dx.doi.org/10.1029/97RS02744

32. Meissner, T. and Wentz, F.J. (2004) The Complex Dielectric Constant of Pure and Sea Water from Microwave Satellite Observations. IEEE Transactions on Geoscience and Remote Sensing, 42, 1836-1849. http://dx.doi.org/10.1109/TGRS.2004.831888

33. Cox, C.S. and Munk, W.H. (1954) Statistics of the Sea Surface Derived from Sun Glitter. Journal of Marine Research, 13, 198-227.

34. Pierson, W. and Moskowitz, L. (1964) A Proposed Spectral form for Fully Developed Wind Sea Based on the Similarity Theory of S. A. Kitaigorodskii. Journal of Geophysical Research, 69, 5181-5190. http://dx.doi.org/10.1029/JZ069i024p05181

35. Bjerkaas, A.W. and Riedel, F.W. (1979) Proposed Model for the Elevation Spectrum of a Wind-Roughened Sea Surface. Technical Report APL-TG-1328-1-31, Applied Physics Laboratory, The Johns Hopkins University, Laurel, 31.

36. Donelan, M.A., Hamilton, J. and Hui, W.H. (1985) Directional Spectra of Wind Generated Waves. Philosophical Transactions of the Royal Society of London. Series A, 315, 509-562.

37. Apel, J.R. (1994) An Improved Model of the Ocean Surface Wave Vector Spectrum and Its Effects on Radar Backscatter. Journal of Geophysical Research, 99, 16269-16291.http://dx.doi.org/10.1029/94JC00846

38. Elfouhaily, T.M., Chapron, B., Katsaros, K. and Vandermark, D. (1997) A Unified Directional Wave Spectrum for Long and Short Wind-Driven Waves.

Journal of Geophysical Research: Oceans, 102, 15781-15796. http://dx.doi.org/10.1029/97JC00467

39. Kudryavtsev, V.N., Hauser, D., Caudal, G. and Chapron, B. (2003) A Semi-Empirical Model of the Normalized Radar Cross-Section of the Sea Surface: 1. Background Model. Journal of Geophysical Research, 108.

40. Contreras, R.F. and Plant, W. (2006) Surface Effect of Rain on Microwave Backscatter from the Ocean: Measurements and Modeling. Journal of Geophysical Research: Oceans, 111, C08019.

41. Hwang, P. and Wang, D.W. (2004) An Empirical Investigation of Source Term Balance of Small Scale Surface Waves. Geophysical Research Letters, 31, L15301. http://dx.doi.org/10.1029/2004GL020080

42. Hwang, P. (2008) Observations of Swell Influence on Ocean Surface Roughness. Journal of Geophysical Research: Oceans, 113, C12024. http://dx.doi.org/10.1029/2008JC005075

43. Hwang, P. (2011) A Note on the Ocean Surface Roughness Spectrum. Journal of Atmospheric and Oceanic Technology, 28, 436-443. http://dx.doi.org/10.1175/2010JTECHO812.1

44. Ulaby, F.T., Moore, R.K. and Fung, A.K. (1986) Microwave Remote Sensing: Active and Passive. Vol. 2. Radar Remote Sensing and Surface Scattering and Emission Theory. Ch. 12, Artech House Publishers, Norwood, 962-966.

45. Smith, S. (1988) Coefficients for Sea Surface Wind Stress, Heat Flux, and Wind Profiles as a Function of Wind Speed and Temperature. Journal of Geophysical Research: Oceans, 93, 15467-15472.

46. Hsu, S.A., Meindl, E.A. and Gilhousen, D.B. (1994) Determining the Power-Law Wind-Profile Exponent under Near-Neutral Stability Conditions at Sea. Journal of Applied Meteorology, 33, 757-765. http://dx.doi.org/10.1175/1520-0450(1994)033<0757:DTPLWP>2.0.CO;2

47. Carvajal, G.K., Eriksson, L.E.B. and Ulander, L.M.H. (2014) Retrieval and Quality Assessment of Wind Velocity Vectors on the Ocean with C-Band SAR. IEEE Transactions on Geoscience and Remote Sensing, 52, 479-492.

CHAPTER 8

OPTIMAL PATTERN SYNTHESIS OF LINEAR ANTENNA ARRAY USING GREY WOLF OPTIMIZATION ALGORITHM

Prerna Saxena and Ashwin Kothari

INTRODUCTION

Antenna arrays play a significant role in modern wireless communication systems because of their advantages over single antenna element such as high directivity, high gain, and beam steering capabilities. These characteristics of antenna arrays make them suitable for applications like beam forming and null positioning [1]. Null positioning in order to eliminate the unwanted signals which cause interference is probably one of the most important goals of array synthesis [2, 3]. Moreover, producing a directive beam at a particular direction while keeping the side lobe level (SLL) small to avoid interference with other radiating sources is another important aim of array synthesis.

Extensive study of linear antenna array synthesis has been reported in the literature [4–6]. For linear array, SLL minimization and null placement can be achieved either by optimizing the excitation amplitude and phase while maintaining uniform spacing as that of conventional array or by optimizing the element spacing while assuming uniform amplitude and phase excitation. Various evolutionary algorithms such as genetic algorithm (GA) [7, 8], simulated annealing (SA) [9], particle swarm optimization (PSO) [10–13], ant colony optimization (ACO) [14], cat swarm optimization (CSO) [15], invasive weed optimization (IWO) [16, 17], Taguchi's method [18], biogeography based optimization [19], artificial bees colony algorithm [20], firefly algorithm [21], cuckoo search algorithm [22], differential search algorithm [23], and backtracking search optimization algorithm [24] have been successfully applied for optimization of linear arrays.

The main aim of this paper is to introduce the grey wolf optimization (GWO) algorithm [25] to the electromagnetics and antenna community. GWO is a metaheuristic algorithm inspired by the leadership hierarchy and hunting mechanism of grey wolves. It was proposed by Mirjalili et al. in 2014 [25]. GWO has been applied to solve practical optimization problems in engineering [25] such as tension/compression spring design, welded beam design, pressure vessel design, and optical buffer design. GWO has also been used in areas like allocation of static synchronous compensator (STATCOM) devices on power system grid [26] and to solve economic dispatch problems [27, 28] and so forth. In [29], an improved grey wolf optimizer for training q-Gaussian Radial Basis Functional-link nets was proposed.

However, this is the first time that GWO is being applied to antenna array optimization to the best of the authors' knowledge. In this paper, GWO has been applied to linear

Citation: Prerna Saxena and Ashwin Kothari, "Optimal Pattern Synthesis of Linear Antenna Array Using Grey Wolf Optimization Algorithm," International Journal of Antennas and Propagation, vol. 2016, Article ID 1205970, 11 pages, 2016. doi:10.1155/2016/1205970.

antenna arrays for optimal pattern synthesis in the following two ways: by optimizing the antenna positions while assuming uniform excitation and by optimizing the antenna current amplitudes while assuming spacing and phase as that of uniform array. GWO is used to achieve an array pattern with minimum side lobe level (SLL) along with placement of deep nulls in the specified directions. In addition, GWO is applied for the minimization of the first side lobe nearest to the main beam (near side lobe minimization) while simultaneously controlling the other side lobes.

This section has presented a brief introduction to linear antenna array, the GWO and its applications in optimization problems, and the main objective of this paper. The rest of the paper is organized as follows: the linear antenna array geometry, configuration, and array factor equations are discussed in Section 2. Section 3presents an elaborate description of the grey wolf optimization algorithm along with a flowchart outlining the steps of GWO implementation. Various examples for linear array synthesis along with the GWO optimized antenna positions and currents as well as the corresponding array patterns are presented in Section 4. The validation of the obtained results by comparison with other nature-inspired evolutionary algorithms is also presented in this section. Section 5 presents the conclusion.

LINEAR ANTENNA ARRAY

A linear antenna array of $2N$ isotropic elements placed symmetrically along the x-axis is considered in this work, as illustrated in Figure 1.

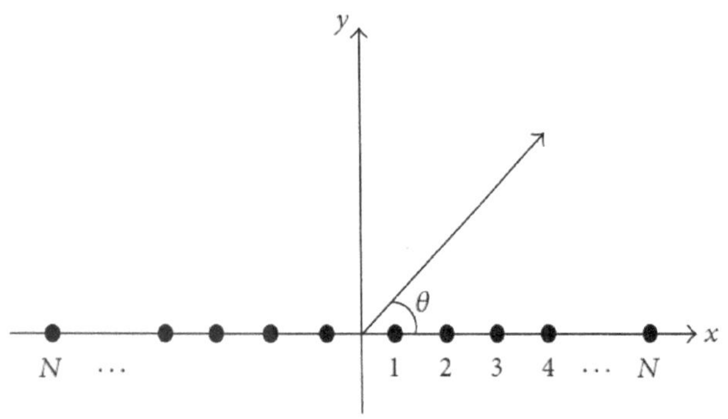

Figure 1: Antenna array geometry.

Due to symmetry, the array factor (AF) [1, 10] in the azimuth plane is given by

$$\mathrm{AF}(\theta) = 2\sum_{n=1}^{N} I_n \cos\left(kx_n \cos(\theta) + \psi_n\right), \tag{1}$$

where I_n, ψ_n, and x_n are the excitation amplitude, phase, and position of nth element in the array. k is the wave number and is given by $2\pi/\lambda$ and θ is the azimuth angle.

For antenna element position optimization, it is assumed that the antenna array is subjected to uniform amplitude and phase excitation; that is, $I_n = 1$ and $\psi_n = 0$. Thus, the AF in (1) gets modified to

$$\mathrm{AF}(\theta) = 2\sum_{n=1}^{N} \cos\left(kx_n \cos(\theta)\right). \tag{2}$$

The proper placement of antennas in the linear array is very important because if antennas are placed too close to each other, this leads to mutual coupling effects and if antennas are placed too far away, this leads to grating lobes. Thus, conditions (3) must be satisfied for antenna position optimization:

$$\left|x_i - x_j\right| > 0.25\lambda,$$

$$\min\{x_i\} > 0.125\lambda; \quad i = 1, 2, \ldots, N,\ i \neq j, \tag{3}$$

where x_j is the antenna position adjacent to the antenna position x_i and $\{x_i\}$ is the set of all antenna positions.

The array factor for antenna current amplitude optimization is given by (4). Uniform phase, that is, $\psi_n = 0$, and the interelement spacing (x_n) of $\lambda/2$ as that of uniform array are assumed:

$$\mathrm{AF}(\theta) = 2\sum_{n=1}^{N} I_n \cos\left(kx_n \cos(\theta)\right). \tag{4}$$

GREY WOLF OPTIMIZATION (GWO) ALGORITHM

GWO is a metaheuristic algorithm which mimics the social hierarchy and hunting mechanism of grey wolves. The hierarchy of grey wolves is as follows: the leaders are male and female, called alphas (α). The second level in the hierarchy comprises the beta (β) wolves followed by delta (δ) wolves and the lowest ranking grey wolves are the omegas (ω). The hierarchy of grey wolves is mathematically modelled by considering α as the fittest solution followed by the second and third best solutions, β and δ, respectively. The rest of the candidate solutions are assumed to be ω.

The main phases of grey wolf hunting are as follows [30]:

(1) Tracking, chasing, and approaching the prey.

(2) Pursuing, encircling, and harassing the prey until it stops moving.

(3) Attack towards the prey.

The encircling behavior is mathematically modelled by (5) as in [25]:

$$\vec{D} = \left|\vec{C} \cdot \overrightarrow{X_p(t)} - \overrightarrow{X(t)}\right|,$$

$$\vec{X}(t+1) = \overrightarrow{X_p}(t) - \vec{A} \cdot \left(\vec{D}\right), \tag{5}$$

where t indicates the current iteration, $\vec{A}$ and $\vec{C}$ are coefficient vectors, $\overrightarrow{X_p}$ is the position vector of the prey, and $\vec{X}$ indicates the position vector of a grey wolf. The vectors $\vec{A}$ and $\vec{C}$ are calculated using

$$\vec{A} = 2\vec{a} \cdot \overrightarrow{r_1} - \vec{a},$$

$$\vec{C} = 2 \cdot \overrightarrow{r_2}, \tag{6}$$

where components of $\vec{a}$ are linearly decreased from 2 to 0 over the course of iterations and , $\overrightarrow{r_1}, \overrightarrow{r_2}$ are random vectors in [0, 1].

In the GWO algorithm, the hunting (optimization) is guided by α, β, and δ. The ω wolves follow these three wolves. The α, β, and δ wolves are supposed to have better knowledge about the potential location of prey. Therefore, the first three best solutions are saved and the other search agents update their positions according to the position of the best search agent. The following equations are used for this purpose [25]:

$$\overrightarrow{D_\alpha} = \left|\overrightarrow{C_1} \cdot \overrightarrow{X_\alpha} - \vec{X}\right|,$$

$$\overrightarrow{D_\beta} = \left|\overrightarrow{C_2} \cdot \overrightarrow{X_\beta} - \vec{X}\right|,$$

$$\overrightarrow{D_\delta} = \left|\overrightarrow{C_3} \cdot \overrightarrow{X_\delta} - \vec{X}\right|,$$

$$\overrightarrow{X_1} = \overrightarrow{X_\alpha} - \overrightarrow{A_1} \cdot \left(\overrightarrow{D_\alpha}\right),$$

$$\overrightarrow{X_2} = \overrightarrow{X_\beta} - \overrightarrow{A_2} \cdot \left(\overrightarrow{D_\beta}\right),$$

$$\overrightarrow{X_3} = \overrightarrow{X_\delta} - \overrightarrow{A_3} \cdot \left(\overrightarrow{D_\delta}\right),$$

$$\vec{X}(t+1) = \frac{\overrightarrow{X_1} + \overrightarrow{X_2} + \overrightarrow{X_3}}{3}. \tag{7}$$

By using (7), a search agent updates its position according to α, β, and δ in the n-dimensional search space as depicted in Figure 2(a) [25]. Furthermore, the final position would be in a random place in the search space within a circle which is defined by the positions of α, β, and δ. Hence, α, β, and δ estimate the position of the prey, and other wolves update their positions randomly around the prey.

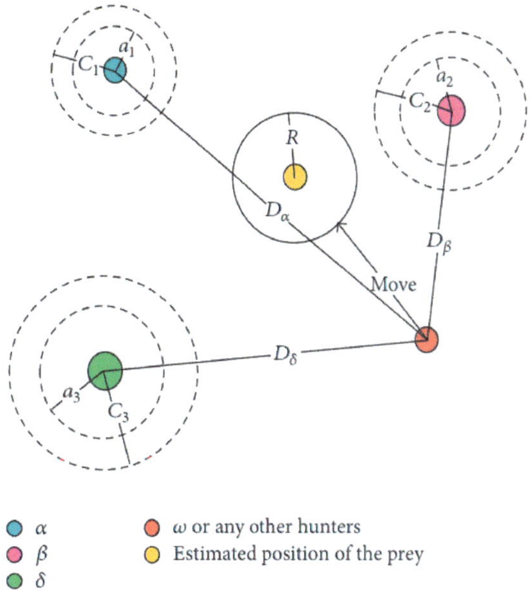

(a)

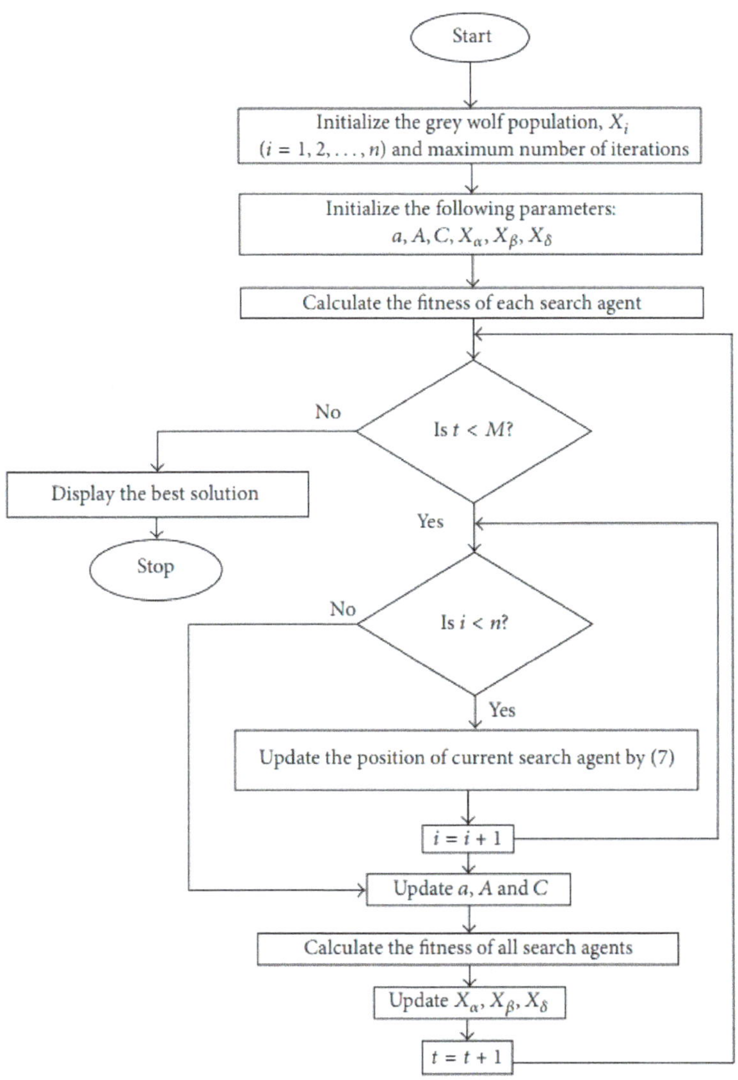

(b)

Figure 2: (a) Position update in GWO [25]. (b) Flowchart of the grey wolf optimization algorithm.

The following steps summarize the GWO algorithm:

1. Create a random population of grey wolves (candidate solutions).
2. α, β, and δ wolves estimate the probable position of the prey over the course of iterations. Each candidate solution updates its distance from the prey.
3. The parameter "$\vec{a}$" is decreased linearly from 2 to 0 in order to emphasize exploration (searching for prey) and exploitation (attacking the prey), respectively. Candidate solutions tend to diverge from the prey when $|\vec{A}|>1$ and converge towards the prey when $|\vec{A}|<1$ so as to avoid stagnation in local solutions.
4. Finally, the GWO algorithm is terminated by the satisfaction of an end criterion. The flowchart of GWO is depicted in Figure 2(b).

RESULTS AND DISCUSSION

In this section, the GWO algorithm is applied to linear antenna array for optimal pattern synthesis. The GWO is implemented on MATLAB® and executed 15 times. The number of iterations for each run is set equal to 1000. All results were obtained using grey wolf population (n) equal to 30.

Antenna Position Optimization

This section presents Examples 1–3 for optimization of the antenna positions (xn) in order to achieve a desired array pattern with minimum SLL and null positioning. Uniform amplitude and phase excitations are assumed; that is, $I_n = 1$ and $\psi_n = 0$.

Peak SLL Minimization

The fitness function used for the minimization of peak SLL is formulated using (8) [13, 19, 22]. Consider

$$\text{Fitness} = \min\left(\max\left(20\log|\text{AF}(\theta)|\right)\right), \tag{8}$$

where $\theta \in$ spatial region in which SLL is to suppressed and AF(θ) is the array factor given by (2).

Example 1 illustrates the design of $2N = 10$ element linear array for achieving minimum SLL in the regions, $\theta = [0°, 74°]$ and $\theta = [106°, 180°]$. The grey wolf optimization algorithm with fitness function given by (8) is used for the determination of optimized element locations, x_n. Uniform amplitude and phase excitations are assumed; that is, $I_n = 1$ and $\psi_n = 0$. The optimized element positions are shown in Table 1 and the array pattern is illustrated in Figure 3. For benchmarking, the peak SLL obtained for this example using the proposed method (GWO) and other nature-inspired optimization techniques is summarized in Table 2. There is an improvement in the reduction of SLL using the proposed approach (GWO) as compared to conventional array as well as arrays optimized using other optimization algorithms such as ACO [14], PSO [15], and CSO [15]. The proposed method (GWO) gives peak SLL of −23.42 dB which is 10.19 dB lower as compared to conventional array. The peak SLL has been lowered from −20.72 dB to −23.42 dB (by 2.7 dB) as compared to PSO optimized array, from −22.66 dB to −23.42 dB as compared to ACO optimized array, and from −22.89 dB to −23.42 dB as compared to CSO optimized array.

Table 1: Optimized positions of the positive half of the 10-element array of Example 1

Method	Optimized element positions				
CSO [15]	0.1516λ	0.4115λ	0.7899λ	1.1048λ	1.6843λ
Proposed	0.1250λ	0.3680λ	0.7522λ	0.9851λ	1.5546λ

Table 2: Optimized peak SLL for 10-element linear array of Example 1

Serial number	Approach	Peak SLL (in dB)
1	Conventional (without optimization) [15]	−13.23
2	PSO [15]	−20.72
3	ACO [14]	−22.66
4	CSO [15]	−22.89
5	Proposed	−23.42

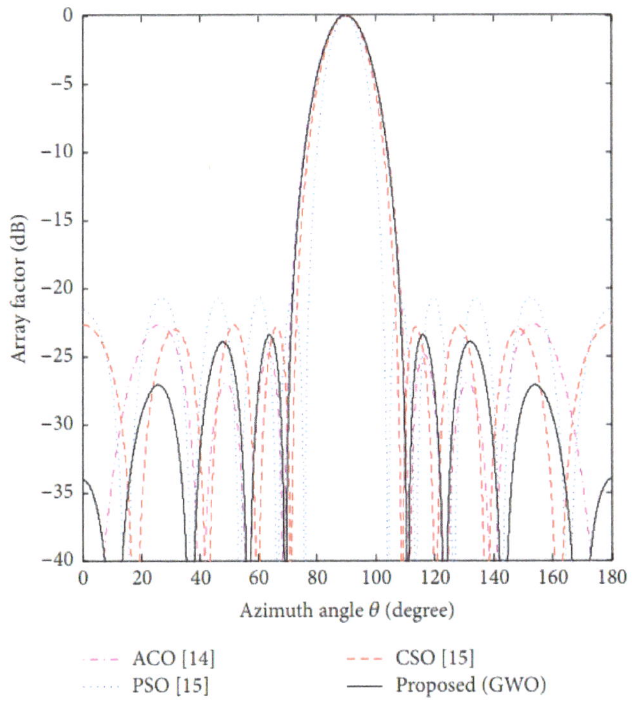

Figure 3: Array pattern for Example 1.

SLL Minimization along with Null Placement

The fitness function as stated in (9) is used for SLL minimization as well as for placement of nulls in desired directions [10, 19]:

$$\text{Fitness} = \sum_i \frac{1}{\Delta\theta_i} \int_{\theta_{li}}^{\theta_{ui}} |\text{AF}(\theta)|^2 \, d\theta + \sum_k |\text{AF}(\theta_k)|^2, \tag{9}$$

where θ_{li} and θ_{ui} are the spatial regions in which SLL is suppressed and $\Delta\theta_i = \theta_{ui} - \theta l_i$. The null direction is given by θ_k. In (9), the first term of the fitness function is for SLL suppression and the second term accounts for the placement of nulls in the desired directions. AF(θ) is the array factor given by (2).

Example 2 illustrates the synthesis of 28-element linear antenna array for achieving SLL minimization in the regions $\theta = [0°, 84°]$ and $\theta = [96°, 180°]$ along with null placement at $\theta = 55°, 57.5°, 60°, 120°, 122.5°$, and $125°$. The fitness function used by the GWO for this example is given by (9). The array pattern is shown in Figure 4 and the optimized positions of the antenna elements are given in Table 3. From Table 3, it is seen that the array optimized using the proposed method (GWO) and CSO [15] optimized array have almost the same length. It is deduced from Figure 4 that the proposed method using GWO enables the placement of nulls, as deep as −84.42 dB at the desired directions.

Table 3: Optimized positions of the positive half of the 28-element array of Example 2

Method	Optimized element positions						
CSO [15]	0.2720λ	0.7547λ	1.1399λ	1.7065λ	2.3287λ	2.8675λ	3.3536λ
	3.7693λ	4.2222λ	4.8991λ	5.4061λ	5.7389λ	6.1564λ	6.7173λ
Proposed	0.2298λ	0.7266λ	1.1376λ	1.7300λ	2.1831λ	2.6710λ	3.0255λ
	3.4867λ	3.7652λ	4.3933λ	4.9039λ	5.3422λ	5.7610λ	6.3378λ

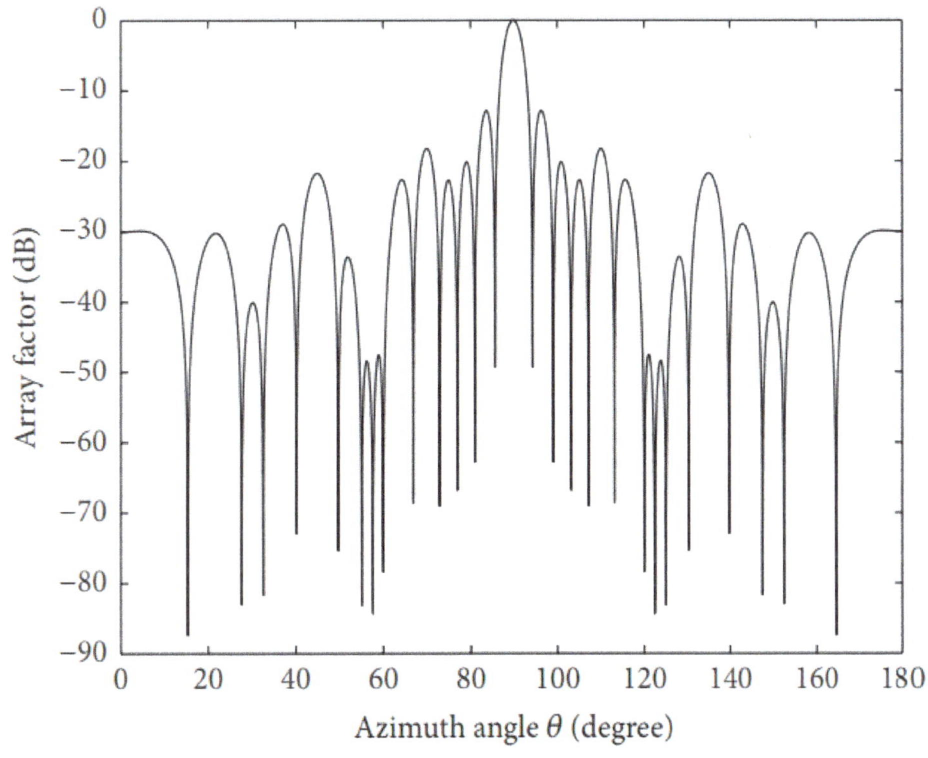

Figure 4: Array pattern for Example 2.

The null depths obtained by the proposed method using GWO at each of the specified directions are summarized in Table 4. The comparative analysis of null depth obtained with the proposed method using GWO and various other state-of-the-art optimization algorithms is shown in Table 5. It is seen that, for this example, the proposed method using GWO places the deepest null of −84.42 dB. There is an improvement of around 34 dB over the null depth obtained by using PSO [15] and ACO [14]. As compared to CSO [15], the proposed approach of using GWO gives improvement of around 19 dB in null depth. However, the SLL obtained is similar for all the above algorithms.

Table 4: Null depths after optimization by GWO algorithm for Example 2

Linear array type	Null depth (in dB)					
	55°	57.5°	60°	120°	122.5°	125°
28-element array	−83.22	−84.42	−78.41	−78.41	−84.42	−83.22

Table 5: Comparative analysis of null depth (in dB) obtained by various optimization algorithms for Example 2

	Method			
	PSO [15]	ACO [14]	CSO [15]	Proposed
Null depth (in dB)	−50	~−50	−65	−84.42

Example 3 illustrates the application of GWO to optimize the antenna element positions for SLL minimization and null placement of 32-element linear antenna array. The fitness function used by the GWO for this example is given by (9). SLL reduction is desired in the spatial regions $\theta = [0°, 85°]$ and $\theta = [95°, 180°]$ whereas nulls are desired at $\theta = 81°$ and $\theta = 99°$ (very close to the first side lobe). The array pattern is shown in Figure 5 and the optimized positions of the antenna elements are given in Table 6. The array optimized by the proposed approach of using GWO has almost the same length as that obtained by CSO [15]. It is seen from Figure 5 that the proposed approach of using GWO enables the placement of nulls as deep as −106 dB at the desired directions ($\theta = 81°$ and $\theta = 99°$), very close to the first side lobe.

Table 6: Optimized positions of the positive half of the 32-element array of Example 3

Method	Optimized element positions			
CSO [15]	0.2883λ	0.6830λ	1.1929λ	1.5199λ
	1.9768λ	2.3247λ	2.6886λ	3.1362λ
	3.4848λ	3.9538λ	4.3822λ	4.9252λ
	5.4817λ	6.2091λ	7.0412λ	7.7500λ
Proposed	0.194307λ	0.740711λ	1.249193λ	1.747565λ
	2.241275λ	2.714261λ	2.999822λ	3.451514λ
	3.753935λ	4.275930λ	4.75λ	5.255635λ
	5.751775λ	6.455911λ	7.25λ	8.0λ

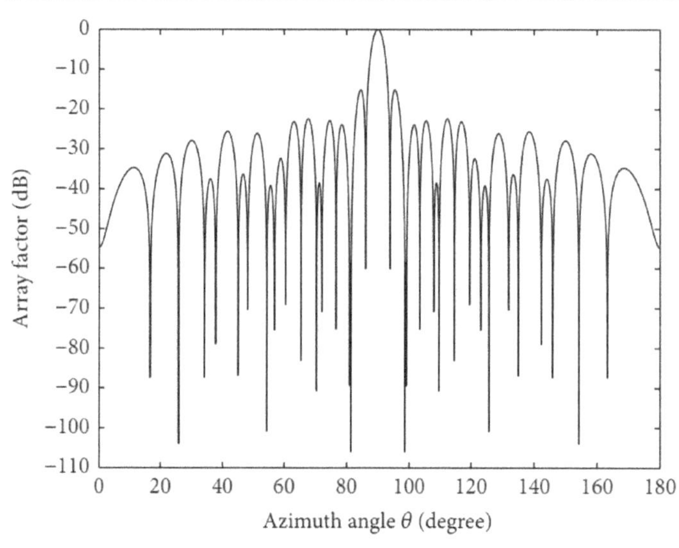

Figure 5: Array pattern for Example 3.

For this example, PSO [15] offers null depth of −60 dB and ACO [14] gives −50 dB nulls whereas CSO [15] places deep nulls of −80 dB as seen from Table 7. However, the proposed approach (GWO) places the deepest null of −106 dB. The first side lobe obtained by GWO is about 3 dB higher than that obtained using CSO [15], but the remaining side lobes are almost similar (some are even lower) to those obtained by using CSO [15].

Table 7: Comparative analysis of null depth (in dB) obtained by various optimization algorithms for Example 3

	Method			
	PSO [15]	ACO [14]	CSO [15]	Proposed
Null depth (in dB)	−60	−50	−80	−106

Near SLL Minimization

Minimization of the first side lobe nearest to the main lobe (near side lobe) by the optimization of antenna positions is illustrated in Example 4. The fitness function mentioned in (10) is used for this purpose [13]. Consider

$$\text{Fitness} = \min\left(\alpha_1 \max\left\{20\log\left|\text{AF}\left(\theta_{\text{AS}}\right)\right|\right\} + \alpha_2 \max\left\{20\log\left|\text{AF}\left(\theta_{\text{NS}}\right)\right|\right\}\right), \tag{10}$$

where $\theta AS = [0°, 76°]$ and $[104°, 180°]$ and $\theta NS = [69°, 76°]$ and $[104°, 111°]$. θNS specifies the near side lobe region and its weight is α_2. The near side lobe region can be controlled by using the fitness function given by (10) and the amount of its reduction is controlled by the weights α_1 and α_2. For illustration of near SLL suppression, design of Example 4 is presented in which a 10-element linear array is optimized using (10) and weights $\alpha_1 = 1$ and $\alpha_2 = 2$. The optimized antenna positions are given in Table 8. It is seen that the length of array optimized by the proposed approach is the same as that of PSO [13] optimized array. The array pattern is shown in Figure 6.

Table 8: Optimized positions of the positive half of the 10-element array of Example 4

Method	**Optimized element positions**				
PSO [13]	0.1685λ	0.5461λ	0.9364λ	1.5107λ	2.2500λ
Proposed	0.1792λ	0.5410λ	0.9354λ	1.5096λ	2.2500λ

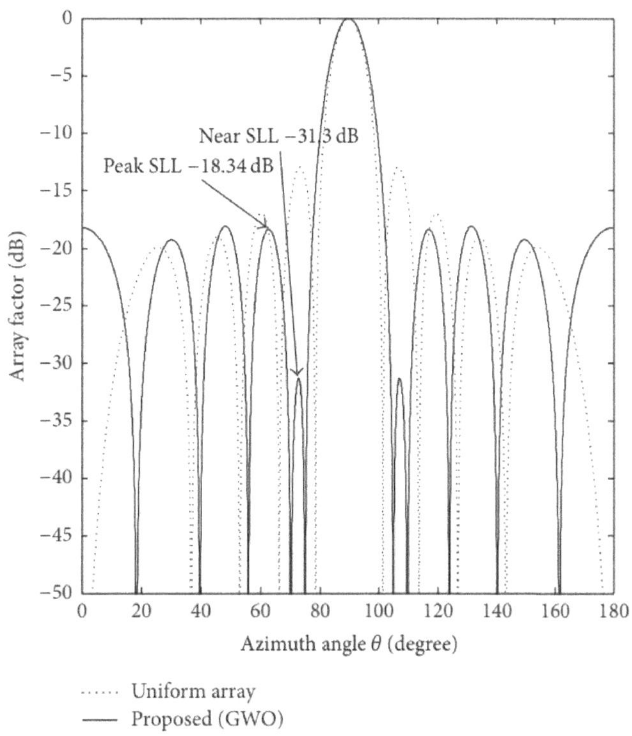

Figure 6: Array pattern for Example 4.

The near SLL obtained by GWO is −31.3 dB which is 18.33 dB lower as compared to uniform array. The peak SLL is found to be −18.34 dB. The benchmarking results are shown in Table 9.

Table 9: Near SLL and peak SLL for Example 4

	Near SLL (dB)	Peak SLL (dB)
Uniform array	−12.97	−12.97
PSO [13]	−31	−18.3
Proposed	−31.3	−18.34

Antenna Current Optimization

In this section, Examples 5–8 are presented for optimization of the antenna current amplitudes (I_n) for optimal pattern synthesis. Handling phase involves a huge complexity as it is not always convenient to implement fractional phase in practice. Hence, uniform phase excitation is assumed; that is, $\psi_n = 0$. The element spacing of $\lambda/2$ is considered which is the same as that for uniform array.

Peak SLL Minimization

This section presents Examples 5 and 6 which illustrate the application of GWO for optimization of antenna current amplitudes in order to minimize the peak SLL in the regions, θ = [0° , 76°] and θ = [104° , 180°]. The fitness function given by (8) is used for both the examples. The optimized current amplitudes obtained by using the proposed method are shown in Table 10. The array patterns for Examples 5 and 6 are illustrated in Figures 7 and 8, respectively. The current amplitude distribution along the array elements is shown in Figures 9(a) and 9(b) for Examples 5 and 6, respectively. The current amplitude is maximum at the center of the array and decreases from the center towards the edge elements. Such an amplitude distribution makes it feasible to use power dividers.

Table 10: Optimized current amplitudes for peak SLL minimization

Linear array type	Method	Optimized current amplitudes
10 elements (Example 5)	PSO [13]	1.0000, 0.9010, 0.7255, 0.5120, 0.4088
	Proposed	1.0000, 0.8962, 0.6963, 0.4935, 0.2964
16 elements (Example 6)	PSO [13]	1.0000, 0.9521, 0.8605, 0.7372, 0.5940, 0.4465, 0.3079, 0.2724
	Proposed	1.0000, 0.9339, 0.8127, 0.6533, 0.4855, 0.3195, 0.1928, 0.1002

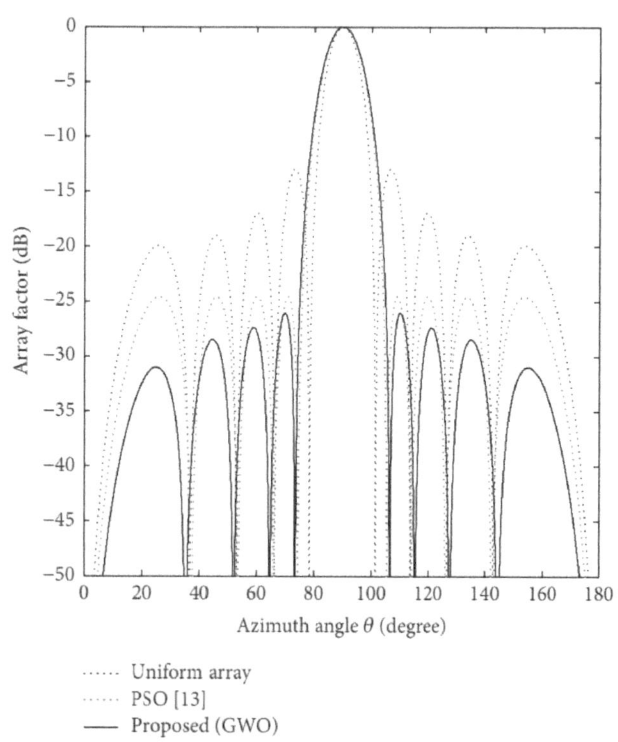

Figure 7: Array pattern for Example 5.

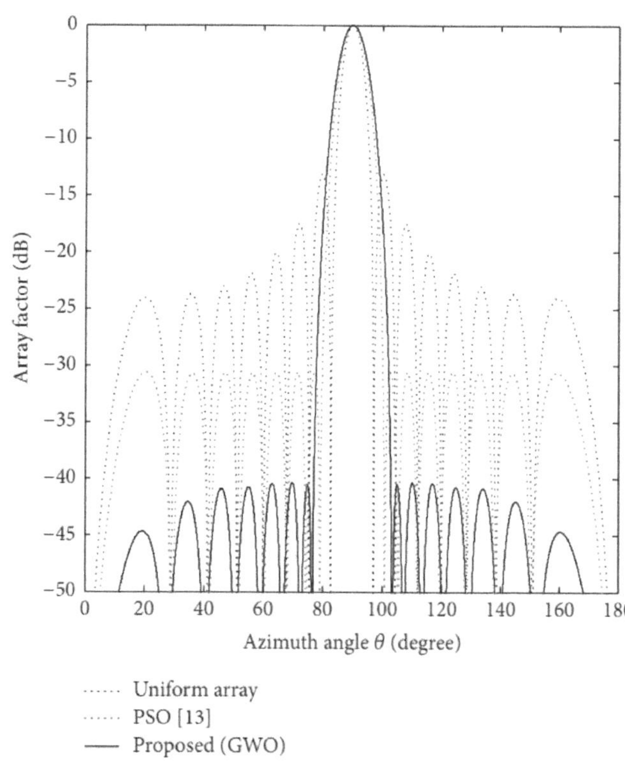

Figure 8: Array pattern for Example 6.

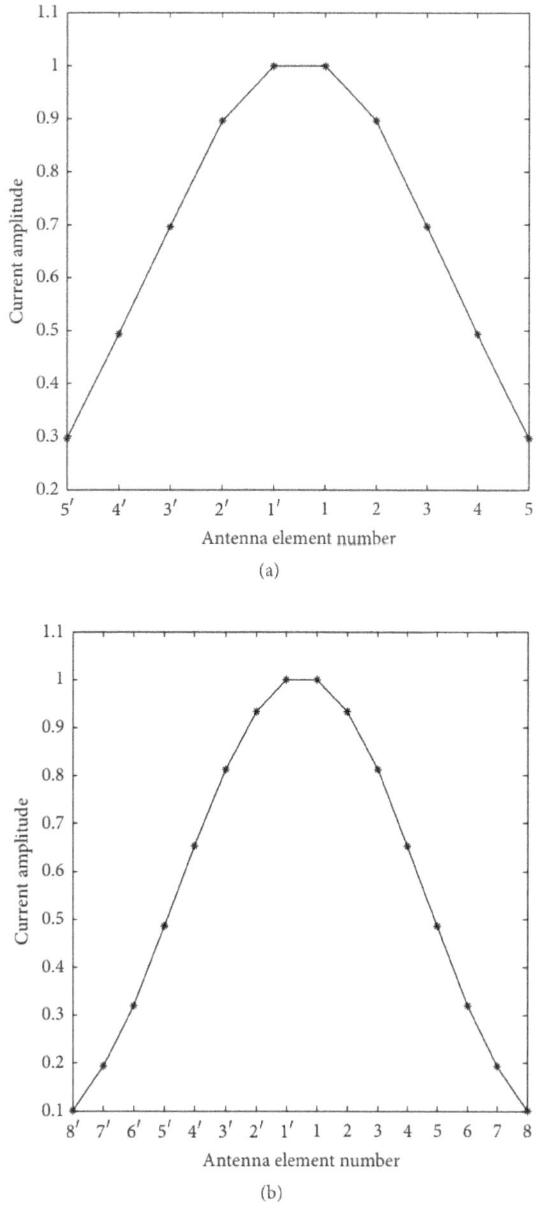

Figure 9: Current distribution for (a) Example 5 and (b) Example 6.

For Example 5, the proposed method (GWO) gives peak SLL of −26.05 dB which is 13.08 dB lower as compared to uniform array. The peak SLL has been lowered from −24.62 dB to −26.05 dB (by 1.43 dB) as compared to PSO [13] optimized array as depicted in Figure 7 and summarized in Table 11.

Table 11: Benchmarking results for peak SLL minimization

	Peak SLL (dB)	
	Example 5	Example 6
Uniform array (without optimization)	−12.97	−13.15
PSO [13]	−24.62	−30.7
Proposed	−26.05	−40.5

The peak SLL obtained by the proposed method for Example 6 is −40.5 dB which is 27.35 dB lower as compared to uniform array. The peak SLL has been lowered from −30.7 dB to −40.5 dB as compared to PSO [13] optimized array. These results are depicted in Figure 8 and summarized in Table 11.

SLL Minimization along with Null Placement

To illustrate this case, Example 7 presents the synthesis of 20-element linear array for achieving SLL minimization in the regions θ = [0°, 82°] and θ = [98°, 180°] along with null placement at θ = 64° , 76°, 104°, and 116° . The fitness function used for this example is given by (9). The optimized current amplitudes are given in Table 12 and the array pattern is shown in Figure 10(a). The current amplitude is maximum at the center of the array and decreases from the center towards the edge elements as shown in Figure 10(b).

Table 12: Optimized current amplitudes of the positive half of the 20-element array of Example 7

Method	Optimized current amplitudes									
BBO [19]	1.0000	0.9747	0.9264	0.8575	0.7022	0.6242	0.4799	0.3607	0.2369	0.1234
Proposed	1.0000	0.9794	0.9254	0.8126	0.7008	0.6000	0.4594	0.3326	0.2133	0.1167

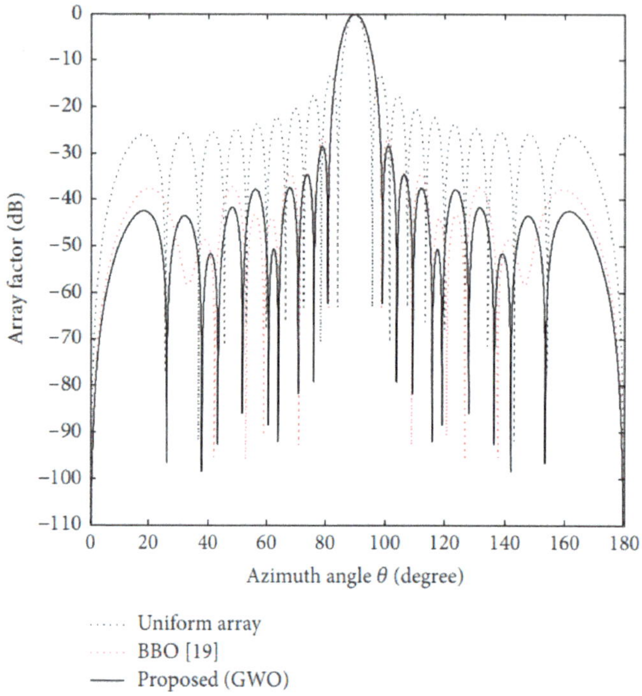

(a)

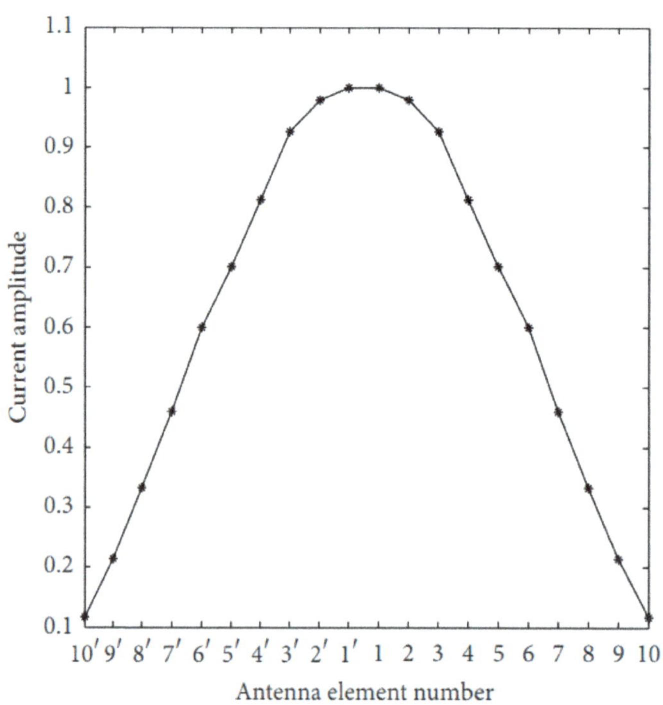

(b)

Figure 10: Example 7: (a) array pattern and (b) current distribution.

The proposed method enables the placement of nulls (as deep as −92.02 dB) in the desired directions. The comparative analysis of peak SLL and null depths obtained by using the proposed method (GWO) and those obtained by uniform array and by using BBO [19] is shown in Table 13.

Table 13: Peak SLL and null depths for Example 7

	Peak SLL (dB)	Null depth (dB)			
		64°	76°	104°	116°
Uniform array	−13.19	−22.7	−17.7	−17.7	−22.7
BBO [19]	−26.71	−79.66	−74.2	−74.2	−79.66
Proposed	−28.44	−92.02	−79.12	−79.12	−92.02

It is seen that the peak SLL obtained by using the proposed approach is −28.44 dB which is 15.25 dB lower than uniform array and 1.73 dB lower than BBO [19] array. Also, the null depths obtained by using the proposed method are much higher than that obtained by using BBO array. The benchmarking results are summarized in Table 13.

Near SLL Minimization

Minimization of the near side lobe of a 10-element array by the optimization of antenna current amplitudes is illustrated in Example 8. The fitness function given by (10) is used for this purpose with θAS = [0°, 76°] and [104°, 180°], θNS = [69° , 76°] and [104°, 111°], and weights α_1 = 1 and α_2 = 2. The optimized antenna current amplitudes are given in Table 14 and the array pattern is shown in Figure 11.

Table 14: Optimized current amplitudes of the positive half of the 10-element array of Example 8

Method	Optimized current amplitudes				
BBO [19]	1.0000	0.8526	0.6586	0.4601	0.5101
Proposed	1.000	0.8169	0.6383	0.4505	0.4505

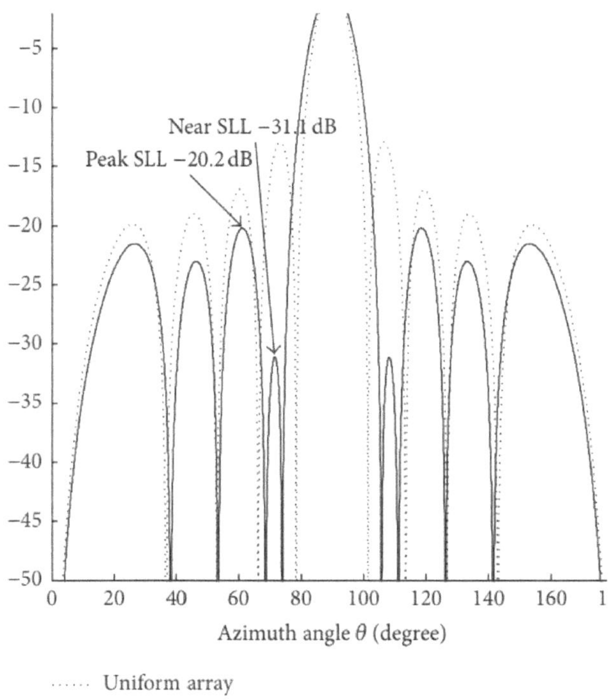

Figure 11: Array pattern for Example 8.

The near SLL obtained by GWO is −31.1 dB which is 18.13 dB lower as compared to uniform array and 2.51 dB lower than BBO [19] optimized array. The peak SLL is −20.2 dB which shows that GWO enables reduction of the near side lobe while simultaneously controlling the other side lobes. The benchmarking results are shown in Table 15.

Table 15: Near SLL and peak SLL for Example 8

Method	Near SLL (dB)	Peak SLL (dB)
Uniform array	−12.97	−12.97
BBO [19]	−28.59	−20.1
Proposed	−31.1	−20.2

Computational Time

The computational time required for GWO to converge to the optimum solution as compared to other state-of-the-art algorithms is shown in Table 16. All the computations are performed using MATLAB on a PC operating at 3 GHz with 2 GB of RAM.

Table 16: Computational time for GWO as compared to other optimization algorithms

Linear antenna array	Computational time for each run (in seconds)		
	PSO [15]	CSO [15]	GWO
10 elements	17	101	26
28 elements	35	246	44

It is seen that GWO converges to the optimum solution much faster than CSO. This is because CSO carries out two processes independently for global (seeking mode) and local (tracing mode) search in each generation which makes it computationally time consuming. However, although GWO has very few parameters to tune and yields improved performance, it takes more time to converge on to the optimum solution as compared to PSO. In PSO, all the particles move through global search and end with local search in the last generation. The momentum effects on particle movement in PSO generally allow faster convergence. On the other hand, in GWO, the adaptive values of parameters $\vec{a}$ and $\vec{A}$ allow a smooth transition between exploration and exploitation. Half of the iterations are devoted to exploration (global search) and the other half are dedicated to exploitation (local search). Furthermore, parameter $\vec{C}$ is used to avoid local optima stagnation, especially in the final iterations. Thus, GWO is more likely to escape locally optimal points and yield a global optimum solution, thus making it more computationally time consuming than PSO as depicted in Table 16.

CONCLUSION

This paper has introduced the GWO as an optimization algorithm to the electromagnetics and antenna community. The application of GWO for optimal pattern synthesis of linear antenna arrays was demonstrated for the first time. GWO was applied to obtain the optimized antenna positions and current amplitudes in order to achieve the desired array pattern with minimum SLL along with null placement in specified directions. Suppression of the first side lobe next to the main lobe (near side lobe) while simultaneously controlling the other side lobes was also demonstrated. The obtained results have been compared with conventional array (nonoptimized) and with arrays optimized using other nature-inspired evolutionary algorithms such as PSO, ACO, CSO, and BBO. The results suggest that optimization of linear antenna arrays using GWO provides considerable enhancements compared to the uniform array and the synthesis obtained from other optimization techniques. This demonstrates the suitability of GWO to be used by the antenna and electromagnetics community for antenna array optimization.

REFERENCES

1. C. A. Balanis, Antenna Theory and Design, John Wiley & Sons, New York, NY, USA, 1997.
2. R. Vescovo, "Null synthesis by phase control for antenna arrays," Electronics Letters, vol. 36, no. 3, pp. 198–199, 2000.
3. K. Güney and A. Akdagli, "Null steering of linear antenna arrays using a modified tabu search algorithm," Progress in Electromagnetics Research, vol. 33, pp. 167–182, 2001.
4. M. H. Er, "Linear antenna array pattern synthesis with prescribed broad nulls," IEEE Transactions on Antennas and Propagation, vol. 38, no. 9, pp. 1496–1498, 1990.
5. M. A. Panduro, D. H. Covarrubias, C. A. Brizuela, and F. R. Marante, "A multi-objective approach in the linear antenna array design," AEU-International Journal of Electronics and Communications, vol. 59, no. 4, pp. 205–212, 2005.
6. D. Bianchi, S. Genovesi, and A. Monorchio, "Design of linear arrays by employing randomly-overlapped subarrays," in Proceedings of the XXXIth URSI General Assembly and Scientific Symposium (URSI GASS ‹14), pp. 1–4, IEEE, Beijing, China, August 2014.
7. Z. Zhang, T. Li, F. Yuan, and L. Yin, "Synthesis of linear antenna array using genetic algorithm to control side lobe level," in Computer Engineering and Networking, pp. 39–46, Springer, 2014.
8. B. Goswami and D. Mandal, "A genetic algorithm for the level control of nulls and side lobes in linear antenna arrays," Journal of King Saud University—Computer and Information Sciences, vol. 25, no. 2, pp. 117–126, 2013.
9. V. Murino, A. Trucco, and C. S. Regazzoni, "Synthesis of unequally spaced arrays by simulated annealing," IEEE Transactions on Signal Processing, vol. 44, no. 1, pp. 119–122, 1996.
10. M. M. Khodier and C. G. Christodoulou, "Linear array geometry synthesis with minimum sidelobe level and null control using particle swarm optimization," IEEE Transactions on Antennas and Propagation, vol. 53, no. 8, pp. 2674–2679, 2005.
11. Y. Li, F. Yang, J. Ouyang, Z. Nie, and H. Zhou, "Synthesis of nonuniform array antennas using particle swarm optimization," Electromagnetics, vol. 30, no. 3, pp. 237–245, 2010.
12. N. Jin and Y. Rahmat-Samii, "Advances in particle swarm optimization for antenna designs: real-number, binary, single-objective and multiobjective implementations," IEEE Transactions on Antennas and Propagation, vol. 55, no. 3 I, pp. 556–567, 2007.
13. M. Khodier and M. Al-Aqeel, "Linear and circular array optimization: a study using particle swarm intelligence," Progress in Electromagnetics Research B, no. 15, pp. 347–373, 2009. ·
14. E. Rajo-Iglesias and Ó. Quevedo-Teruel, "Linear array synthesis using an ant-colony-optimization-based algorithm," IEEE Antennas and Propagation Magazine, vol. 49, no. 2, pp. 70–79, 2007.
15. L. Pappula and D. Ghosh, "Linear antenna array synthesis using cat swarm optimization," AEU—International Journal of Electronics and Communications, vol. 68, no. 6, pp. 540–549, 2014.
16. S. Pal, A. Basak, S. Das, and A. Abraham, "Linear antenna array synthesis with invasive weed optimization algorithm," in Proceedings of the International Conference on Soft Computing and Pattern Recognition (SoCPaR ‹09), pp. 161–166, Malacca, Malaysia, December 2009.
17. S. Karimkashi and A. A. Kishk, "Invasive weed optimization and its features in electromagnetics," IEEE Transactions on Antennas and Propagation, vol. 58, no. 4, pp. 1269–1278, 2010.
18. W.-C. Weng, F. Yang, and A. Z. Elsherbeni, "Linear antenna array synthesis using Taguchi›s method: a novel optimization technique in electromagnetics," IEEE Transactions on Antennas and Propagation, vol. 55, no. 3 I, pp. 723–730, 2007.
19. U. Singh, H. Kumar, and T. S. Kamal, "Linear array synthesis using biogeography based optimization,"Progress in Electromagnetics Research M, vol. 11, pp. 25–36, 2010.
20. B. Basu and G. K. Mahanti, "Fire fly and artificial bees colony algorithm for synthesis of scanned and broad-side linear array antenna," Progress In Electromagnetics Research B, no. 32, pp. 169–190, 2011.
21. M. A. Zaman and M. Abdul Matin, "Nonuniformly spaced linear antenna array design using firefly algorithm," International Journal of Microwave Science and Technology, vol. 2012, Article ID 256759, 8 pages, 2012.
22. M. Khodier, "Optimisation of antenna arrays using the cuckoo search algorithm," IET Microwaves, Antennas & Propagation, vol. 7, no. 6, pp. 458–464, 2013. ·
23. K. Guney, A. Durmus, and S. Basbug, "Antenna array synthesis and failure correction using differential search algorithm," International Journal of Antennas and Propagation, vol. 2014, Article ID 276754, 8 pages, 2014.
24. K. Guney and A. Durmus, "Pattern nulling of linear antenna arrays using backtracking search optimization algorithm," International Journal of Antennas and Propagation, vol. 2015, Article ID 713080, 10 pages, 2015.

25. S. Mirjalili, S. M. Mirjalili, and A. Lewis, "Grey wolf optimizer," Advances in Engineering Software, vol. 69, pp. 46–61, 2014.

26. A. A. M. El-Gaafary, Y. S. Mohamed, A. M. Hemeida, and A.-A. A. Mohamed, "Grey wolf optimization for multi input multi output system," Universal Journal of Communications and Network, vol. 3, no. 1, pp. 1–6, 2015.

27. L. I. Wong, M. Sulaiman, M. Mohamed, and M. Hong, "Grey Wolf Optimizer for solving economic dispatch problems," in Proceedings of the IEEE International Conference on Power and Energy (PECon ‹14), pp. 150–154, Kuching, Malaysia, December 20104.

28. H. M. Song, M. H. Sulaiman, and M. R. Mohamed, "An application of Grey Wolf optimizer for solving combined economic emission dispatch problems," International Review on Modelling and Simulations, vol. 7, no. 5, pp. 838–844, 2014.

29. N. Muangkote, K. Sunat, and S. Chiewchanwattana, "An improved grey wolf optimizer for training q-Gaussian Radial Basis Functional-link nets," in Proceedings of the International Computer Science and Engineering Conference (ICSEC ‹14), pp. 209–214, IEEE, Khon Kaen, Thailand, August 2014.

30. C. Muro, R. Escobedo, L. Spector, and R. P. Coppinger, "Wolf-pack (Canis lupus) hunting strategies emerge from simple rules in computational simulations," Behavioural Processes, vol. 88, no. 3, pp. 192–197, 2011.

CHAPTER 9 STUDY ON THE ROLLOVER CHARACTERISTIC OF IN-WHEEL-MOTOR-DRIVEN ELECTRIC VEHICLES CONSIDERING ROAD AND ELECTROMAGNETIC EXCITATION

Di Tan, Haitao Wang, and Qiang Wang

INTRODUCTION

As one of the most important auto safety problems, vehicle rollover accidents have gained more attention in recent years [1]. Many researchers have conducted significant research on vehicle rollover with a traditional chassis structure [2–4]. However, most of the research is focused on the lateral dynamics and the classic three-degree-of-freedom model is adopted as the basis, without considering the effect of road excitation [5–7]. Taking into account the effect of the road excitation, a seven-degree-of-freedom combined model was developed based on a vehicle with a traditional chassis in [8]. Vehicle rollover was analyzed under certain conditions, and the results indicated that the combined model could reflect the vehicle rollover features more accurately compared with the traditional three-degree-of-freedom model. However, the road input correlation of four wheels was ignored in the study.

For the in-wheel motors (IWMs) electric vehicles (EVs), driving motors are integrated in the wheels directly [9,10]. The tyre runout under different road surface roughness (RSR) excitations and uneven load will cause deformation in the IWM magnet gap. The magnet gap deformation (MGD) can lead to an unbalanced electromagnetic force (EMF), which will influence the vehicle rollover stability [11]. At present, research on IWM driven systems is still in its infancy. Relatively little research has been conducted on topics such as the EMF excitation caused by the RSR and their combined effect on the vehicle rollover. In this paper, the rollover prediction model of an IWM driven EV without a speed reducer is given, including a verified 16-degree-of-freedom coupling dynamic model with the nonlinear tyre model. Then the mathematical model of the excitation source is established which includes the RSR model that considers the input correlation of four wheels, the IWM EMF model that calculates the EMF, and the steering input model. Based on these models, comparative analysis of the vehicle rollover with or without EMF is carried out under the condition of the Fixed Timing Fishhook (FTF) steering and grade B RSR excitation to understand the influence of EMF on the vehicle rollover. In addition, further analysis is conducted to understand the effect of road conditions on the vehicle rollover, such as the road level, vehicle speed, and road adhesion coefficient.

Citation: Di Tan, Haitao Wang, and Qiang Wang, "Study on the Rollover Characteristic of In-Wheel-Motor-Driven Electric Vehicles Considering Road and Electromagnetic Excitation," Shock and Vibration, vol. 2016, Article ID 2450573, 13 pages, 2016. doi:10.1155/2016/2450573

ROLLOVER MODEL OF IWM DRIVEN EV

Rollover Prediction Dynamics Model

16-Degree-of-Freedom Rollover Dynamic Model

A rear IWM driven EV without speed reducer is used for this study. The basic structure of the IWM driven system is shown in Figure 1 [12].

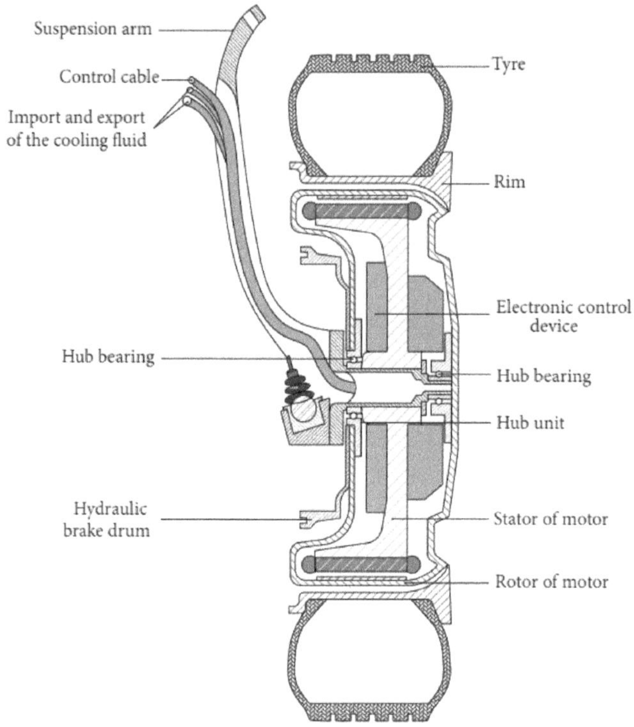

Figure 1: Basic structure of the IWM driven system.

Based on the structure in Figure 1, a dynamic model with 16 degrees of freedom is given, which can reflect the vehicle dynamic characteristics in different directions, as shown in Figure 2. The 16 degrees of freedom include three degrees of freedom for the longitudinal, lateral, and yawing movement of the vehicle, three degrees of freedom for the vertical, pitch, and roll movement of the vehicle body, four degrees of freedom for the rotational movement of the wheels, and six degrees of freedom for the vertical movement of the unsprung mass. In this paper, the model is used to study the rollover characteristics of IWM driven EVs and is termed as a rollover prediction dynamic model.

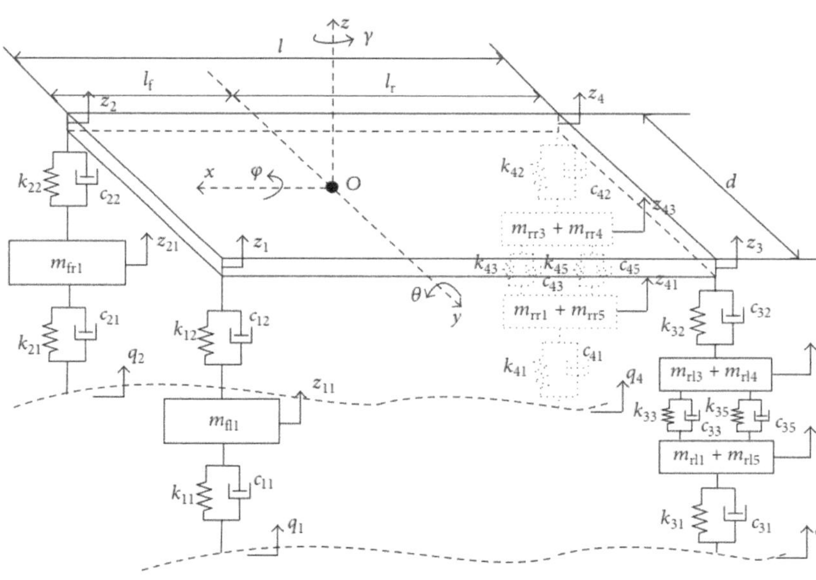

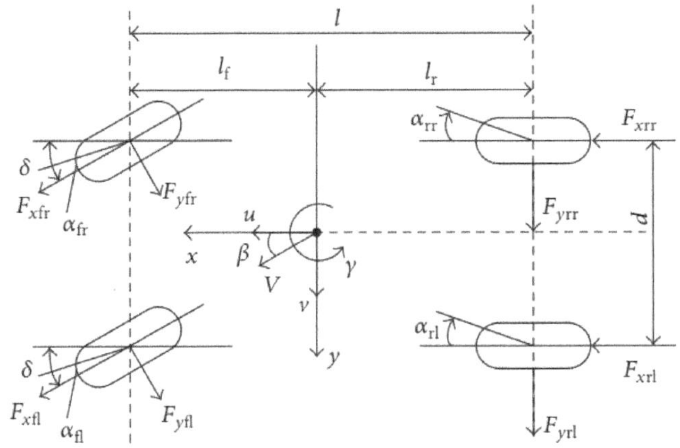

(b) Lateral dynamic model

Figure 2: Rollover prediction dynamic model.

The dynamic equation of the rollover model shown in Figure 2 is given by

$$m\left(\dot{u}-v\gamma+\dot{z}_s\dot{\theta}\right)-m_sh_s\ddot{\theta}-\left(F_{xfl}+F_{xfr}\right)\cos\delta+\left(F_{yfl}+F_{yfr}\right)\sin\delta-F_{xrl}-F_{xrr}=0,$$

$$m\left(\dot{v}+u\gamma-\dot{z}_s\dot{\varphi}\right)+m_sh_s\ddot{\varphi}-\left(F_{xfl}+F_{xfr}\right)\sin\delta-\left(F_{yfl}+F_{yfr}\right)\cos\delta-F_{yrl}-F_{yrr}=0,$$

$$I_z\dot{\gamma}-\left(F_{xfl}\sin\delta+F_{xfr}\sin\delta\right)l_f-\left(F_{yfl}\cos\delta+F_{yfr}\cos\delta\right)l_f+\left(F_{yrl}+F_{yrr}\right)l_r-\left(F_{yfl}\sin\delta-F_{yfr}\sin\delta\right)\frac{d}{2}-\left(F_{xfr}\cos\delta-F_{xfl}\cos\delta+F_{xrr}-F_{xrl}\right)\frac{d}{2}+I_{xz}\ddot{\varphi}=0,$$

$$I_x\ddot{\varphi}-m_sgh_s\varphi+m_sh_sa_y+I_{xz}\dot{\gamma}+\frac{d}{2}\left[k_{12}\left(z_1-z_{11}\right)+c_{12}\left(\dot{z}_1-\dot{z}_{11}\right)+k_{32}\left(z_3-z_{33}\right)+c_{32}\left(\dot{z}_3-\dot{z}_{33}\right)-k_{22}\left(z_2-z_{21}\right)-c_{22}\left(\dot{z}_2-\dot{z}_{21}\right)-k_{42}\left(z_4-z_{43}\right)-c_{42}\left(\dot{z}_4-\dot{z}_{43}\right)\right]=0,$$

$$I_y\ddot{\theta}+m_sa_xh_s-l_f\left[k_{12}\left(z_1-z_{11}\right)+c_{12}\left(\dot{z}_1-\dot{z}_{11}\right)+k_{22}\left(z_2-z_{21}\right)+c_{22}\left(\dot{z}_2-\dot{z}_{21}\right)\right]+l_r\left[k_{32}\left(z_3-z_{33}\right)+c_{32}\left(\dot{z}_3-\dot{z}_{33}\right)+k_{42}\left(z_4-z_{43}\right)+c_{42}\left(\dot{z}_4-\dot{z}_{43}\right)\right]=0,$$

$$m_s\left(\ddot{z}_s+v\dot{\varphi}-u\dot{\theta}\right)+k_{12}\left(z_1-z_{11}\right)+c_{12}\left(\dot{z}_1-\dot{z}_{11}\right)+k_{22}\left(z_2-z_{21}\right)+c_{22}\left(\dot{z}_2-\dot{z}_{21}\right)+k_{32}\left(z_3-z_{33}\right)+c_{32}\left(\dot{z}_3-\dot{z}_{33}\right)+k_{42}\left(z_4-z_{43}\right)+c_{42}\left(\dot{z}_4-\dot{z}_{43}\right)=0,$$

$$m_{fl1}\ddot{z}_{11}+k_{11}\left(z_{11}-q_1\right)+c_{11}\left(\dot{z}_{11}-\dot{q}_1\right)+k_{12}\left(z_{11}-z_1\right)+c_{12}\left(\dot{z}_{11}-\dot{z}_1\right)=0,$$

$$m_{fr1}\ddot{z}_{21}+k_{21}\left(z_{21}-q_2\right)+c_{21}\left(\dot{z}_{21}-\dot{q}_2\right)+k_{22}\left(z_{21}-z_2\right)+c_{22}\left(\dot{z}_{21}-\dot{z}_2\right)=0,$$

$$\left(m_{rl1}+m_{rl5}\right)\ddot{z}_{31}+k_{31}\left(z_{31}-q_3\right)+c_{31}\left(\dot{z}_{31}-\dot{q}_3\right)+k_{33}\left(z_{31}-z_{33}\right)+c_{33}\left(\dot{z}_{31}-\dot{z}_{33}\right)+k_{35}\left(z_{31}-z_{33}\right)+c_{35}\left(\dot{z}_{31}-\dot{z}_{33}\right)=0,$$

$$\left(m_{rl3}+m_{rl4}\right)\ddot{z}_{33}+k_{32}\left(z_{33}-z_3\right)+c_{32}\left(\dot{z}_{31}-\dot{z}_3\right)+k_{33}\left(z_{33}-z_{31}\right)+c_{33}\left(\dot{z}_{33}-\dot{z}_{31}\right)+k_{35}\left(z_{33}-z_{31}\right)+c_{35}\left(\dot{z}_{33}-\dot{z}_{31}\right)=0,$$

$$\left(m_{rr1}+m_{rr5}\right)\ddot{z}_{41}+k_{41}\left(z_{41}-q_4\right)+c_{41}\left(\dot{z}_{41}-\dot{q}_4\right)+k_{43}\left(z_{41}-z_{43}\right)+c_{43}\left(\dot{z}_{41}-\dot{z}_{43}\right)+k_{45}\left(z_{41}-z_{43}\right)+c_{45}\left(\dot{z}_{41}-\dot{z}_{43}\right)=0,$$

$$\left(m_{rr3}+m_{rr4}\right)\ddot{z}_{43}+k_{42}\left(z_{43}-z_4\right)+c_{42}\left(\dot{z}_{43}-\dot{z}_4\right)+k_{43}\left(z_{43}-z_{41}\right)+c_{43}\left(\dot{z}_{43}-\dot{z}_{41}\right)+k_{45}\left(z_{43}-z_{41}\right)+c_{45}\left(\dot{z}_{43}-\dot{z}_{41}\right)=0,$$

$$I_{w_i}\dot{w}_i-T_{mi}+F_{xi}r_t=0. \tag{1}$$

For the research object in this paper is a rear IWM driven EV, the driving torque of the front wheels is 0; that is, $T_{mfl}=T_{mfr}=0$.

When θ and φ are small, the vertical displacement of the join point of vehicle body and the four suspensions can be deduced:

$$z_1=z_s-l_f\theta+\frac{1}{2}d\varphi,$$
$$z_2=z_s-l_f\theta-\frac{1}{2}d\varphi,$$
$$z_3=z_s+l_r\theta+\frac{1}{2}d\varphi,$$
$$z_4=z_s+l_r\theta-\frac{1}{2}d\varphi. \tag{2}$$

The validity of the above dynamic equation has been verified in [12].

Tyre Model

The 16-degree-of-freedom rollover prediction model is a coupling dynamic model which considers the influence of vertical, lateral, and longitudinal dynamics on the vehicle rollover. Therefore, the Magic Formula Tyre Model proposed by Professor Pacejka is used to obtain the nonlinear tyre force accurately. The specific expression of the Magic Formula Tyre Model is [13]

$$y = D\sin\{C\arctan[Bx - E(Bx - \arctan Bx)]\}. \quad (3)$$

Input Model of Excitation Source

RSR Model of Four Wheels

For the rollover prediction model, four random RSR input signals are required for each wheel. Due to the mutual interference of the four-wheel RSR in the space domain, the RSR correlation between the four wheels is considered in the model. The state equation of the four-wheel RSR input model can be expressed as follows [14,15]:

$$\dot{Z}(t) = AZ(t) + B_0 w(t), \quad (4)$$

where

$$Z(t) = [q_1(t)\ \ q_2(t)\ \ q_3(t)\ \ q_4(t)\ \ x_1(t)\ \ x_2(t)],$$

$$\dot{Z}(t) = [\dot{q}_1(t)\ \ \dot{q}_2(t)\ \ \dot{q}_3(t)\ \ \dot{q}_4(t)\ \ \dot{x}_1(t)\ \ \dot{x}_2(t)],$$

$$A = \begin{bmatrix} -2\pi n_c u & 0 & 0 & 0 & 0 & 0 \\ \frac{u}{B}e^{-2\pi n_c d} & -\frac{u}{d} & 0 & 0 & 0 & 0 \\ -12\frac{u}{l} - 2\pi n_c u & 0 & 0 & 0 & 0 & 1 \\ \frac{u}{B}e^{-2\pi n_c d} & -\left(12\frac{u}{l} + \frac{u}{d}\right) & 0 & 0 & 0 & 1 \\ -12\frac{u}{l} & 0 & 0 & 0 & 0 & 1 \\ 72\frac{u^2}{l^2} & 0 & 0 & 0 & -12\frac{u^2}{l^2} & -6\frac{u}{l} \end{bmatrix},$$

$$B_0 = [2\pi n_0\sqrt{G_0 u}\ \ 0\ \ 2\pi n_0\sqrt{G_0 u}\ \ 0\ \ 0\ \ 0]. \quad (5)$$

Based on [16], the value of $n0$ is 0.1 m^{-1}, and nc is 0.01 m^{-1} in this paper.

EMF Model

In this paper, the IWM is the permanent magnet synchronous motor. Ignoring the saturation effect and the core reluctance of the IWM powered by the sine-wave current, the flux density can be calculated as [17, 18]

$$B(\theta_t, t) = F(\theta_t, t)\Lambda(\theta_t, t). \quad (6)$$

(1) Magnet Gap Permeability Λ. For the ideal operating condition of the IWM, the geometric center of the rotor magnetic pole should coincide with that of the stator core. Additionally, the magnet gap length should be distributed along the circumference evenly. However, in reality the tyre runout under different RSR excitations and uneven load will cause the deformation in the IWM magnet gap, as shown in Figure 3.

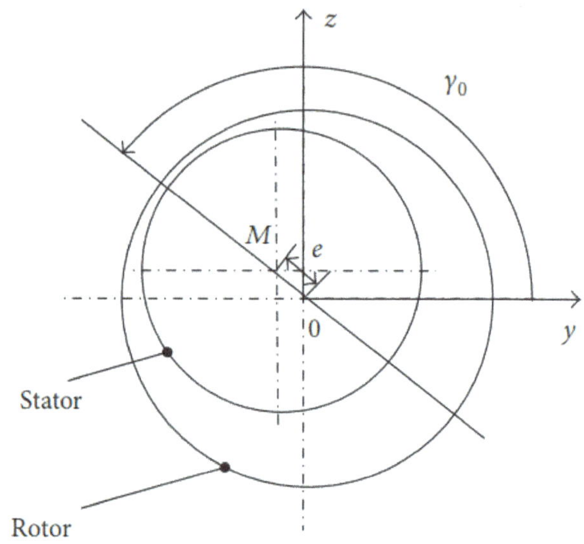

Figure 3: MGD of the IWM.

The length of the even or uneven magnet gap of IWM as shown in Figure 3 at time t can be expressed as

$$g_e(\theta_t, z, t) = g_0(1 - \varepsilon(z)\cos(\theta_t - \omega_t t - \gamma_0)),$$

$$\varepsilon(z) = \frac{e}{g_0}. \quad (7)$$

The Fourier series of the magnet gap permeance is calculated by

$$\Lambda(\theta_t, z, t) = \frac{\mu_0}{g_e(\theta_t, z, t)} = \sum_{\lambda=0}^{\infty}\Lambda_\lambda\cos[\lambda(\theta_t - \omega_t t - \gamma_0)], \quad (8)$$

where

$$\Lambda_\lambda = \begin{cases} \frac{\mu_0}{g_0}\Lambda'_0 = \frac{\mu_0}{g_0}\frac{1}{\sqrt{1-\varepsilon^2}} & \lambda = 0 \\ \frac{\mu_0}{g_0}\Lambda'_\lambda = 2\frac{\mu_0}{g_0}\frac{1}{\sqrt{1-\varepsilon^2}}\left(\frac{1-\sqrt{1-\varepsilon^2}}{\varepsilon}\right)^\lambda & \lambda \geq 1. \end{cases} \quad (9)$$

(2) Magnetomotive Force of the IWM. The magnetomotive force of the IWM is synthesized by the magnetomotive force of the permanent magnet and the stator winding, which can be expressed as

$$F(\theta,t) = F_{s0}(\theta_t,t) + F_{sv}(\theta_t,t) + F_r(\theta_t,t)$$
$$= F_0\cos(\omega_t t - p\theta_t + \varphi_0)$$
$$+ \sum_v F_v \cos(\omega_t t - v\theta_t + \varphi_v)$$
$$+ \sum_\mu F_\mu \cos(\mu\omega_t t - \mu p\theta_t). \quad (10)$$

(3) EMF Model. Substituting (8) and (10) into (6), the flux density is derived as

$$B(\theta_t,t) = \left[F_0\cos(\omega_t t - p\theta_t + \varphi_0) + \sum_v F_v \cos(\omega_t t - v\theta_t + \varphi_v) + \sum_\mu F_\mu \cos(\mu\omega_t t - \mu p\theta_t)\right] \cdot \left\{\frac{\mu_0}{g_0}\left[1 + \sum_{\lambda=1}^{\infty}\frac{\varepsilon^\lambda}{2^{\lambda-1}}\cos(\theta_t - \omega_t t - \gamma_0)\right]\right\}. \quad (11)$$

Based on the Maxwell magnet stress tensor theory, the radial and tangential magnetic force density in a 2D magnetic field under polar coordinates can be calculated by [19, 20]

$$f_r = \frac{1}{2\mu_0}\left(B_r^2 - B_t^2\right),$$
$$f_t = \frac{1}{\mu_0}B_r B_t. \quad (12)$$

Transforming the magnetic force from polar coordinate to Cartesian coordinate, the force acting on the stator and rotor in the z direction can be deduced by integrating the IWM axial gap length l_a as

$$F_z = l_a \int_0^{2\pi}(f_r\sin\alpha - f_t\cos\alpha)\cdot r d\alpha = \frac{rl_a}{2\mu_0} \cdot \int_0^{2\pi}\left(\left(B_t(s,t)^2 - B_r(s,t)^2\right)\sin\alpha - 2B_r(s,t)B_t(s,t)\cos\alpha\right)d\alpha. \quad (13)$$

Steering Input Model

The IWM driven EV used in this study is a front-wheel steering vehicle. The influence of the steering system on the wheel angle is ignored in the process of steering and the steering angle ratio is 15.

ROLLOVER CHARACTERISTICS OF IWM DRIVEN EV

Vehicle Parameters

The main vehicle parameters are listed in Table 1.

Table 1: Vehicle parameters

Symbol	Unit	Value
m	kg	1450
m_s	kg	1160
I_x	kg·m^2	606
I_y	kg·m^2	1800
I_z	kg·m^2	2300
I_{xz}	kg·m^2	45
I_{wi}	kg·m^2	2.5
k_{j1} $(j = 1,2,3,4)$	N/m	200000
c_{j1} $(j = 1,2,3,4)$	N/(m/s)	100
k_{j2} $(j = 1,2)$	N/m	35000
c_{j2} $(j = 1,2)$	N/(m/s)	1500
k_{j2} $(j = 3,4)$	N/m	40000
c_{j2} $(j = 3,4)$	N/(m/s)	1800
$k_{33}, k_{43}, k_{35}, k_{45}$	N/m	8000000
l_f	m	1.2
l_r	m	1.4
d	m	1.44
h_s	m	0.488
r_t	m	0.3

The IWM used in the vehicle is a surface-mount permanent magnet synchronous motor with an outer rotor and an inner stator. The slot and pole combination is 36-slot/12-pole type. The stator winding is a three-phase -type. The basic structure and detailed size of the IWM is shown in Figure 4.

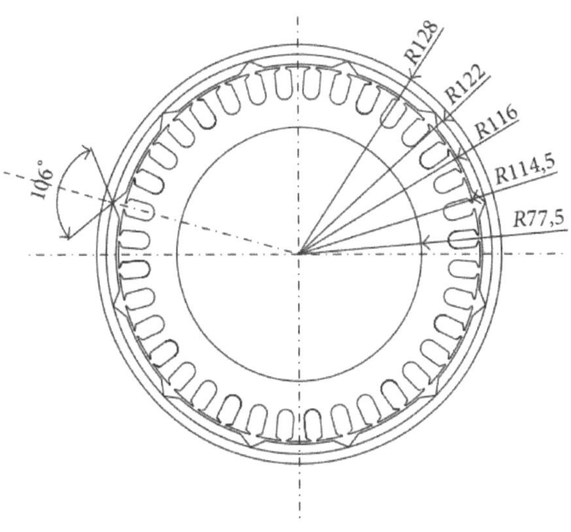

(a) Overall structure

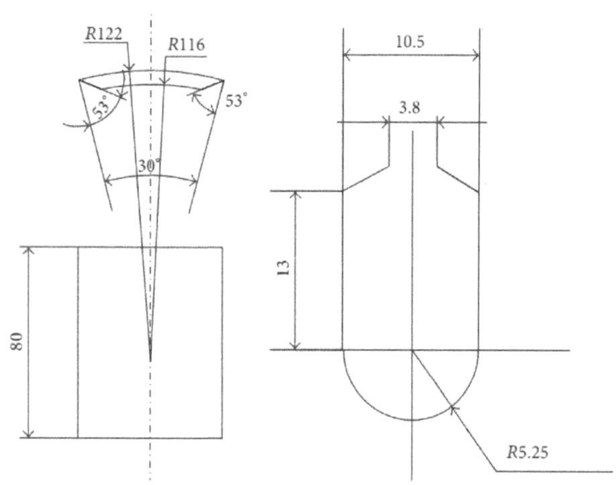

(b) Permanent magnet (c) Slot

Figure 4: Structure of the IWM.

Evaluation Index of Vehicle Rollover

When the vehicle rounds a curve, the greater the vehicle speed is, the greater the lateral acceleration is, but the vertical load of the inside wheels is reduced. In other words, the vertical load is transferred from the inside to outside. When the vertical load of the inside wheel is decreased to 0, the vehicle will not be able to maintain a balance in the roll plane and will begin to roll over. This condition indicates that the vehicle is in an unstable state and that a rollover accident may easily occur. In light of this, a lateral load transfer rate (LTR) is selected as the evaluation index of the vehicle rollover stability. LTR is defined as the ratio of the transferred load from the inside wheel to the outside wheel and the total load [21, 22]; that is,

$$\mathrm{LTR} = \left| \frac{F_{z\mathrm{fl}} + F_{z\mathrm{rl}} - F_{z\mathrm{fr}} - F_{z\mathrm{rr}}}{F_{z\mathrm{fl}} + F_{z\mathrm{rl}} + F_{z\mathrm{fr}} + F_{z\mathrm{rr}}} \right|. \tag{14}$$

The vertical load of the four wheels can be calculated by

$$\begin{aligned}
F_{z\mathrm{fl}} &= \frac{mgl_{\mathrm{r}}}{2l} - ma'_x\frac{h}{2l} - \frac{ma'_y h l_{\mathrm{r}}}{dl} - \frac{K_{\mathrm{f}}\varphi + C_{\mathrm{f}}\dot{\varphi}}{d}\\
F_{z\mathrm{fr}} &= \frac{mgl_{\mathrm{r}}}{2l} - ma'_x\frac{h}{2l} + \frac{ma'_y h l_{\mathrm{r}}}{dl} + \frac{K_{\mathrm{f}}\varphi + C_{\mathrm{f}}\dot{\varphi}}{d}\\
F_{z\mathrm{rl}} &= \frac{mgl_{\mathrm{f}}}{2l} + ma'_x\frac{h}{2l} - \frac{ma'_y h l_{\mathrm{f}}}{dl} - \frac{K_{\mathrm{r}}\varphi + C_{\mathrm{r}}\dot{\varphi}}{d}\\
F_{z\mathrm{rr}} &= \frac{mgl_{\mathrm{f}}}{2l} + ma'_x\frac{h}{2l} + \frac{ma'_y h l_{\mathrm{f}}}{dl} + \frac{K_{\mathrm{r}}\varphi + C_{\mathrm{r}}\dot{\varphi}}{d},
\end{aligned} \tag{15}$$

where

$$\begin{aligned}
a'_x &= a_x - \frac{m_s h_s \ddot{\theta}}{m},\\
a'_y &= a_y + \frac{m_s h_s \ddot{\varphi}}{m}.
\end{aligned} \tag{16}$$

When the vehicle moves in a straight line on the flat and level road, the vertical load of the wheels on both sides is approximately equal, and LTR is close to 0. When the vehicle rounds a curve or the roll angle is unduly large, LTR increases; as LTR is increased to 1, the inside wheel leaves the ground and the vertical load is 0, at which point a rollover accident may likely occur.

Simulation Test and Results Analysis

Simulation Test Conditions

As a typical test condition, FTF steering can be well used to examine the vehicle rollover stability in an extreme manipulation condition such as quick steering wheel correction and opposite rotation [23]. Therefore, the FTF test is considered as the steering maneuver test condition.

To set the FTF test condition, a steering wheel reference value $\Delta\delta$ is determined. Based on [24], the reference value $\Delta\delta$ is 25°. Then, the FTF condition can be set with $\Delta\delta$, as shown in Figure 5. Under the full load condition, the vehicle moves in a straight line at 80 km/h until it reaches a steady state; the driver quickly turns the steering wheel at the speed of 720° /s. Then, when the steering wheel reaches the maximum steering value δ_0 ($\delta_0 = 6.5 \times \Delta\delta$), the driver keeps it for 250 ms and then rotates the steering wheel in the opposite direction with the same speed. As the steering wheel reaches $-\delta_0$, the driver keeps it for 3 s, after which the driver returns the steering wheel to the original position quickly.

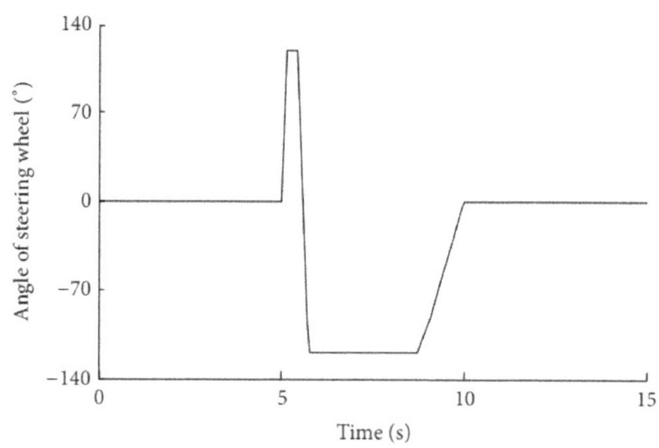

Figure 5: Input of the steering wheel.

Grade B RSR input is calculated using (4), where the vehicle speed u is still kept at 80 km/h.

Contrastive Analysis of the Results

The above FTF steering condition and RSR input are substituted into the mathematical model of vehicle rollover to calculate the vertical load of the wheels with or without EMF. The change in vertical load of the four wheels is obtained, as shown in Figure 6. Then, LTR can be calculated using (14), as shown in Figure 7.

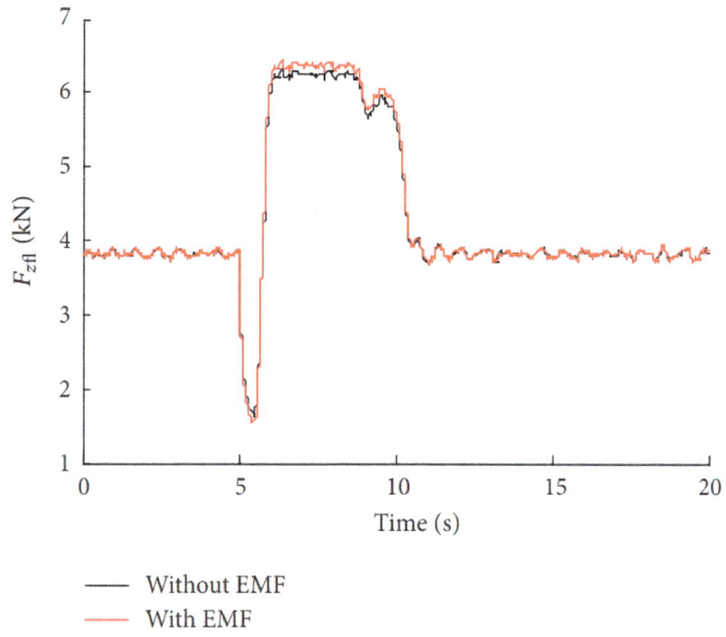
7
6
5
4
3
2
1
F_{zfl} (kN)
0
5
10
15
20
Time (s)
Without EMF
With EMF

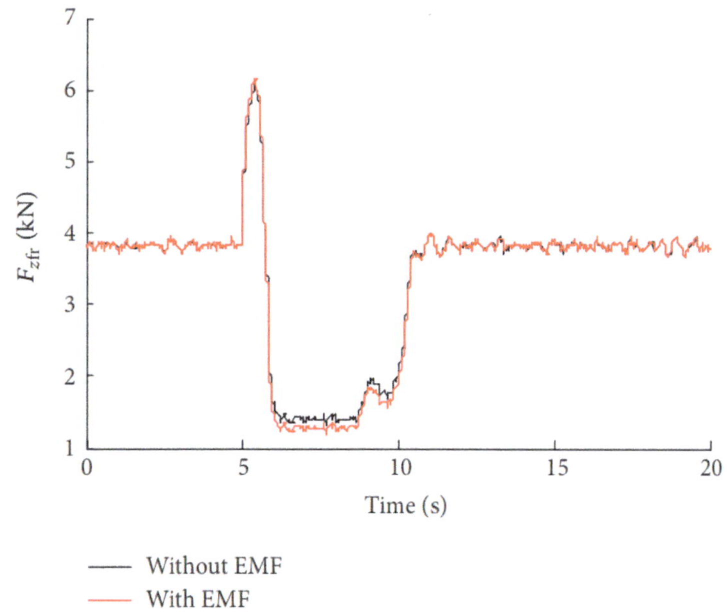
7
6
5
4
3
2
1
F_{zfr} (kN)
0
5
10
15
20
Time (s)
Without EMF
With EMF

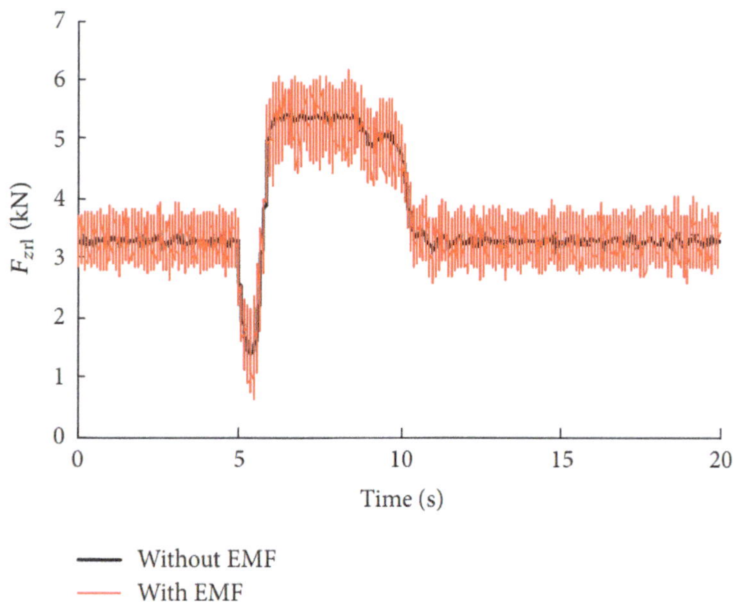
7
6
5
4
3
2
1
0
F_{zrl} (kN)
0
5
10
15
20
Time (s)
Without EMF
With EMF

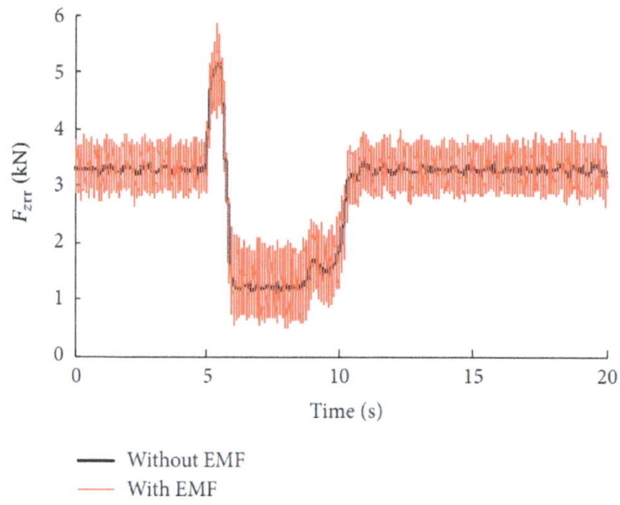

Figure 6: Vertical load of the four wheels.

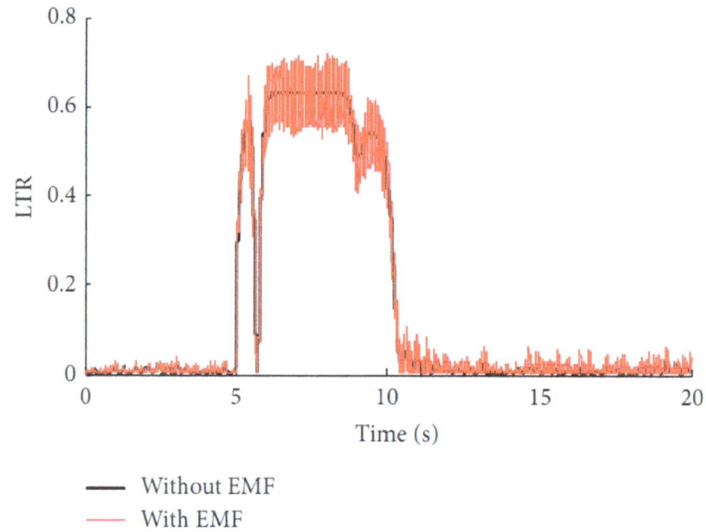

Figure 7: Comparison result of the vehicle LTR with/without EMF.

Because vehicle rollover is a very complex transient process, rollover will occur whenever the LTR is too large. Therefore, the peak value of the transient LTR plays an important role in the evaluation of vehicle rollover risk. Statistical analysis is carried out for the above results to understand the effect of EMF on the response variables more clearly. The maximum, minimum, and change rate for each response variable are listed in Table 2.

Table 2: Statistical analysis of the response variables

		With EMF	Without EMF	Rate of change
F_{zfl} (N)	Maximum	6455	6327	2.02%
	Minimum	1528	1635	6.54%
F_{zfr} (N)	Maximum	6166	6053	1.87%
	Minimum	1202	1328	9.49%
F_{zrl} (N)	Maximum	6152	5431	13.28%
	Minimum	620	1365	54.58%
F_{zrr} (N)	Maximum	5857	5159	13.53%
	Minimum	467	1124	58.45%
LTR	Maximum	0.7184	0.6341	13.29%

According to the comparison results shown in Figures 6 and 7 and Table 2, we can see the condition in which the load transfer of four wheels occurs and the impact on the vertical load of the wheels due to the EMF caused by the RSR is revealed. In particular, this effect can be seen upon the two rear drive wheels, because the EMF generated by the drive motors acts on the corresponding stator and rotor directly and transmits to the rear wheels. With the influence of EMF, the maximum and minimum change rates in vertical load of the left rear wheel are 13.28% and 54.58%, respectively, while that of the change rate without the influence of EMF in vertical load of the left rear wheel can reach 13.53% and 58.45%. Comparison results of the LTR show that LTR has larger fluctuation across the whole simulation test due to the influence of EMF. Additionally, the peak value of LTR is increased from 0.6341 to 0.7184, a relative increase of 13.29% after the consideration of EMF. This indicates that the EMF has some effect on the vertical load transfer and reduces the antirollover ability of the vehicle.

The EMF generated by the two rear IWMs in the simulation test is shown in Figure 8.

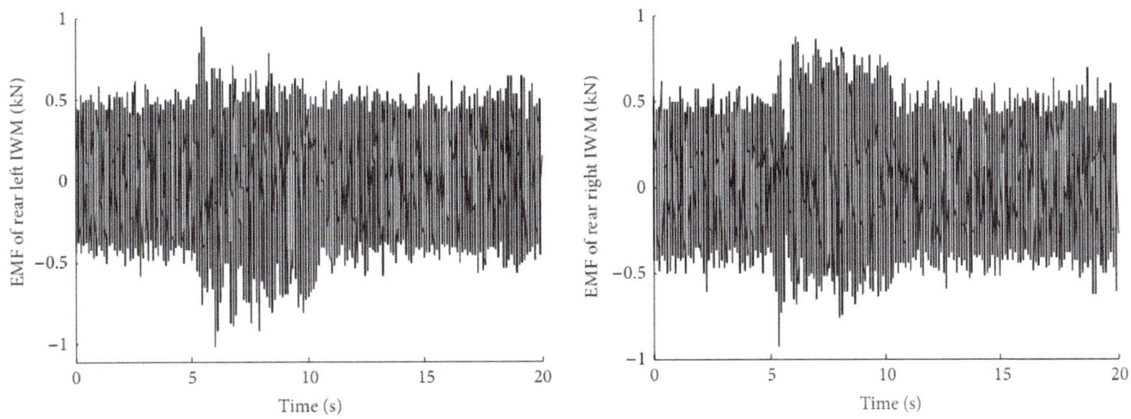

Figure 8: EMF of the IWMs.

As seen from Figure 8, the EMF generated by the left and right IWMs is different and the EMF is increased when the steering wheel is rotated quickly. The reason for this effect lies in the difference of the left and right RSR input and the vertical load transfer under the steering condition, which makes the magnet gap have a different deformation and generates a different amount of EMF. As shown in the figure, the EMF maximum generated by the left rear IWM can reach 1012 N, while the EMF maximum generated by the right rear IWM is 928 N.

Analysis of Road-Influenced Factor

Firstly, the EMF is caused by the road excitation, and different road conditions lead to the generation of different amounts of EMF which will influence the vehicle rollover in varying degrees. Therefore, it is necessary to further study the influence of different road conditions on vehicle rollover. For the RSR input, road level and vehicle speed are two important factors, while the adhesive coefficient is significant for the vehicle steering stability. So, the analysis of road level, vehicle speed, and road adhesion coefficient on the vehicle rollover is carried out in this section.

In the further study, only one influenced factor is changed, whereas the rollover dynamic model, excitation input, and simulation condition are all the same.

Factor of Road Level

Road level is set as grades A, B, and C. That is, the road roughness coefficient is 16×10^{-6}, 64×10^{-6}, and $256 \times 10^{-6}\,m^3$, respectively. A larger road roughness coefficient corresponds to a lager RSR excitation. Then the LTR of different road levels can be obtained, as shown in Figure 9. Vehicle speed is kept at 80 km/h in the simulation.

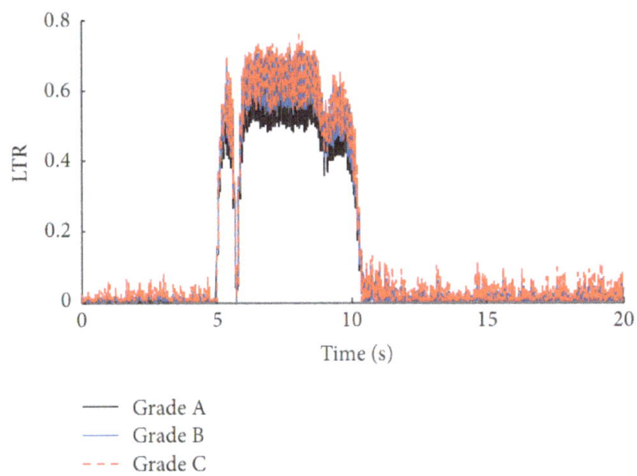

Figure 9: LTR in different road level.

As shown in Figure 9, the value of LTR increases with the increase of the road roughness coefficient. That is, the rollover stability deteriorates with the increase of road level and LTR expresses larger fluctuation when the road roughness coefficient is larger.

Statistical analysis of the simulation results is carried out and the relation of the road level and the peak value of LTR and EMF are obtained, as shown in Figures 10 and 11.

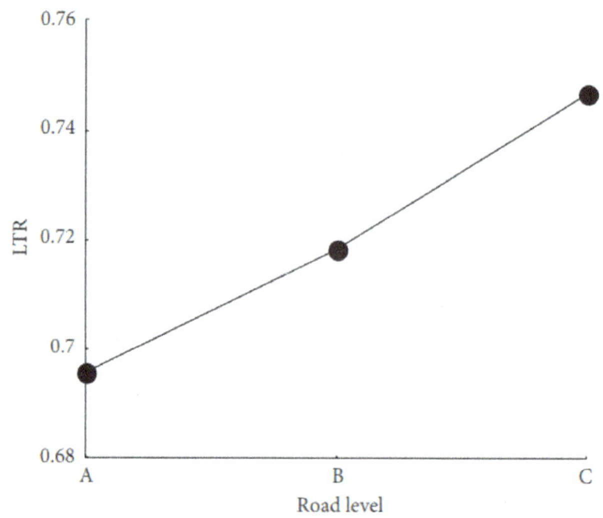

Figure 10: Relation of the road level and peak value of LTR.

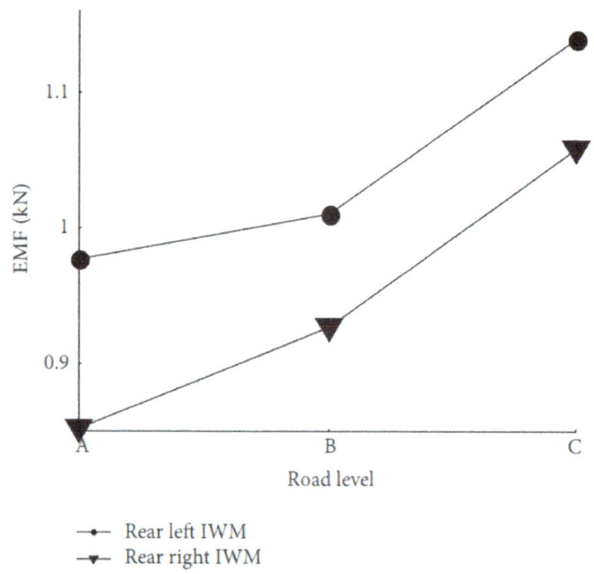

Figure 11: Relation of the road level and peak value of EMF.

As seen from Figure 10, the road level has an effect on the peak value of LTR with a near-linear increasing trend. Additionally, the change rate of LTR can reach 7.34% when the road level is increased from grade A to C. The reason for this effect is that the worse road level makes the RSR excitation increase. The increase in RSR excitation not only causes larger vertical load transfer under the steering condition but further exacerbates the IWM MGD which produces larger EMF and has a certain impact on the LTR. From Figure 11, we can see that the IWM EMF increases with the increase of road level, which further confirms the illustration for the results above.

Factor of Vehicle Speed

In this section, the road adhesion coefficient is 0.6. Vehicle speed is set as 40, 80, and 120 km/h; that is, the rotating speed of IWM is 354, 707, and 1061 rpm. A higher vehicle speed means larger RSR excitation according to (4). Accordingly, the LTR of different vehicle speeds under the grade B FTF condition can be obtained, as shown in Figure 12.

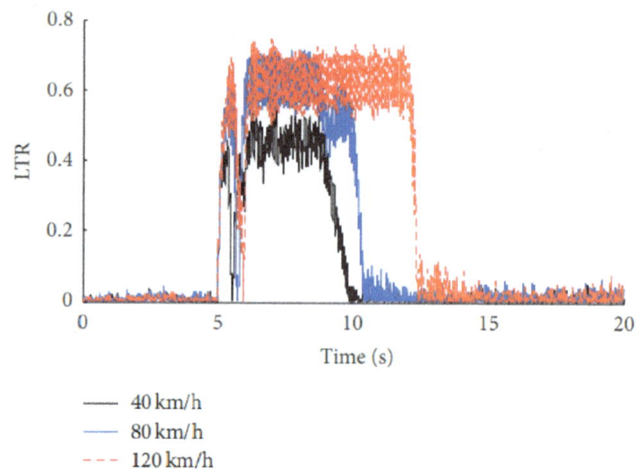

Figure 12: LTR in different vehicle speed.

As shown in Figure 12, the LTR increases with the increase of vehicle speed. When the vehicle speed is 120 km/h, the maximum of LTR can reach up to 0.7577, and the LTR still remains at a higher level even if the steering wheel has been returned to its original position at 10 s. The results indicate that the recovery capability of the straight line driving is decreased with the increase of the vehicle speed in the course of turning. For a certain road level, the available traction is a constant value under the same sideslip angle. When the vehicle speed is increased, larger traction is needed to meet the demand for the longitudinal driving resistance. Therefore, if the vehicle speed reaches a certain level, it is likely that vehicle slide will take place because the smaller amount of traction is not sufficient to meet the demand for the lateral forces and the vehicle will be unable to drive according to a predetermined path.

Similarly, statistical analysis of the simulation results is carried out, and the relation of vehicle speed to the peak value of LTR and EMF is obtained, as shown in Figures 13 and 14.

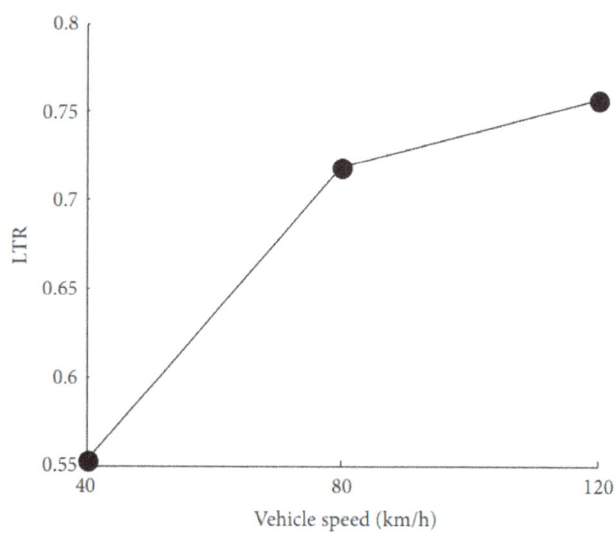

Figure 13: Relation of the vehicle speed and peak value of LTR.

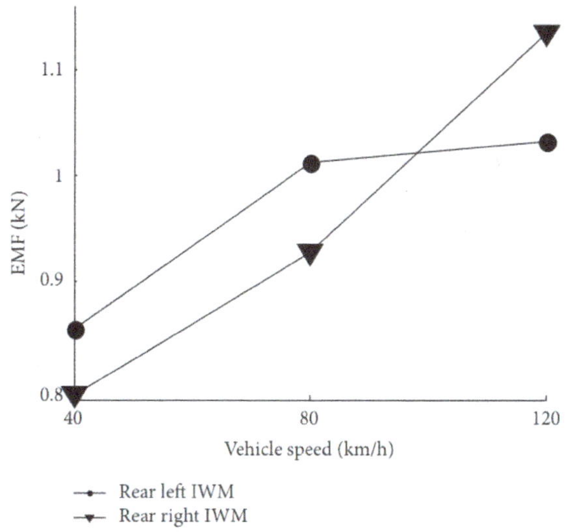

Figure 14: Relation of the vehicle speed and peak value of EMF.

As shown in Figures 13 and 14, the peak value of LTR is increased obviously with the increase of vehicle speed. This is due to the increased vehicle speed which causes the increase of the RSR excitation. The increase in RSR excitation not only influences the size of LTR but further exacerbates the IWM MGD which will produce bigger EMF and have a certain impact on LTR. According to the data of Figure 13, the growth rate of LTR can reach 29.91% when the vehicle speed is increased from 40 to 80 km/h, while the growth rate of LTR is 5.47% as the vehicle speed is increased from 80 to 120 km/h. It follows that the LTR is influenced strongly in the low vehicle speed range. In contrast, there is a slight influence on the LTR in the high speed range.

Factor of Road Adhesion Coefficient

Vehicle speed is kept at 80 km/h, and the road adhesion coefficient is set as 0.3, 0.6, and 0.9. Afterwards, the LTR can be obtained with different adhesion coefficients under the B level FTF steering condition, as shown in Figure 15.

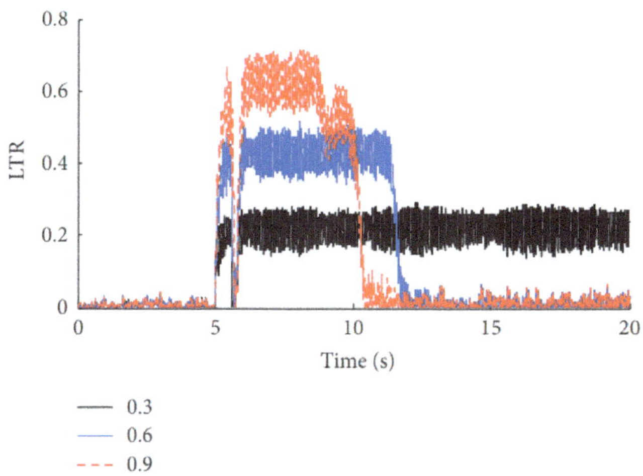

Figure 15: LTR in different road adhesion coefficient.

As shown in Figure 15, the rollover risk (LTR value) of the vehicle is increased with the increase of the road adhesion coefficient. This condition indicates that a lower value of road adhesion coefficient is better for the vehicle rollover safety. However, the result also shows that the lower the road adhesion coefficient is, the worse the vehicles follow the predetermined winding roads. As seen from the figure, when the adhesion coefficient is 0.3, the LTR value always maintains the state before the steering wheel reset and cannot return to the straight line driving even if the steering wheel has been returned to its original position at 10 s. This is because low amounts of traction can only be provided by the road with a lower adhesion coefficient, which is not sufficient to meet the demand for the lateral forces of the vehicle during the rapid turning condition. Although vehicle rollover caused by the curvilinear motion is less likely on a low adhesion road, slippage can easily occur. However, if there are obstacles on the road, it is highly likely that a rollover will occur due to the lateral impact of the vehicle and the obstacles.

The relation of adhesion cohesion to the peak value of LTR and EMF is obtained based on statistical analysis, as shown in Figures 16 and 17.

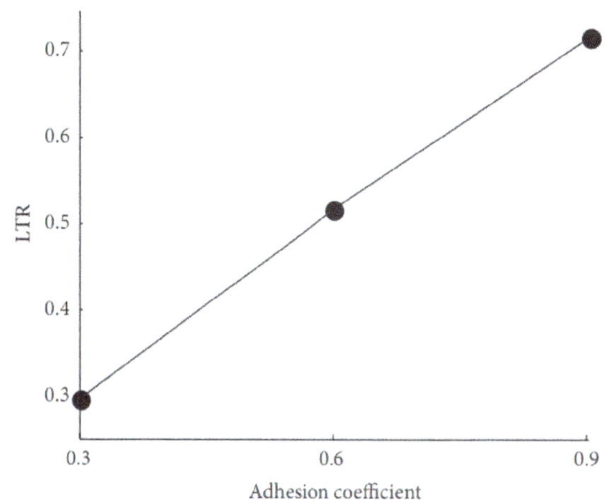

Figure 16: Relation of the adhesion coefficient and peak value of LTR.

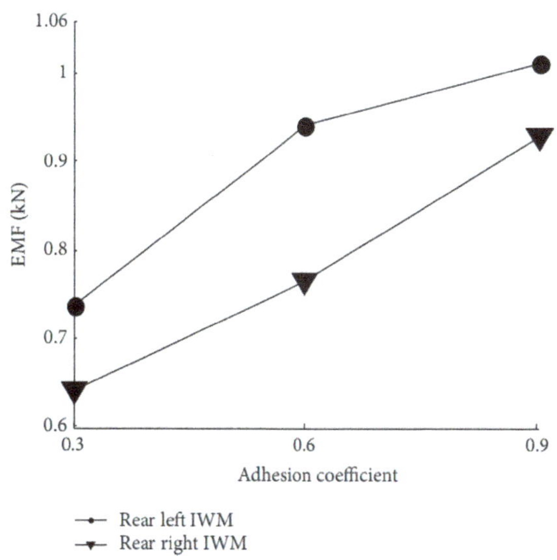

Figure 17: Relation of the adhesion coefficient and peak value of EMF.

As shown in Figure 16, the peak value of LTR is increased linearly with the increase of the vehicle speed. When the adhesion coefficient is increased from 0.3 to 0.9, the corresponding peak value of LTR is increased from 0.2947 to 0.718, and the growth rate of LTR can reach 143.77%. As shown in Figure 17, the road adhesion coefficient also has a great influence on the IWM EMF and a larger adhesion coefficient makes the EMF increase.

DISCUSSION AND CONCLUSIONS

In this paper, a verified 16-degree-of-freedom rollover dynamic model is introduced to study the rollover stability of the IWM driven EVs. Vehicle rollover characteristic with or without EMF under the grade B and FTF steering condition is analyzed. In addition, further study is carried out to understand the vehicle rollover in different road level, vehicle speed, and road adhesion coefficient. Based on this study, some conclusions can be obtained:

- Comparative analysis of the vehicle rollover with and without EMF shows the following: ① The EMF caused by the RSR has a certain impact on the vertical load transfer of the wheels, especially for the drive wheels. ② The LTR has a larger fluctuation across the whole simulation test due to the influence of EMF. Additionally, the EMF makes the LTR increase, which indicates that the EMF has some effect on the vertical load transfer, and reduces the antirollover ability of the vehicle. Therefore, the effect of IWM EMF on the vehicle rollover should be considered in the design of the IWM driven EV. ③ The size of the EMF is related to the RSR and steering input condition after the type and the structure of the electric wheel system are determined. The reason lies in the difference of the left and right RSR input and the vertical load transfer under the steering condition, which makes the magnet gap have different deformation and, therefore, cause different size of EMF.
- Further study of the effect of the road-influenced factors on the vehicle rollover shows that the fluctuation and peak value of the LTR increase with an increase of the road level, vehicle speed, and road adhesion coefficient. That is, the vehicle rollover stability is worse with the increase of the above factors. The reason is that poor road conditions not only cause larger vertical load transfer but further exacerbate the IWM MGD which produces bigger EMF and impacts the load redistribution.
- For the three road-influenced factors, road adhesion coefficient has the greatest effects on the vehicle rollover, followed by the vehicle speed and road level. From the vehicle rollover safety perspective, the road adhesion coefficient should be small. However, if the adhesion coefficient is too small, traction is diminished. Therefore, vehicle slippage will occur because the smaller traction is not sufficient to keep the vehicle running stability. As a result, the road adhesion coefficient should be considered in the design of the vehicle. For the vehicle speed, the recovery capability of straight line driving decreases with the increase of the vehicle speed in the course of turning. If the vehicle speed reaches a certain level, it is likely that vehicle slippage will take place and the vehicle will be unable to drive according to the predetermined path. So, the vehicle speed should be decreased to keep the vehicle stable. Although there is little effect on vehicle rollover by the road lever relative to other factors, it also needs to be considered to improve the vehicle rollover stability in the design.

This paper provides certain theory basis for the design of the IWM driven EVs.

ACKNOWLEDGMENTS

This research is supported by the National Natural Science Foundation of China (Grant no. 51405273) and sponsored by Shandong Province Higher Educational Science and Technology Program (Grant no. J14LB08) and Young Teachers Development Support Program for Shandong University of Technology.

REFERENCES

1. National Highway Traffic Safety Administration, An Analysis of Motor Vehicle Rollover Crashes and Injury Outcomes, NHTSA, Washington, DC, USA, 2007.
2. J. J. Xia, L. Chang, X. M. Hu, et al., "Active control on the side tumbling of heavy vehicle based on model forecast," Transactions of the Chinese Society of Agricultural Engineering, vol. 26, no. 9, pp. 176–180, 2010.
3. R. Kamnik, F. Boettiger, and K. Hunt, "Roll dynamics and lateral load transfer estimation in articulated heavy freight vehicles," Proceedings of the Institution of Mechanical Engineers Part D: Journal of Automobile Engineering, vol. 217, no. 11, pp. 985–997, 2003. · ·
4. Y. I. Ryu, D. O. Kang, S. J. Heo, and J. H. In, "Rollover mitigation for a heavy commercial vehicle,"International Journal of Automotive Technology, vol. 11, no. 2, pp. 283–287, 2010. · ·
5. H. Imine, A. Benallegue, T. Madani, and S. Srairi, "Rollover risk prediction of heavy vehicle using high-order sliding-mode observer: experimental results," IEEE Transactions on Vehicular Technology, vol. 63, no. 6, pp. 2533–2543, 2014. · ·
6. H. Imine, L. M. Fridman, and T. Madani, "Steering control for rollover avoidance of heavy vehicles,"IEEE Transactions on Vehicular Technology, vol. 61, no. 8, pp. 3499–3509, 2012. · ·
7. X. W. Song, J. Li, and Y. Wang, "A refined vehicle rollover model and its application," Automotive Engineering, vol. 31, no. 10, pp. 971–974, 2009.
8. W. N. Bao and S. B. Hu, "Vehicle rollover simulation analysis considering road excitation," Transactions of the Chinese Society of Agricultural Engineering, vol. 31, no. 2, pp. 59–65, 2015. · ·
9. G. Nagaya, "In-wheel Motor System," US Patent, Patent No. US7287611 B2, 2007.
10. A. Watts, A. Valance, A. Whitehead, et al., "The technology and economics of in-wheel-motors," SAE2010-01-2307, SAE International, 2010.

11. Y. T. Luo and D. Tan, "Study on the dynamics of the in-wheel motor system," IEEE Transactions on Vehicular Technology, vol. 61, no. 8, pp. 3510–3518, 2012. · ·

12. C. Lu, D. Tan, and X. Y. Gang, "Parameter sensitivity analysis of coupling dynamics of in-wheel motor driving electric vehicle," Science Technology and Engineering, vol. 15, no. 28, pp. 78–87, 2015.

13. H. B. Pacejka, Tyre and Vehicle Dynamics, Butterworth Heinemann, Oxford, UK, 2002.

14. L. J. Zhang and T. X. Zhang, "Study on general model of random inputs of the vehicle with four wheels correlated in time domain," Journal of Agricultural Machinery Engineering, vol. 36, no. 12, pp. 29–31, 2005.

15. L. J. Zhang, Random Vibration of Vehicle, Northeastern University Press, Shenyang, China, 2007.

16. F. Yu and Y. Lin, Vehicle System Dynamics, China Machine Press, Beijing, China, 2005.

17. A. M. El-Refaie, T. M. Jahns, and D. W. Novotny, "Analysis of surface permanent magnet machines with fractional-slot concentrated windings," IEEE Transactions on Energy Conversion, vol. 21, no. 1, pp. 34–43, 2006. · ·

18. J. Wang, Z. P. Xia, S. A. Long, and D. Howe, "Radial force density and vibration characteristics of modular permanent magnet brushless AC machine," IEE Proceedings: Electric Power Applications, vol. 153, no. 6, pp. 793–801, 2006. · ·

19. K. F. Rasmussen, "Analytical prediction of magnetic field from surface mounted permanent magnet motor," in Proceedings of the IEEE International Electric Machines and Drives Conference (IEMDC ‹99), pp. 34–36, Seattle,. Wash, USA, May 1999. · ·

20. Z. Q. Zhu, D. Howe, and C. C. Chan, "Improved analytical model for predicting the magnetic field distribution in brushless permanent-magnet machines," IEEE Transactions on Magnetics, vol. 38, no. 1, pp. 229–238, 2002. · ·

21. Y. H. Lee and S. J. Yi, "Rollover prevention for sport utility vehicle using fuzzy logic controller," in ICMIT 2005: Control Systems and Robotics, vol. 6042 of Proceedings of SPIE, pp. 1–6, The International Society for Optical Engineering, Chongqing, China, September 2005.

22. Y. Kim, G. Kim, and J. Lim, "Sensitivity analysis of commercial vehicle parameters on dynamic rollover," in Proceedings of the 13th IPC on Automotive Engineering, pp. 51–62, Gyeongju, Republic of Korea, 2005.

23. H. Aleksander, B. Todd, and M. John, "Detection of vehicle rollover,," SAE 2004-01-1757, 2004.

24. G. J. Forkenbrock and W. R. Grrott, "An experimental examination of J-turn and Fish-hook maneuvers that may induce on road, untripped light vehicle roller," SAE 2003-01-1008, 2003.

CHAPTER 10 PLANT RESPONSES TO HIGH FREQUENCY ELECTROMAGNETIC FIELDS

Alain Vian, Eric Davies, Michel Gendraud and Pierre Bonnet

INTRODUCTION

High frequency electromagnetic fields (HF-EMF, i.e., frequencies from 300 MHz to 3 GHz, wavelengths from 1 m to 10 cm) are mainly human-produced, nonionizing electromagnetic radiations that do not naturally occur in the environment, excluding the low amplitude VHF (very high frequency) cosmic radiation. HF-EMF are increasingly present in the environment [1] because of the active development of wireless technology, including cell phones, Wi-Fi, and various kinds of connected devices. Since living material is not a perfect dielectric, it readily interferes with HF-EMF in a way that depends upon several parameters, including (but not restricted to) its general shape, the conductivity and density of the tissue, and the frequency and amplitude of the EMF. The interaction between the living material and the electromagnetic radiation may (or not) induce an elevation of the tissue temperature, thus defining the thermal (versus nonthermal) associated metabolic responses. In the case of a thermal response, the resulting heat dissipation is normalized with the specific absorption rate (SAR) index. This has led to considerable research efforts to study the possible biological effects due to exposure to HF-EMF. While the vast majority of these studies have focused on animals and humans because of health concerns, with contradictory or nonconclusive results [2], numerous experiments have also been performed on plants. Plants are outstanding models compared to animals to conduct such investigations: they are immobile and therefore keep a constant orientation in the EMF and their specific scheme of development (high surface area to volume ratio) makes them ideally suited to efficiently intercept EMF [3]. It is also quite easy in plants to achieve genetically stable plant lines through the selection of species that favor asexual reproduction [4] or self-pollination [5]. Furthermore, metabolic mutants are easily available for several species and constitute invaluable tools to understand the way the EMF signal is transduced [6]. Indeed, several reports have pointed out that plants actually perceive HF-EMF of even small amplitudes and transduce them into molecular responses and/or alterations of their developmental scheme [3–9]. The way that HF-EMF interact with plants remains essentially unanswered. However, since EMF evoke a multitude of responses in plants, they might be considered as a genuine environmental stimulus. Indeed, EMF exposure alters the activity of several enzymes, including those of reactive oxygen species (ROS) metabolism [7], a well-known marker of plant responses to various kinds of environmental factors. EMF exposure also evokes the expression of specific genes previously implicated in plant responses to

Citation: Alain Vian, Eric Davies, Michel Gendraud, and Pierre Bonnet, "Plant Responses to High Frequency Electromagnetic Fields," BioMed Research International, vol. 2016, Article ID 1830262, 13 pages, 2016. doi:10.1155/2016/1830262.

wounding [5, 8] and modifies the development of plants [9]. Furthermore, these responses are systemic insofar as exposure of only a small region of a plant results in almost immediate molecular responses throughout the plant [6]. These responses were abolished in the presence of calcium chelators [6] or inhibitors of oxidative phosphorylation [10] which implies the involvement of ATP pools. In the present review, we describe exposure devices, SAR determination methods, and biological responses (at both the cellular/molecular and whole plant levels) observed after plant exposure to EMF. We focused this review on radiated (i.e., EMF that are emitted through an antenna) HF-EMF (mainly within the range of 300 MHz–3 GHz) and consequently will not address the biological effects of static magnetic fields (SMF), extremely low frequency electromagnetic fields (ELF), or HF current injection, since their inherent physical properties are dramatically different from those of high frequencies. Therefore, the HF-EMF we consider should be viewed through the prism of classical electromagnetism: macroscopic electrodynamics phenomena described in terms of vector and scalar fields.

EXPOSURE SYSTEMS AND DOSIMETRY

HF-EMF are a combination of an electric field and a magnetic field governed by Maxwell's equations. At high frequency, these vector quantities are coupled and obey wave equations whether for propagating waves or for standing waves. In vacuum, the former travel at the speed of light ($\approx 3\ 10^8\,\mathrm{m\,s^{-1}}$) and have the structure of a plane wave (Figure 1(a)). In other media, the speed decreases and the spatial distribution for the electric and the magnetic fields are generally arbitrary (thus not being a plane wave). The latter, which do not propagate but vibrate up and down in place, appear in some particular conditions (e.g., bounded medium like metallic cavity) and play important roles in many physical applications (resonator, waveguide, etc.).

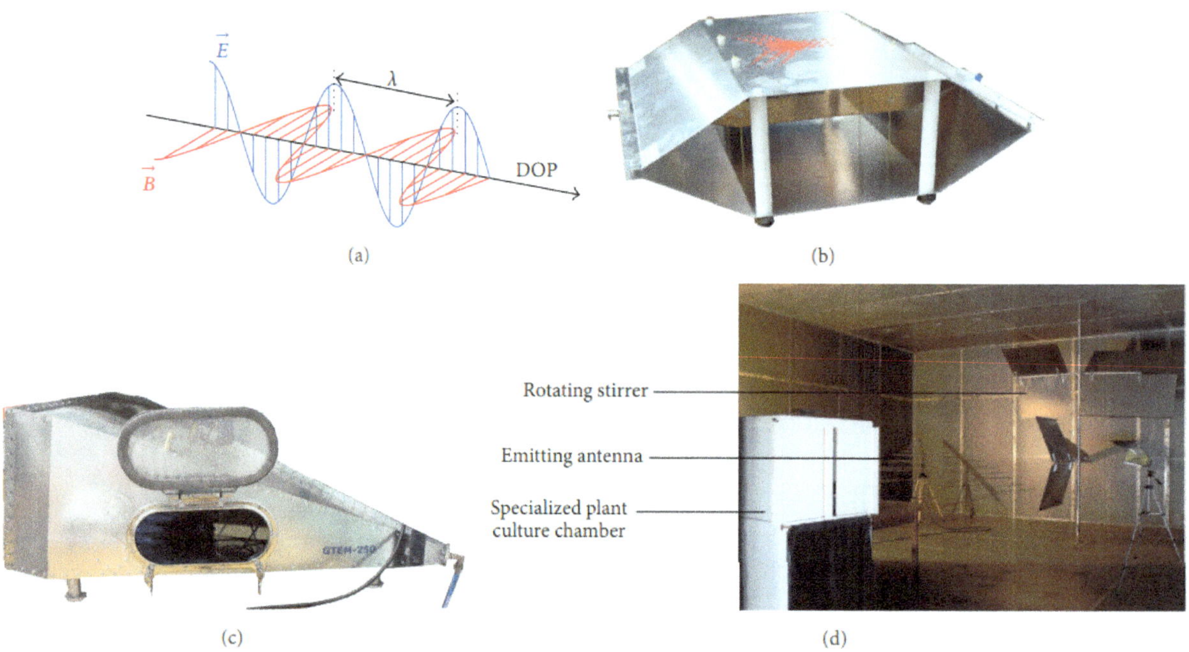

Figure 1: Electromagnetic wave and experimental set-up. (a) Schematic representation of an electromagnetic plane wave showing the transverse and space varying electric (E) and magnetic field (B). The wavelength (λ) is the distance between two crests. DOP: direction of propagation. (b) A TEM cell (transverse electromagnetic cell). (c) A GTEM cell (gigahertz transverse electromagnetic cell). (d) MSRC (mode stirred reverberation chamber). Note the double-sided metallic walls, the emitting antenna, the rotating stirrer, and the specialized culture chamber that stands in the "working volume" where the electromagnetic field characteristics have been extensively characterized.

In both cases, HF-EMF are characterized by an amplitude of the electric (E) or magnetic (H) components (measured in volts or amperes per meter), a frequency f (number of cycles per second of the wave quantity, measured in hertz), and a wavelength λ (distance between wave crests, measured in meters). These properties are related through the following equation:

$$\lambda = \frac{c}{f} = c \times T, \tag{1}$$

where c is the speed of the wave in the considered medium and T is the period of the wave (time between successive wave crests, measured in seconds). The wavelength λ is then the distance traveled by the wave during a period T.

The electromagnetic power density associated with an electromagnetic wave (measured in watts per square meter) is obtained by a vector product between the electric and magnetic field vectors (namely, the Poynting vector) for every point in space. The total HF-EMF power crossing any given surface is derived from Poynting's theorem [11]. For an incident plane wave in vacuum, the time-averaged electromagnetic power P_i (measured in watts) illuminating a surface of 1 m^2 orthogonal to the direction of propagation is given by the following equation:

$$P_i = \frac{E^2}{2 \times Z_0}, \tag{2}$$

where Z_0 is the characteristic impedance of free vacuum space (377 Ω).

The absorbed electromagnetic power (Pd), converted to heat by Joule effect in a volume (V) and averaged over a time period, is given by (3) for an electrically and magnetically linear material that obeys Ohm's law (conductivity σ in siemens per meter):

$$Pd = \iiint_v \frac{\sigma \times E^2}{2} dV. \quad (3)$$

Diversity of Exposure Devices

Due to the wide variety of electromagnetic waves, physicians developed a lot of electromagnetic exposure facilities, mainly for electromagnetic compatibility (EMC) test purposes. Some of these devices are used for plant exposure to HF-EMF.

HF-EMF exposure set-up is usually made up with the following two basic elements:

- HF source (radio frequency generator, Gunn oscillator) associated with a radiating element (antenna, strip-line) and
- a structure that allows the propagation of EM waves and the exposure of the sample.

The simplest exposure set-up relies on the use of standard cell phones as a source of HF-EMF [12, 13] radiating in an open-area test site. While this apparatus has the advantage of being simple and economical, it poses many limitations that may compromise the quality of the exposure. Indeed, these communication devices are operated with different protocols that may modify or even interrupt the emitted power. Also, the biological samples are placed in the immediate neighborhood of the antenna, which is a region where the electromagnetic field is not completely established (near-field conditions) and therefore is difficult to measure; this situation may constitute an issue for bioelectromagnetics studies. These apparatuses are nowadays used only in a small proportion of studies. Moreover, the use of open-area test sites exposes the biological samples to the uncontrolled electromagnetic ambient environment. The use of shielded rooms is a good solution to overcome this issue. Indeed, anechoic chambers provide shielded enclosures, which are designed to completely absorb reflected electromagnetic waves. However, these facilities are often large structures requiring specific equipment and costly absorbers to generate an incident plane wave (far-field illumination) and are consequently seldom used for plant exposure [14, 15].

In contrast, numerous studies are based upon dedicated apparatus of relatively small volume (Figure 1(b)), namely, the transverse electromagnetic (TEM) cell [16]. TEM cells are usually quite small (about 50 cm long × 20 cm wide) and therefore only allow the use of seeds or seedlings as plant models. Many TEM cells are based upon the classic "Crawford cell" [17]. They consist of a section of rectangular coaxial transmission line tapered at each end to adapt to standard coaxial connectors. A uniform plane wave of fixed polarization and direction is generated in the sample space for experiments between the inner conductor (septum) and the upper metallic wall. Because this cost-efficient device is enclosed, high amplitude EMF can be developed with relatively little injected power. Under some conditions, two parallel walls of the TEM cell can be removed (therefore constituting the so-called open TEM cell) without dramatically compromising the performances. This configuration is adequate to allow plant lighting. Special attention must still be paid to the relative position of the samples in the system since the disposition of the different organs within the EMF could severely affect the efficiency of the plant samples' coupling with the electromagnetic field. The main TEM cell limitation is that the upper useful frequency is bound by its physical dimensions limiting the practical size of samples at high frequency.

The gigahertz transverse electromagnetic (GTEM) cell has emerged as a more recent EMF emission test facility (Figure 1(c)) [18]. It is a hybrid between an anechoic chamber and a TEM cell and could therefore be considered as a high frequency version of the TEM cell. The GTEM cell comprises only a tapered section, with one port and a broadband termination. This termination consists of a 50 Ω resistor board for low frequencies and pyramidal absorbers for high frequencies. This exposure device removes the inherent upper frequency limit of TEM cell while retaining some of its advantages (mainly the fact that no antenna set-up is required and the fact that high field strength could be achieved with low injected power).

Waveguides are another kind of screened enclosures that are seldom used in plant exposure [19, 20]. These classical and easy to use exposure devices generate traveling waves along the transmission coordinate and standing waves along the transverse coordinates. In contrast to the TEM cell, waveguides do not generate uniform plane waves but rather allow the propagation of more complex EMF, namely, propagation modes. Each mode is characterized by a cutoff frequency below which the mode cannot propagate. When the ends of the waveguide are short-circuited, a so-called resonant cavity is constituted, from which a recent large facility, originally designed for EMC studies, namely, the mode stirred reverberation chamber (MSRC, Figure 1(d)), is based. While this equipment is expensive and technically difficult to set up, it is the state of the art in terms of electromagnetic field characteristics, allowing the establishment of an isotropic and homogeneous field in a volume large enough to hold a dedicated plant culture chamber (either transparent or shielded toward EMF [6]). This latter characteristic permits experiments on large plants that are kept in an adequate controlled environment [6]. Our group pioneered the use of this facility, based on judicious combinations of standing waves patterns in a complex screened enclosure, in plant bioelectromagnetics studies [8] and extensively described the MSRC functionality [21]. Finally, each exposure set-up may differ in concept, polarization, frequency, or incident power but these setups always need to be optimally designed and based on well-understood physical concepts in order to assess well-controlled HF-EMF exposure conditions (homogeneity, repeatability, reproducibility, etc.).

Different Types of Exposure Signals

From each of the previous exposure devices, two very different types of EMF can be used to expose plants. The most commonly encountered mode is the continuous wave (CW) mode, in which the biological samples are continuously exposed for a specific duration to an EMF of given frequency and amplitude (rarely more than a few dozen V m^{-1}). The second mode is the pulsed electromagnetic field (PEMF) mode, in which the biological samples are subjected to several series of discontinuous pulses of ultrashort duration EMF (within the range of μs to ns) and usually of very high amplitude (up to several hundred kV m^{-1}). This last kind of exposure [22, 23] is seldom used because of the scarcity and great complexity of the equipment needed to generate the EMF and the difficulty to design the dedicated antennae able to deliver such ultrashort power surges [24].

The HF-EMF could also be modulated (i.e., varied in time at a given, usually much lower frequency). Only a few studies explicitly addressed modulation effect on biological responses. Răcuciu et al. [25] exposed maize caryopses to low levels (7 dBm), 900 MHz RF field, for 24 h in either continuous wave (CW), amplitude modulated (AM), or frequency modulated (FM) modes. They found that 12-day-old plant lengths were reduced by about 25% in modulated EMF (AM or FM type) compared to control (unexposed samples), while CW exposure had an opposite (growth stimulation) effect, suggesting that EMF modulation actually modifies biological responses.

Dosimetry

In order to compare the biological effects observed in different exposure conditions, the National Council on Radiation Protection and Measurements officially introduced in 1981 an EMF exposure metric, the specific absorption rate (SAR). The formal definition of this basic dosimetry (the amount of dose absorbed) is "the time derivative of the incremental energy absorbed (dW) by (dissipated in) an incremental mass contained in a volume (dV) of a given density ρ." From this definition and (3), the SAR (measured in W kg^{-1}) is given by the following equation:

$$\mathrm{SAR} = \frac{d}{dt}\left(\frac{dW}{\rho \times dV}\right) = \frac{\sigma \times E^2}{2 \times \rho}. \qquad (5)$$

SAR is the power absorbed by living tissue during exposure to CW-EMF (this quantity does not apply to PEMF mode because of the very short duration of the pulses that do not cause temperature increase in the samples). SAR can be calculated from the dielectric characteristics of plant tissues at the working frequencies, using (4). While ρ could be easily determined, the value of σ is dependent upon the frequency and is difficult to assess in the range of GHz. It is usually evaluated from the literature [40], since the experimental set-up to measure this parameter at a given frequency (waveguide, open waveguide, and coaxial line technique, e.g., D-Line) is rarely used because of its complex set-up. From the biological heat-transfer equation, the SAR can also be determined using the temperature increase evoked in plant tissue after exposure to EMF, using the following equation:

$$\mathrm{SAR} = C \times \left(\frac{dT}{dt}\right)_{t \to 0}, \qquad (6)$$

where C is the heat capacity (J K^{-1} kg^{-1}, which is available for some tissues in the literature) and dT (measured in Kelvin) is the sample temperature increase corresponding to the elapsed time dt (measured in second) since the beginning of HF-EMF exposure. Either for animals or plants, the SAR measurement is subject to uncertainty [46]. Since the specific heat is frequency independent and the temperature distribution is usually more uniform than the internal electric field, (5) provides, for detectable temperature increases, a better way for SAR estimation.

In animal and human tissue, SAR is determined using dedicated phantoms [47] filled with a special liquid that mimics the dielectric properties of biological fluids. While this approach is adequate in animals, in which the developmental scheme produced volumes, it could not be adapted to most plant organs (e.g., leaves) that have a high surface area to volume ratio [3] but could be used in fruits and tuberous structures. In contrast, surface temperature can be easily assessed with dedicated instruments (e.g., Luxtron® fiber optic temperature probe) and used to feed (5) [45]. The SAR can also be determined using the differential power method based on the measurement of power absorption (reviewed in [48]) that takes place in the absence or presence of biological samples [39]. The SAR is then calculated by dividing the absorbed power by the mass of the living material.

BIOLOGICAL RESPONSES

Biological responses should be considered as reporters of, and evidence for, the plant's ability to perceive and interact with EMF. These responses can take place at the subcellular level, implying molecular events or modification of enzymatic activities, or at the level of the whole plant, taking the form of growth modification. Tables 1–3 summarize some work reporting HF-EMF effects observed at the scale of the whole plant, biochemical processes, or gene regulation, respectively.

Table 1: Metabolic pathways affected after plant exposure to HF-EMF radiations

Enzymes or metabolites	Metabolic pathways	Organisms	Exposure conditions	Response to EMF
Phenylalanine ammonia-lyase	Phenylpropanoids	Phaseolus vulgaris	N/A (PEMF)	Synergistic action with growth regulators in cultured cells [26]
Polyphenol oxidase	Polyphenols	Vigna radiata	900 MHz, up to 4 h, 8.55 μW cm^{-2}	8.5-fold increase [27]

α- and β-amylases	Starch metabolism	Vigna radiata	900 MHz, up to 4 h, 8.55 μW cm^{-2}	2.5- and 15-fold increase for α- and β-amylases, respectively [27]
α- and β-amylases, starch phosphorylases	Starch metabolism	Zea mays	1800 MHz, up to 4 h, 332 mW m^{-2}	2-fold increase for amylases. −73% for starch phosphorylases [28]
Water soluble sugars	Sugar metabolism	Phaseolus vulgaris	900 MHz, 4 h	2-fold reduction in soluble sugars [12]
Acid and alkaline invertases	Sucrose metabolism	Zea mays	1800 MHz, up to 4 h, 332 mW m^{-2}	1.8- and 2.6-fold increase for acid and alkaline forms, respectively [28]
Malate and NADP isocitrate dehydrogenases, glucose-6P dehydrogenase	Krebs cycle, pentose phosphate pathway	Plectranthus	900 MHz, 1 h	Lower activity (−10 to −30%) at the end of the stimulus and then a 2-fold increase 24 h later [29]
ATP content and adenylate energy charge (AEC)	Energetic metabolism	Solanum lycopersicon	900 MHz, 10 min, 5 V m^{-1}	Drop of ATP content (30%) and AEC (0.8 to 0.6) 30 min after the stimulus [10]
MDA content, H_2O_2, superoxide dismutase, catalase, guaiacol peroxidase, glutathione reductase, ascorbate peroxidase	Lipid peroxidation-oxidative metabolism	Vigna radiata	900 MHz, 8.55 μW cm^{-2}	All oxidative metabolism markers increased (2-fold to 5-fold) [7]
MDA and H_2O_2 content, catalase, ascorbate peroxidase	Lipid peroxidation	Lemna minor	400 and 900 MHz, 2 to 4 h, 10 to 120 V m^{-1}	MDA and H_2O_2 content, catalase and ascorbate peroxidase activities increased (10–30%) [30]
Peroxidases	Oxidative metabolism	Vigna radiata, Lemna minor	900 MHz, 1 to 4 h, 8.55 μW cm^{-2} or 41 V m^{-1}	Peroxidase activities increased [18, 27]
MDA, oxidized and reduced glutathione, NO synthase	Oxidative metabolism-NO metabolism	Triticum aestivum	2.45 GHz, 5 to 25 s, 126 mW mm^{-2}concomitantly with NaCl treatment	Exposure to EMF reduced the oxidative response of plants to high salt treatment [31]
Protein metabolism-DNA damage	Oxidative protein and DNA damage (comet assay)	Nicotiana tabacum	900 MHz, 23 V m^{-1}	Carbonyl content and tail DNA value increased (1.8-fold and 30%, resp.) [30]
Protein metabolism	Protein content	Phaseolus vulgaris, Vigna radiata, Triticum aestivum	Cell phone, 4 h	Drop in protein content inPhaseolus (71%) and Vigna(57%) [27, 32] and Triticum [13]

Amino acid metabolism	Proline accumulation	Zea mays, Vigna radiata	940 MHz, 2 days Cell phone, 2 h, 8.55 $\mu W\,cm^{-2}$	1.8- and 5-fold increase in Z. mays [33] and V. radiata [7], respectively
Global terpene emission	Monoterpene metabolism	Petroselinum crispum, Apium graveolens, Anethum graveolens	900–2400 MHz, 70–100 $mW\,m^{-2}$	Enhanced emission of terpene compounds [34]

Table 2: Genes whose expression is altered after plant exposure to HF-EMF

Gene	Organism	Function	Exposure conditions	Response to EMF exposure
lebZIP1	Solanum lycopersicon, whole plant	Transcription factor	900 MHz, 5 $V\,m^{-1}$, CW in a MSRC	Increase (3-fold to 4-fold) [6,8]
lebZIP1	Solanum lycopersicon, whole plant	Transcription factor	Cell phone	Increase (3-4-fold) [35]
cam	Solanum lycopersicon, whole plant	Ca^{2+} signal transduction	900 MHz, 5 $V\,m^{-1}$, CW in a MSRC	Increase (5-fold) [5, 10]
cdpk	Solanum lycopersicon, whole plant	Ca^{2+} signal transduction	900 MHz, 5 $V\,m^{-1}$, CW in a MSRC	Increase (5-fold) [10]
cmbp	Solanum lycopersicon, whole plant	mRNA metabolism	900 MHz, 5 $V\,m^{-1}$, CW in a MSRC	Increase (6-fold) [5]
pin2	Solanum lycopersicon, whole plant	Proteinase inhibitor	900 MHz, 5 $V\,m^{-1}$, CW in a MSRC	Increase (4.5-fold [5] and 2.5-fold) [6]
pin2	Solanum lycopersicon, whole plant	Proteinase inhibitor	900 MHz, cell phone	Increase (2-fold) [35]
At4g26260	Arabidopsis thaliana, cell suspension culture	Similar to myo-inositol oxygenase	1.9 GHz, 8 $mW\,cm^{-2}$	Decrease (0.3-fold) [36]
At3g47340	Arabidopsis thaliana, cell suspension culture	Glutamine-dependent asparagine synthetase	1.9 GHz, 8 $mW\,cm^{-2}$	Decrease (0.4-fold) [36]
At3g15460	Arabidopsis thaliana, cell suspension culture	Brix domain protein	1.9 GHz, 8 $mW\,cm^{-2}$	Decrease (0.5-fold) [36]
At4g39675	Arabidopsis thaliana, cell suspension culture	Expressed protein	1.9 GHz, 8 $mW\,cm^{-2}$	Increase (1.5-fold) [36]
At5g10040	Arabidopsis thaliana, cell suspension culture	Expressed protein	1.9 GHz, 8 $mW\,cm^{-2}$	Increase (1.4-fold) [36]
AtCg00120	Arabidopsis thaliana, cell suspension culture	ATPase alpha subunit (chloroplast)	1.9 GHz, 8 $mW\,cm^{-2}$	Increase (1.4-fold) [36]

Table 3: Morphogenetic responses observed after plant exposure to HF-EMF

Plant species	Exposure conditions	Responses to HF-EMF exposure and references
Raphanus sativus	Gunn generator 10.5 GHz, 14 mW, exposure of seeds and hypocotyls	Germination inhibition (45%), reduction of hypocotyl elongation (40%) [37]
Lens culinaris	Cell phone, 1800 MHz (1 mW), exposure of dormant seeds	Reduction of seedlings' root growth (60%) and mitotic index (12%). Abnormal mitosis increased (52%) [38]
Vigna radiata	Cell phone, 900 MHz, 8.55 $\mu W\,cm^{-2}$	Rhizogenesis (root number and length) severely affected [7]
Vigna radiata	Cell phone, 900 MHz, 8.55 $\mu W\,cm^{-2}$	Inhibition of germination (50%), hypocotyl (46%), and root growth (59%). Dry weight reduced by 43% [27]

Phaseolus aureus, Vigna radiata	Cell phone, 4 h exposure	Root and stem elongations severely affected (−44 and −39%, resp.) [12, 32]
Vigna radiata, Lablab purpureus	1.8 GHz, 0.48–1.45 $mW\,cm^{-2}$	Reduction of height and fresh weight [39]
Zea mays	1 GHz, 1 to 8 h, 0.47 $W\,cm^{-2}$	Reduced growth of 12-day-old plants (about 50% after 8 h of exposure) [40]
Zea mays	1800 MHz, 4 h, 332 $mW\,m^{-2}$	Reduced growth of roots and coleoptiles (16 and 22%, resp.) [28]
Vigna radiata, Triticum aestivum	Cell phone, 900 MHz, 4 h exposure	Growth reduction (21 and 50%) in Vigna and Triticum, respectively [13]
Triticum aestivum, Cicer arietinum, Vigna radiata, Vigna aconitifolia	Klystron-based EMF generator, 9.6 GHz, 1 dBm to 3.5 dBm	Growth and biomass reduction [41]
Vigna radiata, Ipomoea aquatica	425 MHz, 2 h, 1 mW	Growth stimulation of primary root [16]
Glycine max	900 MHz, 5.7 to 41 $V\,m^{-1}$	Inhibition of epicotyl and/or root growth, depending on exposure set-up [9]
Lemna minor	400–1900 MHz, 23 to 390 $V\,m^{-1}$, whole plant exposure	Growth slowed down, at least in the first days following exposure [18]
Trigonella foenum-graecum, Pisum sativum	900 MHz, 0.5–8 h	Increased root size, nodule number, and size [42]
Hibiscus sabdariffa	Resulting field from a GSM base antenna (not measured)	Reduction of flower bud abscission with increasing distances from the antenna [43]
Linum usitatissimum	Cell phone or Gunn generator (105 GHz), 2 h	Production of epidermic meristems under calcium deprivation condition [44]
Rosa hybrida	900 MHz, 5–200 $V\,m^{-1}$, whole plant exposure in MSRC	Delayed and reduced (45%) growth of secondary axes [45]

Cellular and Molecular Level

Numerous reports [4, 7, 33] indicate an increase in the production of malondialdehyde (MDA, a well-known marker of membrane alteration) along with ROS metabolism activation after exposing plants to HF-EMF (Table 1). Membrane alteration and ROS metabolism activation are likely to establish transduction cascades that enable specific responses. Indeed, the critical role of calcium, a crucial second messenger in plants, has long been pointed out [6, 10]: the responses (e.g., changes in calm-n6, lecdpk-1, and pin2 gene expression) to EMF exposure are severely reduced when plants are cultivated with excess of calcium or in the presence of calcium counteracting agents (Figure 2) such as chelators (EGTA and BAPTA) or a channel blocker ($LaCl_3$). The importance of calcium in the establishment of the plant response is also highlighted by the fact that early gene expression associated with EMF exposure involves at least 2 calcium-related products (calmodulin and calcium-dependent protein kinase) [5, 10]. This response is also energy-dependent: an important drop (30%, Figure 3) in ATP content and adenylate energy charge (AEC) occurs after HF-EMF exposure [10]. It is not clear for now if the AEC drop is the consequence of altered membranes allowing passive ATP exit or if higher consumption of ATP occurred because of increased metabolic activity. Indeed, it is well known that a drop in AEC stimulates the catabolic enzymatic pathways through allosteric modulations. Nevertheless, inhibiting ATP biosynthesis with the decoupling agent carbonyl cyanide m-chlorophenyl hydrazone (CCCP) abolished plant responses to EMF exposure [10]. Nitric oxide (NO) is another signaling molecule that is tightly related to environmental factors' impact on plants [49]. NO rapidly increases after various kinds of stimuli including drought stress or wounding. Chen et al. [50] recently demonstrated the increased activity of nitric oxide synthase and accumulation of NO after exposing caryopses of wheat for 10 s to high power 2.45 GHz EMF. Similarly, Qiu et al. [51] showed in wheat that the tolerance to cadmium evoked by microwave pretreatment was abolished by the addition of 2-(4-carboxyphenyl)-4,4,5,5-tetramethylimidazoline-1-oxyl-3-oxide (carboxy-PTIO), an NO scavenger, suggesting that microwave-induced NO production was involved in this mechanism. Taken together, these results advocate for the EMF induction of NO synthase. However, these studies used high power EMF (modified microwave oven) as stimulating tool and the fact that a temperature increase of the sample was the cause of NO increase is not excluded. To our knowledge, the involvement of NO has not yet been demonstrated after low power (i.e., nonthermal) EMF exposure. Furthermore, well-known actors of plant responses to environmental stimuli are also involved: the tomato mutants sitiens and JL-5 for abscisic (ABA) or jasmonic (JA) acids biosynthesis, respectively, display normal responses (accumulation of stress-related transcripts) when whole plants are exposed to EMF [6]. In contrast, very rapid distant responses to local exposure that occur in the wild plants (Figure 4(a)) are impaired in sitiens ABA mutant (Figure 4(b)) and JL-5mutants, highlighting the existence of a transmitted signal (whose genesis and/or transmission is dependent on ABA and JA) in the whole plant after local exposure [6]. The nature of this signal is still unknown, but very recent work has demonstrated that membrane potential is affected after exposure to EMF [14]. It could therefore be hypothesized that electrical signals (action potential and/or variation potential) could be the transmitted signal, strongly implying that HF-EMF is a genuine environmental factor.

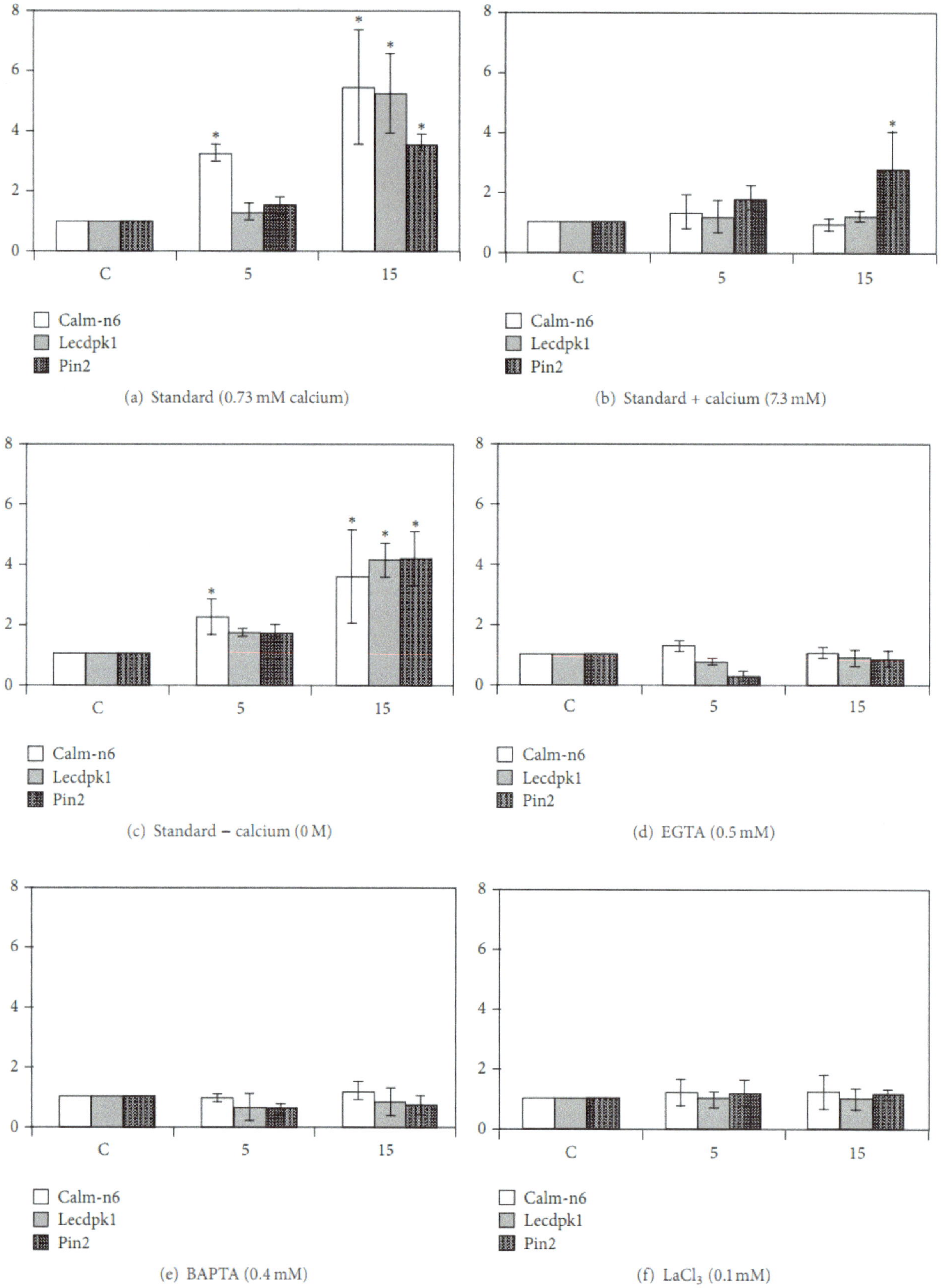

Figure 2: Effect of calcium concentration in culture medium on calm-n6 (calmodulin), pin2 (proteinase inhibitor), and lecdpk1 (calcium-dependent protein kinase) transcript accumulation in response to the HF-EMF exposure. (a) Standard medium (0.73 mM of calcium). (b) Ten-fold extra calcium (7.3 mM). (c) No calcium (0 mM). (d) No calcium (0 mM) with 0.5 mM of EGTA. (e) No calcium (0 mM) with 0.4 mM of BAPTA (1,2-bis(o-aminophenoxy)ethane-N,N,N′,N′-tetraacetic acid), a specific Ca^{2+} chelator. (f) No calcium (0 mM) with 0.1 mM $LaCl_3$. Bars

represent mean values ± SE from at least three independent experiments. An asterisk over the bars states the significant differences according to the one-sided Mann-Whitney U test. Reproduced from [10], with permission.

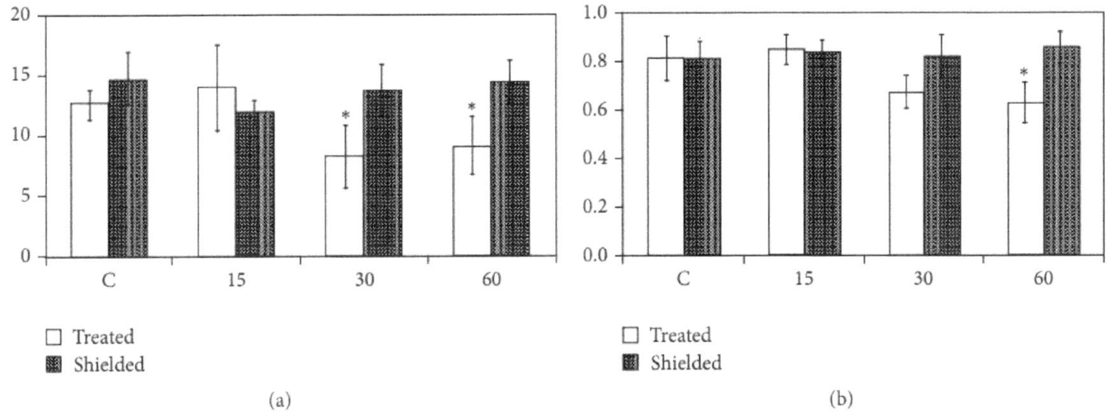

Figure 3: ATP concentration and adenylate energy charge (AEC) changes after HF-EMF exposure ($5\,V\,m^{-1}$, 10 min) in a mode stirred reverberation chamber. C: control, unexposed plants. 15, 30, and 60: time (min) after the end of HF-EMF exposure. (a) ATP concentration (pmol mg^{-1} Prot.). (b) Adenylate energy charge (ratio). Bars represent mean values ± SE from at least three independent experiments. An asterisk over the bars states the significant differences according to the one-sided Mann-Whitney U test. Reproduced from [10], with permission.

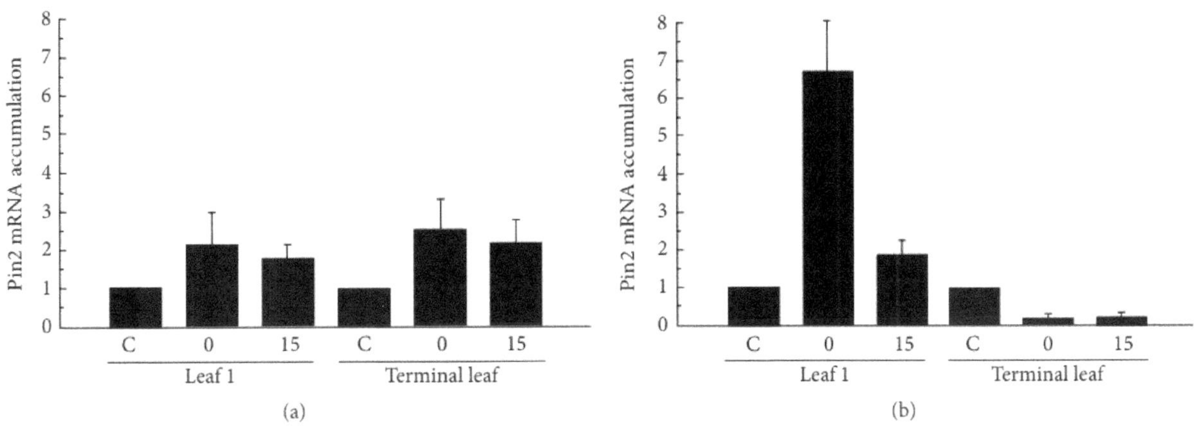

Figure 4: Local and systemic responses after HF-EMF exposure. (a) Systemic response after local exposure to HF-EMF in wild type. Local and distant responses after stimulation of leaf 1 (with the rest of the plant being protected from the EMF). The stimulated tissue (leaf 1) and distant one (terminal leaf) both displayed responses (accumulation of pin2 transcript). HF-EMF exposure: $5\,V\,m^{-1}$, 10 min. (b) Impairment of distant response after exposure to HF-EMF in sitiens (ABA deficient) mutant. The stimulated tissue (leaf 1) displays the response to EMF exposure (accumulation of pin2 transcript), while the response in the distant tissue (terminal leaf) is impaired. HF-EMF exposure: $5\,V\,m^{-1}$, 10 min. Reproduced from [6], with permission.

Alterations of Enzymatic Activities

Table 1 summarizes some of the enzymatic activities that are modified after exposing plants to HF-EMF. As previously noted, ROS metabolism is very often activated after plant exposure to EMF. Enzymatic activities such as peroxidase, catalase, superoxide dismutase, and ascorbate peroxidase have twofold to fourfold increase [4, 7, 18, 27, 33]. The question remains open to determine if this could be the consequence of a direct action of EMF on living tissue. Indeed, the very low energy that is associated with the EMF at these frequencies makes them nonionizing radiations. Side effects of elevated ROS metabolism are also noted: H_2O_2 production [4, 7], MDA increases [4, 7, 33], and protein damage [30]. An increase in polyphenol oxidase [27] and phenylalanine ammonia-lyase [26] may indicate stress responses linked to an increased lignification, a common response of plants to environmental stress.

Protein content is reduced in Vigna and Phaseolus [27, 32] as well as in Triticum [13]. It is not yet known if the decrease in protein content results from an increase in protein degradation and/or a decrease in protein synthesis, but this may constitute a stimulating field of investigation, since evidence shows that mRNA selection from translation occurs after plant exposure to HF-EMF [10]. Hydrolytic enzymatic activities (α- and β-amylases and invertases) responsible for the production of soluble sugar increase in germinating seeds after exposure to HF-EMF [12, 28, 32], while the starch phosphorylase activity, phosphorolytic and potentially reversible, is diminished. In contrast, HF-EMF exposure causes a drop of soluble sugar that may be related to the inhibition of Krebs cycle and pentose phosphate pathway in Plectranthus (Lamiaceae) leaves after exposure to 900 MHz EMF [29], suggesting that seeds and adult leaves respond in a different way to HF-EMF exposure. The accumulation of proline, reported by several authors [7, 33], and an increase in terpenoid emission and content in aromatic plants [34] are also classical responses of plants to environmental stresses.

Modification of Gene Expression

While numerous reports focused on enzymatic activities alterations after exposure to EMF, only a few studies concentrate on gene expression modifications (Table 2). Tafforeau et al. [44] demonstrated using Gunn generator (105 GHz) several reproducible variations in 2D gel electrophoresis profiles, showing that gene expression is likely to be altered by the exposure treatment. Jangid et al. [52] provided indirect proof (RAPD profiles) suggesting that high power microwave irradiation (2450 MHz, 800 W cm^{-2}) modifies gene expression in Vigna aconitifolia, while these results do not exclude a possible thermal effect of microwave treatment. Arabidopsis thaliana suspension-cultured cells exposed to HF-EMF (1.9 GHz, 8 mW cm^{-2}) showed differential expression of several genes (p values < 0.05) compared to the control (unexposed) condition in microarray analysis [36]. Most of them are downregulated (while At4g39675, At5g10040, and AtCg00120 displayed a slight increase; see Table 2). However, the RT-PCR p value lowers the significance of these variations and these authors consequently concluded the absence of HF-EMF effect on plant gene expression. In contrast, short duration, high frequency, low amplitude EMF exposure (10 min, 900 MHz, 5 V m^{-1}) performed on whole 3-week-old tomatoes in MSRC [5, 6, 8, 10] demonstrated altered expressions of at least 5 stress-related genes (Table 2), suggesting that whole plants are more sensitive to HF-EMF than cultured cells. These experiments have been independently replicated by Rammal et al. [35], using a longer exposure period and a far less sophisticated exposure set-up (cell phone). Stress responses of plants quite often display a biphasic pattern [53]: a very rapid increase in transcript accumulation that lasts 15–30 min, followed by a brief return to basal level, and then a second increase (after 60 min). This pattern was observed after tomato exposure to EMF so we questioned the meaning of the early and late population of transcripts in terms of physiological significance by measuring their association to polysomes (which reflects their putative translation to proteins). We found that the early (0–15 min) mRNA population was only faintly associated with polysomes, yet being poorly translated, while the late mRNA population (60 min) is highly associated with polysomes [10]. This result strongly suggests that only the late mRNA population may have a physiological importance since it is the only one to be efficiently translated into proteins.

Whole Plant Level

The biochemical and molecular modifications observed after plant exposure to EMF and described in the previous paragraphs might induce morphogenetic alterations of plant development. Indeed, an increasing number of studies report modifications of plant growth after exposure to HF-EMF (Table 3). These treatments are effective at different stages of plant development (seeds, seedlings, or whole plants) and may affect different organs or developmental processes including seeds germination and stem and root growth, indicating that biological samples of even small sizes (a few mm) are able to perceive HF-EMF. Seed exposure to EMF generally results in a reduced germination rate [27, 37, 39], while in other cases germination is unaffected [42] or even stimulated [16]. The seedlings issued from EMF-exposed seeds displayed reduced growth of roots and/or stem [13, 28, 32, 37–39, 41] but rarely a stimulatory effect [16]. This point strongly differs from exposure to static magnetic fields or extremely low frequency EMF, in which the stimulatory effects on growth are largely predominant [54]. Ultrashort pulsed high power EMF (PEMF, 4 μs, 9.3 GHz, 320 kV m^{-1}) also tends to stimulate germination of seeds of radish, carrot, and tomato and increase plant height and photosynthetic surface area in radish and tomato [20] and roots of tobacco seedlings [22]. These different effects of PEMF compared to HF-EMF on plants may be related to their fundamental difference in terms of physical properties. Exposure to HF-EMF of seedlings or plants (rather than seeds) also generally resulted in growth inhibition [9,18, 27, 28, 39]. Singh et al. [7] showed that rhizogenesis (root number and length) is severely affected in mung bean after exposure to cell phone radiation, possibly through the activation of several stress-related enzymes (peroxidases and polyphenol oxidases). Akbal et al. [38] showed that root growth was reduced by almost 60% in Lens culinaris seeds exposed in the dormant state to 1800 MHz EMF radiation. Concomitantly, these authors reported an increase in ROS-related enzymes, lipid peroxidation, and proline accumulation, with all of these responses being characteristic of plant responses to stressful conditions. Afzal and Mansoor [13] investigated the effect of a 72 h cell phone exposure (900 MHz) on both monocotyledonous (wheat) and dicotyledonous (mung bean) plants seeds: germination was not affected, while the seedlings of both species displayed growth inhibition, protein content reduction, and strong increase in the enzymatic activities of ROS metabolism. It is however worth noting that growth of mung bean and water convolvulus seedlings exposed at a lower frequency (425 MHz, 2 h, 1 mW) is stimulated because of higher elongation of primary root [11], while duckweed (Lemna minor, Araceae) growth was significantly slowed down not only by exposure at a similar frequency (400 MHz, 4 h, 23 V m^{-1}) but also after exposure at 900 and 1900 MHz for different field amplitudes (23, 41, and 390 V m^{-1}) at least in the first days following the exposure [18]. Surducan et al. [15] also found stimulation of seedling growth in bean and maize after exposure to EMF (2.452 GHz, 0.005 mW cm^{-2}). Senavirathna et al. [55] studied real-time impact of EMF radiation (2 GHz, 1.42 W m^{-2}) on instantaneous growth in the aquatic plant, parrot's feather (Myriophyllum aquaticum, Haloragaceae), using nanometer scale elongation rate fluctuations. These authors demonstrated that EMF-exposed plants displayed reduced fluctuation rates that lasted for several hours after the exposure, strongly suggesting that plants' metabolism experienced a stressful situation. It is worth noting that the exposure did not cause any plant heating (as measured using sensitive thermal imaging). Some other kind of morphological changes also occurred after plant exposure to HF-EMF: induction of epidermal meristems in flax [44], flower bud abscission [43], nitrogen-fixation nodule number increase in leguminous [42], or delayed reduced growth of secondary axis in Rosa [45].

These growth reductions may be related to a lower photosynthetic potential since Răcuciu et al. [40] showed that exposing 12-day-old maize seedlings to 0.47 W kg^{-1} 1 GHz EMF induces a drop in photosynthetic pigment content: the diminution was especially important in chlorophyll a, which was reduced by 80% after 7 h of exposure. Ursache et al. [56] showed that exposure of maize seedlings to microwave ($1\,mW\,cm^{-2}$, 10.75 GHz) also caused a drop in chlorophyll a and b content. Similarly, Hamada [57] found a decrease in chlorophyll content in 14-day-old seedlings after exposing the caryopses for 75 min at 10.5 GHz. Kumar et al. showed a 13% decrease in total chlorophyll after 4 h exposure of maize seedlings to 1800 MHz ($332\,mW\,m^{-2}$). These modifications may be related to abnormal photosynthetic activity, which relies on many parameters, including chlorophyll and carotenoid content. Senavirathna et al. [58] showed that exposing duckweeds to 2–8 GHz, $45–50\,V\,m^{-1}$ EMF induced changes in the nonphotosynthetic quenching, indicating a potential stressful condition. Three aromatic species belonging to Apiaceae family (Petroselinum crispum, Apium graveolens, and Anethum graveolens) strongly respond to global system for mobile communications radiation (GSM, 0.9 GHz, $100\,mW\,cm^{-2}$) or wireless local area network (WLAN, 2.45 GHz, $70\,mW\,cm^{-2}$) exposure by decreasing the net assimilation rate (over 50%) and the stomatal conductance (20–30%) [34].

CONCLUSION AND FUTURE PROSPECTS

An increasing number of reports highlight biological responses of plants after exposure to HF-EMF at the molecular and the whole plant level. The exposure conditions are, however, far from being standardized and illustrate the diversity of exposure conditions employed. However, future work should avoid exposure in near-field conditions (i.e., in immediate vicinity of the emission antenna) where the field is instable and difficult to characterize. Similarly, the use of communication devices (i.e., cell phones) should be avoided as emission sources since it may be difficult to readily control the exposure conditions because of built-in automation that may overcome the experimental set-up. The use of specialized devices (TEM cells, GTEM cell, waveguides, MSRC, etc.) in which a precise control of exposure condition can be achieved is highly preferable.

Shckorbatov [59] recently reviewed the possible interactions mechanisms of EMF with living organisms. While the classical targets (interaction with membranes, free radicals, and intracellular regulatory systems) have all been observed in plants, a convincing interpretation of the precise mechanism of HF-EMF interaction with living material is still needed. Alternative explanation (i.e., electromagnetic resonance achieved after extremely high frequency stimulation which matches some kind of organ architecture) has also been proposed for very high frequency EMF (several dozen GHz) [60]. However, the reality of this phenomenon in vivo (studied for now only through numerical simulations) and its formal contribution to the regulation of plant development have not yet been experimentally established. Amat et al. [61] proposed that light effects on plants arose not only through chromophores, but also through alternating electric fields which are induced in the medium and able to interact with polar structures through dipole transitions. The possible associated targets (ATP/ADP ratio, ATP synthesis, and Ca^{2+} regulation) are also those affected by exposure to HF-EMF [10]. It could therefore be speculated that HF-EMF may use similar mechanisms. The targeted pathways, especially Ca^{2+}metabolism, are well known to modulate numerous responses of plants to environmental stress. While deeper understanding of plant responses to HF-EMF is still needed, these treatments may initiate a set of molecular responses that may affect plant resistance to environmental stresses, as already demonstrated in wheat for $CaCl_2$ [62] or UV [63] tolerances, and constitute a valuable strategy to increase plant resistance to environmental stressful conditions.

ACKNOWLEDGMENTS

The authors would like to gratefully acknowledge the LRC BIOEM ("Laboratoire de Recherche Conventionné"—LRC no. 002-2011-CEA DAM/Institut Pascal) for its financial support.

REFERENCES

1. A. Balmori, "Electromagnetic pollution from phone masts. Effects on wildlife," Pathophysiology, vol. 16, no. 2-3, pp. 191–199, 2009. · ·
2. H. Kleinlogel, T. Dierks, T. Koenig, H. Lehmann, A. Minder, and R. Berz, "Effects of weak mobile phone—electromagnetic fields (GSM, UMTS) on well-being and resting EEG," Bioelectromagnetics, vol. 29, no. 6, pp. 479–487, 2008. · ·
3. A. Vian, C. Faure, S. Girard et al., "Plants respond to GSM-like radiations," Plant Signaling & Behavior, vol. 2, no. 6, pp. 522–524, 2007. · ·
4. M. Tkalec, K. Malarić, and B. Pevalek-Kozlina, "Exposure to radiofrequency radiation induces oxidative stress in duckweed Lemna minor L.," Science of the Total Environment, vol. 388, no. 1–3, pp. 78–89, 2007.
5. D. Roux, A. Vian, S. Girard et al., "Electromagnetic fields (900 MHz) evoke consistent molecular responses in tomato plants," Physiologia Plantarum, vol. 128, no. 2, pp. 283–288, 2006.
6. É. Beaubois, S. Girard, S. Lallechere et al., "Intercellular communication in plants: evidence for two rapidly transmitted systemic signals generated in response to electromagnetic field stimulation in tomato," Plant, Cell and Environment, vol. 30, no. 7, pp. 834–844, 2007. · ·
7. H. P. Singh, V. P. Sharma, D. R. Batish, and R. K. Kohli, "Cell phone electromagnetic field radiations affect rhizogenesis through impairment of biochemical processes," Environmental Monitoring and Assessment, vol. 184, no. 4, pp. 1813–1821, 2012. · ·
8. A. Vian, D. Roux, S. Girard et al., "Microwave irradiation affects gene expression in plants," Plant Signaling and Behavior, vol. 1, no. 2, pp. 67–69, 2006. · ·
9. M. N. Halgamuge, S. K. Yak, and J. L. Eberhardt, "Reduced growth of soybean seedlings after exposure to weak microwave radiation from GSM 900 mobile phone and base station," Bioelectromagnetics, vol. 36, no. 2, pp. 87–95, 2015. · ·
10. D. Roux, A. Vian, S. Girard et al., "High frequency (900 MHz) low amplitude ($5\,V\,m^{-1}$) electromagnetic field: a genuine environmental stimulus that affects transcription, translation, calcium and energy charge in tomato," Planta, vol. 227, no. 4, pp. 883–891, 2008. · ·
11. J. A. Stratton, Electromagnetic Theory, McGraw-Hill, New York, NY, USA, 1941.

12. V. P. Sharma, H. P. Singh, R. K. Kohli, and D. R. Batish, “Mobile phone radiation inhibits Vigna radiata(mung bean) root growth by inducing oxidative stress,” Science of the Total Environment, vol. 407, no. 21, pp. 5543–5547, 2009. · ·

13. M. Afzal and S. Mansoor, “Effect of mobile phone radiations on morphological and biochemical parameters of Mung bean (Vigna radiata) and wheat (Triticum aestivum) seedlings,” Asian Journal of Agricultural Sciences, vol. 4, no. 2, pp. 149–152, 2012.

14. M. D. H. J. Senavirathna and T. Asaeda, “Radio-frequency electromagnetic radiation alters the electric potential of Myriophyllum aquaticum,” Biologia Plantarum, vol. 58, no. 2, pp. 355–362, 2014. · ·

15. E. Surducan, V. Surducan, A. Butiuc-Keul, and A. Halgamagy, “Microwaves irradiation experiments on biological samples,” Studia Univesitas Babes-Bolyai Biologia, vol. 58, no. 1, pp. 83–98, 2013.

16. P. Jinapang, P. Prakob, P. Wongwattananard, N. E. Islam, and P. Kirawanich, “Growth characteristics of mung beans and water convolvuluses exposed to 425-MHz electromagnetic fields,” Bioelectromagnetics, vol. 31, no. 7, pp. 519–527, 2010. · ·

17. M. L. Crawford, “Generation of standard EM fields using TEM transmission cells,” IEEE Transactions on Electromagnetic Compatibility, vol. 16, no. 4, pp. 189–195, 1974. ·

18. M. Tkalec, K. Malarić, and B. Pevalek-Kozlina, “Influence of 400, 900, and 1900 MHz electromagnetic fields on Lemna minor growth and peroxidase activity,” Bioelectromagnetics, vol. 26, no. 3, pp. 185–193, 2005. · ·

19. L. M. Liu, F. Garber, and S. F. Cleary, “Investigation of the effects of continuous-wave, pulse- and amplitude-modulated microwaves on single excitable cells of chara corallina,” Bioelectromagnetics, vol. 3, no. 2, pp. 203–212, 1982. · ·

20. A. Radzevičius, S. Sakalauskiene, M. Dagys et al., “The effect of strong microwave electric field radiation on: (1) vegetable seed germination and seedling growth rate,” Zemdirbyste, vol. 100, no. 2, pp. 179–184, 2013. · ·

21. S. Lalléchère, S. Girard, D. Roux, P. Bonnet, F. Paladian, and A. Vian, “Mode Stirred Reverberation Chamber (MSRC): a large and efficient tool to lead high frequency bioelectromagnetic in vitro experimentation,” Progress in Electromagnetics Research B, vol. 26, pp. 257–290, 2010. · ·

22. G. Cogalniceanu, M. Radu, D. Fologea, and A. Brezeanu, “Short high-voltage pulses promote adventitious shoot differentiation from intact tobacco seedlings,” Electro- and Magnetobiology, vol. 19, no. 2, pp. 177–187, 2000. ·

23. K. Dymek, P. Dejmek, V. Panarese et al., “Effect of pulsed electric field on the germination of barley seeds,” LWT—Food Science and Technology, vol. 47, no. 1, pp. 161–166, 2012. · ·

24. B. Cadilhon, L. Pécastaing, S. Vauchamp, J. Andrieu, V. Bertrand, and M. Lalande, “Improvement of an ultra-wideband antenna for high-power transient applications,” IET Microwaves, Antennas and Propagation, vol. 3, no. 7, pp. 1102–1109, 2009. · ·

25. M. Răcuciu, S. Miclăuş, and D. E. Creangă, “Non-thermal, continuous and modulated rf field effects on vegetal tissue developed from exposed seeds,” in Electromagnetic Field, Health and Environment, A. Krawczyk, R. Kubacki, S. Wiak, and C. Lemos Antunes, Eds., vol. 29 of Studies in Applied Electromagnetics and Mechanics, pp. 142–148, IOS Press, Amsterdam, The Netherlands, 2008.

26. D. B. Jones, G. P. Bolwell, and G. J. Gilliatt, “Amplification, by pulsed electromagnetic fields, of plant growth regulator induced phenylalanine ammonia-lyase during differentiation in suspension cultured plant cells,” Electromagnetic Biology and Medicine, vol. 5, no. 1, pp. 1–12, 1986. · ·

27. V. P. Sharma, H. P. Singh, D. R. Batish, and R. K. Kohli, “Cell phone radiations affect early growth of vigna radiate (Mung Bean) through biochemical alterations,” Zeitschrift für Naturforschung C, vol. 65, no. 1-2, pp. 66–72, 2010. ·

28. A. Kumar, H. P. Singh, D. R. Batish, S. Kaur, and R. K. Kohli, “EMF radiations (1800 MHz)-inhibited early seedling growth of maize (Zea mays) involves alterations in starch and sucrose metabolism,”Protoplasma, pp. 1–7, 2015. ·

29. M. Kouzmanova, M. Dimitrova, D. Dragolova, G. Atanasova, and N. Atanasov, “Alterations in enzyme activities in leaves after exposure of Plectranthus Sp. plants to 900 MHZ electromagnetic field,”Biotechnology & Biotechnological Equipment, vol. 23, supplement 1, pp. 611–615, 2009. ·

30. S. Radic, P. Cvjetko, K. Malaric, M. Tkalec, and B. Pevalek-Kozlina, “Radio frequency electromagnetic field (900 MHz) induces oxidative damage to DNA and biomembrane in tobacco shoot cells (Nicotiana tabacum),” in Proceedings of the IEEE/MTT-S International Microwave, pp. 2213–2216, IEEE, Honolulu, Hawaii, USA, June 2007. · ·

31. Y.-P. Chen, J.-F. Jia, and Y.-J. Wang, “Weak microwave can enhance tolerance of wheat seedlings to salt stress,” Journal of Plant Growth Regulation, vol. 28, no. 4, pp. 381–385, 2009. · ·

32. V. P. Sharma, G. Singh, and R. K. Kohli, “Effect of mobile phone EMF on biochemical changes in emerging seedlings of Phaseolus aureus Roxb,” The Ecoscan, vol. 3, no. 3-4, pp. 211–214, 2009.

33. H. Zare, S. Mohsenzadeh, and A. Moradshahi, “Electromagnetic waves from GSM a mobile phone simulator and abiotic stress in Zea mays L,” Journal of Nutrition & Food Sciences, vol. S11, p. 3, 2015. ·

34. M.-L. Soran, M. Stan, Ü. Niinemets, and L. Copolovici, “Influence of microwave frequency electromagnetic radiation on terpene emission and content in aromatic plants,” Journal of Plant Physiology, vol. 171, no. 15, pp. 1436–1443, 2014. · ·

35. M. Rammal, F. Jebai, H. Rammal, and W. H. Joumaa, “Effects of long term exposure to RF/MW radiations on the expression of mRNA of stress proteins in Lycopersicon esculentum,” WSEAS Transactions on

Biology and Biomedicine, vol. 11, pp. 10–14, 2014.

36. J. C. Engelmann, R. Deeken, T. Müller, G. Nimtz, M. R. G. Roelfsema, and R. Hedrich, "Is gene activity in plant cells affected by UMTS-irradiation? A whole genome approach," Advances and Applications in Bioinformatics and Chemistry, vol. 1, pp. 71–83, 2008. ·

37. A. Scialabba and C. Tamburello, "Microwave effects on germination and growth of radish (Raphanus sativus L.) seedlings," Acta Botanica Gallica, vol. 149, no. 2, pp. 113–123, 2002. ·

38. A. Akbal, Y. Kiran, A. Sahin, D. Turgut-Balik, and H. H. Balik, "Effects of electromagnetic waves emitted by mobile phones on germination, root growth, and root tip cell mitotic division of Lens culinarisMedik," Polish Journal of Environmental Studies, vol. 21, no. 1, pp. 23–29, 2012. ·

39. Y. C. Chen and C. Chen, "Effects of mobile phone radiation on germination and early growth of different bean species," Polish Journal of Environmental Studies, vol. 23, no. 6, pp. 1949–1958, 2014. ·

40. M. Răcuciu, C. Iftode, and S. Miclaus, "Inhibitory effects of low thermal radiofrequency radiation on physiological parameters of Zea mays seedlings growth," Romanian Journal of Physics, vol. 60, no. 3-4, pp. 603–612, 2015.

41. L. Ragha, S. Mishra, V. Ramachandran, and M. S. Bhatia, "Effects of low-power microwave fields on seed germination and growth rate," Journal of Electromagnetic Analysis and Applications, vol. 3, no. 5, pp. 165–171, 2011. ·

42. S. Sharma and L. Parihar, "Effect of mobile phone radiation on nodule formation in the leguminous plants," Current World Environment Journal, vol. 9, no. 1, pp. 145–155, 2014. ·

43. A. O. Oluwajobi, O. A. Falusi, and N. A. Zubbair, "Flower bud abscission reduced in hibiscus sabdariffa by radiation from GSM mast," Environment and Pollution, vol. 4, no. 1, pp. 53–57, 2015. ·

44. M. Tafforeau, M.-C. Verdus, V. Norris et al., "Plant sensitivity to low intensity 105 GHz electromagnetic radiation," Bioelectromagnetics, vol. 25, no. 6, pp. 403–407, 2004. · ·

45. A. Grémiaux, S. Girard, V. Guérin et al., "Low-amplitude, high-frequency electromagnetic field exposure causes delayed and reduced growth in Rosa hybrida," Journal of Plant Physiology, vol. 190, pp. 44–53, 2016.

46. V. J. Berdinas-Torres, Exposure›s systems and dosimetry of large-scale in vivo studies [Ph.D. thesis], Swiss Federal Institute of Technology, Zürich, Switzerland, 2007.

47. M. J. Van Wyk, M. Bingle, and F. U. C. Meyer, "Antenna modeling considerations for accurate SAR calculations in human phantoms in close proximity to GSM cellular base station antennas,"Bioelectromagnetics, vol. 26, no. 6, pp. 502–509, 2005. · ·

48. S. Miclăuş and M. Răcuciu, "A dosimetric study for experimental exposures of vegetal tissues to radiofrequency fields," in Electromagnetic Field, Health and Environment, A. Krawczyk, R. Kubacki, S. Wiak, and C. Lemos Antunes, Eds., vol. 29 of Studies in Applied Electromagnetics and Mechanics, pp. 133–141, IOS Press, Amsterdam, The Netherlands, 2008.

49. J. León, M. C. Castillo, A. Coego, J. Lozano-Juste, and R. Mir, "Diverse functional interactions between nitric oxide and abscisic acid in plant development and responses to stress," Journal of Experimental Botany, vol. 65, no. 4, pp. 907–921, 2014. · ·

50. Y.-P. Chen, J.-F. Jia, and X.-L. Han, "Weak microwave can alleviate water deficit induced by osmotic stress in wheat seedlings," Planta, vol. 229, no. 2, pp. 291–298, 2009. · ·

51. Z.-B. Qiu, J.-L. Guo, M.-M. Zhang, M.-Y. Lei, and Z.-L. Li, "Nitric oxide acts as a signal molecule in microwave pretreatment induced cadmium tolerance in wheat seedlings," Acta Physiologiae Plantarum, vol. 35, no. 1, pp. 65–73, 2013. · ·

52. R. K. Jangid, R. Sharma, Y. Sudarsan, S. Eapen, G. Singh, and A. K. Purohit, "Microwave treatment induced mutations and altered gene expression in Vigna aconitifolia," Biologia Plantarum, vol. 54, no. 4, pp. 703–706, 2010. · ·

53. A. Vian, C. Henry-Vian, and E. Davies, "Rapid and systemic accumulation of chloroplast mRNA-binding protein transcripts after flame stimulus in tomato," Plant Physiology, vol. 121, no. 2, pp. 517–524, 1999. · ·

54. J. A. Teixeira da Silva and J. Dobránszki, "Magnetic fields: how is plant growth and development impacted?" Protoplasma, 2015. ·

55. M. D. H. J. Senavirathna, T. Asaeda, B. L. S. Thilakarathne, and H. Kadono, "Nanometer-scale elongation rate fluctuations in the Myriophyllum aquaticum (Parrot feather) stem were altered by radio-frequency electromagnetic radiation," Plant Signaling and Behavior, vol. 9, no. 4, Article ID e28590, 2014. · ·

56. M. Ursache, G. Mindru, D. E. Creangă, F. M. Tufescu, and C. Goiceanu, "The effects of high frequency electromagnetic waves on the vegetal," Romanian Journal of Physics, vol. 5, no. 1-2, pp. 133–145, 2009.

57. E. A. M. Hamada, "Effects of microwave treatment on growth, photosynthetic pigments and some metabolites of wheat," Biologia Plantarum, vol. 51, no. 2, pp. 343–345, 2007. · ·

58. M.D.H.J.Senavirathna,A.Takashi,andY.Kimura,"Short-duration exposure to radiofrequency electromagnetic radiation alters the chlorophyll fluorescence of duckweeds (Lemna minor),"Electromagnetic Biology and Medicine, vol. 33, no. 4, pp. 327–334, 2014. · ·

59. Y. Shckorbatov, "The main approaches of studying the mechanisms of action of artificial electromagnetic fields on cell," Journal of Electrical & Electronic Systems, vol. 3, no. 2, 2014. ·

60. A. M. Pietak, "Structural evidence for electromagnetic resonance in plant morphogenesis," BioSystems, vol. 109, no. 3, pp. 367–380, 2012. · ·

61. A. Amat, J. Rigau, R. W. Waynant, I. K. Ilev, and J. J. Anders, "The electric field induced by light can explain cellular responses to electromagnetic energy: a

hypothesis of mechanism," Journal of Photochemistry and Photobiology B: Biology, vol. 82, no. 2, pp. 152–160, 2006. · ·

62. Z. Qiu, J. Li, Y. Zhang, Z. Bi, and H. Wei, "Microwave pretreatment can enhance tolerance of wheat seedlings to $CdCl_2$ stress," Ecotoxicology and Environmental Safety, vol. 74, no. 4, pp. 820–825, 2011. · ·

63. Y.-P. Chen, "Microwave treatment of eight seconds protects cells of Isatis indigotica from enhanced UV-B radiation lesions," Photochemistry and Photobiology, vol. 82, no. 2, pp. 503–507, 2006.

CHAPTER 11 EVOLUTIONARY ALGORITHM GEOMETRY OPTIMIZATION OF OPTICAL ANTENNAS

Ramón Díaz de León-Zapata, Gabriel González, Efrén Flores-García, Ángel Gabriel Rodríguez, and Francisco Javier González

INTRODUCTION

Printed circuit antennas which have been extensively used in the radio frequency (RF) spectrum have also been used to detect electromagnetic radiation at optical and infrared frequencies. The use of these types of antennas at optical frequencies provides several advantages over traditional optical and infrared detectors; among these advantages are low profile, low cost, faster response times, compatibility with integrated circuit technology, and wavelength and polarization selectivity [1, 2].

Even though several RF antenna designs have been successfully used at optical and infrared frequencies [3], the design of antennas at optical frequencies by using traditional RF antenna design rules can result in nonoptimized antennas due to the dispersion of material properties that can be neglected at lower frequencies [4, 5].

Evolutionary algorithms imitate nature, where all living organisms possess specific genetic material which contains information about each organism that can be transferred to new generations via reproduction. The other organism involved in reproduction also transfers some of its characteristics [6]. These characteristics are encoded in genes stored in chromosomes, which together constitute the genetic material known as a genotype [7]. Genes are modified during the characteristic transfer process as a consequence of the crossover between maternal and paternal chromosomes. Mutation may also occur, altering the information contained within the genes of a given chromosome. Although the newly created individual possesses the information of its parents, the combination of two different organisms makes the individual unique. This organism begins life in an environment that is not significantly different from that of its parents. The new organism develops in a manner that allows it to survive and transfer its genome, permitting the species to persist in a given environment. An individual that cannot adapt to its environment will struggle to survive and transfer its genes to new generations.

These evolutionary processes can be used to optimize the solution of nonanalytical problems assuming that the environment is defined based on known values and characteristics. The evolved individual or the whole evolved population can constitute the potential solution to a given problem [8–11].

An adequate mathematical function must be chosen to define the fitness of any given individual representing how well adapted they are to their environment. Exchange of

Citation: Ramón Díaz de León-Zapata, Gabriel González, Efrén Flores-García, Ángel Gabriel Rodríguez, and Francisco Javier González, "Evolutionary Algorithm Geometry Optimization of Optical Antennas," International Journal of Antennas and Propagation, vol. 2016, Article ID 3156702, 7 pages, 2016. doi:10.1155/2016/3156702 genetic materials and mutations will occur during the genetic crossover process. Thus, an optimal solution will be created that best suits a given environment.

These algorithmic techniques allow the exploration of unconventional geometries or combination of materials that can be used in the design of nanophotonic circuits and devices for diverse applications [12].

In this work, a genetic algorithm (GA) is used to obtain the geometry that optimally concentrates the electromagnetic field of a dipole-type nanoantenna at a resonance frequency of 500 THz, which can be varied based on the nanostructure dimensions. The nanoantenna radiation pattern was obtained via computational simulations and compared to a classical dipole geometry.

METHOD

The most basic and general type of genetic algorithm called "Holland Genetic Algorithm" proposed by Henry Holland [13] was used. The flow diagram of the genetic algorithm is presented in Figure 1. It starts with an initial random population's chromosomes (population is equivalent to the geometry coordinates and the chromosomes are their values in this work). A loop is implemented to evaluate repeatedly the fitness of chromosomes in the population where only those with better health survive and can combine their chromosomes to produce new individuals. Some genetic operators like random mutation are inserted at this point only to certain percentage of the new individuals, just like in biological specimens. The loop ends when a stopping criterion is achieved; it can be the number of iterations, level of health, and so forth, and this final stage has the best chromosome (or solution).

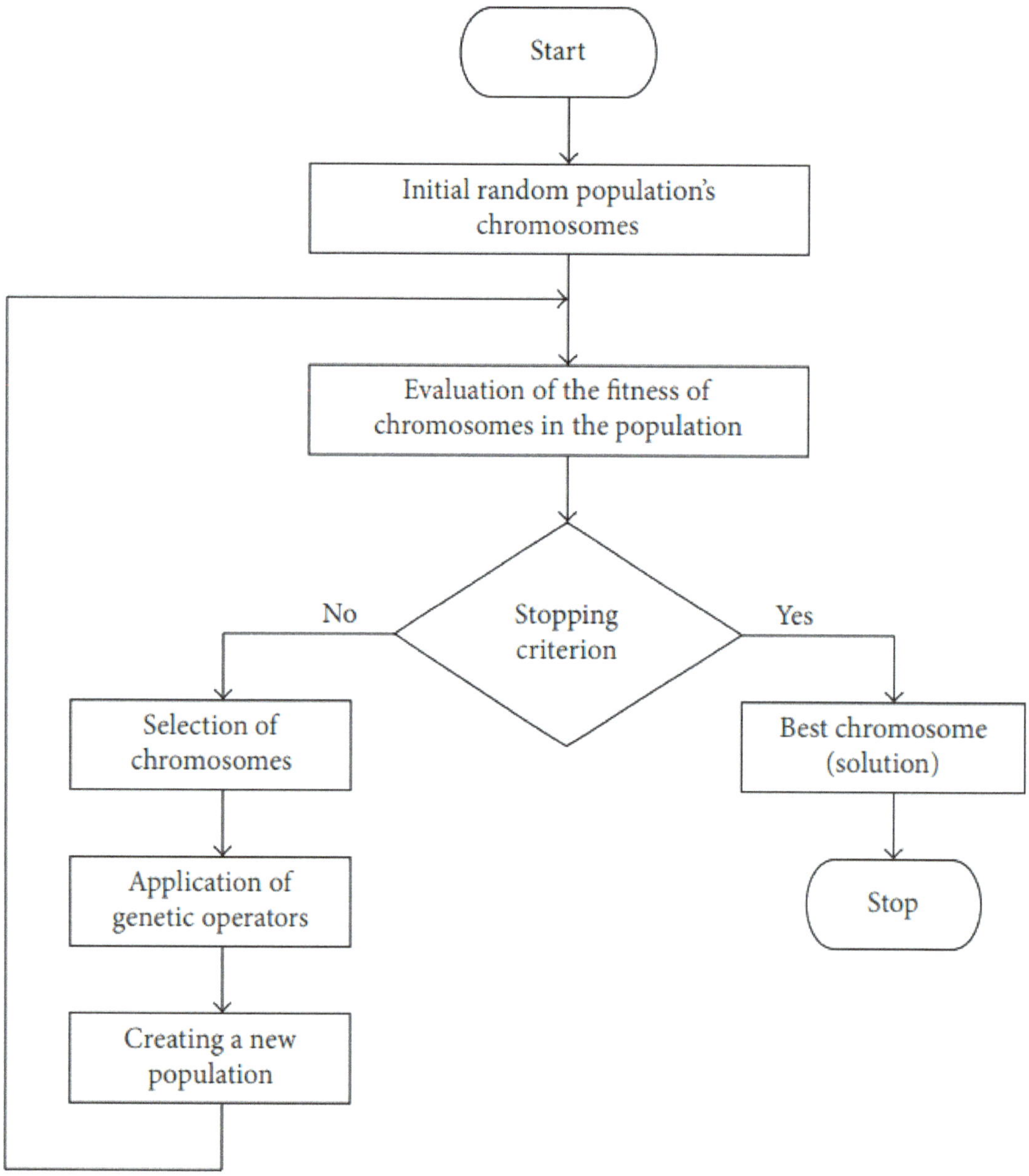

Figure 1: Flow diagram of the genetic algorithm.

Various studies, such as [8, 11], have applied this algorithm to multiple problems, where the resulting optimized geometries are different from classical macroscopic antennas, as shown in Figure 2.

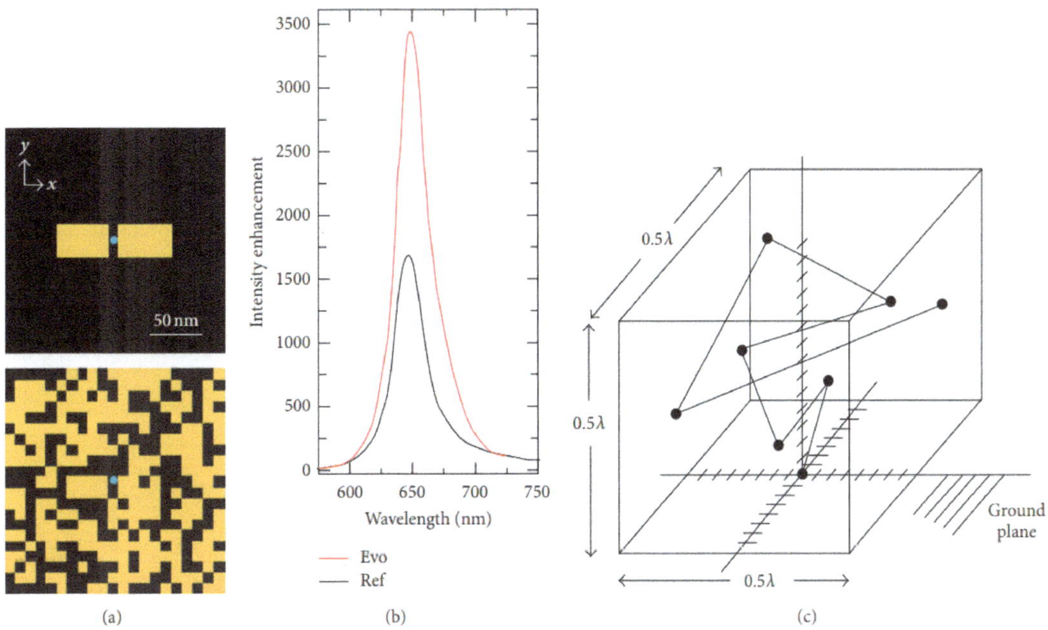

Figure 2: (a) Top: reference classical dipole. Bottom: geometry generated by the genetic algorithm. (b) Efficiency comparison between the reference antenna and that generated by the genetic algorithm, taken from [11]. (c) Geometric model obtained using the evolutionary algorithm, taken from [8].

Figure 2(a) illustrates a significant electromagnetic field concentration increase, but the active antenna area also increased. Therefore, the obtained geometry efficiencies cannot be directly compared due to the effective antenna area increase. Figure 2(c) represents a series of macroscopic antenna simplifications, which are not applicable at the nanoscale. Thus, this final geometry does not offer a viable alternative for the type of nanoantennas that we wish to analyze and eventually fabricate.

The authors assume a specific preestablished antenna area based on the desired resonance frequency. The geometry is then modified using the evolutionary algorithm as a malleable material. No empty regions are added, and a solid shape is sought which achieves a maximum electromagnetic field concentration at the center of the nanostructure.

This algorithm has been successfully applied to multiple problems; the results show that the resulting optimized geometries are different from classical RF antennas [8, 11].

The genetic algorithm was programmed in MATLAB where a link was made to COMSOL Multiphysics where the electromagnetic simulations were performed and the results were returned to MATLAB where the fitness function was evaluated. Iterative changes or adjustments are then made to optimize the solution, and the new proposed nanostructure is analyzed using COMSOL until a fixed number of iterations was achieved.

The genetic algorithm performs the nanostructure analysis required to suggest a geometry that approaches the optimal design conditions, assuming a (two-dimensional) flat nanostructure. The solution space is constrained to dimensions near the optical wavelength for an antenna irradiated by a normal incident electromagnetic wave. The main parameters required by the genetic algorithm, such as the dimensions of the simulation space (maximum size of the antenna), number of chromosomes (number of individuals to be analyzed), chromosome resolutions (quality), overlap between generations (mortality index), mutation rate (as a percentage), and mating rules, must be input by the user.

Figure 3 illustrates the MATLAB and COMSOL processes, which together comprise the genetic algorithm.

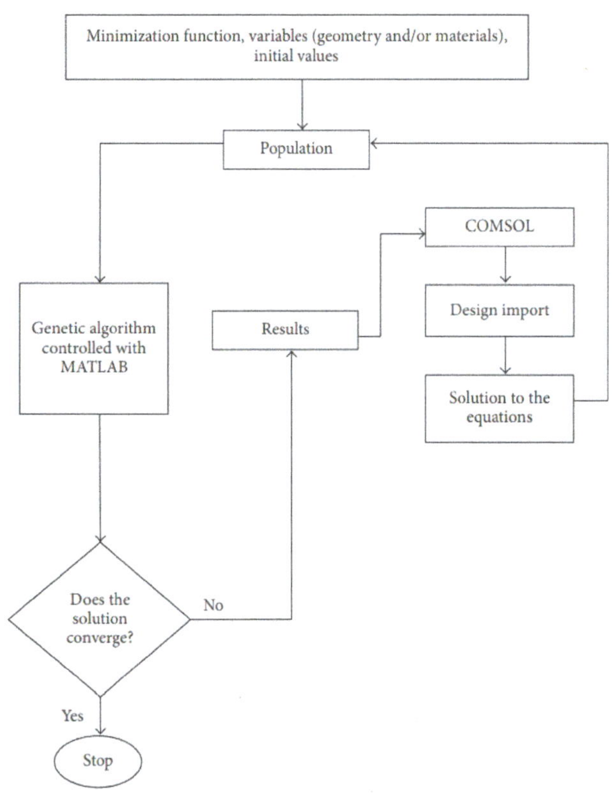

Figure 3: Flow diagram of the application process, noting the links between the COMSOL and MATLAB software packages.

The Bézier curve control points, which are modeled as chromosomes, are used to obtain the optimized geometry of the dipole-type nanoantenna. A total of 10 lines are set, including 4 with two Bézier control points, 4 with only one, and 2 straight lines with no control points. The lines with no control points represent the section where the electromagnetic field is applied to the nanoantenna. The algorithm execution stops if the average loss in the electric field is zero or if the iterative limit is reached, which is based on the number of generations [14].

The fitness function [15] is shown in (1). The function should accept a vector, whose length is the number of independent variables, and return a scalar. This function must accept a vector of length 24, corresponding to the variables (points in x and y coordinates) that conform the geometry (which includes the straight and Bézier lines), and return a scalar equal to the value of the function that represents the value of the electromagnetic power loss density for such geometry. By minimizing this equation (minimizing losses), we can find the strongest electromagnetic field. As an application example, the optimized fitness function calculates the minimum loss (optimization function) of the electromagnetic field at a frequency of 500 THz and is a parameter that COMSOL submits to MATLAB, which then evaluates it using the genetic algorithm function:

$$F_{\text{Fitness}} = \min\left(\frac{1}{2}\operatorname{Re}\left(\mathbf{J}_{\text{tot}} \cdot \mathbf{E}_{\text{tot}}\right)\right), \tag{1}$$

where J_{tot} is the total electric current density over the whole geometry shown in

$$\mathbf{J}_{\text{tot}} = \sigma\mathbf{E}. \tag{2}$$

σ is the electrical conductivity and E_{tot} is the electric field over the whole geometry shown in

$$\mathbf{E}_{\text{tot}} = \left(\mu_0 \int \left(\mathbf{J} + \varepsilon_0 \frac{\partial \mathbf{E}}{\partial t}\right) da\right). \tag{3}$$

The refractive index in the optical regime of operation in the nanostructured materials (metamaterials) becomes a function of frequency [5].

A geometric model with 12 Bézier curve control points, plus 11 fixed points which represent the geometry limits (noting the first point is also the last one to have a closed geometry), is found during the first iteration, based on steady-state initial conditions, 100 chromosomes, a 50% intergeneration overlap, and a 1% mutation rate. In addition, a single crossover point was used, which can occur between any pair of segments in the chromosome with equal probability. Figure 4 shows the superposition of the geometry as it evolves. Figure 5plots the electric field power loss as it approaches zero. Values after the fourth population generation, which rapidly decrease from $1.3e^{-7}$ nW/m^2 to $8e^{-15}$ nW/m^2 , cannot be detected due to scale limitations.

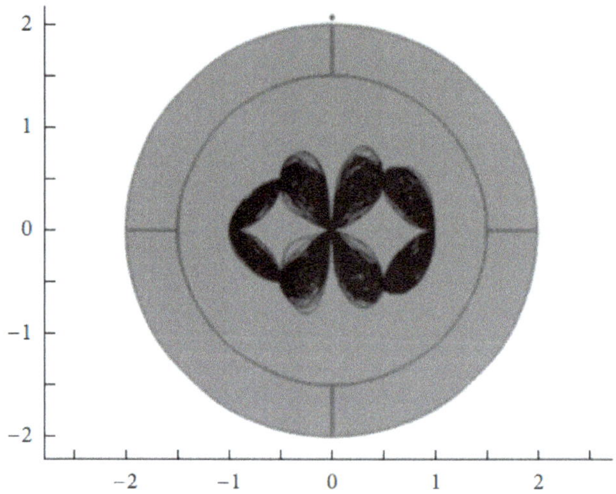

Figure 4: Lines demonstrating the geometry modification process due to mutations when applying the genetic algorithm.

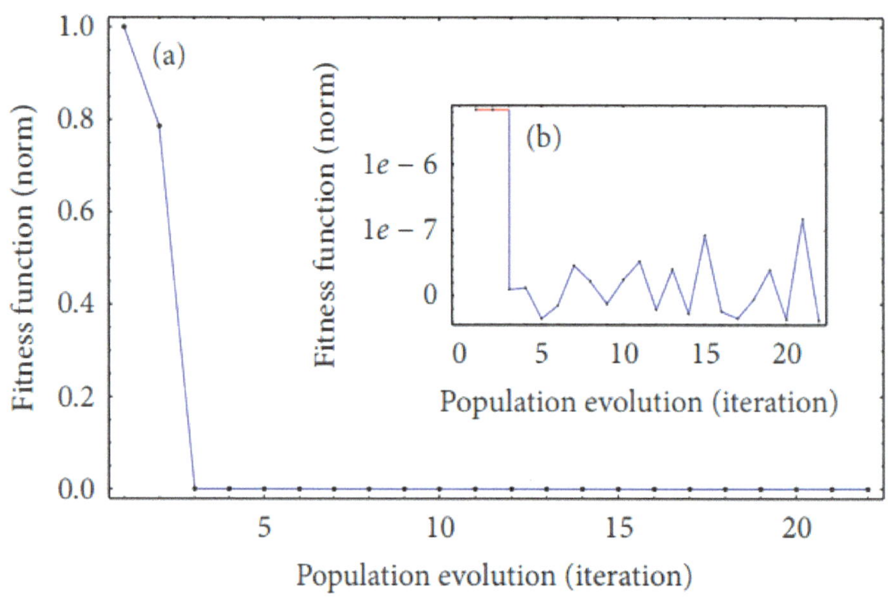

Figure 5: (a) Plot showing the trend toward zero of the electromagnetic field loss after each new population generation. (b) Data trends for a zoomed portion of the main plot.

The geometry obtained at the end of the genetic algorithm execution is shown in Figure 6(a). Figure 6(b) illustrates the data point at which the maximum optical energy absorption-emission is obtained in the terahertz range.

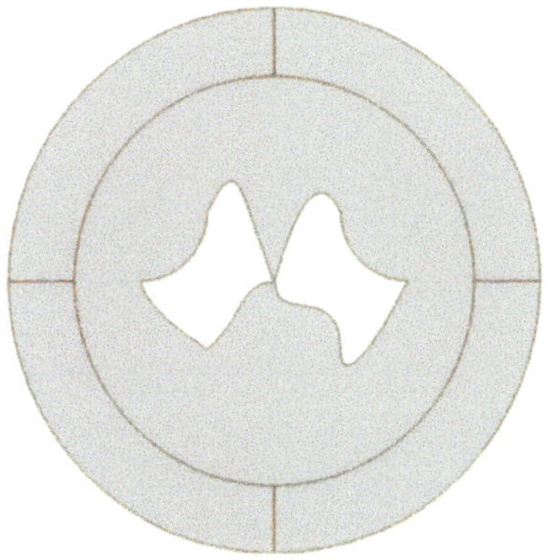

(a)

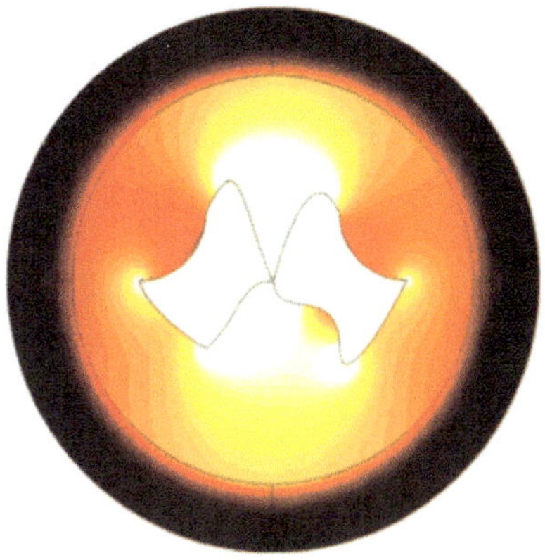

(b)

Figure 6: (a) Geometry obtained after the genetic algorithm optimization function application. (b) Finite element method simulation of the nanostructure electromagnetic radiation pattern.

According to the results obtained by [8, 11] and the resulting geometry in Figure 6(b), the geometries obtained by the genetic algorithms represent improvements over the classical or conventional radiofrequency antenna geometries. These geometries were obtained at reduced computer costs and over a shorter processing time compared with traditional analytical processes or trying wide classical RF geometries one by one.

RESULTS

Many studies use analogies between nanostructure geometries and conventional radiofrequency macroscopic antenna geometries based on the assumption that their behaviors can be extrapolated to optical frequencies as in [16]. However, the proposed computational model demonstrated that nanoantenna geometries require further study. The nanoantenna shape analyzed in this study does not feature a conventional geometry, such as those utilized for the radiofrequency range. Figure 7 shows a comparison between the classical dipole (a), the dipole generated by the algorithm after the first iteration (b), and the final geometry, which is based on the lowest electromagnetic field loss at the dipole center (c). Figure 7(a) is a theoretical dipole which fully covers the area proposed to evolve our geometry, demonstrating that increasing area does not increase the electromagnetic field even at their (Figure 7(a)) own resonant frequency.

(a) (b) (c)

Figure 7: Electromagnetic field concentration comparison according to the geometry. Panel (a) shows a classical dipole; (b) shows the first iteration of the genetic algorithm; and (c) shows the final geometry, which is based on the maximum electromagnetic field concentration at the center of the nanostructure.

Figure 8 compares the electromagnetic field intensities of the classical dipole and the geometry generated by the genetic algorithm. The results obtained via the genetic algorithm represent an optimal concentration, and both geometries encompass the same effective area. Compared to Figure 2(b), Figure 8 shows an increase in the electromagnetic signal and the effective bandwidth, with the difference that the area remained the same; as an added advantage, the GA design presented in this work is simpler than the one shown in Figure 2(a); therefore, the fabrication process would also be simpler. The peaks of both plots represent the dipole resonance frequencies. From Figure 8, with an increase from 0.4 for non-GA dipole to 0.93167 for GA dipole at resonant frequency in the antenna response (a difference of about 0.536), we found that the geometries obtained by the genetic algorithm provided a 33% better result than the classic dipole geometry and with the bandwidth for non-GA dipole being about 39 THz (from 481 to 520 THz) and for the GA dipole about 48.7 THz (from 475.6 to 524.4 THz) it was increasing the bandwidth by 25%.

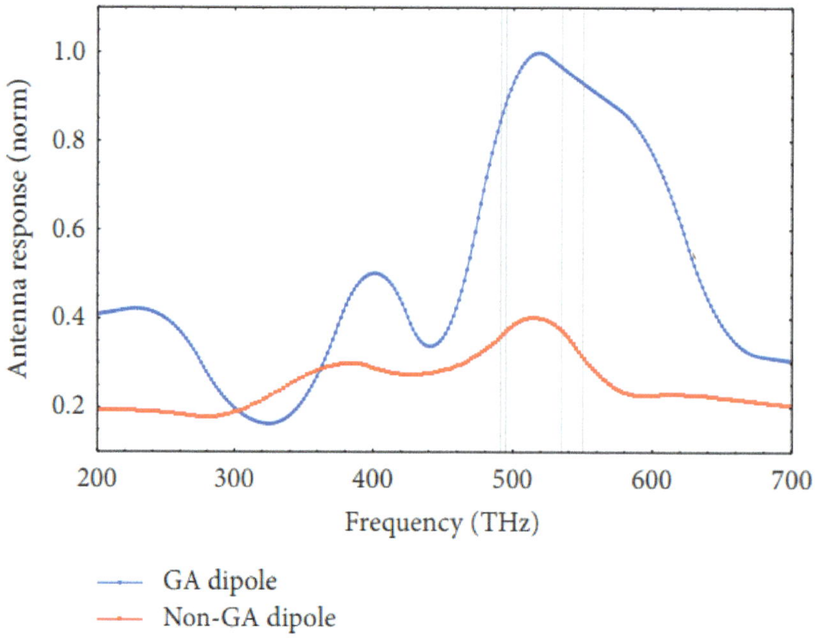

Figure 8: Electromagnetic field concentration comparison between the classical dipole geometry and that generated by the genetic algorithm.

CONCLUSIONS

The classical or conventional dipole-type antenna geometries (in the radiofrequency regime) do not encompass the maximum electromagnetic field concentration (in the optical regime). This is due to the intrinsic differences between electron and photon behaviors. In addition, many macroscopic antenna assumptions and simplifications are not applicable for nanoscale optical frequency regimes because the electromagnetic wavelength is comparable to or even shorter than the antenna dimensions. Thus, a new refractive index function is introduced which defines the electromagnetic field behavior at such frequencies.

The proposed alternative genetic algorithm was applied to improve dipole geometry while accounting for the nanoscopic scale properties of these structures. Our results demonstrate that the final nanoantenna shape is significantly different than the classical case in the context of providing the optimal electromagnetic field concentration.

The results of this study will be used in future nanostructure fabrication and characterization studies using two materials with different Seebeck coefficients (one positive and one negative). The materials will generate maximum heating in the region of interest, producing a direct electric current [17] that can be stored in batteries for subsequent use. This will contribute to the creation of renewable energy devices.

ACKNOWLEDGMENTS

This work was supported by the "Cátedras CONACYT" program, Project 32 of "Centro Mexicano de Innovación en Energía Solar" from Fondo Sectorial CONACYT-Secretaría de Energía-Sustentabilidad Energética and by the National Laboratory Program from CONACYT through the Terahertz Science and Technology National Lab (LANCYTT). Ramón Díaz de León-Zapata thanks Tecnológico Nacional de México for the "Licencia por Beca Comisión 2015" scholarship and the "Materiales y Dispositivos Optoelectrónicos de la Universidad Autónoma de San Luis Potosí" academic group for their support and collaboration.

REFERENCES

1. F. J. González, B. Ilic, J. Alda, and G. D. Boreman, "Antenna-coupled infrared detectors for imaging applications," IEEE Journal on Selected Topics in Quantum Electronics, vol. 11, no. 1, pp. 117–120, 2005.
2. F. J. Gonzalez, Wiley Enciclopedia of Electrical and Electronics Engineering, Edited by J. Webster, John Wiley & Sons, New York, NY, USA, 2015.
3. F. J. González, "Thermal-impedance simulations of antenna-coupled microbolometers," Infrared Physics and Technology, vol. 48, no. 3, pp. 223–226, 2006.
4. C. Fumeaux, M. A. Gritz, I. Codreanu, W. L. Schaich, F. J. González, and G. D. Boreman, "Measurement of the resonant lengths of infrared dipole antennas," Infrared Physics & Technology, vol. 41, no. 5, pp. 271–281, 2000.
5. F. J. González, J. Alda, J. Simón, J. Ginn, and G. Boreman, "The effect of metal dispersion on the resonance of antennas at infrared frequencies," Infrared Physics and Technology, vol. 52, no. 1, pp. 48–51, 2009.
6. C. R. Reeves and J. E. Rowe, Genetic Algorithms: Principles and Perspectives, vol. 20, Kluwer Academic Publishers, Boston, Mass, USA, 2003. View at MathSciNet
7. M. Mitchell, An Introduction to Genetic Algorithms, Massachusetts Institute of Technology, Cambridge, Mass, USA, 1999.
8. L. S. Derek, "Antenna design using genetic algorithms," in Proceedings of the Genetic and Evolutionary Computation Conference (GECCO ‹02), New York, NY, USA, July 2002.
9. D. E. Goldberg, Genetic Algorithms in Search, Optimization and Machine Learning, Kluwer Academic, Boston, Mass, USA, 1989.
10. L. Rutkowski, Computational Intelligence, Polish Scientific, Berlin, Germany, 2008.

11. T. Feichtner, O. Selig, M. Kiunke, and B. Hecht, "Evolutionary optimization of optical antennas," Physical Review Letters, vol. 109, no. 12, Article ID 127701, 5 pages, 2012.

12. S. Chakraborty, U. Mukherjee, and K. Anchalia, "Circular micro-strip(Coax feed) antenna modelling using FDTD method and design using genetic algorithms: a comparative study on different types of design techniques," in Proceedings of the International Conference on Computer and Communication Technology (ICCCT ‹14), pp. 329–334, Allahabad, India, September 2014.

13. M. Mitchel, An Introduction to Genetic Algorithms, MIT Press, 1999.

14. B. Hahn and D. Valentine, Essential Matlab, Academic Press, New York, NY, USA, 2013.

15. Y. K. Zhong, S. M. Fu, N. P. Ju, P. Y. Chen, and A. Lin, "Experimentally-implemented genetic algorithm (Exp-GA): toward fully optimal photovoltaics," Optics Express, vol. 23, no. 19, pp. A1324–A1333, 2015.

16. E. Briones, A. Cuadrado, J. Briones et al., "Seebeck nanoantennas for the detection and characterization of infrared radiation," Optics Express, vol. 22, no. 21, pp. A1538–A1546, 2014.

17. B. Mora-Ventura, R. Díaz de León, G. García-Torales, J. L. Flores, J. Alda, and F. J. González, "Responsivity and resonant properties of dipole, bowtie, and spiral Seebeck nanoantennas," Journal of Photonics for Energy, vol. 6, no. 2, Article ID 024501, 2016.

CHAPTER 12 MICROSTRUCTURE AND ELECTROMAGNETIC PROPERTIES OF NI-ZN-CO FERRITE UP TO 20 GHZ

Charalampos Stergiou

INTRODUCTION

Among ferrimagnetic oxides, the cubic ferrite of Mn-Zn type has the largest production and market share due to their superior behavior under ac magnetic fields at frequencies roughly reaching 1 MHz. However, for the growing electronic, telecommunication, and power applications at higher frequencies other types of magnetic oxides, such as Ni-Zn cubic and hexagonal ferrite, are preferably employed. These materials possess higher electrical resistivity and magnetocrystalline anisotropy, which ensure sufficiently low power losses and high magnetic response in this frequency range [1, 2]. In particular in the group of Ni-Zn spinel ferrite types, by introducing Co as a substituting element the operation band extends to even higher frequencies, which actually reflects the persistence of moderate magnetic permeability values up to a higher cutoff frequency limit. This effect basically occurs on account of the strong positive contribution of cobalt to the ferrite anisotropy, which originates in the unquenched orbital moment of Co^{2+} cation in the octahedral crystal sites [3]. The induced anisotropy, even by low Co doping, inhibits the domain wall displacement and intensifies the ferromagnetic resonance, which may be exploited in the design of Ni-Zn ferrite with low loss above 1 MHz [4–7].

Except for defining the frequency region of operation through compositional control, the developed microstructure of the ceramic is also known to affect the electromagnetic (EM) performance of ferrite. In fact, advanced tailoring of the morphological features through adjustment of the processing parameters is one of the most significant technical objectives in ferrite industry over time [1, 8, 9]. Therefore, various recently published reports examine the strong relation between grain size, porosity, density, and magnetic properties of Ni-Zn ferrite at high frequencies, with a view of forming certain material design directions [10–14].

Nevertheless, there is a deficiency in the respective literature focusing on the Ni-Zn ferrite with high Co concentration, which is enlarged by the necessity for knowledge of both complex magnetic permeability $\mu*(f)$ and dielectric permittivity $\varepsilon *(f)$ in order to design the final passive components. To this end, we probe into the impact of heat treatment and microstructure on the microwave EM properties of Corich Ni-Zn spinels by varying the top temperatures during the prefiring and sintering stages of the ceramic synthesis route.

MATERIALS AND METHODS

In the present investigation, polycrystalline samples with the chemical composition $Ni_{0.25}Zn_{0.25}Co_{0.5}Fe_2O_4$ were synthesized

Citation: Charalampos Stergiou, "Microstructure and Electromagnetic Properties of Ni-Zn-Co Ferrite up to 20 GHz,"Advances in Materials Science and Engineering, vol. 2016, Article ID 1934783, 7 pages, 2016. doi:10.1155/2016/1934783.

by the conventional solid state reaction route. Specifically, stoichiometric amounts of the precursor oxides NiO (Alfa Aesar, >99.0%), ZnO (Merck, >99.0%), Co_2O_3 (Riedel-de Haën, >95.0%), and Fe_2O_3 (Riedel-de Haën, >99.0%) were mixed for 3 hours in distilled water, dried overnight, and calcined in air for 5 hours. The prefired powders were ball milled for 3 hours, dried, and subsequently granulated with the addition of polyvinyl alcohol solution. The granulated powders were then pressed under axial compaction (80 MPa) to form disc specimens of 12 mm diameter and were finally sintered in air for 10 hours. Table 1 depicts the applied variations of annealing temperatures and the respective sample designations.

Table 1: Experimental variables under test

Sample	Prefiring temperature T_P (°C)	Sintering temperature T_S (°C)
S1 (REF)	1000	1250
S2	750	1250
S3	1000	1200

Both the prefired and the sintered materials were pulverized and subjected to phase and crystal structure identification with the X-ray powder diffraction method (Seifert, XRD 3003 TT, CuKa). The microstructure of the developed ferrite was examined with scanning electron microscopy (JEOL, SEM JSM6300), from which the median grain sizes Dv50 were calculated by the linear intercept technique. With regard to the static magnetic characterization, the specific saturation magnetization (σ_S) and coercive field (H_c) were obtained from hysteresis loops recorded up to 10 kOe at room temperature by using a vibrating-sample magnetometer (Quantum Design, VSM VersaLab 3T). Moreover, the materials' constitutive EM properties (complex permittivity $\varepsilon*$ and permeability $\mu*$) were measured in the frequency range 10 MHz–20 GHz by means of impedance and vector network analysis (Agilent, E4991A and HP, 8720C) and by applying the Nicolson-Ross-Weir algorithm [15]. To this effect, the sintered discs were precisely ground to form ring samples with outer diameter of 7 mm, inner diameter of 3 mm, and height of 1.5 mm. Finally, the prepared ferrite was evaluated with reference to its ability to attenuate either the EM waves reflected from a conducting plane or the waves propagating through it. Specifically, the return losses (RL) were calculated from the respective overall reflection coefficient (Γ) in decibels according to the formula $RL = -20 \cdot \log(|\Gamma|)$.

RESULTS AND DISCUSSION

Structural Properties

The investigation with XRD technique of the powders prefired at 750°C and 1000°C has resulted in the patterns displayed in Figure 1. We thus observe that prefiring at 750°C results in a mixture of the desired Ni-Zn ferrite (ICDD-PDF no. 87-2337) and the cation oxides Fe_2O_3 (ICDD-PDF no. 33-0664), Co_3O_4 (ICDD-PDF no. 09-0418), and NiO (ICDD-PDF no. 89-7130), whereas after prefiring at 1000°C only the spinel ferrite phase is present. Due to this compositional variation, the pressibility of the two ceramic powders is altered, with the higher press density PD under constant pressure being achieved by the pure ferrite powder (Table 2). The lower PD of the multiphase material is attributed to its pressibility, as the theoretical densities of the cation oxides (Fe_2O_3, Co_3O_4, and NiO) are similar or even higher than that of ferrite. With regard to the sintered materials, despite their different heat treatment history described in Table 1, all the diffraction peaks were indexed according to a single phase of spinel ferrite (ICDD-PDF no. 87-2337) with inverse spinel structure (S.G.: Fd-3m), as it is depicted in Figure 2. In particular for the S1 sample, by applying the Rietveld analysis of the XRD data the cell constant was refined to a = 8.3918 Å [16]. Beyond the similarities in crystal structure of the sintered ferrite, the respective SEM images reveal significant microstructural variations between them (Figures3 (a)–3(c)). Thus, we notice that the samples S2 and S3 with lower prefiring and/or sintering temperatures exhibit increased total porosity volume and suppressed grain growth, compared to that of sample S1. These variations are easily spotted by comparing the median grain size Dv50, the sintered density SD, and the respective relative density, which are cited in Table 2. In conclusion, it seems that the pores formed due to the insufficient pressibility of multiphase composition (lower T_p), are preserved during sintering, and finally yield the lowest SD. On the other hand, the lower sintering temperature T_S leads to incomplete densification process and a porous polycrystalline ceramic with the finest grains.

Table 2: Morphology-related properties of the sintered ferrite

Sample	Press density PD (g/cm^3)	Median grain size $Dv50$ (μm)	Sintered density SD (g/cm^3)	Relative density (%)
S1	2.94 ± 0.02	3.4 ± 0.2	5.08 ± 0.01	94.1
S2	2.73 ± 0.02	3.0 ± 0.1	4.82 ± 0.02	89.3
S3	2.94 ± 0.02	2.1 ± 0.2	4.93 ± 0.01	91.3

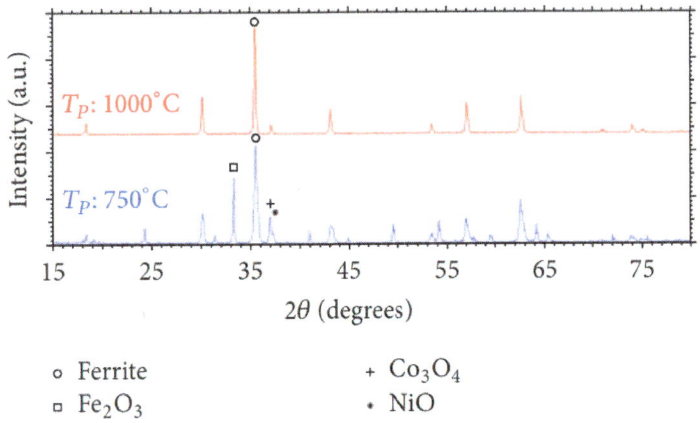

Figure 1: X-ray diffraction patterns of the prefired samples.

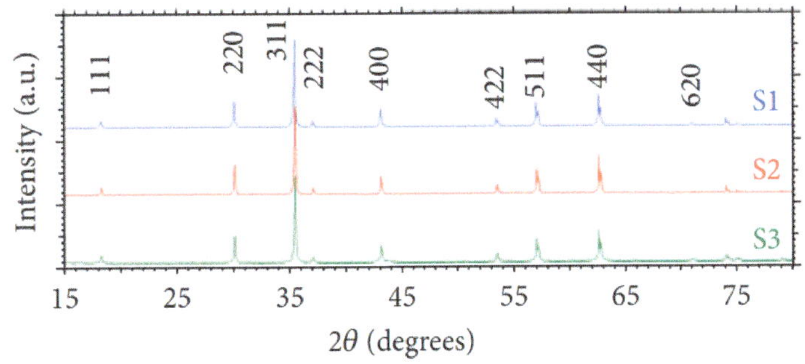

Figure 2: X-ray diffraction patterns of the sintered samples.

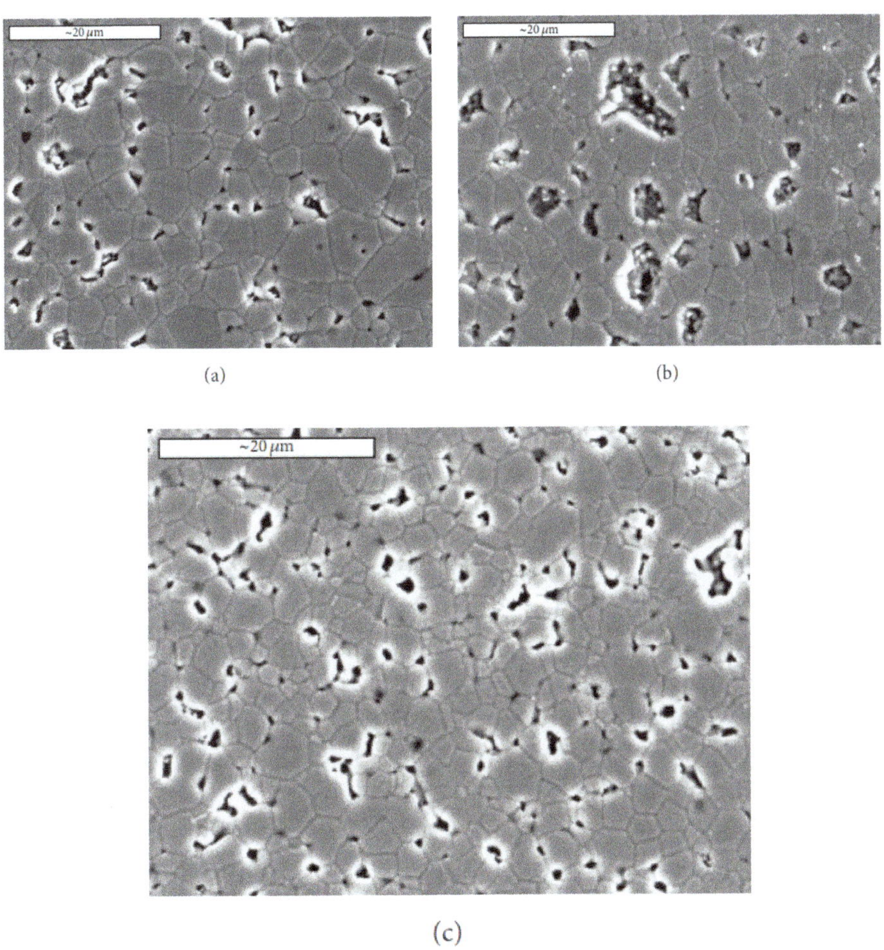

Figure 3: SEM images of the samples (a) S1, (b) S2, and (c) S3. The scale bar length is approximately 20 μm.

Static Magnetic Properties

In order to examine the impact of microstructure on the static magnetic properties of the spinel ferrite, we have extracted the values of the specific saturation magnetization M_S and coercive field H_C from the respective hysteresis loops under dc magnetic field (Table 3). Taking into account the measurement standard error of less than 1%, the magnetization M_S, expressing the magnetic moments per unit mass (emu/g), is practically unaffected by the morphological features of different samples with a fixed composition and crystal structure. Concerning the coercive field, due to its strong dependence on the grain size, it is increased in sample S3 with the lowest Dv50 (2.1 μm). Actually, the increased volume concentration of boundaries in finer grains enhances the pinning possibility of magnetic domain walls, thereby raising H_C as the measure of the irreversible changes in magnetization.

Table 3: Static magnetic properties of the sintered ferrite

Sample	Spec. saturation magnetization M_S(emu/g)	Coercive field H_C (Oe)
S1	90.5	20.1
S2	89.4	20.4
S3	89.5	29.0

Electromagnetic Properties

The dispersion of complex permeability with frequency, $\mu*(f) = \mu'(f) - j\mu''(f)$, was measured in the range from 10 MHz to 20 GHz and the variations of the respective real and imaginary parts are separately shown in Figure 4. The recorded curves include two dynamic magnetization mechanisms, which affect different frequency regions. Specifically, due to the multidomain nature of this polycrystalline ferrite, both the domain wall motion (relaxation-type) and the ferromagnetic resonance occur. The magnetization contribution of the domain wall bulging by its nature appears at much lower frequencies than the ferromagnetic resonance [17]. From the $\mu''(f)$ spectra of Figure 4, we derive that the characteristic frequency of domain wall relaxation is increased from 550 MHz in S1 to 650 MHz in the sample with low T_P (S2) and 850 MHz in the sample with low T_S (S3). The observed decline of the domain wall process and its movement towards higher frequency are consistent with the shrinkage of the median grain size of the samples when heat treated at lower temperatures [18, 19]. Similarly, the ferromagnetic resonance peak at higher frequencies takes place at 6.55 GHz in the S1 ferrite and shifts to 7 GHz in S2 and 8.05 GHz in S3. Since the resonance frequency f_R is proportional to an effective internal field, the higher f_R corresponds with a stronger internal field due to the demagnetization effects rising from the porosity and the finer grains [20, 21]. The shift of the spectrum towards higher frequencies is accompanied by the reduction of the respective permeability levels, which is in accordance with fundamental Snoek's law defining the product $(\mu' - 1) \times f_R$ [3].

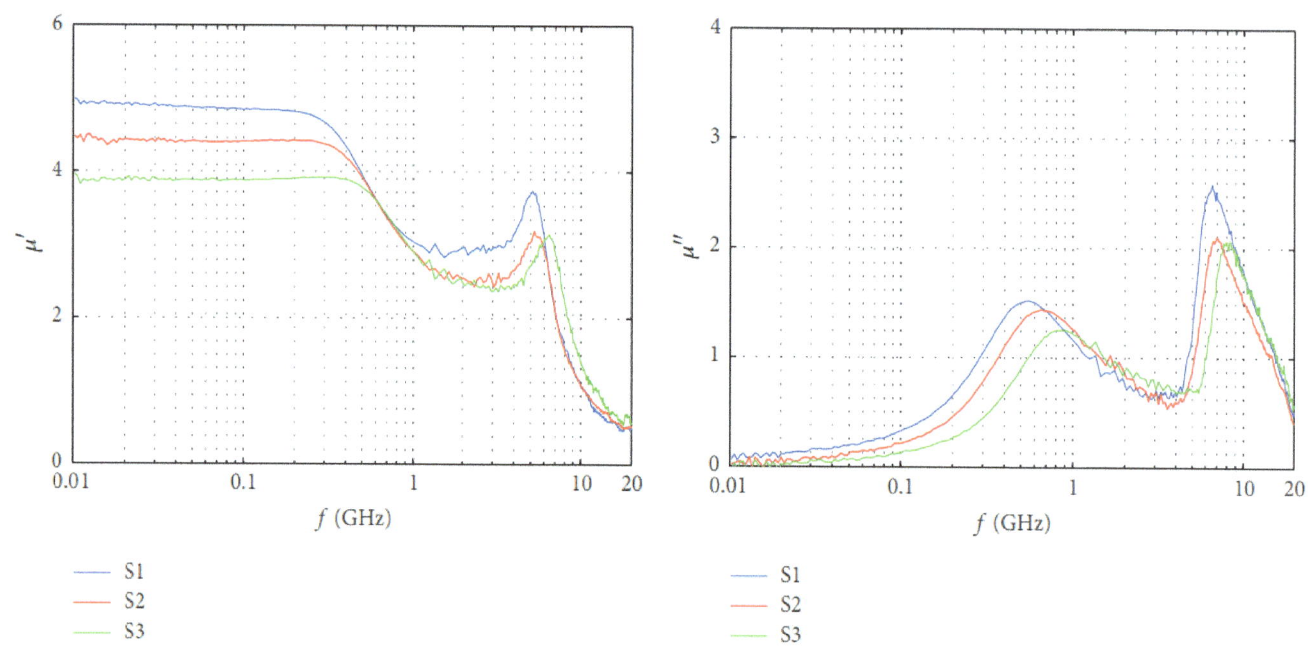

Figure 4: Complex permeability spectra of the sintered ferrite samples.

It is also important to comment on the appearance of permeability dispersion characteristics at unusually high frequencies for cubic ferrite. This seems apparent controversy with the established notion of the existence of a cutoff limit for spinel applications far below 1 GHz. However, although it is not observed in the most common types of Mn-Zn and Ni-Zn ferrite, the occurrence of resonance peaks above 1 GHz has been reported in Co-rich Ni-ferrite and Ni-Zn ferrite [22, 23]. This crucial impact of Co on the permeability spectrum stems from its inherent strongly positive contribution to the first-order anisotropy constant.

With regard to the dielectric behavior of the prepared materials, the complex permittivity, $\varepsilon*(f) = \varepsilon'(f) - j\varepsilon''(f)$, was measured in the range of 1 to 20 GHz. We focus our investigation on the real part ε', displayed in Figure 5, as the differences in the imaginary part ε'' of the samples lie within the confidence interval of the measurement. Thus, in the monitored frequency window, ε'' curves are indistinct and are decreasing monotonically with frequency from 0.5 to 0.1, whereas ε' curves show some noticeable variations. In fact, we get that sintering at 1200°C results in the lowest permittivity values of the S3 sample. Above 1 GHz the dielectric response of materials is ascribed to the mechanism of orientational polarization and is thus related to the dipole concentration [24]. Therefore, the minimum ε' value of the sample S3 sintered at the lowest *TS* is most likely the aftereffect of low Fe^{2+} content in combination with the inferior density.

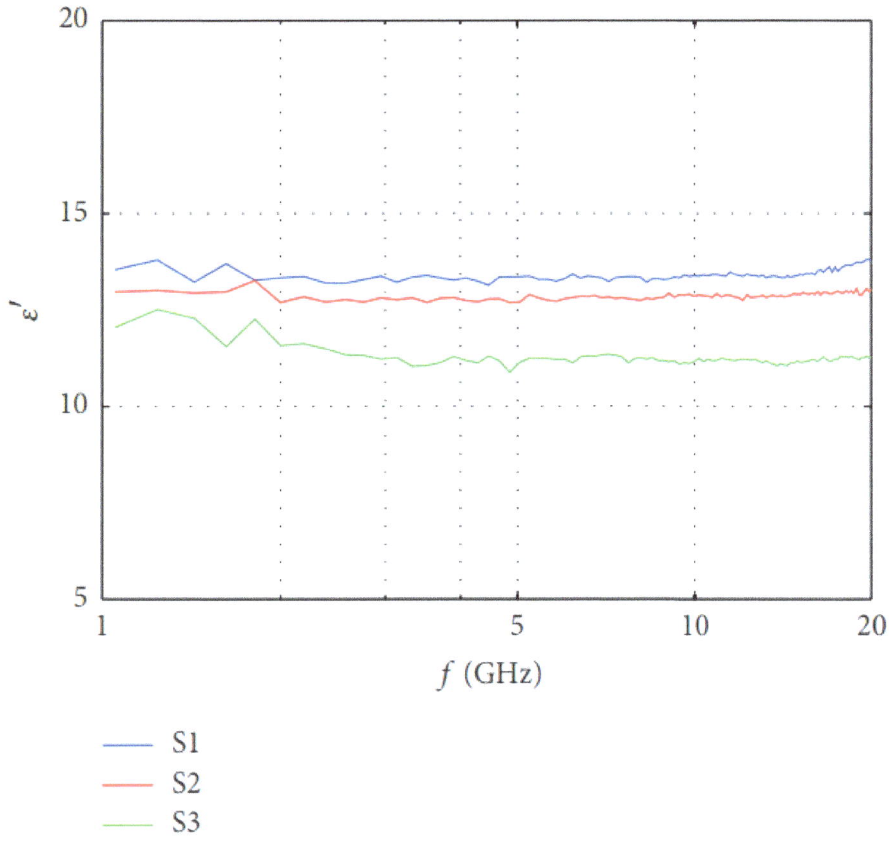

Figure 5: Spectra of real part of permittivity of the sintered ferrite samples.

Absorber Performance

By using the measured $\mu*(f)$ and $\varepsilon*(f)$ data, we have created the contour maps of Figure 6, which depict the return losses (RL) in dB of the different produced materials for a metal-backed single-layer configuration. RL are represented as a function of frequency (*f*) and material thickness (*d*). In these maps we notice narrow zones of high RL values, which coincide with the $\lambda_g/4$ curves, where $\lambda_g = \lambda/\mathrm{Re}\{\sqrt{\varepsilon^*\mu^*}\}$ is the wavelength in the material. We thus conclude that the dominant mechanism for reflection reduction in the tested ferrite is the destructive interference, when the layer thickness is an odd multiple of quarter wavelength. As only small differences are identified in the electromagnetic properties of the produced samples, the RL plots of Figure 6 possess similar qualitative characteristics. However, we should notice the shift of the high RL peaks to higher frequency region for samples S2 and S3, in correspondence with the permeability spectrum shift. That displacement of the high return loss performance to higher frequencies in S2 and S3 is more clearly discerned in Figure 7, where we plot the maximum achievable return losses at each frequency point, for material thickness up to 5 mm. The represented curves actually constitute an envelope of the RL(*f*) curves obtained for different *d* values. From this projection of the maps on the frequency axis we derive that by altering the heat treatment of the spinel ferrite we have moved the major peaks with magnitude above 30 dB from 6.2 GHz to 7.9 GHz. Still, the strong suppression of reflectance is achieved by ferrite layers with thickness of less than 2 mm.

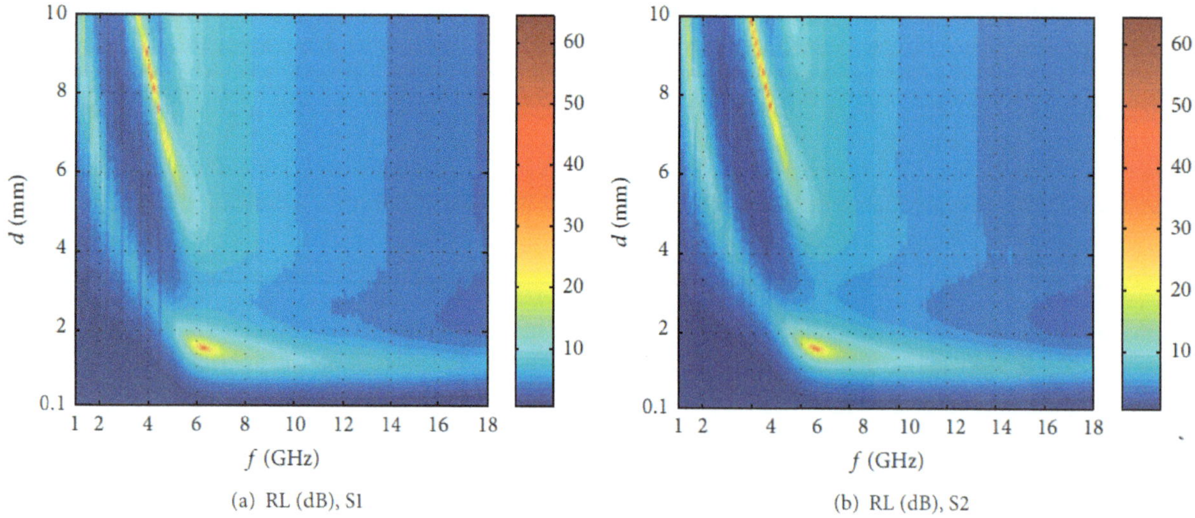

Figure 6: Return losses RL(f, d) of metal-backed (a) S1, (b) S2, and (c) S3 ferrite.

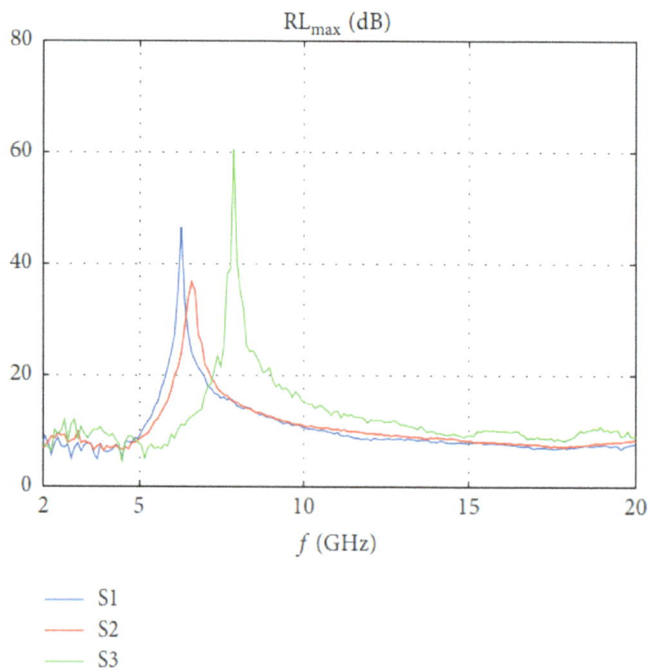

Figure 7: Envelopes of the RL(f, d) curves of the S1, S2, and S3 ferrite. Thickness d ranges up to 5 mm.

CONCLUSIONS

We have investigated the impact of heat treatment during the production of cubic ferrite $Ni_{0.25}Zn_{0.25}Co_{0.5}Fe_2O_4$ on its microwave electromagnetic properties. By varying the top temperature values during prefiring (T_P: 750°C, 1000°C) and sintering (T_S: 1200°C, 1250°C), single-phase spinel ferrite with similar crystal structure and different microstructure was obtained. Specifically, the combination of high T_P and T_S gives rise to high-density magnetic ceramics, whereas low T_P forms large pores and low T_S leads to the finest grains. To this end, the latter conditions of low T_S result in the maximum coercive field HC, according to the dc hysteresis loops. In respect of the ac magnetic properties, the high Co content has moved the characteristic frequencies of dynamic magnetization processes to unusually high values, due to the strong magnetocrystalline anisotropy. That is, the domain wall relaxation and ferromagnetic resonance occur, respectively, above 550 MHz and 6.55 GHz in the complex permeability spectrum $\mu*(f)$. By decreasing any of the two temperatures, T_P or T_S, the resultant smaller grain size and higher porosity shift the permeability spectrum to even higher frequencies. The lower sintering temperature also affects the microwave dielectric properties by decreasing permittivity in the 1–20 GHz range, which indicates the reduced concentration of dipoles. These variations of permeability and permittivity drive the zero-reflection conditions from a metal surface from 6.2 GHz to 7.9 GHz. The underlying mechanism is the quarter wavelength phase cancelation of multiple reflections, which creates the conditions for narrowband EM wave absorption. Although the changes of temperatures T_P and T_S induce significant microstructural dissimilarities, we finally conclude that sintering temperature is a more critical parameter in affecting the microwave electromagnetic properties of Ni-Zn-Co ferrite ceramics.

REFERENCES

1. V. A. Brabers, "Ferrimagnetic insulators," in Handbook of Magnetism and Advanced Magnetic Materials, John Wiley & Sons, New York, NY, USA, 2007.
2. M. Wu and A. Hoffmann, Recent Advances in Magnetic Insulators—From Spintronics to Microwave Applications, Elsevier, Academic Press, San Diego, Calif, USA, 2013.
3. J. Smit and H. P. J. Wijn, Ferrites, Philips Technical Library, Eindhoven, Netherlands, 1959.
4. I. Mikami, "Role of induced anisotropy in magnetic spectra of cobalt-substituted nickel-zinc ferrites,"Japanese Journal of Applied Physics, vol. 12, no. 5, pp. 678–693, 1973.
5. A. Lucas, R. Lebourgeois, F. Mazaleyrat, and E. Labouŕ, "Temperature dependence of spin resonance in cobalt substituted NiZnCu ferrites," Applied Physics Letters, vol. 97, no. 18, Article ID 182502, 2010.
6. A. Lucas, R. Lebourgeois, F. Mazaleyrat, and E. Laboure, "Temperature dependence of core loss in cobalt substituted Ni-Zn-Cu ferrites," Journal of Magnetism and Magnetic Materials, vol. 323, no. 6, pp. 735–739, 2011.
7. E. Brun, R. Lebourgeois, and E. Laboure, "Low permeability ferrite with low losses at high frequency,"Journal of the Japan Society of Powder and Powder Metallurgy, vol. 61, supplement 1, pp. S227–S230, 2014.
8. A. Goldman, Modern Ferrite Technology, Van Nostrand Reinhold, New York, NY, USA, 1990.
9. A. J. Moulson and J. M. Herbert, Electroceramics. Materials, Properties, Applications, John Wiley & Sons, Chichester, UK, 2nd edition, 2003.
10. C. A. Stergiou and V. Zaspalis, "Analysis of the complex permeability of NiCuZn ferrites up to 1 GHz with regard to Cu content and sintering temperature," Ceramics International, vol. 40, no. 1, pp. 357–366, 2014.
11. J.-L. Mattei, L. Huitema, P. Queffelec et al., "Suitability of Ni-Zn ferrites ceramics with controlled porosity as granular substrates for mobile handset miniaturized antennas," IEEE Transactions on Magnetics, vol. 47, no. 10, pp. 3720–3723, 2011.
12. X. Wu, S. Yan, W. Liu, Z. Feng, Y. Chen, and V. G. Harris, "Influence of particle size on the magnetic spectrum of NiCuZn ferrites for electromagnetic shielding applications," Journal of Magnetism and Magnetic Materials, vol. 401, pp. 1093–1096, 2016.
13. C. Clausell, A. Barba, L. Nuño, and J. C. Jarque, "Effect of average grain size and sintered relative density on the imaginary part – μ'' of the complex magnetic permeability of $(Cu_{0.12}Ni_{0.23}Zn_{0.65})Fe_2O_4$ system,"Ceramics International, vol. 42, no. 3, pp. 4256–4261, 2016.
14. I. R. Ibrahim, M. Hashim, R. Nazlan et al., "Grouping trends of magnetic permeability components in their parallel evolution with microstructure in $Ni_{0.3}Zn_{0.7}Fe_2O_4$," Journal of Magnetism and Magnetic Materials, vol. 355, pp. 265–275, 2014.
15. J. Baker-Jarvis, M. D. Janezic, B. F. Riddle, et al., "Measuring the permittivity and permeability of lossy materials: solids, liquids, metals, building materials and negative-index materials," NIST Technical Note1536, National Institute of Standards and Technology, US Department of Commerce, 2005.
16. C. Stergiou and G. Litsardakis, "Structural and magnetic properties of yttrium and lanthanum-doped Ni-Co and Ni-Co-Zn spinel ferrites," AIP Conference Proceedings, vol. 1627, pp. 117–122, 2014.
17. R. Valenzuela, Magnetic Ceramics, Cambridge University Press, Cambridge, UK, 1994.
18. S. Chikazumi, Physics of Ferromagnetism, Oxford University Press, Oxford, UK, 2nd edition, 1997.
19. D. Jiles, Introduction to Magnetism and Magnetic Materials, CRC Press, Boca Raton, Fla, USA, 2nd edition, 1998.
20. H. Igarashi and K. Okazaki, "Effects of porosity and grain size on the magnetic properties of NiZn ferrite," Journal of the American Ceramic Society, vol. 60, no. 1-2, pp. 51–54, 1977. ·
21. R. Valenzuela, "The temperature behavior of resonant and non-resonant microwave absorption in Ni-Zn

ferrites," in Electromagnetic Waves, V. Zhurbenko, Ed., chapter 18, InTech, Rijeka, Croatia, 2011.

22. G.-J. Yin and S.-B. Liao, "Permeability spectra of Co substituted Ni ferrites," IEEE Transactions on Magnetics, vol. 27, no. 6, pp. 5459–5461, 1991.

23. J.-L. Mattei, E. Le Guen, and A. Chevalier, "Dense and half-dense NiZnCo ferrite ceramics: their respective relevance for antenna downsizing, according to their dielectric and magnetic properties at microwave frequencies," Journal of Applied Physics, vol. 117, no. 8, Article ID 084904, 2015.

24. M. T. Sebastian, Dielectric Materials for Wireless Communication, Elsevier, Amsterdam, Netherlands, 2008.

CHAPTER 13

REDUCTION EMI OF BLDC MOTOR DRIVE BASED ON SOFTWARE ANALYSIS

Navid Mousavi, Tohid Rahimi, and Homayoun Meshgin Kelk

INTRODUCTION

Among different types of motors, the DC motors, due to their high starting torque, are still being used in many industries, especially in the aviation industry. Because of the brushes and mechanical commutation, high vibration may occur in DC motors. It should be mentioned that the brushless DC (BLDC) motors have the advantages of DC motors. For instance, due to the electrical commutation mechanisms, the BLDC motors also produce less vibration and acoustic noise. Therefore, the BLDC motors have been recognized as suitable alternatives for DC motors. The developments of technology in semiconductor devices have caused these devices to switch high currents or voltages at short time intervals [1]. Historically, the problem of electromagnetic interference (EMI) in the development and construction process was less regarded and discussed. In general, a product is designed based on the basic concepts of electromagnetic compatibility (EMC) and then tested from the standpoint of the EMC. If the product needs an additional filter or requires to be redesigned for meeting the EMI standards, the time and costs of the process of making the final product will significantly increase. For AC drives, these filters are not only accompanied with high costs, but also they occupy a large space. In order to reduce such costs and avoid their reconstruction, engineers must design strategies based on the EMI from the start of the production. Due to high dv/dt and di/dt, the EMI issue is caused by the flow of the parasitic current through different parts of the drive system [2]. EMI noise is divided into two parts: conductive and radiative noise. Radiation noise is emitted in the form of electromagnetic waves to space that is not discussed in this paper. The conductive electromagnetic noise flows through the common DC current or through parasitic elements and ground plane. The conductive EMI is divided into two types: common mode (CM) and different mode (DM). CM noise flows in the same direction on elements and returns via ground plane. DM noise flows in a direction on one element and returns via other elements. High dv/dt is responsible for CM noise occurrence [3–5]. However, high di/dt is responsible for DM noise occurrence [6, 7].

Many researches about the EMI reduction and prevention of leakage current, especially for electric motors used in industries based on using the passive and active filters, have been centralized [8–14]. To implement simple and effective methods to overcome the EMI, Passive filters with large volume and weight system will be an indispensable part of the motor drive. References [15, 16] have proposed a process for the design of DC side filter to overcome noise. Similar to the process of designing filters for the DC side, a design process

Citation: Navid Mousavi, Tohid Rahimi, and Homayoun Meshgin Kelk, "Reduction EMI of BLDC Motor Drive Based on Software Analysis," Advances in Materials Science and Engineering, vol. 2016, Article ID 1497360, 9 pages, 2016. doi:10.1155/2016/1497360.

of AC side filter in [17] is presented. The mentioned references investigated proposed methods based on simple concepts of EMI emission based on laboratory samples or high-frequency models derived from practical tests. In [18], performance and size of the EMI filter with respect to the core material of inductance as a part of the filter structure are discussed. In [19], a new active common mode filter for inverter fed motors is presented. However, active filters require mainly a separate DC supply. In [5], a simple and new method is presented to design passive filters to reduce EM emissions in BLDC motor drives. With a focus on manufacturing technology, in [20], the EMI in Si insulated gate bipolar transistor (IGBT) and silicon carbide (SiC) junction field-effect transistor (JFET) based motor drives are compared. In fact, EMI resulted from the two types of power switches with new manufacturing technologies which are compared.

Hysteresis current control is widely used due to the simplicity of implementation and accurate and rapid response [21]. Meanwhile, in some applications, hysteresis controllers are used to improve performance and transient response of electric motors. In [22, 23], the hysteresis control methods are used to reduce torque ripple of the electric motors.

In this paper, a new process is presented in order to design filters based on the leakage current and the frequency components of the DC link current. This process can determine the size of the passive elements and their placements in BLDC drive structure. The analysis of electromagnetic noise and presentation of the proposed method based on the frequency spectrum of the DC link current and based on the leakage current from the motor to the ground network are done. The proposed method presents a new process based on the filtering method to overcome EMI. To cover the requirement analysis, the Maxwell software is used. The EMI reduction methods are applied to the high-frequency model of the BLDC motor drive. These models can be founded based on the experimental results or finite element analysis. In addition, the noise analysis is carried out in the case of hysteresis control.

CHARACTERISTICS OF THE MOTOR

The proposed methods are investigated on a 33-watt, 10,000 rpm BLDC motor drive. In order to implement the motor drive, the Maxwell circuit editor is used. According to the rotor position, the firing pulses of the power switches are ordered. Since the studied BLDC motor is of the outer permanent magnet type, the self- and mutual-inductance variations are small.

It should be mentioned that one of the aims of the present paper has been to figure out the magnitude of the leakage current studding. Therefore, the effect of the inductance variations must be applied. The characteristics of the motor are summarized in Table 1. The winding pattern of the motor is illustrated in Figure 1. The winding of each phase covers the whole circumference of the motor.

Table 1: Characteristics of the motor.

Variable	Value	Unit
Resistance per phase	233.37	mΩ
Back-EMF constant	0.0156	V·s/rad
Torque constant	0.0154	Nm/A

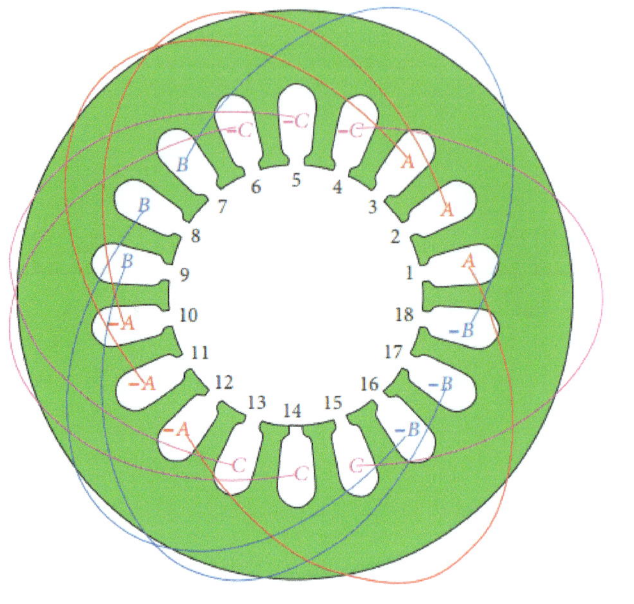

Figure 1: Motor winding scheme studied.

THE HIGH-FREQUENCY MODEL IN THE CIRCUIT EDITOR

The overall high-frequency model of the motor-drive system is shown in Figure 2. Each switch as well as each phase has high-frequency elements [24]. The parasitic capacitances of one phase can be connected to a scope. The waveform of this current is chosen as a waveform of the parasitic current. The high-frequency model of each phase is exhibited in Figure 3. The coils, named as L_{a1}, L_{a2}, and L_{a3}, provide the connection between the circuit editor and Maxwell software. The time-varying behaviors of the inductances and the back-EMF voltages appeared by means of the mentioned coils.

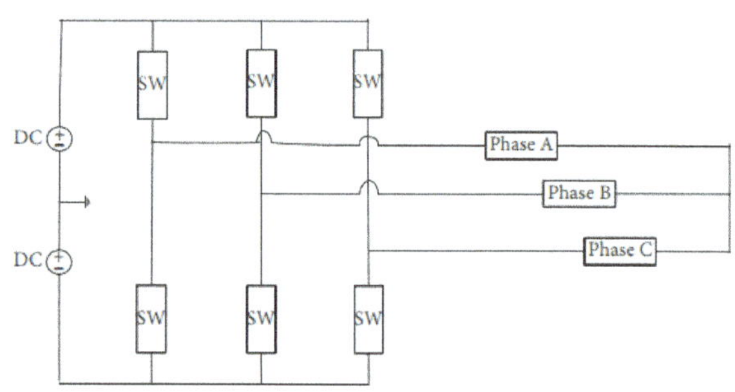

Figure 2: The overall high-frequency model.

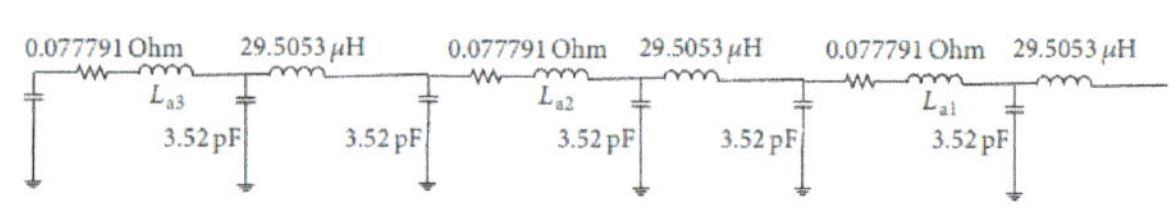

Figure 3: The high-frequency model of each phase.

THE PROPOSED PROCEDURE

First, the motor-drive system without the use of filters is simulated in order to show the initial EMI level and the maximum amplitude of the leakage current. The high levels of the leakage current, sudden spikes in the motor terminal voltage waveform, and severe variations in the DC link current confirmed the necessity of using filters. The terminal voltage waveform, the phase current, the DC link current, the leakage current (motor to ground) waveform, and the FFT of the DC link current are shown in Figures 4–8, respectively. As observed in Figure 4, severe variations in line voltage have emerged. According to the concepts explained in the introduction section, the high variations in the BLDC motor terminal voltage are the reason for the ground leakage current.

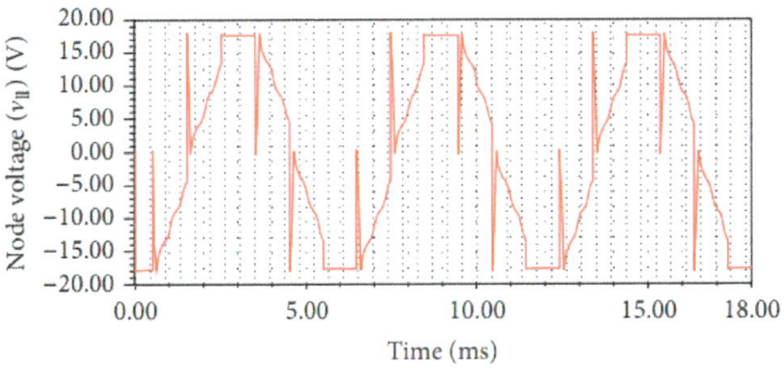

Figure 4: The line-to-line voltage waveform (without filter).

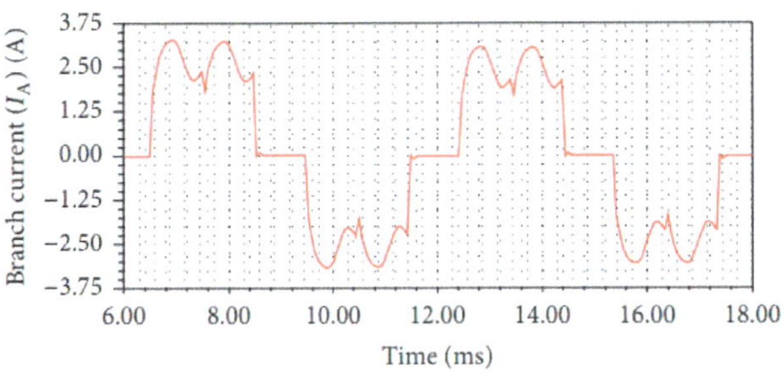

Figure 5: The phase current waveform (without filter).

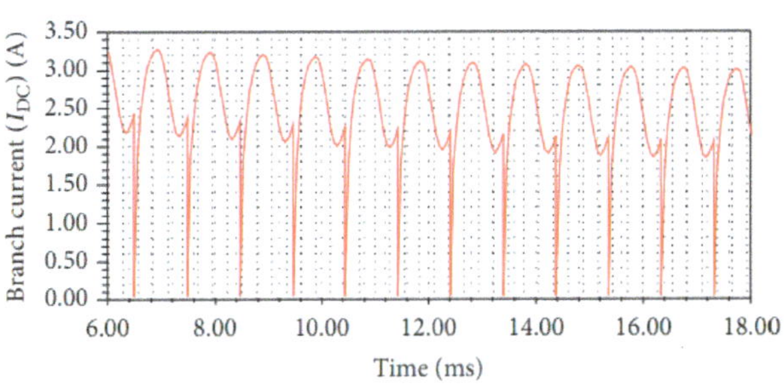

Figure 6: The DC link current (without filter).

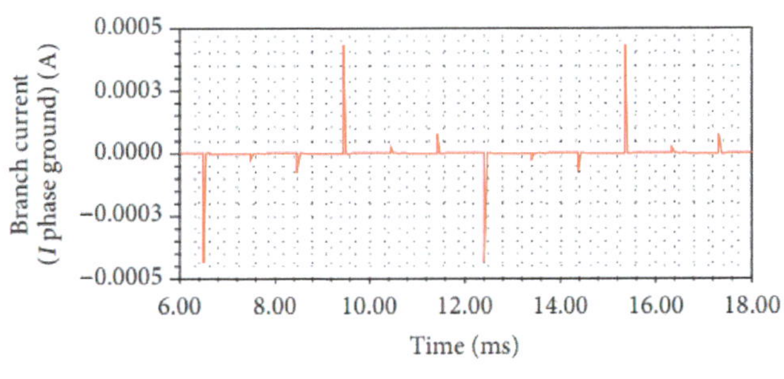

Figure 7: The leakage current (without filter).

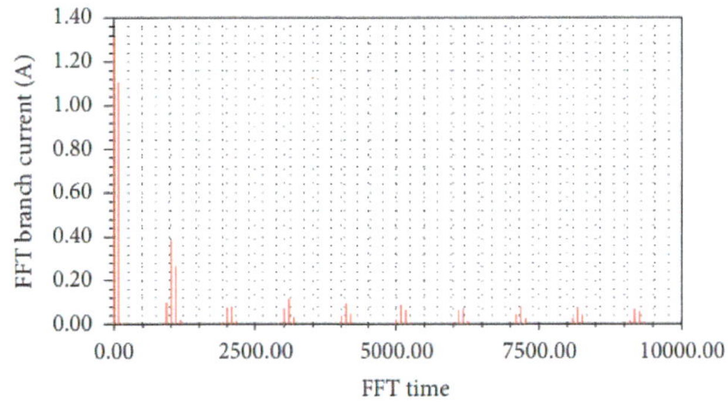

Figure 8: The FFT of DC link current (without filter).

The high-frequency components of the DC side current (see Figure 8) provide the conducted EMI. The conducted EMI progressed through the common DC bus and has an adverse effect on the other electrical systems that are fed from the common DC bus. This problem is more important in satellite and submarine systems. The conducted noise is the cause of desert effect on communication and radio systems. The high-frequency components of the DC link current can be reduced by using LC filter. The design of LC filter in the DC line is based on the spectrum of the DC link current. The LC filter weakens the high component of the DC link current. It is commonly used for low impedance sources and high impedance loads. In this study, the ideal voltage source with low impedance and high input impedance of the inverter is used. Therefore, LC filter would be a suitable choice. Equation (1) shows the LC filter design process:

$$2000 = \frac{1}{2\pi\sqrt{LC}}$$

$$C = 10\,\mu\text{F},$$

$$L = 640\,\mu\text{H}$$

$$f_c = \frac{1}{2\pi\sqrt{640\,\mu\text{H} * 10\,\mu\text{F}}} = 1989.43\text{ Hz}. \qquad (1)$$

Attenuation of the high-frequency components for frequencies above the cut-off frequency of the LC filter is effective. Since the standard capacitors and inductors are used to design filter, the exact cut-off frequency of the LC filter cannot be obtained. To filter the high-frequency components, especially above 2000 Hz, the cut-off frequency that is obtained equal to 1989.43 Hz is considered. The value of the cut-off frequency of the LC filter is about 2000 Hz. In practice, the self has series resistance. If the value of inductance is large, as a result, the resistance of the inductor will increase and the voltage drop will occur. If the value of inductance is 640 μH, the equivalent resistance of the chosen choke will have a relatively high value. Therefore, a choke, named SN203-640M-5.0AV, is selected. The choke has 64 μH inductance and 23 mΩ equivalent resistance. Equation (2) shows the new LC filter design process:

$$2000 = \frac{1}{2\pi\sqrt{LC}}$$

$$C = 100\,\mu\text{F},$$

$$L = 64\,\mu\text{H}$$

$$f_c = \frac{1}{2\pi\sqrt{64\,\mu\text{H} * 100\,\mu\text{F}}} = 1989.43\text{ Hz}. \tag{2}$$

Small changes appeared in the line-to-line voltage waveform, the phase current waveform, and the leakage current waveform. The LC filter weakened the high component of the DC source current, as seen in Figure 9. The simulation results showed that the inverter input current has high di/dt and therefore the MOSFET parasitic capacitance can conduct the high component of the current. Thus, it seemed necessary to add another passive element to the DC line. In order to reduce the input ripple current of the inverter, a series inductor is used. The series inductance must be chosen in such a way to lower the inverter input current ripple and prevent the effective voltage drop resulted from the inherent resistance of the inductor. According to the available inductors, a 160 μH inductor (with series resistance of 32 mΩ) was selected according to the engineering experience. In this case, the input current ripple decreased from 2.5 to 0.7 ampere. The inductor, series with the inverter input, increased the reliability of the drive system. It reduces the amplitude of the inrush current, caused by short circuit of switches; as a result, the damage to the common DC bus, from which the other electronic and electrical systems are fed, will be less.

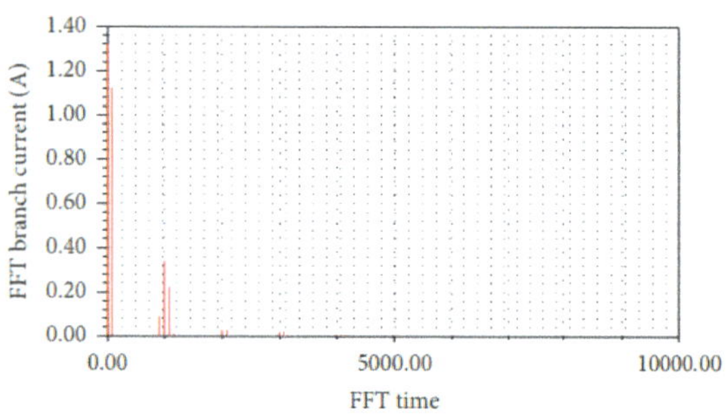

Figure 9: The FFT of DC link current (with LC filter).

The line-to-line voltage, the leakage current, the FFT analysis of the DC link current, and the inverter input current are shown in Figures 10–13, respectively.

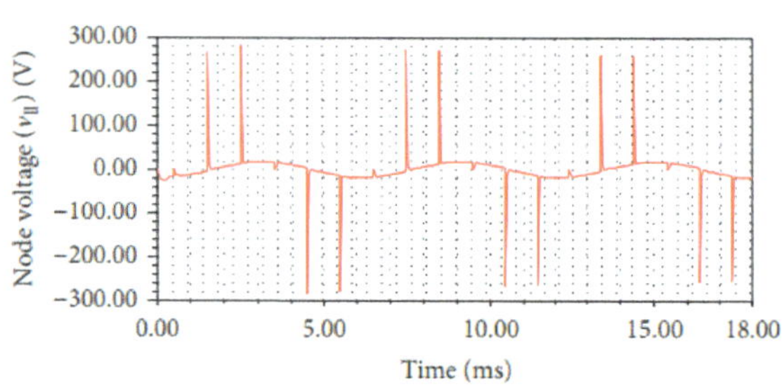

Figure 10: The line-to-line voltage waveform (with LCL filter).

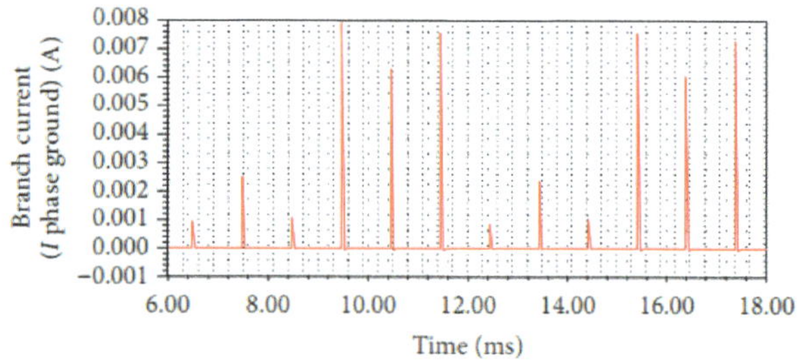

Figure 11: The leakage current (with LCL filter).

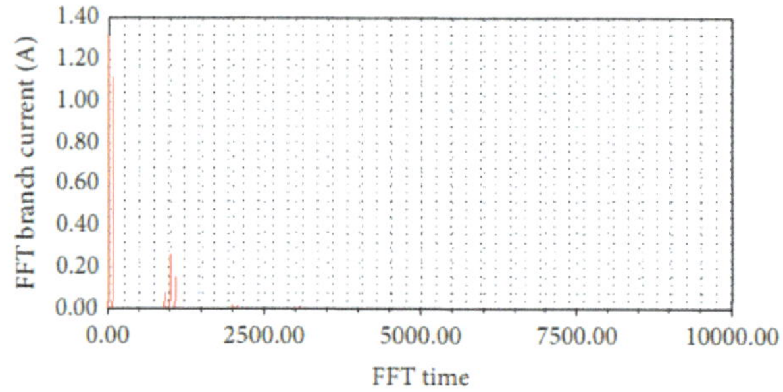

Figure 12: The FFT of DC link current (with LCL filter).

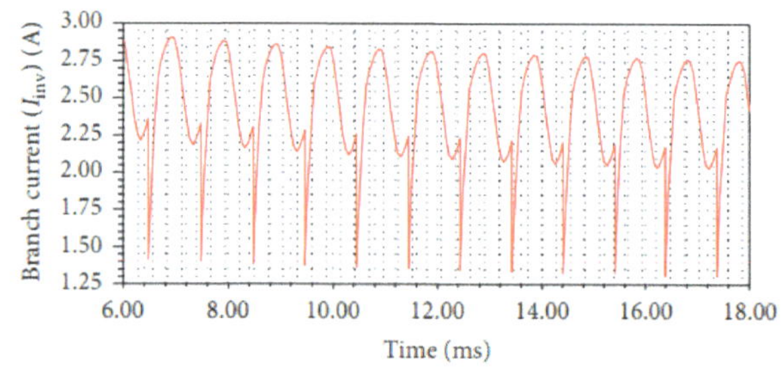

Figure 13: The inverter input current (with LCL filter).

Due to the high equivalent inductance, the maximum amplitude of the leakage current increased significantly. It is necessary to use the capacitor bank in front of the AC line. In fact, by adding a capacitor bank, upon cessation of the motor phase current, the phase current flowed through the capacitor bank. Therefore, severe spikes in the voltage waveform due to the reaction of the inductance were reduced or eliminated. The capacitance value of the capacitor bank must be in such a way that does not cause interference with motor functions in the operating frequency motor (the impedance value of the capacitor bank must be high enough in the operating frequency motor) and the resonance frequency, as a result of the motor capacitor bank, must be far enough from the switching frequency. The amplitude of the current, drawn by the motor and the capacitor bank, may be of a high value due to the resonance. The equivalent circuit of the motor and the capacitor bank are shown in Figure 14 at different times. The phase motor is series with the capacitor and is the disconnected phase. It is required to adjust the value of the capacitor in such a way that the impedance of the middle branch for several multiples of the resonance would be too high (i.e., (3) must be satisfied). In addition, the impedance of the parallel branch consisting of two-series capacitors must be large enough (i.e., (4) must be satisfied). These requirements are applied in order to remove the effective current, drawn by the capacitor bank:

$$10 * 166.667 < \frac{1}{2\pi * \sqrt{(119 * 10^{-6}) * C}}$$

$$C < 76.6264\,\mu\text{F} \tag{3}$$

$$\frac{1}{0.5 * \pi * 166.67 * C} > 20\sqrt{(0.233373^2) + (166.67 * 2\pi * 119 * 10^{-6})}$$

$$C < 160\,\mu\text{F}. \tag{4}$$

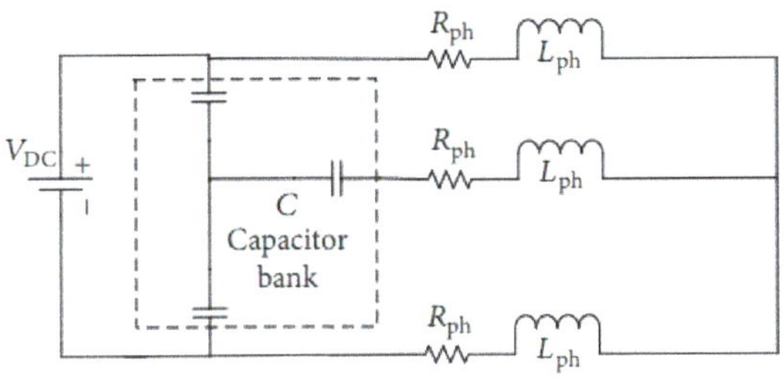

Figure 14: The equivalent circuit of the motor by the capacitor bank.

On the other hand, the resonant frequency of the capacitor must be far enough from the operating frequency of motor:

$$10 * 166.667 < \frac{1}{2\pi * \sqrt{(2 * 119 * 10^{-6}) * 0.5C}}$$

$$C < 76.6264\,\mu\text{F}. \tag{5}$$

Due to the application of the capacitor banks, the input current of the inverter will experience jumps. If the values of the capacitors are small, there will be higher amplitude jumps. It is essential to select an appropriate capacitor bank. For this purpose, a 20 μF capacitor bank has been considered. The line-to-line voltage waveform, the leakage current, the FFT analysis of the DC link current, the inverter input current, the 3-phase currents, and the current waveform of a power switch are obtained by finite element analysis. The mentioned results can be seen in Figures 15–20, respectively. A remarkable decrease in voltage jumps of the line voltage waveform is obvious. Due to the reduction in the amplitude of the jumps, the earth leakage current from the motor to the ground plane was significantly reduced, as can be observed in (7).

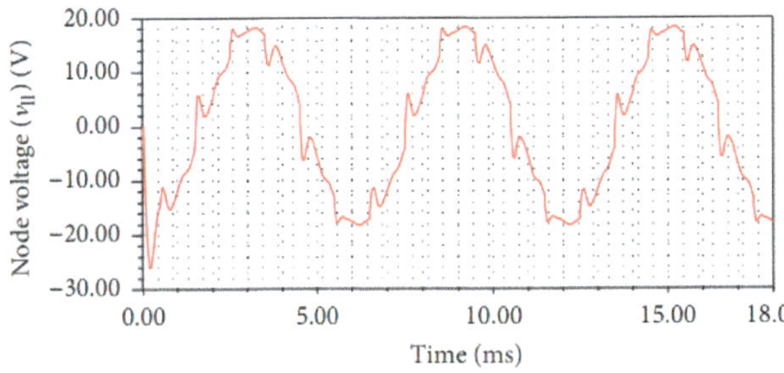

Figure 15: The line-to-line voltage waveform (with LCL filter and capacitor bank).

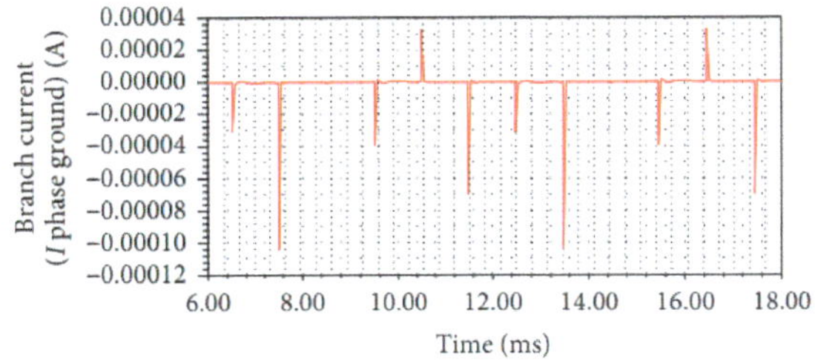

Figure 16: The leakage current (with LCL filter and capacitor bank).

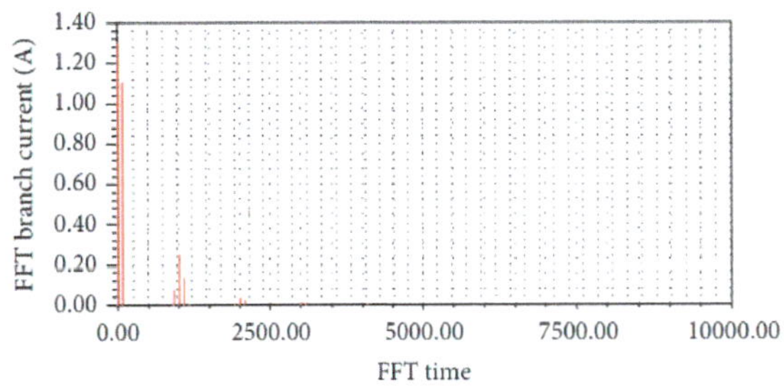

Figure 17: The FFT of DC link current (with LCL filter and capacitor bank).

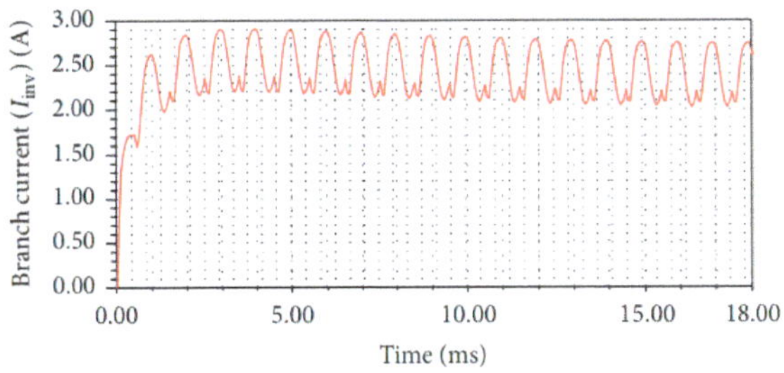

Figure 18: The inverter input current (with LCL filter and capacitor bank).

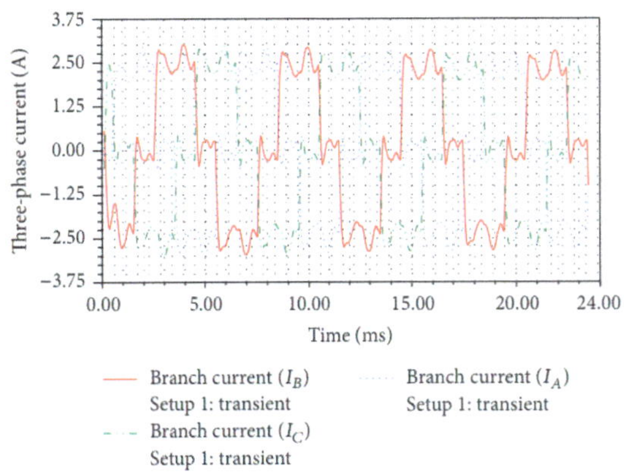

Figure 19: The 3-phase motor currents (with LCL filter and capacitor bank).

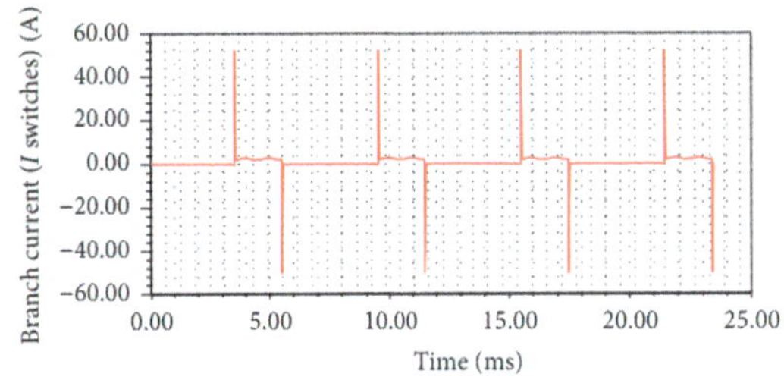

Figure 20: The current waveform of a switch (with LCL filter and capacitor bank).

The amplitude of the leakage current was reduced from 400 μA to 80 μA. The high-frequency components of the DC link current were effectively reduced. The three-phase current waveforms are still in the shape of a square. High di/dt at the switching time could cause damage to the power switches. Hence, it is essential to protect the power switches against high di/dt with the aid of a proper circuit. An RLD snubber is chosen, because the snubber circuits can reduce the leakage current. Due to the application of passive filters, its effect on the EMI reduction is small. The circuit of the selected snubber can be seen in Figure 21:

$$L_s < \frac{P_n(=33\,\text{w})}{8000 * (2\pi * 166.67) * 2.2^2} = 0.814\,\text{nH}$$

$$L_s = 100\,\text{nH} \quad (6)$$

$$\frac{L_s}{R_s} < 2T_s$$

$$R_s > 0.5 * \frac{100 * 10^{-9}}{5 * 10^{-6}} = 0.01\ \text{ohm} \quad (7)$$

$$R_s = 200\ \text{mohm}. \quad (8)$$

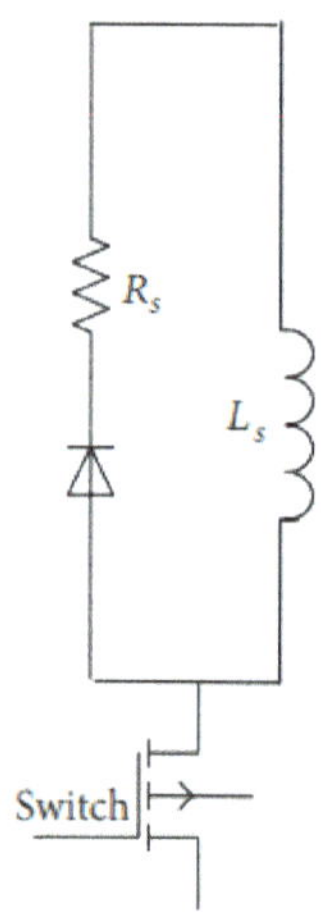

Figure 21: The overall structure of RLD snubber.

The amplitude of the current jumps was reduced from 25 (A) to 10.5 (A). In order to design the snubber, (6) to (8) are employed. The leakage current and a power switch current are shown in Figures 22 and 23. The snubber circuit affected the leakage reduction. By comparing Figures 16 and 22, it can be perceived that the maximum amplitude of the leakage current has been reduced from 100 μA to 70 μA.

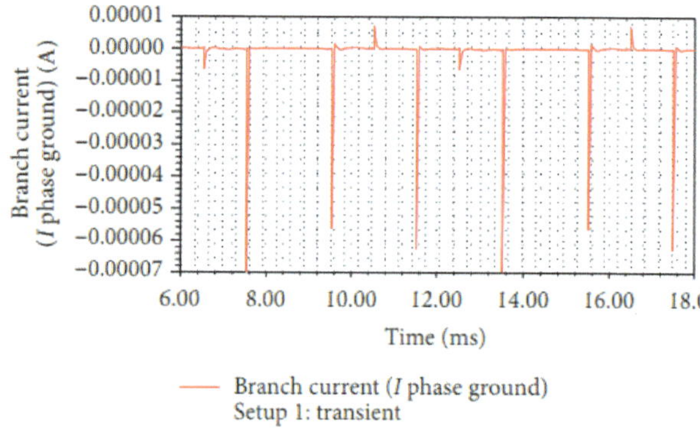

Figure 22: The leakage current (with LCL filter, capacitor bank, and RLD snubber).

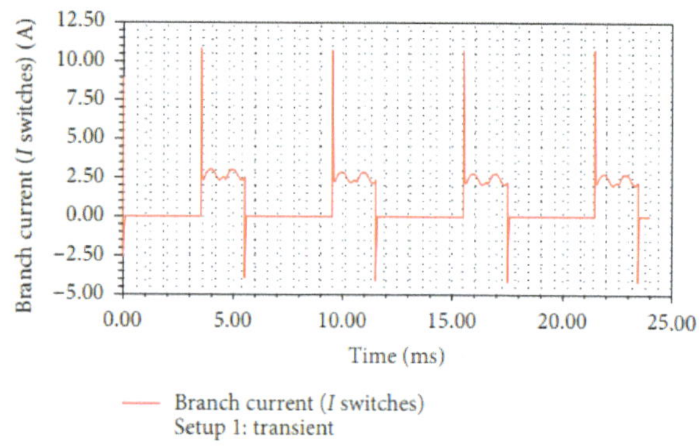

Figure 23: The current waveform of a switch (with LCL filter, capacitor bank, and RLD snubber).

Finally, the overall structure of the drive system by the passive filter is shown in Figure 24. Each power switch is series with a snubber circuit. The DC line filter is placed in front of the DC link. The capacitor bank is in parallel with the terminals of the motor. Finally, the overall structure of the drive system by the passive filters is shown in Figure 24.

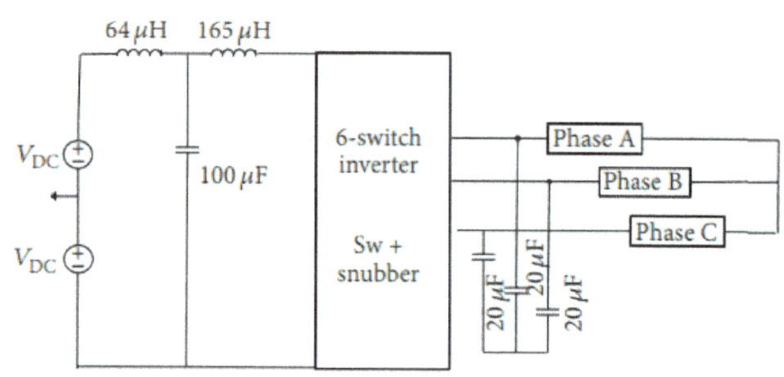

Figure 24: The overall structure of BLDC motor drive (with LCL filter, capacitor bank, and RLD snubber).

THE EMI OF BLDC DRIVE WITH HYSTERESIS CONTROL

The hysteresis control is widely used in controlling the AC motor in order to achieve fast dynamics, protect over current situations, and maintain the current in the form similar to the square waveform. In this control strategy, the measured load currents are compared with the references by using hysteresis comparators. Each comparator determines the switching state of the corresponding inverter leg, such that the load currents are imposed to remain within the hysteresis band and hence the frequency switching rises more than the traditional control state. The increase in the switching frequency increases the motor inductance reaction and consequently sudden changes in the line voltage, which cause the leakage current, are increased every 360° of the operating drive system. The finite element results showed that the hysteresis control has caused high dv/dt in the line-to-line voltage waveform and a high density leakage current. The line-to-line voltage, the leakage current, and the three-phase current are shown in Figures 25–27.

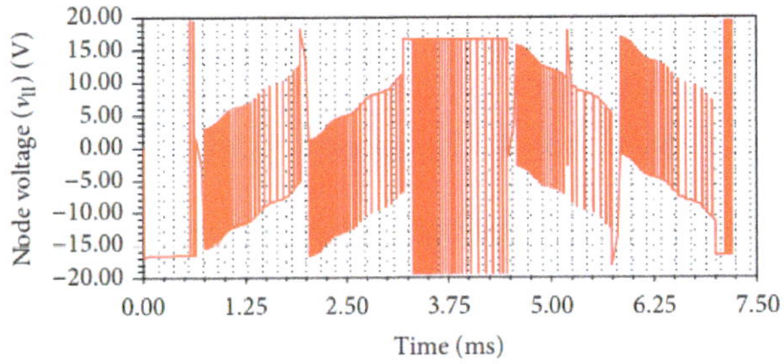

Figure 25: The line-to-line voltage waveform in hysteresis control mode (without filter).

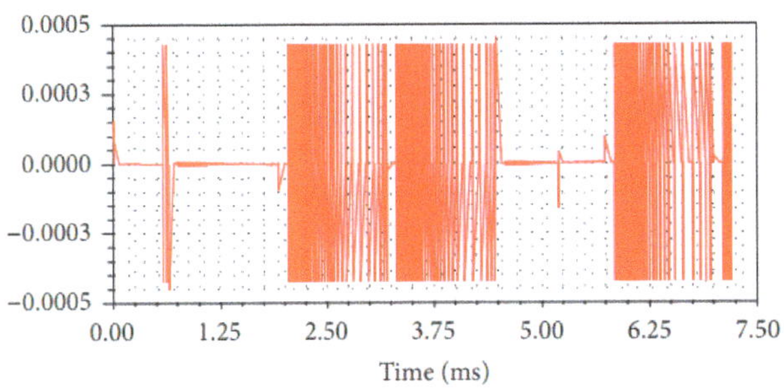

Figure 26: The leakage current in hysteresis control mode (without filter).

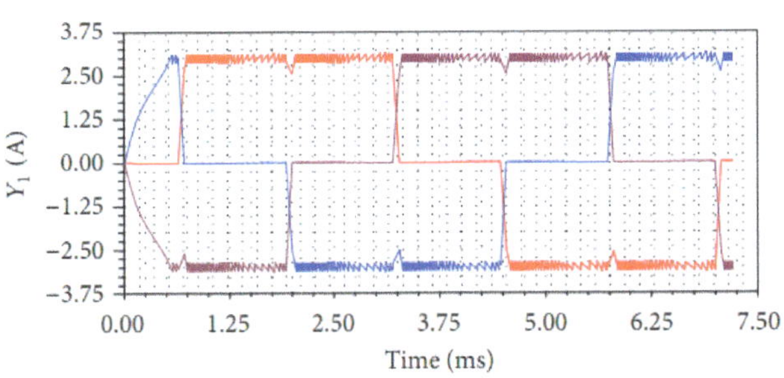

Figure 27: Three-phase current in hysteresis control mode (without filter).

Due to the increased switching frequencies, the use of inductance in the inverter input significantly increased dv/dt. The inductance is series with the inverter input. Therefore, it was avoided in the hysteresis control mode. The designed LC filter and the capacitor bank are used to overcome the EMI. The line-to-line voltage and the leakage current can be seen in Figures 28 and 29. Due to the employment of the capacitor bank, the large voltage changes in rapid succession are reduced more effectively and so the maximum amplitude and density of the leakage current were reduced as well. The maximum amplitude of the leakage current was reduced from 450 to 180 μA.

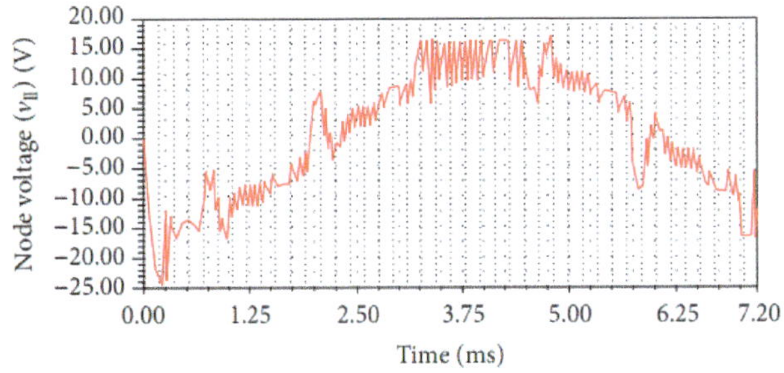

Figure 28: The line-to-line voltage waveform in hysteresis control mode (with filter).

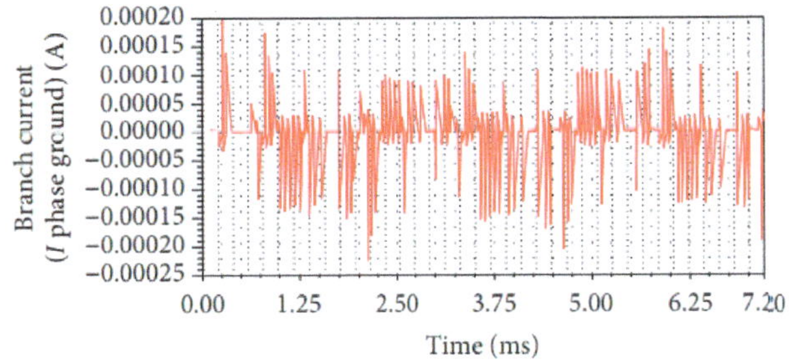

Figure 29: The leakage current in hysteresis control mode (with filter).

The use of a hysteresis controller increased the high-frequency components, as can be observed in Figure 30. The high-frequency components up to 75 kHz can be seen in Figure 30. When the designed filter is used in BLDC system, the high-frequency components were effectively weakened, especially for above 10 kHz as seen in Figure 31.

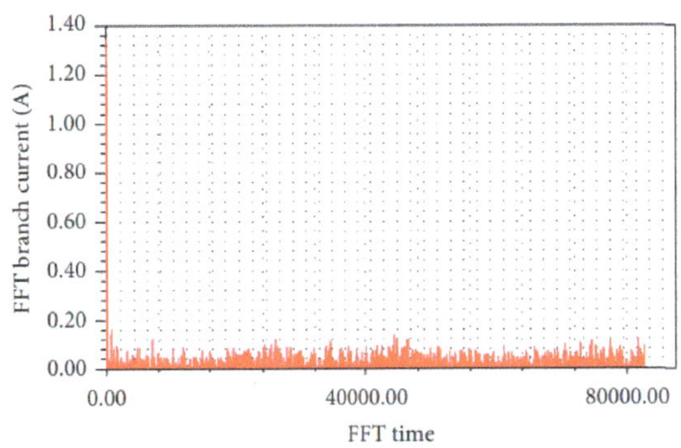

Figure 30: The FFT of DC link current in hysteresis control mode (without passive filter).

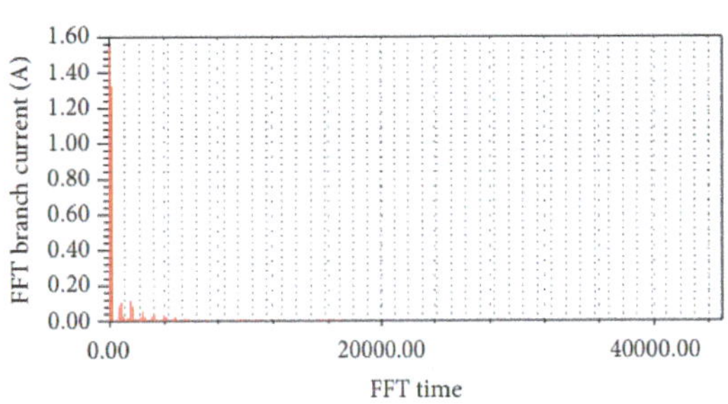

Figure 31: The FFT of DC link current in hysteresis control mode (with LCL filter and capacitor bank).

CONCLUSION

In this paper, the proposed method has been implemented in order to overcome the electromagnetic noise on the basis of the high-frequency model of the motor-drive system using Maxwell's software package and the concepts of the EMI, and the structure and amounts of the proposed filter were extracted. In fact, the type and amount of passive elements and their placement were systematically investigated. At each step, the frequency spectra of the current and the leakage current were set as the criteria for assessing the impact of the strategies for reducing the noise.

Based on these two factors, the elements, placement, and the values of the selected passive elements in each step were determined. According to the results from the obtained filter structure, there is a good potential in reducing electromagnetic noise.

Control methods, as the current hysteresis control, which increased the switching frequency, caused higher electromagnetic interference. It should be expressed that, in these cases, using appropriate filters is more necessary, and they have confirmed their effectiveness.

REFERENCES

1. G. L. Skibinski, R. J. Kerkman, and D. Schlegel, "EMI emissions of modern PWM ac drives," IEEE Industry Applications Magazine, vol. 5, no. 6, pp. 47–80, 1999.
2. J.-S. Lai, X. Huang, E. Pepa, S. Chen, and T. W. Nehl, "Inverter EMI modeling and simulation methodologies," IEEE Transactions on Industrial Electronics, vol. 53, no. 3, pp. 736–744, 2006. ·
3. L. Ran, S. Gokani, J. Clare, K. J. Bradley, and C. Christopoulos, "Conducted electromagnetic emissions in induction motor drive systems part I: time domain analysis and identification of dominant modes," IEEE Transactions on Power Electronics, vol. 13, no. 4, pp. 757–767, 1998. ·
4. F. Costa, E. Laboure, A. Puzo, and A. Pons, "Influence of the driver circuits in the generation and transmission of EMI in a power converter: effects on its electromagnetic susceptibility," European Power Electron Journal, vol. 5, no. 1, pp. 35–44, 1995.
5. T. Rahimi, S. Y. Khangah, and B. Yousefi, "Reduction EMI due to di/dt and dv/dt DC and AC sides of BLDC motor drive," in Proceedings of the 5th Annual International Power Electronics, Drive Systems and Technologies Conference (PEDSTC '14), pp. 428–433, Tehran, Iran, February 2014.
6. Y. Tang, H. Zhu, B. Song, J. S. Lai, and C. Chen, "EMI experimental comparison of PWM inverters between hard- and soft-switching techniques," in Proceedings of the IEEE Power Electronics on Transportation Conference, pp. 71–77, IEEE, October 1998.
7. H. Zhu, J.-S. Lai, A. R. Hefner Jr., Y. Tang, and C. Chen, "Analysis of conducted EMI emissions from PWM inverter based on empirical models and comparative experiments," in Proceedings of the 30th Annual IEEE Power Electronics Specialists Conference (PESC '99), pp. 861–867, Charleston, SC, USA, July 1999.
8. S. Ogasawara and H. Akagi, "Modeling and damping of high-frequency leakage currents in PWM inverter-fed AC motor drive systems," IEEE Transactions on Industry Applications, vol. 32, no. 5, pp. 1105–1114, 1996. ·
9. F. Wang, "Motor shaft voltages and bearing currents and their reduction in multilevel medium-voltage PWM voltage-source-inverter drive applications," IEEE Transactions on Industry Applications, vol. 36, no. 5, pp. 1336–1341, 2000. ·
10. J.-S. Lai, X. Huang, S. Chen, and T. W. Nehl, "EMI characterization and simulation with parasitic models for a low-voltage high-current AC motor drive," IEEE Transactions on Industry Applications, vol. 40, no. 1, pp. 178–185, 2004. ·
11. T. Rahimi and S. H. Hosseini, "EMI consideration of high reliability electrical power subsystem (EPS) of satellite," in Proceedings of the 6th Annual International Power Electronics, Drive Systems, and Technologies Conference (PEDSTC '15), pp. 412–417, Tehran, Iran, February 2015. ·
12. H. Akagi, H. Hasegawa, and T. Doumoto, "Design and performance of a passive EMI filter for use with a voltage-source PWM inverter having sinusoidal output voltage and zero common-mode voltage," IEEE Transactions on Power Electronics, vol. 19, no. 4, pp. 1069–1076, 2004. ·
13. S. Ogasawara, H. Ayano, and H. Akagi, "An active circuit for cancellation of common-mode voltage generated by a PWM inverter," IEEE Transactions on Power Electronics, vol. 13, no. 5, pp. 835–841, 1998. ·
14. S. Wang, Characterization and cancellation of high-frequency parasitics for EMI filters and noise separators in power electronics applications [Ph.D. thesis], Department of Electrical Engineering, Virginia Polytechnic Institute and State University, Blacksburg, Va, USA, 2005.
15. R. Lai, Analysis and design for a high power density three-phase AC converter using SiC devices [Ph.D. thesis], Department of Electrical Engineering, Virginia Polytechnic Institute and State University, Blacksburg, Va, USA, 2008.
16. Y. Y. Maillet, High-density discrete passive EMI filter design for Dc-fed motor drives [M.S. thesis], Department of Electrical Engineering, Virginia Polytechnic Institute and State University, Blacksburg, Va, USA, 2008.
17. A. Carrubba, M. C. Di Piazza, G. Tine, and G. Vitale, "Evaluation of common mode disturbance mitigation devices in AC motor drives through HF modeling," in Proceedings of the IEEE International Symposium on Industrial Electronics, pp. 2315–2320, Montreal, Canada, July 2006. ·
18. Y. Maillet, R. Lai, S. Wang, F. Wang, R. Burgos, and D. Boroyevich, "High-density EMI filter design for dc-fed motor drives," IEEE Transactions on Power Electronics, vol. 25, no. 5, pp. 1163–1172, 2010. ·
19. M. C. Di Piazza, M. Luna, and G. Vitale, "EMI reduction in DC-fed electric drives by active common-mode compensator," IEEE Transactions on Electromagnetic Compatibility, vol. 56, no. 5, pp. 1067–1076, 2014. ·
20. X. Gong and J. A. Ferreira, "Comparison and reduction

of conducted EMI in SiC JFET and Si IGBT-based motor drives," IEEE Transactions on Power Electronics, vol. 29, no. 4, pp. 1757–1767, 2014. ·

21. Hysteresis Control Circuit Diagram, Electrical and Computer Engineering Department, University of Missouri-Rolla, Rolla, Mo, USA, 2002.
22. R. Carlson, M. Lajoie-Mazenc, and J. C. D. S. Fagundes, "Analysis of torque ripple due to phase commutation in brushless DC machines," IEEE Transactions on Industry Applications, vol. 28, no. 3, pp. 632–638, 1992. ·
23. W. Chang-hee, S. Joong-Ho, and I. Choy, "Commutation torque ripple reduction in brushless DC motor drives using a single DC current sensor," IEEE Transactions on Power Electronics, vol. 19, no. 2, pp. 985–990, 2004.
24. A. Carrubba, M. C. Di Piazza, G. Tine, and G. Vitale, "Evaluation of common mode disturbance mitigation devices in AC motor drives through HF modelling," in Proceedings of the IEEE International Symposium on Industrial Electronics, pp. 2315–2320, Montreal, Canada, July 2006. ·

CHAPTER 14

INFLUENCE OF GROWTH RATE AND MAGNETIC FIELD ON MICROSTRUCTURE AND PROPERTIES OF DIRECTIONALLY SOLIDIFIED AG-CU EUTECTIC ALLOY

Xiaowei Zuo , Congcong Zhao , Lin Zhang and Engang Wang

Keywords: Ag-Cu eutectic alloy; magnetic field; growth rate; eutectic lamellar spacing; microhardness; strength

INTRODUCTION

Eutectic alloys are important in casting, soldering and welding processes due to their low melting temperatures, good forming flexibility, and excellent mechanical properties [1]. These eutectic alloys, with lamellar, fibrous, or rod structures, can be produced by directional solidification. These structures are strongly dependent on the solidification parameters, which are determined by the material properties. The relationship between the eutectic microstructure and solidification conditions has been studied extensively, and a general quantitative equation was first proposed by Jackson and Hunt (J-H model) [2,3], which was based on a solution of the diffusion problem at the solid/liquid interface. They predicted that the product of the growth rate V and the square of the eutectic lamellar spacing (ELS) λ should be constant, i.e., $\lambda^2 V$ = const., with the hypothesis that the interfacial undercooling was at a minimum as the eutectics were produced [2]. They concluded that the ELS is one of the most crucial microstructural parameters for eutectic alloys, and the growth rate is capable producing variations in both the liquid diffusion coefficient and the convection, changing the ELS correspondingly.

The external magnetic field (MF) has a similar effect on the eutectic solidification. Extensive experimental studies have been carried out to quantify the changes to the eutectic microstructure after subjecting alloys to MF during solidification. Li et al. [4] found that, at a low growth rate, MF decreased the eutectic spacing of the directionally solidified (DS) Al-Al_2Cu eutectic alloy, and they posited that the change was a result of the decrease of the liquid diffusion coefficient induced by the external MF. Liu et al. [5] found that an applied MF during solidification of Al-Si alloys decreased the lamellar spacing of the eutectics by decreasing the diffusion at the solid/liquid interface. Li et al. [6,7] found that the ELS in Cu-Ag alloys was decreased after applying a 12-T MF. Li et al. [8], however, found that the application of a static MF to semi-continuous casting Al-Fe alloys increased the ELS due to the initiation of the secondary arms. Therefore, the effect of an external MF on the ELS is not entirely clear, and the properties of materials solidified with MF are rarely studied.

In this paper, Ag-Cu, a representative eutectic system, was chosen for study, partly due to its wide applications in both DC and pulsed high-field magnets [9,10,11,12,13]. In previous publications [7,14,15,16,17], we introduced a MF during conventional solidification to fabricate high-performance Cu-

Citation: Zuo, Congcong Zhao, Lin Zhang and Engang Wang, Influence of Growth Rate and Magnetic Field on Microstructure and Properties of Directionally Solidified Ag-Cu Eutectic Alloy, doi:10.3390/ma9070569.

Ag alloys, and the results showed that the ELS of hypoeutectic Cu-Ag alloys decreased due to the external MF application during solidification. The purpose of the present work is to investigate how the solidification growth rate and a MF applied in directional solidification influences the ELS of Ag-Cu alloys. Meanwhile, DS Ag-Cu alloys were prepared via cold rolling for comparison, to illustrate the differences between the resulting properties of Ag-Cu alloys.

MATERIALS AND METHODS

Ag-28%Cu (weight percent, abbreviated as Ag-Cu in the following texts) eutectic master alloys were prepared by vacuum induction melting under argon atmosphere using oxygen-free Cu and electrolytic Ag rods. The polished sample, with a diameter of 9.5 mm and a height of 120 mm, was placed in a quartz crucible with a diameter of 10 mm. Together they were loaded in a directional solidification apparatus. The schematic diagram of the directional solidification equipment is illustrated in Figure 1, where a horizontal static MF up to 1.12 Tesla (T) was subjected to the liquid-solid interface of the molten metal. Solidification of the Ag-Cu eutectic alloy was carried out at a constant temperature gradient (4 K/mm) with different growth rates (50, 100, and 200 μm/s) under different horizontal MFs (0, 0.8, and 1.12 T). The detailed experimental protocol can be found in previous works [14,15]. The specimens with a thickness of 5 mm machined from the as-DS Ag-Cu alloys were then cold rolled to a total reduction of 94% without any heat treatment. The reduction is defined as $100 \times (t_0 - t)/t_0\%$, where t_0 is the initial thickness and t is the thickness after cold rolling. Differential scanning calorimetry (DSC, Shimadzu Corporation, Kyoto, Japan) analysis was carried out on the rolled samples under an argon atmosphere with a Shimadzu DSC-60. The calibration procedures for the DSC tests have been reported elsewhere previously [18].

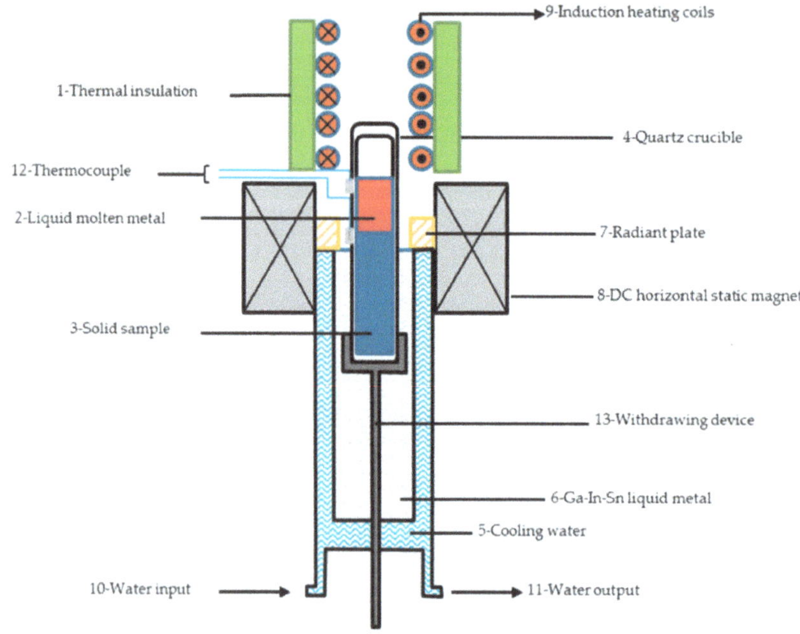

Figure 1. Schematic diagram of the directional solidification apparatus with a horizontal static MF.

The as-DS and as-rolled specimens were sectioned, polished, and etched in a solution of 120 mL H_2O, 20 mL HCl, and 5 g $FeCl_3$. The microstructure was observed by scanning electron microscope. The ELS of Ag-Cu eutectic alloys, λ, was measured on the longitudinal sections in 4–6 regions of the samples and analyzed quantitatively by the linear intercept method [19]. Microhardness measurements were performed with a standardized Vickers device using a 100 g load with a dwelling time of 10 s on the polished surfaces. The electrical resistivity of the samples was measured by the standard DC four-probe technique. The electrical conductivity was determined by using the standard conversion equation (IACS = the International Annealed Copper Standard, and 100% IACS = 1.7241 μΩ·cm). The tensile samples were machined to standard rectangular shape (width: 2 mm, gage length: 4 mm, thickness: less than 0.1 mm) according to the guidelines in ASTM E8-04, and tensile tests were performed on a MTS machine at a rate of $10^{-3}\cdot s^{-1}$ at room temperature.

RESULTS AND DISCUSSION

The effects of both the growth rate and the external MF on microstructural evolution of as-DS Ag-Cu eutectic alloys are shown in Figure 2. It indicates that the as-DS eutectic Ag-Cu alloys primarily consist of lamellar eutectic colonies, where dark lamellar structures are Cu phases, and light ones are Ag matrix phases [20,21]. With increasing growth rate (Figure 2a–c), the lamellar Cu gradually aligns along the growth direction and the eutectic spacing between Cu and Ag is refined. Some researchers [22] found that the crystallographic growth direction of Ag was <110>. Thus, we assumed that the increasing growth rate aligned Cu lamellar along the <110> crystallographic direction. The ELS of as-DS Ag-Cu eutectic alloys were statistically analyzed and the results are summarized in Figure 2f. The results show that the ELS of as-DS Ag-Cu eutectic alloys are about 200–320 nm (Figure 2a–c). With increasing growth rate in the absence of a MF, the ELS of Ag-Cu alloys was gradually decreased by 32% from 293 to 200 nm. This confirmed our previous thoughts that an increased growth rate would refine the ELS.

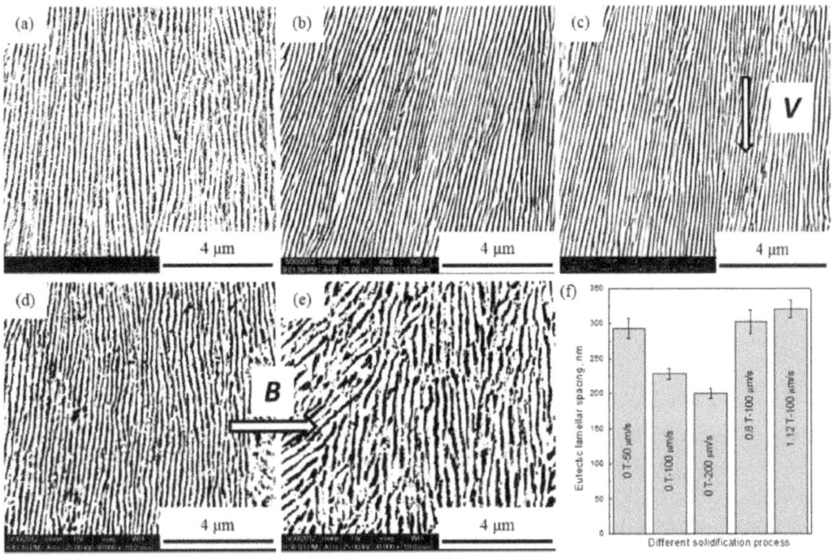

Figure 2. Microstructure of as-DS Ag-Cu eutectic alloys with different MFs (B) and different growth rates (V) at the constant temperature gradient of 4 K/mm. (**a**) 0 T and 50 μm/s; (**b**) 0 T and 100 μm/s; (**c**) 0 T and 200 μm/s; (**d**) 0.8 T and 100 μm/s; (**e**) 1.12 T and 100 μm/s; (**f**) the quantitative results of the ELS.

The effect of the growth rate, V, on the average ELS, λ, was previously studied in numerous eutectic lamellar systems and the experimental results generally follow the J-H model $\lambda^2 V$ = constant [2]. The relationship between the ELS of Ag-Cu eutectic alloy and the growth rate agreed well with the J-H model, and the constant was found to be 5.8 $\mu m^3/s$ (Figure 3). We hypothesized that as the growth rate increased, the enriched solute in front of the lamellae increased. This resulting depression in front of Ag or Cu lamellae promoted the nucleation and subsequent growth of another Cu or Ag lamellae, which consequently decreased the ELS. The constant was different from that of 14 $\mu m^3/s$, which was reported in previous works of Ag-Cu eutectic alloys by Cline et al. [22]. This deviation may have resulted from the refining effects from the melt. Intermedia induction heating (label 9 in Figure 1) instead of resistive heating was used to melt the DS samples, which induced electromagnetic forces during the melt. This force promoted the melt convection, which is called the electromagnetic stirring [23]. This might be responsible for the finer ELS in Ag-Cu eutectic alloy [22].

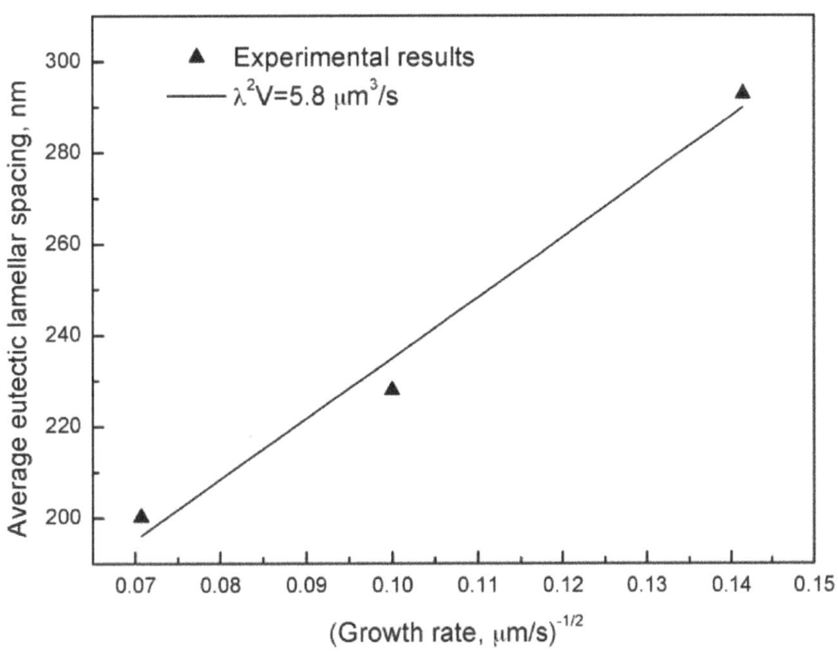

Figure 3. The ELS of as-DS Ag-Cu alloys as a function of the inverse square root of growth rate.

Even with an increase of the imposed MF (Figure 2b,d,e), for all tested growth conditions, the eutectic microstructure remained lamellar. However, the lamellar eutectic structure gradually degenerated, previously demonstrated in other work [4,24]. The external MF also made the lamellar Cu phases coarsen and distribute discontinuously. However, as the MF was increased to 0.8 T (Figure 2f) the ELS is increased by 33%, from 228 nm to 303 nm, but the ELS only changes slightly as the MF is increased further to 1.12 T. MF changed the growth direction of lamellar Cu (Figure 2d,e). In our previous work, we found that utilizing MF in solidification of DS Cu-6 wt %Ag alloy tilted the growth direction of Cu dendrite against the heat flow [15], and the angle between growth direction and heat flow direction agreed well with the equation proposed by Pocheau et al. [25]. We believe that the MF had an analogous impact on Cu lamellae and changed the growth direction of the Cu lamellar phase. We conclude that the imposed MF coarsened the eutectic lamellar structure and changed the growth direction of lamellar eutectics.

While the external MF and the growth rate during directional solidification affected the microstructures of Ag-Cu alloys, this influence was inversely related to the effect of the growth rates on the ELS and the lamellar structures. We hypothesized that the changes might be related to the effect of the MF on the liquid diffusion coefficient. The growth of the eutectic structure strongly depended on the solute diffusion in front of the liquid-solid interface [1]. The inter-lamellar spacing increased proportionally to the decreasing square root of the growth rate. Therefore, the changes of the eutectic inter-lamellar spacing were thought to be a result of the indirect effect of the MF on the growth rate in terms of diffusion. On the other hand, the melt convection affected the diffusion, resulting in the change to the ELS. There were two main mechanisms, which were recognized as effects between the MF and the electrical conductive melt. One was the electromagnetic damping effect, where an induced Lorentz force between MF and melt always tended to weaken convection, thereby damping the natural convection. When applying a MF to the directional solidification of eutectics, the decreasing convection slows down the solute diffusion at the solid-liquid interface, developing solute enriched layers, which in turn result in the decrease of the ELS. Another mechanism is the so-called thermoelectromagnetic convection (TEMC) effect [26,27,28]. During solidification, a thermoelectric current arises at the liquid-solid interface owing to the Seebeck effect. In the presence of an external MF, the direction of the electric current is changed and it is not uniformly parallel to the direction of the MF, which causes Lorentz forces within the solid-liquid zone, which might drive some motion. The influence of the high external MF on the melt convection was formulated by the Hartmann number. Applying the scale of the interlamellar spacings (about 0.2 μm) and the relative physical parameters [26], the Hartmann number is about 0.001 when imposed by an order of magnitude of 1 T MF. This indicated that the thermosolutal convection in molten metal, which was damped by the 1 T MF was negligible. However, at the scale of the interlamellar spacing, the TEMC was dominant and promoted the melt convection, which was consistent with previous reports [29,30]. Such weak convective flow gave rise to an increase of the ELS since the convection produced a slight shift of the solute concentration profile at the interface and a decrease of the eutectic growth undercooling [31]. Increasing convection was capable of increasing the atom diffusion in front of the liquid–solid interface, and coarsening the ELS of the Ag-Cu eutectic alloy. Therefore, we concluded that at the scale of the interlamellar spacing, TEMC rather than the electromagnetic damping effect was dominant, thus increasing the ELS of the Ag-Cu eutectic. Moreover, the increasing convection disturbed the preferential growth directions of the lamellar Cu phases, thus making lamellar Cu deviate from the preferred growth direction, which tended to align the orientation of Cu phases along the pulling direction.

The microhardness and the electrical conductivity were measured and plotted as a function of the ELS in Figure 4a. The microhardness and the ultimate tensile strength (UTS), shown in Table 1, were decreased with increasing ELS. This was attributed to the Hall-Petch relationship between the microhardness and the characteristic microstructure, indicating that the increasing characteristic spacing decreased the strength [32]. The variations of the microhardness with the interlamellar spacing at a constant temperature gradient (G = 4 K/mm) are shown in Figure 4b. The value of hardness increased by increasing the value of $\lambda^{-0.5}$. The relationship between hardness and the exponent function was obtained by using linear regression analysis, and the result was given as:

$$\text{Hardness} = 7.2 + 47.7\,\lambda^{-0.5} \qquad (1)$$

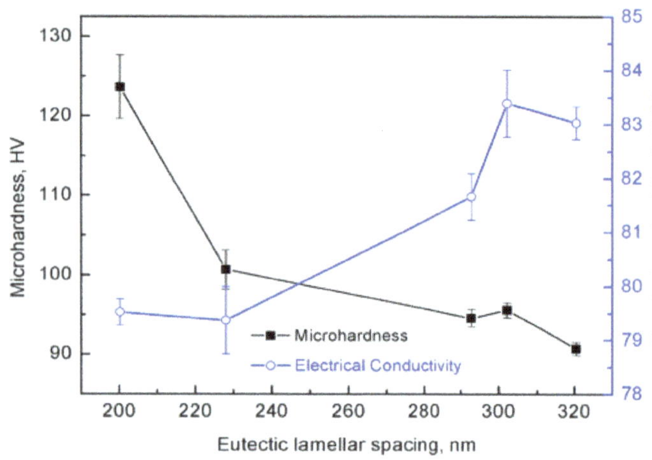

(a)

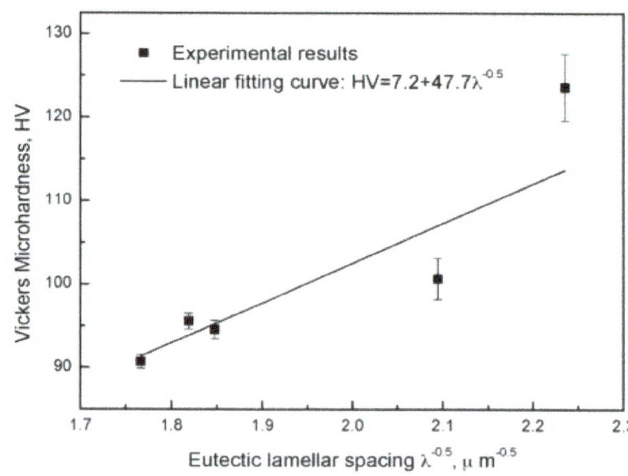

(b)

Figure 4. (**a**) Microhardness and electrical conductivity plotted as the ELS of DS Ag-Cu eutectic alloys with different MFs and growth rates; (**b**) Variation of microhardness as a function of the reciprocal square root of average interlamellar spacing for as-DS Ag-Cu alloy.

Table 1. The mechanical and electrical properties of as-DS and as-rolled Ag-Cu alloys.

Specimen	Elongation	UTS	Yield Strength	Conductivity
	%	(M· Pa)	(M· Pa)	%IACS
0 T-As-DS	10.4%	349.1	310.9	79.4%
0.8 T-As-DS	14.0%	338.8	319.2	83.4%
1.12 T-As-DS	12.2%	263.9	261.0	83.0%
0 T-As-rolled	2.7%	854.8	818.3	62.88%
0.8 T-As-rolled	3.0%	949.5	884.8	53.36%
1.12 T-As-rolled	2.3%	874.5	840.4	56.68%

Microstructure analysis (Figure 2) shows that the ELS decreases with decreasing growth rate, and, by increasing MF. Microhardness is increased with decreasing growth rate and by increasing MF according to Hall-Petch equation [32]. However, Figure 4a shows that the electrical conductivity of Ag-Cu alloys increases gradually with increasing ELS. Generally, the total resistivity of the metallic material is described by the sum of the individual contributions from phonons, impurities, dislocations, and interfaces [33]. Contributions from phonons, impurities, and scattering are assumed to be negligible if there are no other factors except the spacing. The effect of the size contribution to the resistance is defined as [31]:

$$\rho=\rho_0[1+0.75(1-p)l_0/\lambda] \quad (2)$$

where the l_0 is the mean free path of the electron in Ag at room temperature and p is the specularity factor which is in the range of 0.8–0.95 [12,34]. The external MF was incapable of changing ρ_0, p or l_0 because these were intrinsic physical factors of the elements [12,34]. As the ELS λ increased, electrical resistivity decreased and electrical conductivity increased, which was consistent with our results in Figure 4a.

Relative mechanical and electrical parameters of Ag-Cu alloys are presented in Table 1. It indicates that the external MF decreases the UTS of as-cast Ag-Cu alloys. With respect to the as-rolled samples, however, the sample with 0.8 T displays an increased UTS by 8% compared to that without MF.

DSC results displayed in Figure 5 reveal that there are exothermal reactions during the heating process, which indicate the release of deformation energy because there are no peaks found in the as-cast sample. The sample with 0.8 T MF released more latent heat during the heating profile. Higher UTS was believed to be a result of the higher deformation energy, which confirmed our mechanical results (Table 1). Generally, electrical conductivity and UTS are trade-offs, which is consistent with the fact that the sample with 0.8 T MF in Table 1 has the lowest electrical conductivity. The lower conductivity might be a result of more interface scattering resistivity, which is found in the sample with higher strength.

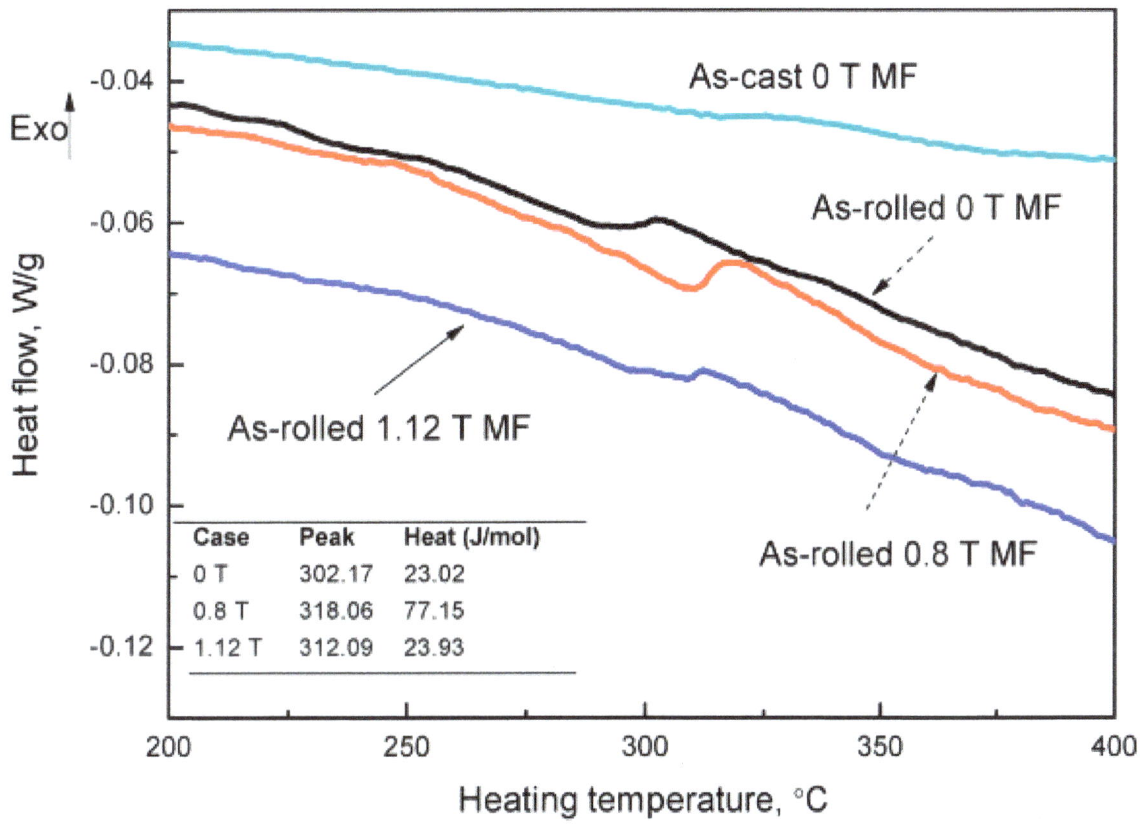

Figure 5. DSC curves of as-rolled Ag-Cu alloys at a heating rate of 10 K/min.

The external MF changed the <110> growth direction and coarsened the ELS. During deformation, Cu and Ag have <110> textures [13]. The change in the growth direction might increase the barriers to form this kind of texture, thus increasing the strength. On the other hand, the coarsening ELS definitely decreased the strengthening according to Hall-Petch equation. The competitive strengthening between the ELS and the change in growth direction controlled the overall strength of the as-rolled Ag-Cu alloy. The change in the growth direction might be dominant, which was thought to be one of the reasons that the as-rolled sample with 0.8 T MF had a higher strength than the 0 T sample.

CONCLUSIONS

The influences of growth rate and external MFs up to 1.12 T on the ELS, mechanical, and electrical properties of DS Ag-Cu alloys were investigated.

(1) The relationship between the measured lamellar spacing of as-DS Ag-Cu eutectic alloys and growth rate followed the J-H model and the constant was found to be 5.8 $\mu m^3/s$.

(2) The external MF changed the growth direction of the Ag-Cu lamellar eutectic and coarsened the spacing. It decreased the microhardness and UTS, but increased the electrical conductivity.

(3) The UTS of the as-rolled Ag-Cu sample with 0.8 T increased by 8% than that without MF. DSC results showed that higher deformation energy was released in the as-rolled sample with 0.8 T MF.

(4) The competitive strengthening between the coarsening ELS and the change in growth direction affected the strength. In our case, the change in growth direction might play a more important role in the strength of DS Ag-Cu alloys.

ACKNOWLEDGMENTS

The authors gratefully acknowledge the financial supports from the National Natural Science Foundation of China (Nos. 51474066 and 51004038), the Research Fund for the Doctoral Program of Higher Education (No. 20120042110008), the Fundamental Research Funds for the Central Universities (No. L1509004), and the 111 Project of China (No. B07015). A portion of this work was performed at the National High Magnetic Field Laboratory (NHMFL), which is supported by National Science Foundation Cooperative Agreement No. DMR-1157490 and the State of Florida. The authors are grateful to Ke Han from NHMFL for their supports on this work, Rongmei Niu from NHMFL for their help with DSC calibration, mechanical properties tests and rolling.

Author Contributions

Xiaowei Zuo, Congcong Zhao, and Lin Zhang conceived and designed the experiments; Xiaowei Zuo and Congcong Zhao performed the experiments; Xiaowei Zuo and Congcong Zhao analyzed the data; Xiaowei Zuo and Engang Wang supervised the project and Xiaowei Zuo wrote the paper. All authors have reviewed the manuscript.

REFERENCES

1. Kurz, W.; Fisher, D.J. *Fundamentals of Solidification*, 4th ed.; Trans Tech Publications Ltd.: Acdermannsdorf, Switzerland, 1998; p. 305.
2. Jackson, K.A.; Hunt, J.D. Lamellar and rod eutectic growth. *Trans. Metall. Soc.* 1966, *236*, 1129–1142.
3. Seetharaman, V.; Trivedi, R. Eutectic growth: Selection of interlamellar spacings. *MTA* 1988, *19*, 2955–2964.
4. Li, X.; Ren, Z.M.; Fautrelle, Y. Effect of a high axial magnetic field on the microstructure in a directionally solidified Al-Al_2Cu eutectic alloy. *Acta Mater.* 2006, *54*, 5349–5360.
5. Liu, T.; Wang, Q.; Zhang, H.W.; Lou, C.S.; Nakajima, K.; He, J.C. Effects of high magnetic fields on solidification microstructure of Al-Si alloys. *J. Mater. Sci.* 2011, *46*, 1628–1634.
6. Li, G.M.; Liu, Y.; Su, Y.; Wang, E.G.; Han, K. Influence of high magnetic field on as-cast structure of Cu-25 wt %Ag alloys. *China Foundry* 2013, *10*, 162–166.
7. Li, G.M.; Wang, E.G.; Zhang, L.; Zuo, X.W.; He, J.C. Effect of high magnetic field on solidified structure, drawn structure and electrical conductivity of Cu-25Ag alloys. *Rare Metal Mater. Eng.* 2012, *41*, 701–706.
8. Li, L.; Zhang, Y.D.; Esling, C.; Zhao, Z.H.; Zuo, Y.B.; Zhang, H.T.; Cui, J.Z. Formation of twinned lamellas with the application of static magnetic fields during semi-continuous casting of Al-0.24 wt %Fe alloy. *J. Cryst. Growth* 2009,*311*, 3211–3215.
9. Davy, C.A.; Ke, H.; Kalu, P.N.; Bole, S.T. Examinations of Cu-Ag composite conductors in sheet forms. *IEEE Trans. Appl. Supercond.* 2008, *18*, 560–563.
10. Han, K.; Toplosky, V.; Goddard, R.; Lu, J.; Niu, R.M.; Chen, J. Impacts of heat treatment on properties and microstructure of Cu16at%Ag conductors. *IEEE Trans. Appl. Supercond.* 2011, *22*, 6900204.
11. Zhao, C.C.; Zuo, X.W.; Wang, E.G.; Niu, R.M.; Han, K. Simultaneously increasing strength and electrical conductivity in nanostructured Cu–Ag composite. *Mater. Sci. Eng. A* 2016, *652*, 296–304.
12. Zuo, X.W.; Zhao, C.C.; Niu, R.M.; Wang, E.G.; Han, K. Microstructural dependence of magnetoresistance in CuAg alloy solidified with high magnetic field. *J. Mater. Process. Technol.* 2015, *224*, 208–212.
13. Han, K.; Vasquez, A.A.; Xin, Y.; Kalu, P.N. Microstructure and tensile properties of nanostructured Cu-25 wt %Ag. *Acta Mater.* 2003, *51*, 767–780.
14. Zuo, X.W.; Guo, R.; An, B.L.; Zhang, L.; Wang, E.G. Microstructure, hardness and electrical resistivity of directionally solidified Cu-6%Ag alloy under a transverse magnetic field. *Acta Metall. Sin.* 2016, *52*, 143–150.
15. Zuo, X.W.; Guo, R.; Zhao, C.C.; Zhang, L.; Wang, E.G.; Han, K. Microstructure and properties of Cu-6 wt %Ag composite thermomechanical-processed after directionally solidifying with magnetic field. *J. Alloys Compd.* 2016, *676*, 46–53.
16. Zuo, X.W.; Han, K.; Zhao, C.C.; Niu, R.M.; Wang, E.G. Microstructure and properties of nanostructured Cu28 wt

%Ag microcomposite deformed after solidifying under a high magnetic field. *Mater. Sci. Eng. A* 2014, *619*, 319–327.

17. Zuo, X.W.; Zhao, C.C.; Wang, E.G.; Zhang, L.; Han, K.; He, J.C. Microstructure evolution of the proeutectic Cu dendrites in diamagnetic Cu-Ag alloys by electromagnetic suppressing convection. *J. Low Temp. Phys.* 2013, *170*, 409–417.
18. Niu, R.M.; Han, K.; Su, Y.F.; Salters, V.J. Atomic-scale studies on the effect of boundary coherency on stability in twinned Cu. *Appl. Phys. Lett.* 2014, *104*, 4.
19. Fan, J.L.; Li, X.Z.; Su, Y.Q.; Guo, J.J.; Fu, H.Z. Effect of growth rate on microstructure parameters and microhardness in directionally solidified Ti-49Al alloy. *Mater. Des.* 2012, *34*, 552–558.
20. Tian, Y.Z.; Zhang, Z.F. Bulk eutectic Cu-Ag alloys with abundant twin boundaries. *Scr. Mater.* 2012, *66*, 65–68.
21. Tian, Y.Z.; Li, J.J.; Zhang, P.; Wu, S.D.; Zhang, Z.F.; Kawasaki, M.; Langdon, T.G. Microstructures, strengthening mechanisms and fracture behavior of Cu-Ag alloys processed by high-pressure torsion. *Acta Mater.* 2012, *60*, 269–281.
22. Cline, H.E.; Lee, D. Strengthening of lamellar vs. Equiaxed Ag-Cu eutectic. *Acta Metall.* 1970, *18*, 315–323.
23. Griffiths, W.D.; McCartney, D.G. The effect of electromagnetic stirring during solidification on the structure of Al-Si alloys. *Mater. Sci. Eng. A* 1996, *216*, 47–60.
24. Li, X.; Fautrelle, Y.; Ren, Z.M. Morphological instability of cell and dendrite during directional solidification under a high magnetic field. *Acta Mater.* 2008, *56*, 3146–3161.
25. Pocheau, A.; Deschamps, J.; Georgelin, M. Dendrite growth directions and morphology in the directional solidification of anisotropic materials. *JOM* 2007, *59*, 71–76.
26. Lehmann, P.; Moreau, R.; Camel, D.; Bolcato, R. A simple analysis of the effect of convection on the structure of the mushy zone in the case of horizontal Bridgman solidification-comparison with experimental results. *J. Cryst. Growth*1998, *183*, 690–704.
27. Li, X.; Fautrelle, Y.; Ren, Z.M. Influence of an axial high magnetic field on the liquid-solid transformation in Al-Cu hypoeutectic alloys and on the microstructure of the solid. *Acta Mater.* 2007, *55*, 1377–1386.
28. Li, X.; Fautrelle, Y.; Gagnoud, A.; Du, D.F.; Wang, J.; Ren, Z.; Nguyen-Thi, H.; Mangelinck-Noel, N. Effect of a weak transverse magnetic field on solidification structure during directional solidification. *Acta Mater.* 2014, *64*, 367–381.
29. Zhang, T.; Ren, W.; Dong, J.; Li, X.; Ren, Z.; Cao, G.; Zhong, Y.; Deng, K.; Lei, Z.; Guo, J. Effect of high magnetic field on the primary dendrite arm spacing and segregation of directionally solidified superalloy DZ417G. *J. Alloys Compd.*2009, *487*, 612–617.
30. Li, X.; Ren, Z.; Wang, J.; Han, Y.; Sun, B. Influence of a weak static magnetic field on the primary dendrite arm spacing of a directionally solidified Ni-based superalloy. *Mater. Lett.* 2011, *67*, 205–209.
31. Ma, D.; Jie, W.Q.; Li, Y.; Ng, S.C. Effect of weak convection on lamellar spacing of eutectics. *Acta Mater.* 1998, *46*, 3203–3210.
32. Freudenberger, J.; Lyubimova, J.; Gaganov, A.; Witte, H.; Hickman, A.L.; Jones, H.; Nganbe, M. Non-destructive pulsed field CuAg-solenoids. *Mater. Sci. Eng. A* 2010, *527*, 2004–2013.
33. Verhoeven, J.D.; Downing, H.L.; Chumbley, L.S.; Gibson, E.D. The resistivity and microstructure of heavily drawn Cu-Nb alloys. *J. Appl. Phys.* 1989, *65*, 1293–1301.
34. Heringhaus, F. *Quantitative Analysis of the Influence of the Microstructure of Strength, Resistivity, and Magnetoresistance of Eutectic Ag-Cu*; Verlag Shaker: Aachen, Germany, 1998; p. 168.

CHAPTER 15 COMPUTATIONAL ELECTROMAGNETIC METHODS FOR GRAPHENE MODELING

Yu Shao, Jing Jing Yang, and Ming Huang

INTRODUCTION

Graphene consists of a monolayer of carbon atoms arranged on a two-dimensional honeycomb lattice which is made up of hexagons. Since its first discovery in 2004, graphene has attracted tremendous research interest in various fields due to its distinctive properties [1–3]. Graphene is a rising star not only in material science and condensed matter physics, but also in the electronic and photonic device communities [4].

Graphene has unique electronic band structure, and the electrons in it behave as massless Dirac fermion [5, 6]. Graphene acquires a pronounced electric field effect which means carrier concentration can be tuned by electrostatic gating. It has been shown that, in bilayer graphene, the electronic gap between conduction bands and the valence can be tuned between zero and midinfrared energies, which makes bilayer graphene the only known semiconductor with a tunable energy gap [7, 8]. Zhang et al. [9] presented a widely tunable electronic bandgap in electrically gated graphene, and they realized a gate-controlled, continuously tunable bandgap of up to 250 meV. Large-area graphene films of the order of centimeters on copper substrates were realized by chemical vapor deposition using methane [10], which facilitated the fabrication of graphene transistors. A good review of graphene transistors in both logic and radiofrequency applications is provided in [11].

Besides electron-device application, graphene has also been recognized as a novel optical material for photonic application. Compared to traditional metal, graphene exhibits several favorable properties. Particularly, surface plasmon polaritons (SPPs), which are electromagnetic waves coupled charge excitations on the surface of a conductor, can be excited in graphene. The plasmons in graphene are tightly confined and the volumes of SPPs in graphene can be 10^6 times smaller than those in free space [12]. This property leads to strong light-matter interaction in graphene [13]. Additionally, the dielectric properties of graphene can be electrically or chemically tuned by changing the charge carrier density and Fermi energy [13–15]. Brar et al. [12] created and probed plasmons in graphene with $\lambda_p \leq \lambda_0/100$ and resonant energies as high as 310 meV for 15 nm nanoresonators. Fei et al. [16] showed common graphenes/SiO_2/Si back-gated structure support propagating surface plasmons and they altered both the amplitude and the wavelength of these plasmons by varying the gate voltage. The graphene-based plasmonics may enable the novel optic devices working in different frequency ranges—from terahertz to the visible range—with extremely high speed, low driving

Citation: Yu Shao, Jing Jing Yang, and Ming Huang, "Computational Electromagnetic Methods for Graphene Modeling," International Journal of Antennas and Propagation, vol. 2016, Article ID 7478621, 9 pages, 2016. doi:10.1155/2016/7478621.

voltage, low power consumption, and compact sizes [17]. Vakil and Engheta in [18] showed that graphene can be made into a one-atom-thick platform for infrared metamaterials and transformation optical devices by manipulating spatially inhomogeneous and nonuniform conductivity patterns. Ju et al. [19] used metamaterial made up of periodic graphene microribbon arrays for terahertz plasmon excitations and demonstrated that the plasmon resonance can be tuned over a broad terahertz frequency range by adjusting microribbon width and electrostatic doping.

Our group has been dedicated to the study of molecular sensors for a long time. Francescato et al. [20] presented a platform for broadband molecular spectroscopy based on the propagation of strongly confined antibonding plasmons supported by graphene sandwiches. This novel scheme measures the absorption spectrum of the molecule and shows a broadband capability and high sensitivity. Recently, Yang et al. [21] proposed a cylindrical graphene plasmon waveguide and investigated its application in molecular sensing.

The terahertz antenna based on graphene has promising application in wireless communications in nanosystems. The tunable property of graphene facilitates the designing of reconfigurable antenna. Huang et al. [22] presented a beam reconfigurable antenna based on a switchable high-impedance surface using single-layer graphene. The graphene-based antennas which have broad wavelength tuning range are proposed in [23, 24].

In this paper, the electromagnetic simulation of graphene is introduced and some widely used computational electromagnetic methods for graphene modeling are reviewed. The advantages and drawbacks of each method are discussed and numerical examples of these methods are given to illustrate their performance and application.

ELECTROMAGNETIC SIMULATION OF GRAPHENE

To investigate the electromagnetic property of graphene, Maxwell equations need to be solved to simulate the wave propagation in graphene. In most cases, Maxwell equations have no analytical solutions except for a few simple canonical problems. Therefore, numerical methods are crucial to understand the wave guiding and scattering by graphene. There are three popular computational electromagnetic methods to simulate graphene: finite-difference time-domain method (FDTD), finite element method (FEM), and method of moment (MoM). Each method has its own advantages and disadvantages depending on the specific problem. Many commercial software packages based on these methods are available, such as FDTD Solutions, CST, HFSS, COMSOL, IE3D, and FEKO. All the software packages can be used to model graphene; however, none of these are developed specifically for graphene which makes them less efficient in dealing with this monoatomic layer device.

Graphene is modeled as a thin surface with complex conductivity which can be tuned by external electric field bias or chemical doping. The surface conductivity of graphene is commonly described by Kubo formula [32]:

$$\cdot \int_0^{\infty} \varepsilon \left(\frac{\partial f_d(\varepsilon)}{\partial \varepsilon} - \frac{\partial f_d(-\varepsilon)}{\partial \varepsilon} \right) d\varepsilon$$

$$- \int_0^{\infty} \frac{f_d(-\varepsilon) - f_d(\varepsilon)}{(\omega - j2\Gamma)^2 - 4(\varepsilon/\hbar)^2} d\varepsilon \Bigg], \tag{1}$$

where Γ is the phenomenological scattering rate, e is the electron charge, ε is the energy state, and $\hbar$ denotes the reduced Plank constant. The first term of (1) is corresponding to the intraband contribution while the second term is due to the interband contribution. That is,

$$\sigma_g = \sigma_{\text{intra}} + \sigma_{\text{inter}}. \tag{2}$$

The intraband term in (2) can be evaluated by

$$\sigma_{\text{intra}} = \frac{e^2 k_B T}{\pi \hbar^2 (2\Gamma + j\omega)} \left[\frac{\mu_c}{k_B T} + 2 \ln \left(e^{-\mu_c / k_B T} + 1 \right) \right] \tag{3}$$

and the interband term in (2) is approximated by

$$\sigma_{\text{inter}} = -\frac{je^2}{4\pi\hbar} \ln \left(\frac{2|\mu_c| - (\omega - j2\Gamma)\hbar}{2|\mu_c| + (\omega - j2\Gamma)\hbar} \right) \quad (k_B T \ll |\mu_c|), \tag{4}$$

where μ_c is the chemical potential, T is the temperature, and k_B is the Boltzmann constant.

In electromagnetic simulation, the 2D graphene surface is usually approximated by a 3D dielectric slab whose 3D conductivity is evaluated by the following equation:

$$\sigma_{3D} = \frac{\sigma_g}{d}, \tag{5}$$

where d is the thickness of the graphene layer. The accuracy of 3D dielectric slab model degrades as d increases. It can be shown that, using the 2D conductivity, the SPP wavelength is [25]

$$\lambda_{\text{SPP}} = \lambda_0 \left(1 - \left(\frac{2}{\eta_0 \sigma} \right)^2 \right)^{-0.5}. \tag{6}$$

The 3D dielectric slab can support even and odd modes. The odd mode has the wavelength of

$$\lambda_{\text{odd}} = 2\pi \left(-\frac{2}{d} \coth^{-1} \varepsilon_{3D} \right)^{-1} \tag{7}$$

and (6) can be approximated by (7) only if the following three conditions are satisfied [25]:

$$\frac{d}{\lambda_{\text{SPP}}} \ll 1,$$

$$|\sigma| \ll \frac{2}{\eta_0},$$

$$\left| \frac{\sigma}{d} \right| > 2\omega\varepsilon_0. \tag{8}$$

Figure 1 shows the relative error of using the dielectric slab model for graphene with respect to normalized d and σ; more details can be found in [25].

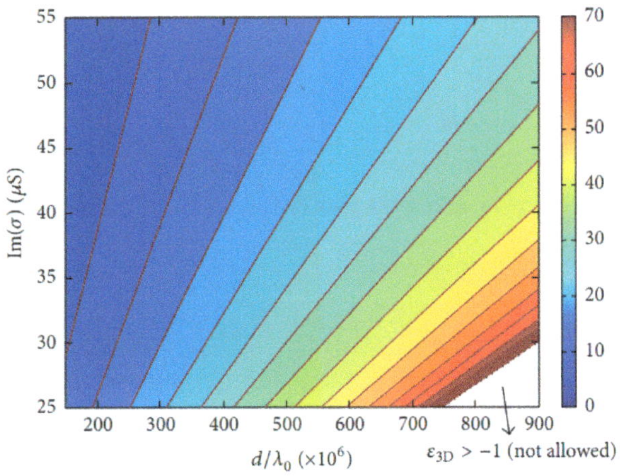

Figure 1: Relative error of the dielectric slab model with respect to normalized d and σ [25].

FINITE-DIFFERENCE TIME-DOMAIN METHOD

FDTD is time-domain method which has specific advantage in transient problem and wideband problem. The broadband results can be obtained in one simulation. Compared to frequency-domain method, it takes less computing memory and simulation time. Additionally, FDTD is an iterative scheme which eliminates solving large linear system; hence, it is simple, robust, and easy to implement.

There are three general approaches to model graphene in FDTD: (I) using standard FDTD method with very fine mesh within the graphene sheet; (II) using the subcell FDTD method; (III) using surface resistive boundary condition.

In the first approach, graphene is modeled as thin layer with finite thickness and the surface conductivity is transformed to volume permittivity. According to the Courant-Friedrichs-Lewy (CFL) stability condition of FDTD, fine mesh leads to small time step, so this approach is both memory- and time-consuming which reduces its efficiency. To alleviate the time step constraint, some unconditionally stable FDTD methods are developed for graphene simulation. In [33], the locally one-dimensional (LOD) FDTD is applied for the efficient simulation of graphene device. They demonstrated that the LOD-FDTD can be 60% faster than standard FDTD with reasonable accuracy. Wang et al. [26] proposed an unconditionally stable one-step leapfrog alternating-direction-implicit (ADI) FDTD to study the SPPs in graphene structure. In Figure 2, they showed the SPPs propagating along the spiral waveguide. The one-step ADI-FDTD method gives accurate result which is comparable to conventional FDTD. They also showed that the conventional FDTD took 1.7×10^5s for this problem, while ADI-FDTD only took 0.8×10^5s with CFLN = 10.

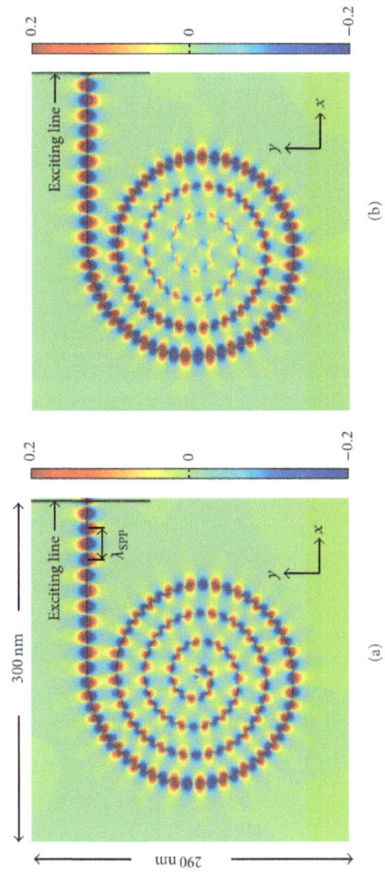

Figure 2: The normalized E_y component of guided SPPs along the spiral graphene waveguide at time t=1.9258 ps calculated by (a) conventional FDTD and (b) the proposed improved ADI-FDTD with CFLN = 10 [26].

In the second approach, graphene is treated as thin layer occupying a fraction of FDTD cell [27]. The Yee cell for the subcell FDTD method is shown in Figure 3, where E_z component is split into $E_{Z,i}$ and $E_{z,o}$ in cells occupying graphene [27]. However, this scheme is complex in mathematics and needs special type of PML to model infinitely thin sheets [34].

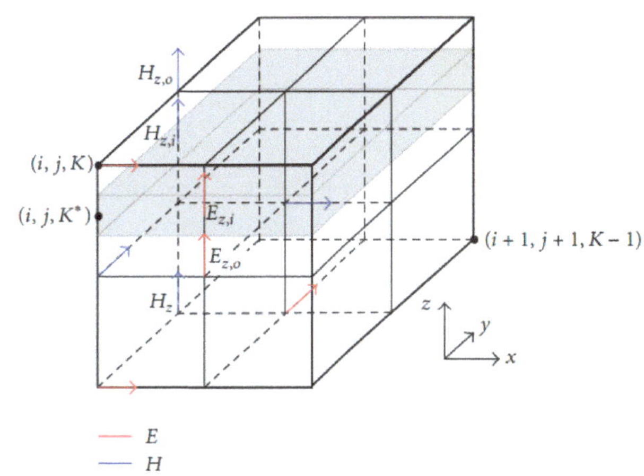

Figure 3: Modified Yee cell for subcell FDTD method at the location of graphene (shaded area) [27].

In the third approach, graphene is modeled as a zero-thickness conductive sheet over which the fields satisfy the surface boundary condition. Nayyeri et al. [28, 35] presented this scheme in detail. As shown in Figure 4(a), a conductive sheet locates at K+1/2. The tangential component of H field is discontinuous over the sheet, so H_x is split into ${}^1H_x$ and ${}^2H_x$, which satisfy the following discretized Faraday law:

$$\mu_1 \frac{{}^1H_x^{n+1/2} - {}^1H_x^{n-1/2}}{\Delta t} = \frac{E_y^n(K+1/2) - E_y^n(K)}{\Delta x/2}, \tag{9a}$$

$$\mu_1 \frac{{}^2H_x^{n+1/2} - {}^2H_x^{n-1/2}}{\Delta t} = \frac{E_y^n(K+1) - E_y^n(K+1/2)}{\Delta x/2}. \tag{9b}$$

The surface boundary condition is

$$ {}^2H_x - {}^1H_x = \sigma_s E_y \tag{10}$$

from which $E_y^n(K+1/2)$ can be written as

$$E_y^n\left(K+\frac{1}{2}\right) = \frac{1}{2\sigma_s}\left[\left({}^2H_x^{n+1/2} + {}^2H_x^{n-1/2}\right) - \left({}^1H_x^{n+1/2} + {}^1H_x^{n-1/2}\right)\right]. \tag{11}$$

Substituting (11) into (9a) and (9b), ${}^1H_x$ and ${}^2H_x$ can be derived as

$${}^1H_x^{n+1/2} = \frac{1}{1-c_1c_2}\left(F_1^n + c_1F_2^n\right), \tag{12a}$$

$${}^2H_x^{n+1/2} = \frac{1}{1-c_1c_2}\left(F_2^n + c_2F_1^n\right). \tag{12b}$$

The expressions of c_1, c_2, F_1^n, and F_2^n are shown in [28]. Figure 4(b) is a 3D FDTD cell containing conductive sheet at K+1/2. In this case, tangential components of magnetic field (H_x, H_y) and normal component of electric field (E_z) are discontinuous over the sheet, so they are split into two parts on the two sides of the surface. Their updating equations can be derived in a similar way to 1D case and are presented in [28].

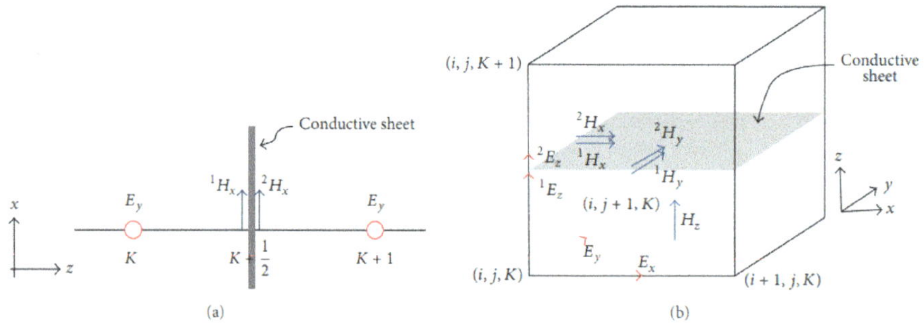

Figure 4: (a) 1D FDTD cell including a conductive sheet at grid K+1/2. (b) 3D FDTD cell with a conductive sheet at grid K+1/2 [28].

The surface boundary condition approach has been validated [28]. Figure 5(a) shows a 2D problem in which an infinite line source radiates above an infinite graphene sheet. The pattern of E_z at the wavelength of 100 µm is shown in Figure 5(b). A strong agreement between the proposed FDTD result and that derived from semianalytic method is obtained. In addition, unlike the subcell FDTD method, the classical PML is applicable to the surface boundary condition scheme.

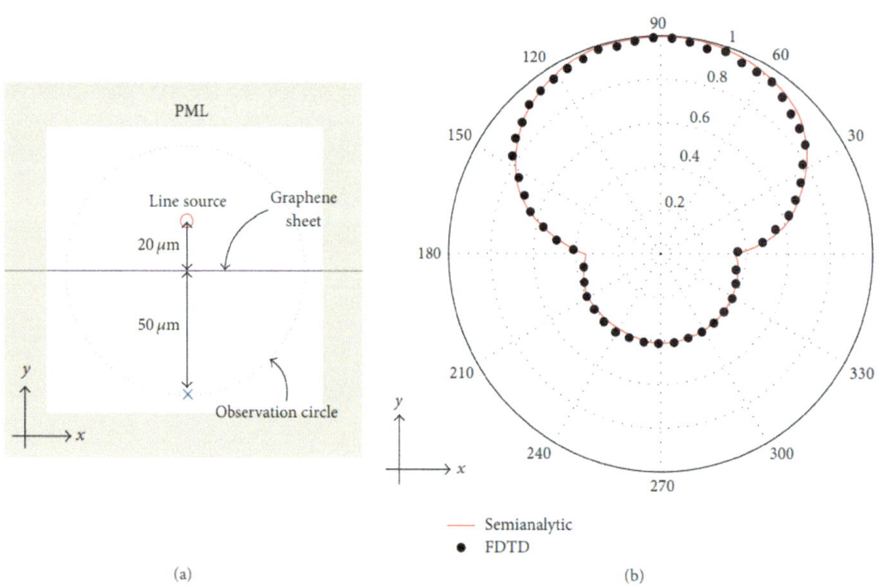

Figure 5: (a) Line source scattering by a graphene sheet. (b) Normalized pattern of E_z at the wavelength of $\lambda_0 = 100\,\mu m$ [28].

In time-domain simulation, the conductivities of graphene shown in (2)–(4) need to be converted from frequency domain into time domain. The intraband conductivity, which has a Drude-like expression, can be easily converted into time domain. However, the interband conductivity has a complex logarithmic form which needs a vector fitting technique. A summation of partial fractions in terms of complex conjugate pole-residues is applied to approximate the conductivity. The detailed description of the fitting technique can be found in [35–37].

FINITE ELEMENT METHOD

FEM is a full-wave numerical technique for electromagnetic boundary-value problems. The basic principle of this method is to discretize the whole computing domain with a finite number of subdomains in which the unknown function is expanded by simple interpolation functions with unknown coefficients. Then, a system of algebraic equations of these unknown coefficients is obtained by using Ritz variational or Galerkin's method. Finally, by solving this linear system, the approximated solution of the entire domain can be obtained.

In FDTD, the computing space is discretized by orthogonal grid; this staircase approximation will reduce the modeling fidelity when it comes to complex geometry, while in FEM, where the triangles or tetrahedral elements are applied, arbitrary geometries can be modeled accurately. As a result, FEM has advantage in complex and inhomogeneous problems. Furthermore, FEM is a frequency domain method which makes it efficient in dealing with narrow-band problems.

Brar et al. [12] solved Maxwell's equation by FEM and modeled graphene as a thin sheet with 0.1 nm thickness; the results suggested that graphene can increase light-matter interactions at infrared energies. Software packages such as HFSS and COMSOL are based on FEM, and they are used widely in graphene simulation. Recently, we used COMSOL to investigate the transmission properties of a cylindrical graphene plasmon waveguide [21]. Tamagnone et al. [29] simulated a reconfigurable graphene antenna with HFSS; the structure of the antenna is shown in Figure 6(a); the input impedance of the antenna can be tuned by changing the chemical potential as shown in Figure 6(b).

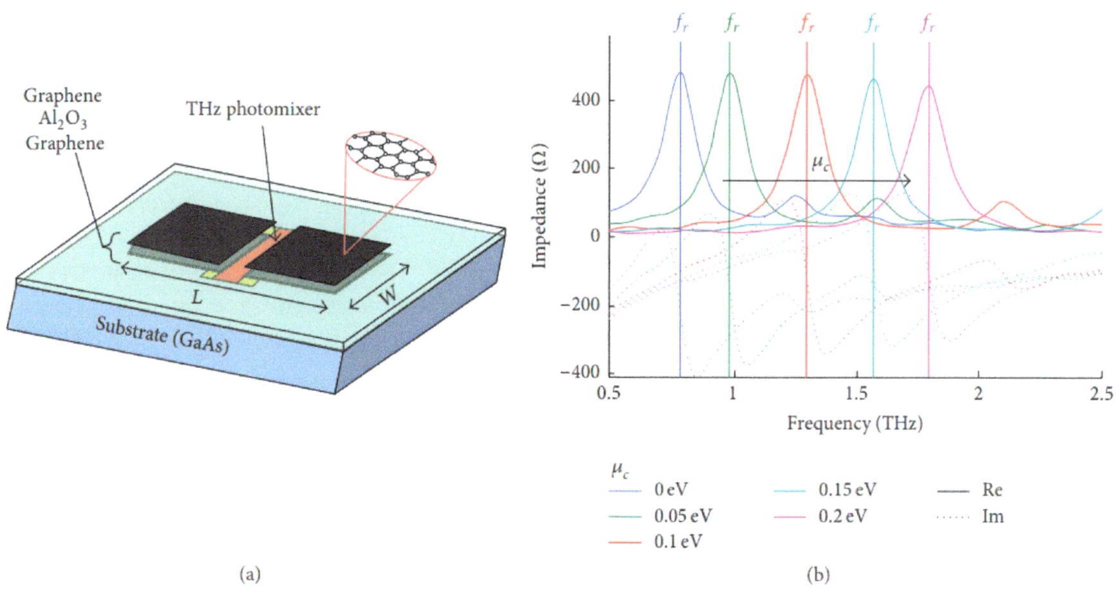

Figure 6: (a) Structure of a reconfigurable graphene antenna. (b) Input impedance of the antenna with respect to chemical potential [29].

METHOD OF MOMENTS

In contrast to FDTD and FEM which solve differential equations, MoM is a technique used to solve electromagnetic surface or volume integral equations in the frequency domain. In MoM, the quantities of interest are not the fields but the electromagnetic sources (surface or volume current), so only the surface of the geometry needs to be discretized. The surface current is discretized into wire segments and/or surface patches. A linear system can be constructed by the method of weighted residuals and the results of the linear equations give the surface current. The far-field result can be derived from the surface current by Green's function. Because it only requires calculating the boundary values instead of the values throughout the space, MoM is highly efficient for electrically large objects and is widely used in solving radiation and scattering problems. However, when applied to complex inhomogeneous cases, it will be very computationally expensive and less efficient. The analytical expressions of dyadic Green's function for graphene are derived in [38]. Shapoval et al. [39] proposed integral equations based on surface-impedance boundary condition to analyze plane wave scattering and absorption by graphene-strip gratings. The method of moments for graphene nanoribbons was developed in [40–44], in which the issue of nonlocality of graphene conductivity was taken into account. Nonlocal effect arises from spatial dispersion of graphene which is nonnegligible when dealing with slow modes supported by graphene nanoribbons. The spatially dispersive intraband conductivity tensor was derived in [41].

Software packages such as IE3D and FEKO are based on MoM. IE3D is applied to simulate the microwave propagation in a coplanar waveguide over graphene from 40 MHz to 110 GHz [45]. Cabellos-Aparicio et al. [30] used FEKO to study the radiated power of a graphene plasmonic antenna fed by photoconductive source. The antenna structure is shown in Figure 7(a) and the radiated power with respect to frequency is shown in Figure 7(b). The detailed parameters of the photoconductive antenna can be found in [30].

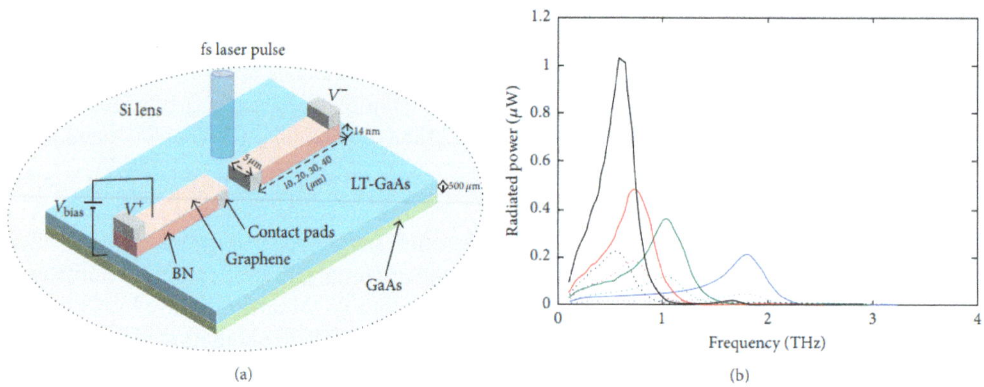

Figure 7: (a) Structure of graphene RF plasmonic antenna fed with a photoconductive material. (b) Radiated power with respect to frequency for graphene antennas with lengths 10 µm (blue), 20 µm (green), 30 µm (red), and 40 µm (black) [30].

DISCONTINUOUS GALERKIN TIME-DOMAIN METHOD

The graphene involved problems are often multiscale. Graphene is monoatomic and its thickness is much smaller than wavelength, so it is an electrically fine structure. In contrast, the substrate belongs to electrically coarse structure because its dimension is much greater than wavelength. As we have mentioned before, time-domain methods have the advantage that the broadband characterization can be obtained with only a single simulation. However, FDTD method has a serious efficiency problem in multiscale simulation because it requires high discretization density to model electrically fine structure due to its cartesian grid. The finite element time-domain (FETD) method is capable of modeling complex and fine structures and achieving high-order accuracy with high-order basis functions. The major drawback is a global linear system of equations that needs to be solved at each time step. The multiscale problems usually contain a large number of unknowns; FETD will be very computationally expensive in this case. Discontinuous Galerkin time-domain (DGTD) method is promising in multiscale problems [46]. DGTD allows for domain decomposition. A multiscale structure is divided into several subdomains and each subdomain can be discretized separately. All operations in DGTD are local, so large global matrix is split into several smaller matrices. Unlike FETD, the matrices of DGTD are inverted and stored before time marching, and different time integration scheme can be used in different subdomain. Additionally, DGTD is naturally adapted to parallel computing. Recently, DGTD has been used in nanophotonics field and considered as a viable alternative to the well-established FDTD and FETD methods [47].

Li et al. [31] proposed DGTD method with resistive boundary condition for the electromagnetic analyzing of graphene. They also applied this method to the magnetized graphene from microwave to THz range, where the anisotropic and disperse surface conductivity is involved [48]. Figure 8(a) shows the reflection Γ_R, transmission Γ_T, and absorption Γ_A coefficients of an infinitely large graphene sheet under the illumination of a normally incident plane wave; the results of DGTD agree very well with the theoretical data. Figure 8(b) is the normalized extinction cross section of a freestanding graphene patch; good consistency is achieved between DGTD and integral equation method. More numerical examples can be found in [31].

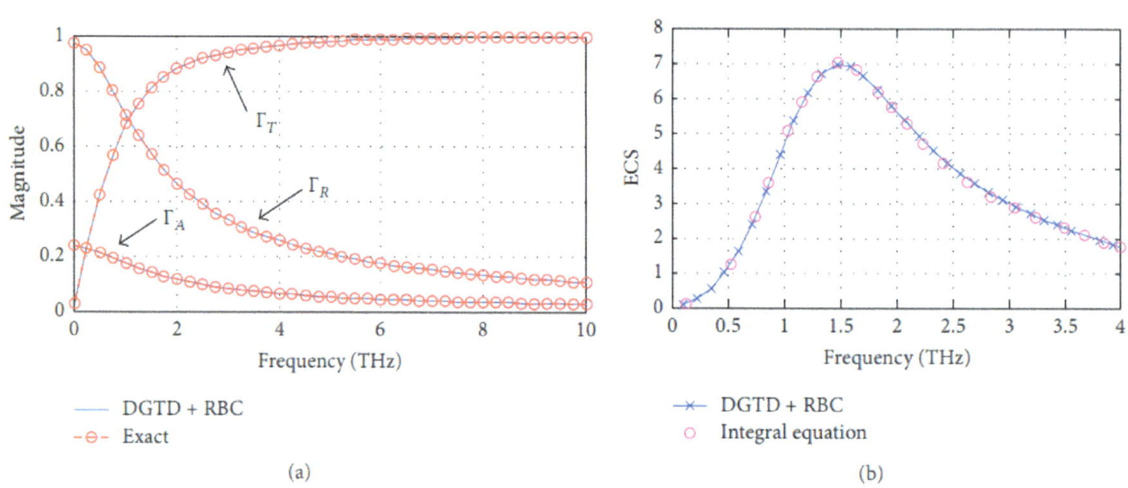

Figure 8: (a) The magnitude of reflection Γ_R, transmission Γ_T, and absorption Γ_A coefficients calculated by DGFD as well as the theoretical value. (b) Normalized extinction cross section of a graphene patch calculated by DGTD and integral equation method [31].

CONCLUSION

Due to its intriguing properties, graphene is used more and more widely in electronic and photonic community. This paper gives a brief review of application of graphene, as well as the computational methods which can be used to investigate its electromagnetic properties. Although a number of famous commercial software packages are available for graphene simulation, none of them are developed specifically for graphene; therefore, they suffer from inefficiency. Each computational method has its own advantages and drawbacks; one should choose the appropriate method according to the specific problem. FDTD is simple and easy to implement; however, it is less efficient in modeling fine structure. FEM is flexible in geometry modeling but solving large global system makes it computationally expensive. DGTD is very suitable for multiscale problem and we believe it will be more and more widely used in modeling and designing graphene-assisted device.

ACKNOWLEDGMENTS

This work was supported by the National Natural Science Foundation of China (Grants nos. 61161007, 61261002, 61461052, and 11564044), the Specialized Research Fund for the Doctoral Program of Higher Education (Grants nos. 20135301110003 and 20125301120009), China Postdoctoral Science Foundation (Grants nos. 2013M531989 and 2014T70890), and the Key Program of Natural Science of Yunnan Province (Grants nos. 2013FA006 and 2015FA015).

REFERENCES

1. K. S. Novoselov, A. K. Geim, S. V. Morozov et al., "Electric field effect in atomically thin carbon films,"Science, vol. 306, no. 5696, pp. 666–669, 2004.
2. A. K. Geim and K. S. Novoselov, "The rise of graphene," Nature Materials, vol. 6, no. 3, pp. 183–191, 2007.
3. C. Lee, X. Wei, J. W. Kysar, and J. Hone, "Measurement of the elastic properties and intrinsic strength of monolayer graphene," Science, vol. 321, no. 5887, pp. 385–388, 2008.
4. P. Avouris, "Graphene: electronic and photonic properties and devices," Nano Letters, vol. 10, no. 11, pp. 4285–4294, 2010.
5. K. S. Novoselov, A. K. Geim, S. V. Morozov et al., "Two-dimensional gas of massless Dirac fermions in graphene," Nature, vol. 438, no. 7065, pp. 197–200, 2005.
6. A. H. Castro Neto, F. Guinea, N. M. R. Peres, K. S. Novoselov, and A. K. Geim, "The electronic properties of graphene," Reviews of Modern Physics, vol. 81, no. 1, pp. 109–162, 2009.
7. T. Ohta, A. Bostwick, T. Seyller, K. Horn, and E. Rotenberg, "Controlling the electronic structure of bilayer graphene," Science, vol. 313, no. 5789, pp. 951–954, 2006.
8. E. V. Castro, K. S. Novoselov, S. V. Morozov et al., "Biased bilayer graphene: semiconductor with a gap tunable by the electric field effect," Physical Review Letters, vol. 99, no. 21, Article ID 216802, 2007.
9. Y. Zhang, T.-T. Tang, C. Girit et al., "Direct observation of a widely tunable bandgap in bilayer graphene,"Nature, vol. 459, no. 7248, pp. 820–823, 2009.
10. X. Li, W. Cai, J. An et al., "Large-area synthesis of high-quality and uniform graphene films on copper foils," Science, vol. 324, no. 5932, pp. 1312–1314, 2009.
11. F. Schwierz, "Graphene transistors," Nature Nanotechnology, vol. 5, no. 7, pp. 487–496, 2010.
12. V. W. Brar, M. S. Jang, M. Sherrott, J. J. Lopez, and H. A. Atwater, "Highly confined tunable mid-infrared plasmonics in graphene nanoresonators," Nano Letters, vol. 13, no. 6, pp. 2541–2547, 2013.
13. F. H. L. Koppens, D. E. Chang, and F. J. G. de Abajo, "Graphene plasmonics: a platform for strong light–matter interactions," Nano Letters, vol. 11, no. 8, pp. 3370–3377, 2011.
14. J. Chen, M. Badioli, P. A. Gonzalez, et al., "Optical nano-imaging of gate-tunable graphene plasmons,"Nature, vol. 487, pp. 77–81, 2012.
15. M. Jablan, H. Buljan, and M. Soljačić, "Plasmonics in graphene at infrared frequencies," Physical Review B, vol. 80, no. 24, Article ID 245435, 2009.
16. Z. Fei, A. S. Rodin, G. O. Andreev et al., "Gate-tuning of graphene plasmons revealed by infrared nano-imaging," Nature, vol. 486, no. 7405, pp. 82–85, 2012.
17. A. N. Grigorenko, M. Polini, and K. S. Novoselov, "Graphene plasmonics," Nature Photonics, vol. 6, no. 11, pp. 749–758, 2012.
18. A. Vakil and N. Engheta, "Transformation optics using graphene," Science, vol. 332, no. 6035, pp. 1291–1294, 2011.
19. L. Ju, B. Geng, J. Horng et al., "Graphene plasmonics for tunable terahertz metamaterials," Nature Nanotechnology, vol. 6, no. 10, pp. 630–634, 2010.
20. Y. Francescato, V. Giannini, J. Yang, M. Huang, and S. A. Maier, "Graphene sandwiches as a platform for broadband molecular spectroscopy," ACS Photonics, vol. 1, no. 5, pp. 437–443, 2014.
21. J. Yang, J. Yang, W. Deng, F. Mao, and M. Huang, "Transmission properties and molecular sensing application of CGPW," Optics Express, vol. 23, no. 25, pp. 32289–32299, 2015.
22. Y. Huang, L.-S. Wu, M. Tang, and J. Mao, "Design of a beam reconfigurable thz antenna with graphene-based switchable high-impedance surface," IEEE Transactions on Nanotechnology, vol. 11, no. 4, pp. 836–842, 2012.
23. I. Llatser, C. Kremers, A. Cabellos-Aparicio, J. M. Jornet, E. Alarcón, and D. N. Chigrin, "Graphene-based nano-patch antenna for terahertz radiation," Photonics and Nanostructures—Fundamentals and Applications, vol. 10, no. 4, pp. 353–358, 2012.
24. Y. Yao, M. A. Kats, P. Genevet et al., "Broad electrical tuning of graphene-loaded plasmonic antennas,"Nano Letters, vol. 13, no. 3, pp. 1257–1264, 2013.
25. E. Forati, G. W. Hanson, A. B. Yakovlev, and A. Alù, "Planar hyperlens based on a modulated graphene

monolayer," Physical Review B, vol. 89, no. 8, Article ID 081410, 2014.

26. X.-H. Wang, W.-Y. Yin, and Z. Chen, "Broadband modeling surface plasmon polaritons in optically pumped and curved graphene structures with an improved leapfrog ADI-FDTD method," Optics Communications, vol. 334, pp. 152–155, 2015.

27. G. D. Bouzianas, N. V. Kantartzis, C. S. Antonopoulos, and T. D. Tsiboukis, "Optimal modeling of infinite graphene sheets via a class of generalized FDTD schemes," IEEE Transactions on Magnetics, vol. 48, no. 2, pp. 379–382, 2012.

28. V. Nayyeri, M. Soleimani, and O. M. Ramahi, "Modeling graphene in the finite-difference time-domain method using a surface boundary condition," IEEE Transactions on Antennas and Propagation, vol. 61, no. 8, pp. 4176–4182, 2013.

29. M. Tamagnone, J. S. G. Diaz, J. R. Mosig, and J. P. Carrier, "Reconfigurable terahertz plasmonic antenna concept using a graphene stack," Applied Physics Letters, vol. 101, no. 21, Article ID 214102, 2012.

30. A. Cabellos-Aparicio, I. Llatser, E. Alarcón, A. Hsu, and T. Palacios, "Use of terahertz photoconductive sources to characterize tunable graphene RF plasmonic antennas," IEEE Transactions on Nanotechnology, vol. 14, no. 2, pp. 390–396, 2015.

31. P. Li, L. J. Jiang, and H. Bağcı, "A resistive boundary condition enhanced DGTD scheme for the transient analysis of graphene," IEEE Transactions on Antennas and Propagation, vol. 63, no. 7, pp. 3065–3076, 2015. · ·

32. G. W. Hanson, "Dyadic green's functions for an anisotropic, non-local model of biased graphene," IEEE Transactions on Antennas and Propagation, vol. 56, no. 3, pp. 747–757, 2008.

33. I. Ahmed, E. H. Khoo, and E. Li, "Efficient modeling and simulation of graphene devices with the LOD-FDTD method," IEEE Microwave and Wireless Components Letters, vol. 23, no. 6, pp. 306–308, 2013. · ·

34. X. Yu and C. D. Sarris, "A perfectly matched layer for subcell FDTD and applications to the modeling of graphene structures," IEEE Antennas and Wireless Propagation Letters, vol. 11, pp. 1080–1083, 2012.

35. V. Nayyeri, M. Soleimani, and O. M. Ramahi, "Wideband modeling of graphene using the finite-difference time-domain method," IEEE Transactions on Antennas and Propagation, vol. 61, no. 12, pp. 6107–6114, 2013. · ·

36. H. Lin, M. F. Pantoja, L. D. Angulo, J. Alvarez, R. G. Martin, and S. G. Garcia, "FDTD modeling of graphene devices using complex conjugate dispersion material model," IEEE Microwave and Wireless Components Letters, vol. 22, no. 12, pp. 612–614, 2012.

37. D.-W. Wang, W.-S. Zhao, X.-Q. Gu, W. Chen, and W.-Y. Yin, "Wideband modeling of graphene-based structures at different temperatures using hybrid FDTD method," IEEE Transactions on Nanotechnology, vol. 14, no. 2, pp. 250–258, 2015.

38. A. Y. Nikitin, F. J. Garcia-Vidal, and L. Martin-Moreno, "Analytical expressions for the electromagnetic dyadic green's function in graphene and thin layers," IEEE Journal of Selected Topics in Quantum Electronics, vol. 19, no. 3, Article ID 4600611, 2013.

39. O. V. Shapoval, J. S. Gomez-Diaz, J. Perruisseau-Carrier, J. R. Mosig, and A. I. Nosich, "Integral equation analysis of plane wave scattering by coplanar graphene-strip gratings in the thz range," IEEE Transactions on Terahertz Science and Technology, vol. 3, no. 5, pp. 666–674, 2013.

40. R. Araneo, G. Lovat, and P. Burghignoli, "Graphene nanostrip lines: dispersion and attenuation analysis," in Proceedings of the 16th Workshop on Signal and Power Integrity (SPI '12), pp. 75–78, IEEE, Sorrento, Italy, May 2012.

41. G. Lovat, G. W. Hanson, R. Araneo, and P. Burghignoli, "Semiclassical spatially dispersive intraband conductivity tensor and quantum capacitance of graphene," Physical Review B, vol. 87, no. 11, Article ID 115429, 2013.

42. P. Burghignoli, R. Araneo, G. Lovat, and G. Hanson, "Space-domain method of moments for graphene nanoribbons," in Proceedings of the 8th European Conference on Antennas and Propagation (EuCAP '14), pp. 666–669, IEEE, The Hague, The Netherlands, April 2014. · ·

43. R. Araneo, P. Burghignoli, G. Lovat, and G. W. Hanson, "Modal propagation and crosstalk analysis in coupled graphene nanoribbons," IEEE Transactions on Electromagnetic Compatibility, vol. 57, no. 4, pp. 726–733, 2015.

44. A. Fallahi, T. Low, M. Tamagnone, and J. Perruisseau-Carrier, "Nonlocal electromagnetic response of graphene nanostructures," Physical Review B, vol. 91, Article ID 121405, 2015.

45. M. Dragoman, D. Neculoiu, A. Cismaru et al., "Coplanar waveguide on graphene in the range 40 MHz-110 GHz," Applied Physics Letters, vol. 99, no. 3, Article ID 033112, 2011.

46. J. Chen and Q. H. Liu, "Discontinuous Galerkin Time-Domain methods for multiscale electromagnetic simulations: a review," Proceedings of the IEEE, vol. 101, no. 2, pp. 242–254, 2013.

47. S. Descombes, C. Durochat, S. Lanteri, L. Moya, C. Scheid, and J. Viquerat, "Recent advances on a DGTD method for time-domain electromagnetics," Photonics and Nanostructures—Fundamentals and Applications, vol. 11, no. 4, pp. 291–302, 2013.

48. P. Li and L. J. Jiang, "Modeling of magnetized graphene from microwave to THz range by DGTD with a scalar RBC and an ADE," IEEE Transactions on Antennas and Propagation, vol. 63, no. 10, pp. 4458–4467, 2015. · ·

CHAPTER 16

A HYBRID APPROACH FOR FAST COMPUTATION OF MULTIPLE INCIDENT ANGLES ELECTROMAGNETIC SCATTERING PROBLEMS WITH COMPRESSIVE SENSING AND ADAPTIVE CROSS APPROXIMATION

Guo-hua Wang and Yu-fa Sun

INTRODUCTION

Accurate and efficient analysis of the electromagnetic scattering with multiple incident angles has attracted increasing attention from academia and industry. The method of moment (MoM) [1, 2] together with iterative methods is often used for the accurate analysis of electromagnetic scattering problems, and some fast algorithms like the multilevel fast multipole algorithm (MLFMA) [3], the adaptive integral method (AIM) [4], and the adaptive cross approximation (ACA) [5, 6] are generally applied to reduce both the computation time and the memory requirements for electrically large radiation and scattering problems. But, for the problems with multiple incident angles, the above methods become time-consuming due to the presence of multiple right-hand side (RHS) problems.

An efficient method to accelerate the solution to multiple RHS problems is to remove redundancies in the initial excitation and reduce the number of RHS. In [7, 8], the compressive sensing (CS) theory [9, 10] was introduced into the traditional MoM to solve wide-angle monostatic scattering problem of 2D targets. In [11], the CS technique was used to analyze the multistatic scattering problems. In this paper, a new excitation matrix is firstly constructed with smaller RHS number than traditional MoM, and the measurement values of the induced current coefficients can be calculated by direct method or iterative method. Finally, true current coefficients can be reconstructed by a recovery method such as the orthogonal matching pursuit (OMP) [12] algorithm. However, the number of the measurement values and sparse transformation base in the CS are difficult to determine according to different scattering targets. In [7], to balance the computational complexity and the accuracy, the number of measurements was set to 24. To reduce the number of measurements to a lower level than the CS technique, a hybrid compressive method named CS-ACA is advised. Such method utilizes the low-rank characteristics of the new excitation matrix generated by the CS. The CS technique is firstly used with a reduced number of RHS; then the ACA algorithm is introduced into CS to recompress the new excitation matrix so as to reduce the number of RHS further, so as to realize the efficient calculation of the multiple excitation 3D electromagnetic problems.

Citation: Guo-hua Wang and Yu-fa Sun, "A Hybrid Approach for Fast Computation of Multiple Incident Angles Electromagnetic Scattering Problems with Compressive Sensing and Adaptive Cross Approximation," International Journal of Antennas and Propagation, vol. 2016, Article ID 1416094, 6 pages, 2016. doi:10.1155/2016/1416094.

A HYBRID COMPRESSIVE TECHNIQUE OF THE EXCITATION MATRIX

Compressive Sensing Theory

The CS theory exploits the inherent scarcity or compressibility features of the signal in some transform basis and directly samples the most important information of the signal at the rate far below Nyquist rate [13]. The original signal can be reconstructed accurately by solving an optimization problem.

Let $\mathbf{x} \in \mathbf{R}^R$ represent a discrete signal of length R, which can be expanded as a linear combination of a set of standard orthogonal bases $\mathbf{\Psi} = [\psi_1, \psi_2, \ldots, \psi_R]$:

$$\mathbf{x} = \mathbf{\Psi} \cdot \boldsymbol{\alpha} = \sum_{i=1}^{R} \psi_i \alpha_i, \tag{1}$$

where ψ_i is a set of basis functions in orthogonal basis Ψ and $\boldsymbol{\alpha} = [\alpha_1, \alpha_2, \ldots, \alpha_R]^T$ is the expanded complex coefficients vector of signal. T represents the transpose operation. If x can be represented by combination of only K nonzero entries and K<<R, it can be named K-sparse [14] in Ψ field. Assume that x is one-dimensional signal of length R and M is the number of measurements. By constructing $M \times R$ $(M \approx K\ln(R/K) \ll R)$ -dimensional measurement matrix ϕ and computing M inner products operation, we can obtain the observation signal y, which is the linear measurement of the original signal $\mathbf{x}$. y can also be seen as the projection of a sparse signal in a low dimensional space:

$$\mathbf{y} = \mathbf{\Phi} \cdot \mathbf{x} = \mathbf{\Phi} \cdot \mathbf{\Psi} \cdot \boldsymbol{\alpha} = \mathbf{A}^{CS} \cdot \boldsymbol{\alpha}, \tag{2}$$

where the CS matrix is defined as M×R-dimensional matrix A^{CS}. Once the measurement y has been obtained, a reconstruction algorithm is required to reconstruct the sparse signal α from the measured vector y of length M. As the number of measurements is much smaller than the dimension of the original signal, the reconstruction process becomes an issue of solving the underdetermined equations, which generally has no determined solution. However, if A^{CS} satisfies strict restricted isometry property (RIP) [15] condition, this issue will have a determined solution, and the signal recovery issue can be converted to convex optimization algorithm; then the approximation of α can be obtained by solving the -norm minimization problem:

$$\hat{\boldsymbol{\alpha}} = \min \quad \|\boldsymbol{\alpha}\|_1 \qquad \text{s.t.} \quad \mathbf{y} = \mathbf{A}^{CS} \cdot \boldsymbol{\alpha}. \tag{3}$$

Finally, the original signal can be reconstructed as

$$\hat{\mathbf{x}} = \mathbf{\Psi} \cdot \hat{\boldsymbol{\alpha}}. \tag{4}$$

Application of CS to Multiple Excitation Problems

In this paper, the MoM in the frequency domain is used to solve 3D electromagnetic scattering problems by solving the related electric field integral equations (EFIEs). By using Rao-Wilton-Glisson (RWG) vector basis functions [16] and employing Galerkin's method of moments, the related EFIEs can be transformed into a dense, complex system of linear equations:

$$\mathbf{Z} \cdot \mathbf{J} = \mathbf{V}, \tag{5}$$

where $\mathbf{Z} \in \mathbf{C}^{N\times N}$ is the impedance matrix and N is the number of unknowns. $\mathbf{V} \in \mathbf{C}^{N\times R}$ is the excitation matrix with R plane waves and $\mathbf{J} \in \mathbf{C}^{N\times R}$ is the induced current density matrix. For various angles of incidence, computing responses became time-consuming due to the presence of a multiple RHS problem. In order to reduce the number of RHS and accelerate the solution process, according to CS theory [7–11], new excitation is constructed as follows:

$$\mathbf{V}_m^{CS} = \sum_{r=1}^{R} \phi_{mr} \mathbf{V}(\theta_r), \quad m = 1, 2 \cdots M, \tag{6}$$

in which the new source is formed by random superposition of the incident at different incident angles. Substituting V^{CS} into (5),

$$\mathbf{Z} \cdot \mathbf{J}_m^{CS} = \mathbf{V}_m^{CS}, \quad m = 1, 2 \cdots M, \tag{7}$$

in which $\mathbf{V}_m^{CS}$ is the column vector of new excitation matrix $\mathbf{V}^{CS}$. As the impedance matrix Z is not changed with incident angle, each column vector of $\mathbf{J}_m^{CS}$ can be described as

$$\mathbf{J}_m^{CS} = \sum_{r=1}^{R} \phi_{mr} \mathbf{J}(\theta_r), \quad m = 1, 2 \cdots M. \tag{8}$$

Rewriting $\mathbf{J}_m^{CS}$ and $\mathbf{J}(\theta_r)$ as M-dimensional row vector, (8) can be expressed as a matrix form

$$\mathbf{J}^{CS} = \begin{pmatrix} \mathbf{J}_1^{CS} \\ \mathbf{J}_2^{CS} \\ \vdots \\ \mathbf{J}_M^{CS} \end{pmatrix} = \begin{pmatrix} \phi_{11} & \phi_{12} & \cdots & \phi_{1R} \\ \phi_{21} & \phi_{22} & \cdots & \phi_{2R} \\ \vdots & \vdots & \ddots & \vdots \\ \phi_{M1} & \phi_{M2} & \cdots & \phi_{MR} \end{pmatrix} \begin{pmatrix} \mathbf{J}(\theta_1) \\ \mathbf{J}(\theta_2) \\ \vdots \\ \mathbf{J}(\theta_R) \end{pmatrix} = \mathbf{\Phi} \cdot \mathbf{J} \tag{9}$$

in which ϕ is M×R-dimensional Gaussian random number matrix. Actually, J^{CS} can just be seen as the M-dimensional measurements of true induced current coefficients J. On the other hand, according to CS theory [7–11], J^{CS} can also be written as

$$\mathbf{J}^{CS} = \mathbf{\Phi} \cdot \mathbf{\Psi} \cdot \boldsymbol{\alpha} = \mathbf{A}^{CS} \cdot \boldsymbol{\alpha} \tag{10}$$

in which A^{CS} is defined as CS matrix. Once the expanded complex coefficients matrix α has been calculated by OMP, J can be reconstructed accurately. As the induced current is K-sparse in orthogonal basis Ψ and the number of measurement values M<<R, the number of RHS in linear system equations is reduced.

Recompression of the RHS Excitation Matrix

As stated in previous works [7, 8], the new excitation matrix V^{CS} is formed by random superposition of various plane waves with different directions of incidence, which is of low rank and can be recompressed by a low-rank representation [17] algorithm to reduce the RHS to lower level than that by using CS method.

An efficient technique to achieve a matrix low-rank approximation is ACA algorithm, which has been introduced by Bebendorf in [5]. In 2005, the algorithm has been successfully applied to moderate electrical size scattering and radiation computations [6]. The ACA algorithm is purely algebraic in nature, kernel independent, and relatively easy to implement. Also, the algorithm does not require a priori knowledge of each impedance submatrix.

The new excitation matrix V^{CS} can be well approximated by the product of two full-rank matrixes using ACA method:

$$\mathbf{V}^{CS} \approx \mathbf{V}_U^{CS} \cdot \mathbf{V}_V^{CS}, \tag{11}$$

where $\mathbf{V}_U^{CS} \in \mathbf{C}^{N\times Q}$ and $\mathbf{V}_V^{CS} \in \mathbf{C}^{Q\times R}$ and Q is the effective rank of new excitation matrix which is much smaller than the number of RHS. Induced current density matrix J^{CS} in (9) can be calculated approximately by exploiting the following expression:

$$\mathbf{J}^{CS} \approx \left(\mathbf{Z}^{-1} \cdot \mathbf{V}_U^{CS}\right) \cdot \mathbf{V}_V^{CS}. \tag{12}$$

This expression reduces the computational effort massively, because the number of RHS is equal to the rank Q, which is much smaller than the initial number of RHS in (5).

NUMERICAL RESULTS

To test the performance of the new method, the scattering characteristics of differently shaped objects are presented and compared with those obtained by using the traditional MoM solution. The measurement matrix ϕ in the CS is set to Gaussian random matrix, orthogonal basis Ψ is set to discrete cosine transform matrix, the OMP is used as the recovery algorithm, and the ACA iteration error threshold is set to 1e-4. Both examples are calculated on a PC with 3 GHz Intel Core 2 Duo CPU and 8 GB RAM (only one core is used). To provide a comparison between the original and the recovered current coefficients, the following relative root mean square error (RRMSE) is introduced:

$$\text{RRMSE} = \sqrt{\frac{1}{N}\sum_{i=1}^{N}\frac{\left|I_i^{\text{Recovered}} - I_i^{\text{original}}\right|^2}{\left|I_i^{\text{original}}\right|^2}}. \tag{13}$$

NASA Almond Model

The scattering problem of a 252.3744 mm perfect electric conductor (PEC) Almond [18] at the frequency of 7 GHz is considered first. The surface of the Almond has been discretized by RWG basis functions with average edge length of $\lambda/10$, which requires 12621 unknowns. The model is illuminated by a set of incident plane waves with incident direction of $\varphi \in [0^\circ, 180^\circ]$ and $\theta = 90^\circ$, where the angle interval for φ is 1. The monostatic radar cross section (RCS) is computed by using the CS-ACA method proposed in this paper. The CS technique is firstly applied with the measurement number M=40, and then the new excitation matrix V^{CS} is recompressed by using the ACA, which can further reduce the number of RHS from M=40 to rank Q=12. The backscattering monostatic RCS in horizontally transmitted and horizontal received (HH) field polarization calculated by the CS-ACA method and the traditional MoM are shown in Figure 1.

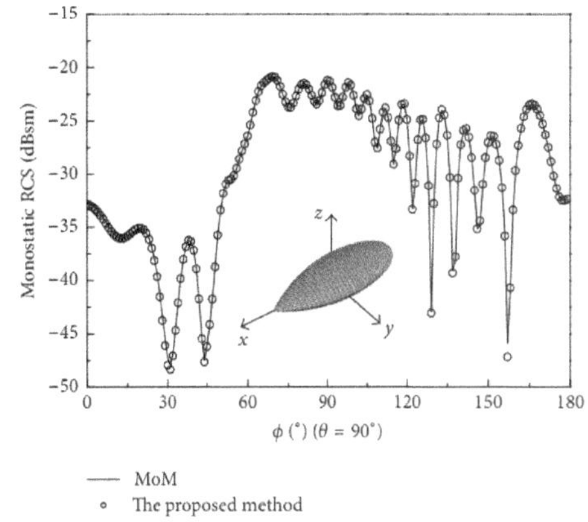

Figure 1: Monostatic RCS of PEC Almond model.

The recovered current coefficients and the true current coefficients (computed by MoM) obtained using a RWG basis with multiple incident angles are shown in Figure 2.

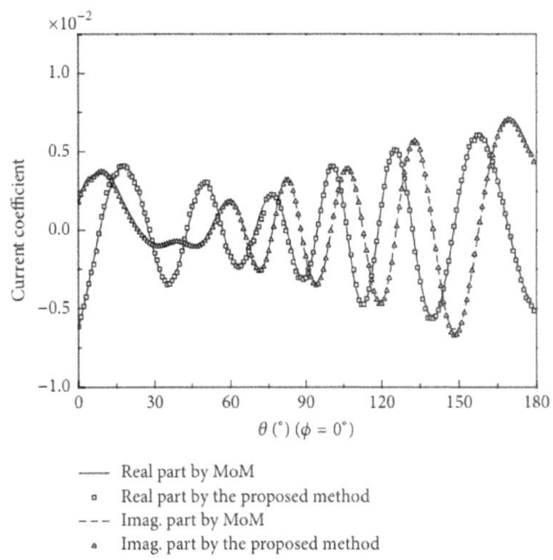

Figure 2: True and recovery current coefficients for Almond model.

As the effective rank of V^{CS} takes different values, the RRMSE varies with it. The relationship between the global RRMSE and the effective rank is presented in Figure 3, from which we can find that the CS-ACA method can reduce the number of RHS effectively with very low RRMSE.

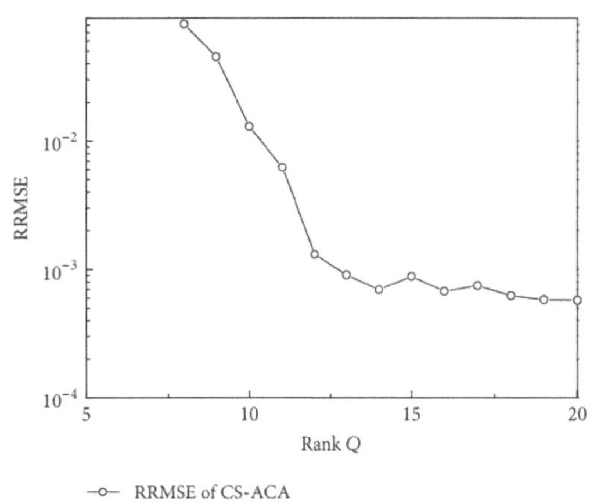

Figure 3: RRMSE of CS-ACA method for Almond.

High numerical accuracy of the current distributions makes it possible to obtain a reliable value of the near-field distribution, as shown in [19], where the field generated by the surface currents lying on a Cartesian quadrant has been assessed in closed analytical form to evaluate the exact field distribution excited by current pulse basis functions. The spatial distribution of the electric field component at a distance $\lambda_0/1000$ above the Almond, computed using the proposed hybrid approach, is shown in Figure 4.

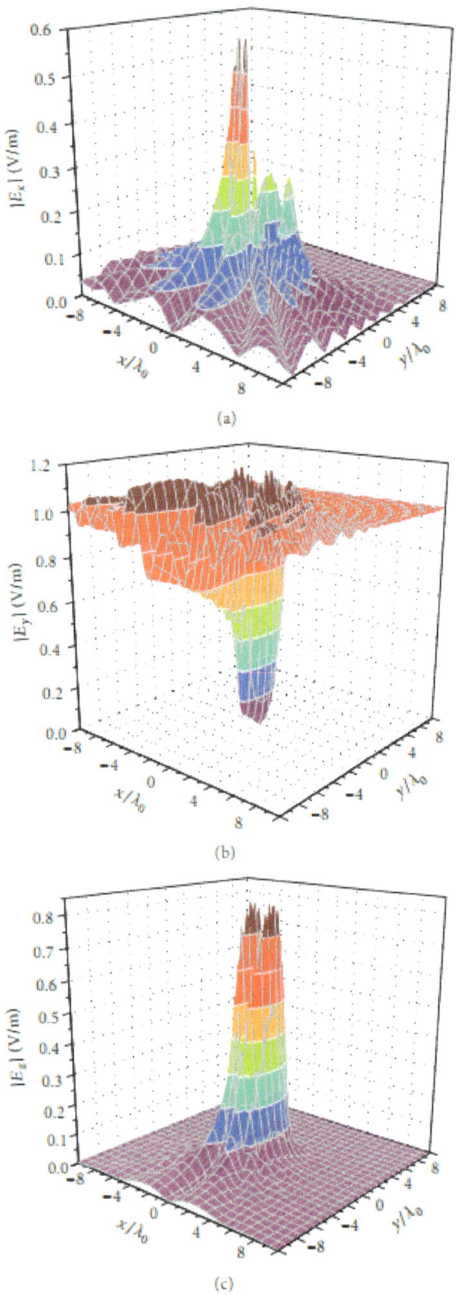

Figure 4: Magnitude of the total electric field $\lambda_0/1000$ above the Almond: (a) x-, (b) y-, and (c) z-component of the electric field. Characteristics of the incident field: E_0=1V/m, $\theta_i = 90^\circ, \varphi_i = 0^\circ$, , and polarization angle $\alpha = 90^\circ$.

PEC Missile Model

The second example is a PEC missile model shown in Figure 5, which has a length of 9.91 m, a wingspan of 6.28 m, and a height of 2.19 m.

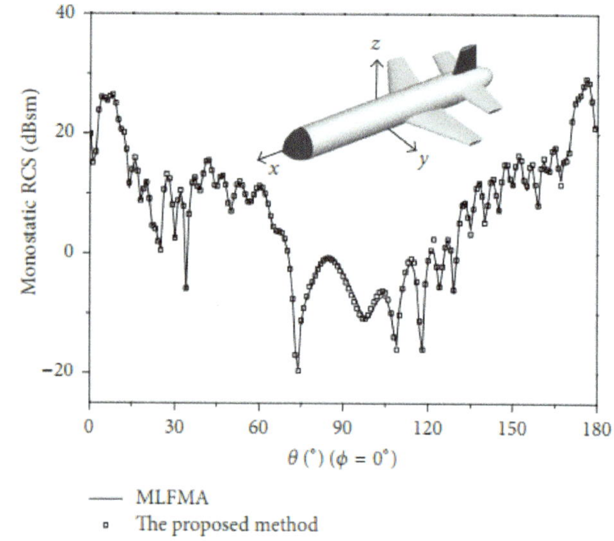

Figure 5: Monostatic RCS of PEC missile model.

The HH polarization backscattering monostatic RCS is computed at the frequency of 300 MHz, which requires 25263 unknowns. The missile model is illuminated by a set of incident plane waves with the incident direction of $\theta \in [0^\circ, 180^\circ]$ and $\varphi = 0^\circ$, where the angle interval for θ is 1°. Firstly, the number of measurements in the CS is set to M=40, and then the new excitation matrix generated by the CS is recompressed by the ACA, which can further reduce the RHS number to Q=14. The HH polarization backscattering monostatic RCS in x-z-plane calculated by the CS-ACA method in this paper and the MLFMA are shown in Figure 4. Recovered and true current coefficients obtained through using a RWG basis are shown in Figure 6, while the relationship between global RRMSE with rank Q of V^{CS} is shown in Figure 7. From this figure, it appears that the RRMSE is still very low when rank Q is greater than 14. From the above examples, we can conclude that the results obtained by the CS-ACA method are very accurate.

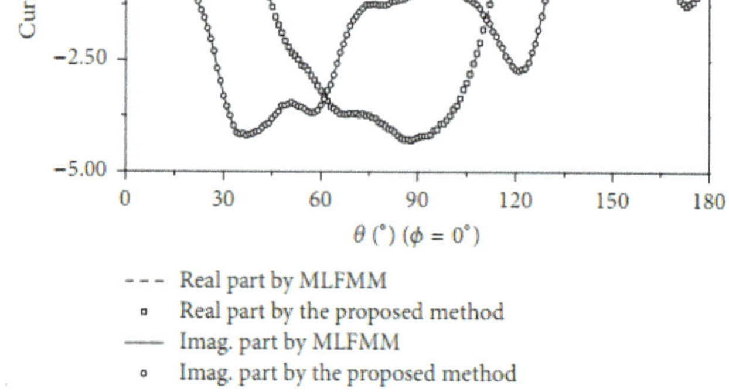

Figure 6: True and recovery current coefficients for missile model.

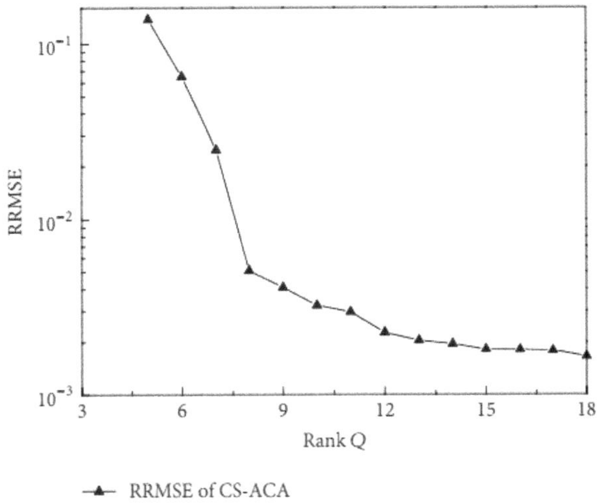

Figure 7: RRMSE for missile model.

Table 1 lists the CPU time requirements of the proposed examples. It can be seen that the CS-ACA method has compressed much more CPU time demand than the conventional MoM implementation.

Table 1: CPU time requirements.

Target	Method	RHS number	Time/s
PEC NASA Almond	MoM	180	1564
	CS-ACA	12	257
PEC missile model	MLFMA	180	765
	CS-ACA	14	436

CONCLUSIONS

An efficient method to analyze the electromagnetic scattering problems with multiple incident angles based on the MoM technique has been presented. A hybrid compressive method using the CS technique and the ACA is introduced to compress the excitation matrix. The numerical results demonstrate that the adopted approach leads to very accurate RCS results. Furthermore, the ratio of the number of RHS between the compressed excitation matrix and the original excitation matrix in the conventional MoM is close to one to fifteen.

ACKNOWLEDGMENTS

This work was supported by the National Natural Science Foundation of China (Grant no. 61172020), the Specialized Research Fund for the Doctoral Program of Higher Education of China (Grant no. 20123401110006), and the Fund for 211 Project of Anhui University, China (Grant no. ZYGG201202).

REFERENCES

1. R. F. Harrington, Field Computation by Method of Moments, Macmillan, New York, NY, USA, 1968.
2. W. C. Gibson, The Method of Moments in Electromagnetics, CRC Press, Boca Raton, Fla, USA, 2007.
3. J. Song, C. C. Lu, and W. C. Chew, "Multilevel fast multipole algorithm for electromagnetic scattering by large complex objects," IEEE Transactions on Antennas and Propagation, vol. 45, no. 10, pp. 1488–1493, 1997.
4. E. Bleszynski, M. Bleszynski, and T. Jaroszewicz, "AIM: adaptive integral method for solving large-scale electromagnetic scattering and radiation problems," Radio Science, vol. 31, no. 5, pp. 1225–1251, 1996.
5. M. Bebendorf, "Approximation of boundary element matrices," Numerische Mathematik, vol. 86, no. 4, pp. 565–589, 2000.
6. K. Zhao, M. N. Vouvakis, and J.-F. Lee, "The adaptive cross approximation algorithm for accelerated method of moments computations of EMC problems," IEEE Transactions on Electromagnetic Compatibility, vol. 47, no. 4, pp. 763–773, 2005. ·
7. M. S. Chen, F. L. Liu, H. M. Du, and X. L. Wu, "Compressive sensing for fast analysis of wide-angle monostatic scattering problems," IEEE Antennas and Wireless Propagation Letters, vol. 10, no. 3, pp. 1243–1246, 2014.
8. X.-Y. Cao, M.-S. Chen, and X.-L. Wu, "Application of compressed sensing to solving 3D electromagnetic scattering problems over a wide angle," in Proceedings of the International Conference on Microwave and Millimeter Wave Technology (ICMMT '12), pp. 720–723, Shenzhen, China, May 2012.
9. R. G. Baraniuk, "Compressive sensing," IEEE Signal Processing Magazine, vol. 24, no. 4, pp. 118–121, 2007.
10. R. G. Baraniuk, V. Cevher, M. F. Duarte, and C. Hegde, "Model-based compressive sensing," Institute of Electrical and Electronics Engineers. Transactions on Information Theory, vol. 56, no. 4, pp. 1982–2001, 2010.
11. L. Carin, D. Liu, W. Lin, and B. Guo, "Compressive sensing for multi-static scattering analysis," Journal of Computational Physics, vol. 228, no. 9, pp. 3464–3477, 2009.
12. T. T. Cai and L. Wang, "Orthogonal matching pursuit for sparse signal recovery with noise," IEEE Transactions on Information Theory, vol. 57, no. 7, pp. 4680–4688, 2011.
13. E. J. Candés, "Compressive sampling," in Proceedings of the International Congress of Mathematicians, pp. 1433–1452, European Mathematical Society, Madrid, Spain, 2006.
14. D. L. Donoho, "Compressed sensing," IEEE Transactions on Information Theory, vol. 52, no. 4, pp. 1289–1306, 2006.
15. E. J. Candés, J. Romberg, and T. Tao, "Robust uncertainty principles: exact signal reconstruction from highly incomplete frequency information," IEEE Transactions on Information Theory, vol. 52, no. 2, pp. 489–509, 2006.
16. S. M. Rao, D. R. Wilton, and A. W. Glisson, "Electromagnetic scattering by surfaces of arbitrary shape,"IEEE Transactions on Antennas and Propagation, vol. 30, no. 3, pp. 409–418, 1982.
17. J. Shaeffer, "Direct solve of electrically large integral equations for problem sizes to 1M unknowns," IEEE

Transactions on Antennas and Propagation, vol. 56, no. 8, pp. 2306–2313, 2008.

18. A. C. Woo, H. T. G. Wang, M. J. Schuh, and M. L. Sanders, "Benchmark radar targets for the validation of computational electromagnetics programs," IEEE Transactions on Antennas and Propagation, vol. 35, no. 1, pp. 84–89, 1993.

19. R. Cicchetti, D. Caratelli, and A. Faraone, "Exact closed-form expression of the electromagnetic field excited by a uniform current distribution lying on a Cartesian quadrant," IEEE Transactions on Antennas and Propagation, vol. 61, no. 4, pp. 2142–2155, 2013.

CHAPTER 17 THE TERAHERTZ CONTROLLED DUPLEX ISOLATOR: PHYSICAL GROUNDS AND NUMERICAL EXPERIMENT

Konstantin Vytovtov, Said Zouhdi, Rostislav Dubrovka, and Volodymyr Hnatushenko

INTRODUCTION

The terahertz domain actively developed in the last decades [1–5]. Various devices have been created and numerous electromagnetic structures have been used in this frequency range. One of the most interesting classes of terahertz devices is the class of nonreciprocal devices [1–10]. Usually, the Faraday rotation effect itself is exploited naturally to achieve desired nonreciprocal performance [1]. The electrically tunable Faraday effect in a HgTe thin film has been reported in [2]. Faraday rotation has also been studied for pump pulse of terahertz radiation [3]. In [4], the heterostructures based on the piezoelectric and semiconductor layers have been proposed for radio frequency applications. Other examples of nonreciprocal properties of nonlinear devices in terahertz range have been presented in [5] where a Faraday isolator operating on the coupled microwave resonators circuit has been proposed. A theoretical study of the reflection of infrared radiation from antiferromagnets and M-type barium hexagonal ferrite using an attenuated total reflection geometry has been presented in [6]. Nonreciprocal devices using attenuated total reflection and thin film magnetic layered structures have been described in [7]. Thus, we can clearly see that this topic is quite relevant in modern microwave science.

In this paper, we consider nonreciprocal properties of an anisotropic structure based on dependence of the reflection coefficient on an incident wave direction and an anisotropy axis orientation, but not on the Faraday effect. The physical grounds and numerical simulations of a controlled bidirectional dual-frequency isolator not based on the Faraday effect are presented. For this, the analytical investigation and numerical calculations of a stratified anisotropic structure are carried out. At first, the elements of the translation matrix are written in the compact analytical form (4) for the general case of an anisotropic medium for the first time. It is important that these elements are expressed in elementary algebraic functions. The obtained form is very useful in subsequent numerical experiments.

In this work, it also is found that the reflection coefficients are different for different incidence angles α and $-\alpha$. Thus, a structure shows nonreciprocal properties.

These properties allow us to build nonreciprocal elements in the terahertz domain. The physical principles of construction of a two-way double-frequency (duplex) isolator are described here. These principles are based on dependence of reflection coefficient on an incidence wave direction and an orientation of

Citation: Konstantin Vytovtov, Said Zouhdi, Rostislav Dubrovka, and Volodymyr Hnatushenko, "The Terahertz Controlled Duplex Isolator: Physical Grounds and Numerical Experiment," International Journal of Microwave Science and Technology, vol. 2016, Article ID 1468508, 7 pages, 2016. doi:10.1155/2016/1468508.

an anisotropy axis. It should be noted that the Faraday effect is also present in these structures and also defines nonreciprocal properties of anisotropic structures. Today, a wide variety of isolators is described in contemporary scientific literature [7–9]. A feature of the presented device is the fact that it is a two-way dual-frequency one. Switching an isolator (changing a direction and a frequency) is implemented by two solenoids. Its tuning is done mechanically by changing an orientation of external magnetic field. Note that here a very complicated geometry of the problem is used to create the isolator. But only this geometry allows us to obtain the pronounced minimums of reflectance for a given incidence angle, as well as a wide range of total reflection.

DESCRIPTION OF THE ISOLATOR SCHEME

A presentation of the offered duplex isolator is shown in Figure 1(a). This device contains two transceiver antennas, two solenoids generating a magnetizing field, and a stratified anisotropic slab (a nonreciprocal element).

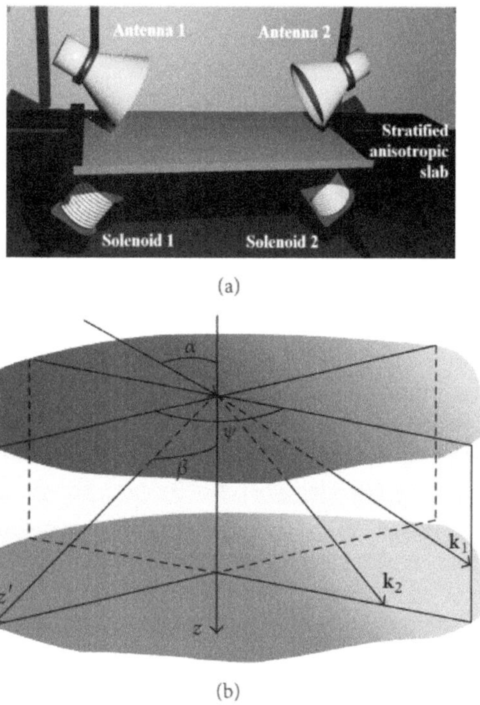

Figure 1: (a) Appearance of the controlled double-frequency bidirectional isolator; (b) geometry of an anisotropy layer.

An orientation of the axes within a homogeneous layer of a stratified anisotropic slab is presented in Figure1(b). Here, z' is an anisotropy axis, z is a normal to a structure, k_1 and k_2 are wave vectors of forward refracted waves within an anisotropic layer, β is an inclination angle, ψ is an angle between an incidence plane and a plane containing an anisotropy axis, and α is an incidence angle. The choice of such a complicated geometry is due to the need to obtain an asymmetric (nonreciprocal) structure, the need to obtain wide range of total reflection, and the need to control the structure. Of course nonreciprocal properties can be observed in the simplest case of orientation of an anisotropy axis along interfaces [10, 11], but in that case total reflection regions are not sufficiently wide and a structure has low selective properties in comparison with the given case. Additionally, here mechanical tuning of an isolator is assumed. And using such a complicate geometry gives us the additional freedom to control the device. Moreover, properties of a structure having such a geometry are poorly understood. Therefore, the results of these investigations have both practical and theoretical significance.

The layers of a stratified structure are described by a scalar permittivity and a dyadic permeability in the gyrotropic form:

$$\mu = \begin{vmatrix} \mu_{xx} & j\mu_{xy} & 0 \\ -j\mu_{xy} & \mu_{xx} & 0 \\ 0 & 0 & \mu_{zz} \end{vmatrix} \tag{1}$$

With

$$\mu_{xx} = 1 + \frac{\omega_M(\omega_0 + j/T)}{(\omega_0 + j/T)^2 - \omega^2},$$
$$\mu_{xy} = \frac{\omega_M \omega}{(\omega_0 + j/T)^2 - \omega^2}. \tag{2}$$

Here, $\omega_M = 4\pi\gamma M, \omega_0 = \gamma H_0, \gamma$ is the gyromagnetic relation, $H_0 = 0.1T$ is an external magnetic field, and T is a relaxation time.

In our numerical investigation, we take into account losses and frequency dispersion of the ferrite material and the elements in (1) are written in the Lanfau-Lifshits form [12]. The dependence of the permeability elements on a frequency for the first and second layers of a period is given in Figure 2. Here, μ'_{xx} and μ'_{xy} are the real parts of the elements, and μ''_{xx} and μ''_{xy} are the imaginary parts of the elements. The resonance frequencies are 1.1 GHz and 1.8 GHz for the first and second layers accordingly.

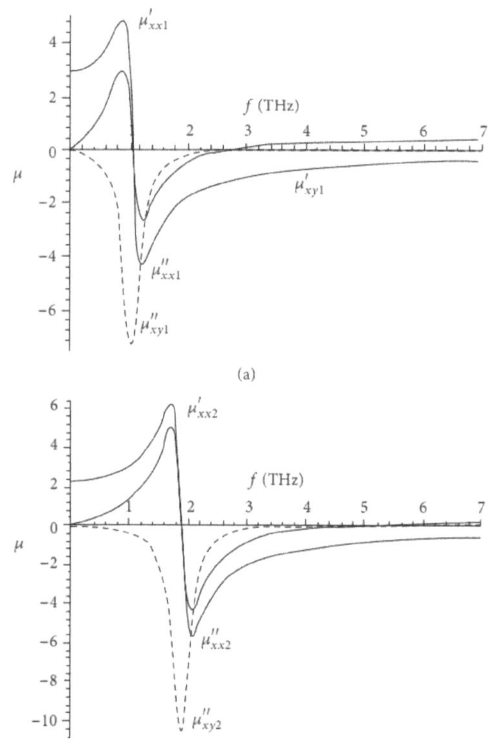

Figure 2: The dependence of permeability elements on frequency for the first (a) and second (b) layers of a period.

METHOD OF CALCULATION

The well-known translation matrix method and the reflection matrix method [13, 14] are used in this study of a stratified anisotropic slab. Both of these methods are accurate and analytical and these can be used in any frequency range for both lossless and lossy media without restrictions. According to the definition, a translation matrix relates tangential field components on both boundaries of a plane-parallel structure:

$$\begin{vmatrix} E_x(z) \\ E_y(z) \\ H_x(z) \\ H_y(z) \end{vmatrix} = \mathbf{M}(z) \begin{vmatrix} E_x(0) \\ E_y(0) \\ H_x(0) \\ H_y(0) \end{vmatrix} \tag{3}$$

Here, M(z) is a 2×2 matrix for an isotropic medium or a 4×4 translation matrix for an anisotropic medium.

The method is described in the literature in detail for both isotropic [13] and anisotropic [14] media. The matrices have been found for a normal orientation of an anisotropy axis, for a tangential orientation of an anisotropy axis, for arbitrary orientation of an anisotropy axis, and for a normal incidence of a plane harmonic wave. However, the expressions for the most general case of an anisotropic medium with an arbitrary orientation of the anisotropy axes under an oblique incidence of a wave have not been obtained in a compact analytical form. In this paper, the translation matrix elements of a homogeneous layer are presented for the first time as

$$M_{ml} = \sum_{i=1}^{4} (-1)^{m+i} \gamma_{mi} \frac{\det \mathbf{G}_{li}}{\det \mathbf{G}} \exp(-jk_i z). \tag{4}$$

Here, G_{li} is the minor of the element γ_{mi} of the 4 × 4-matrix G. Matrix G is defined by the expression

$$\begin{vmatrix} E_x \\ E_y \\ H_x \\ H_y \end{vmatrix} = \begin{vmatrix} \gamma_{11} & \gamma_{12} & \gamma_{13} & \gamma_{14} \\ \gamma_{21} & \gamma_{22} & \gamma_{23} & \gamma_{24} \\ \gamma_{31} & \gamma_{32} & \gamma_{33} & \gamma_{34} \\ \gamma_{41} & \gamma_{42} & \gamma_{43} & \gamma_{44} \end{vmatrix} \begin{bmatrix} A_1 \exp(-jk_1 z) \\ A_2 \exp(-jk_2 z) \\ A_3 \exp(-jk_3 z) \\ A_4 \exp(-jk_4 z) \end{bmatrix}$$

$$= \mathbf{G} \begin{bmatrix} A_1 \exp(-jk_1 z) \\ A_2 \exp(-jk_2 z) \\ A_3 \exp(-jk_3 z) \\ A_4 \exp(-jk_4 z) \end{bmatrix}, \tag{5}$$

where A_i is a constant. Here also K_i is a wavenumber defined by well-known fourth-order dispersion relation [12–16]:

$$k^4 + a_3 k^3 + a_2 k^2 + a_1 k = 0. \tag{6}$$

The coefficients a_i in (6) must be obtained directly from Maxwell equations.

The expression of matrix G in (5) can be obtained directly from Maxwell equations. Actually, expression (5) is a general solution of Maxwell's equations for this case [12–14] (see Appendix). The achieved form of the translation matrix (4) essentially simplifies the solution of the programming problem and improves accuracy of numerical calculations. Moreover, no analytical expression of the translation matrix element has been presented in scientific literature for this complicate case.

The reflection matrix relates the tangential components of incident and reflected waves [15]:

$$\begin{vmatrix} E_{x\,\text{refl}} \\ E_{y\,\text{refl}} \end{vmatrix} = \mathbf{R} \begin{vmatrix} E_{x\,\text{inc}} \\ E_{y\,\text{inc}} \end{vmatrix}. \tag{7}$$

Here $E_{x\,\text{inc}}$ and $E_{y\,\text{inc}}$ are the components of an incident field, $E_{x\,\text{refl}}$ and $E_{y\,\text{refl}}$ are the components of a reflected field, and R is a 2 × 2 reflection matrix.

NONRECIPROCAL PROPERTIES OF ANISOTROPIC STRUCTURE

The main element of the isolator is a stratified anisotropic slab (Figure 1) possessing nonreciprocal properties. In this section, we consider the main physical features of a slab, which ensure its isolator properties. An orientation of the axes for the studied case is shown in Figure 1(b). Angle ψ is given in Figure 1(b). The presented complicated geometry of the problem gives us the interesting features of electromagnetic field in the structure. Such properties cannot appear in a case of more simple geometry. In particular, study of wavenumbers in a homogeneous anisotropic layer suggests that the wavenumbers can be real, imaginary, or complex even in a lossless medium. In the considered case, imaginary parts of wavenumbers indicate existence of a complex wave [16] but not in presence of losses in an anisotropic structure. Indeed a resulting structure of electromagnetic field in a layer is determined by a superposition of all four eigenwaves only, but not each eigenwave separately. In addition, these numbers are not the same for the incidence angles α and −α (Figure 3). The numerical calculations analytical investigations give us the following result (Figure 4): $k_1(\alpha) = -k_4(-\alpha)$, $k_2(\alpha) = -k_3(-\alpha)$, $k_3(\alpha) = -k_2(-\alpha)$, and $k_4(\alpha) = -k_1(-\alpha)$. Thus, we can see nonreciprocal properties of the medium in the considered case.

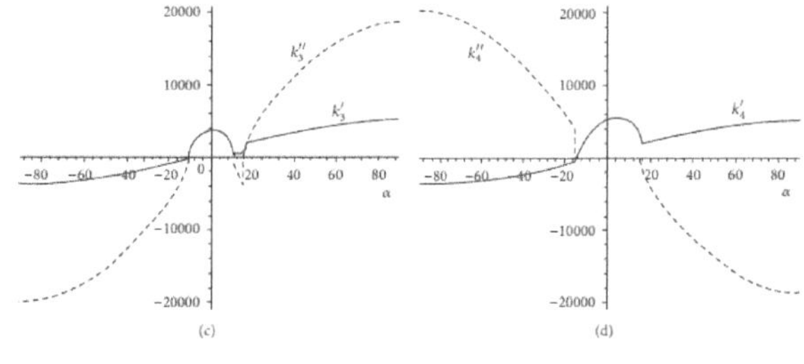

Figure 3: The dependence of the wavenumbers on an incidence angle ($\beta = 60^\circ$, $\psi = 30^\circ$, $f = 10\,\text{THz}$, $\mu_{xx} = 4.9$, $\mu_{xy} = 2.9$, and $\mu_{zz} = 0.$).

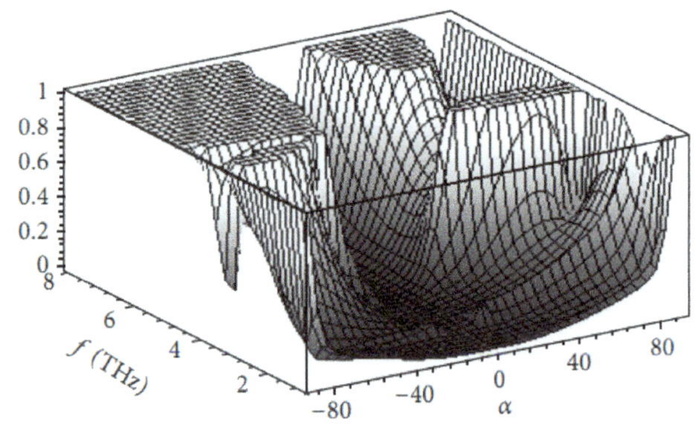

Figure 4: The dependence of the reflection on a frequency and an incidence angle ($\beta = 30^\circ, \psi = 50^\circ$), d_1=mm, and $d_1/d_2 = 1.3125$.

It also is clear (Figure 3) that two of the four wavenumbers can be real and two others can be complex for the certain parameters of the structure ($10^\circ < \alpha < 16^\circ$). It should also be noted that all four wave numbers have positive (negative) real part (Figure 3) in the certain range of the incidence angles ($\alpha > 16^\circ$). Such unusual combinations of wavenumbers are possible only in the case of this complex geometry of the problem. Similarly, it also is verified that the field components are different for the incidence angles α and −α.

Obviously, this kind of wavenumber affects reflection properties of a medium. So now we analyze the dependence of the reflection coefficient on a frequency and an anisotropy axis orientation. One of the possible variants of this dependence is presented in Figure 4. The study shows that there is no regularity in the reflection coefficient dependence on a frequency and an orientation of the anisotropy axis. Here a very wide range of total reflection even for small incidence angles can be seen. Their existence is determined by multiple reflection phenomenon, existence of four eigenwaves with different wavenumbers and propagation angles in each layer, presence of complex waves, and the losses in the medium. The resonant character of the reflection coefficient is determined by the effects described above also. Note that there are no sharp edges of the reflection coefficient in the neighborhood of the unity. It also is shown (Figure 5) that the dependence of the reflection coefficient of a frequency is not symmetric with respect to α = 0. This is due to the asymmetric dependence of wave numbers and the field components on the incidence angle.

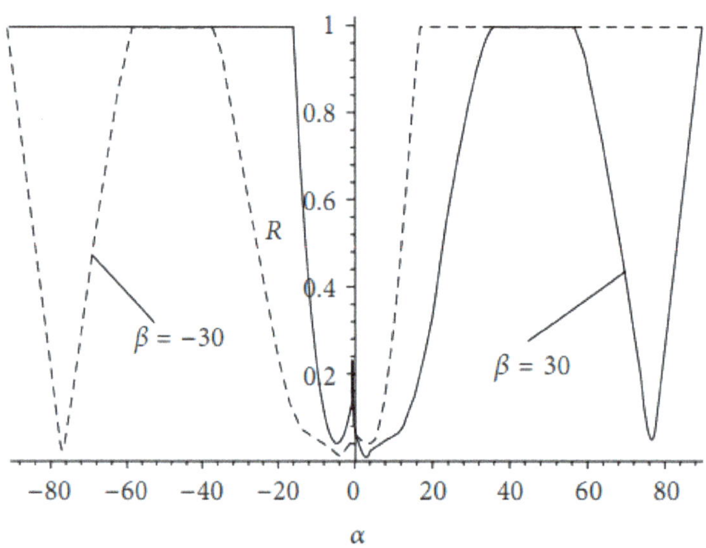

Figure 5: The dependence of the reflection coefficient on the incidence angle for $f = 6.35 \cdot 10^{12}$, $\psi = 50^\circ$, $d_1 = 0.78$ mm, and $d_1/d_2 = 1.3125$; (1) $\beta = 30^\circ$ solid line and (2) $\beta = -30^\circ$ dash line.

OPERATION PRINCIPLE OF THE ISOLATOR

The operation principle of the device is based on an angle dependence of medium properties (in particular, the reflection coefficient) but not on the Faraday effect. Wherein for the certain structure parameters and an orientation of the anisotropy axis (an external magnetic field direction) it is possible to obtain the total reflection at an incidence angle α and the minimal reflection at an incidence angle −α. Analogously, it is possible to choose an orientation of the external magnetic field for the given structure parameters such that it is possible to obtain the total reflection for −α and the minimal reflection for α. Thus, changing a direction of wave propagation (−α or α) can be accomplished by changing a direction of an external magnetic field. Analogously it is possible to find the two frequencies that give such an effect.

Now let us describe, for clarity, the operation principle of a bidirectional double-frequency isolator shown in Figure 1. If a voltage is applied to solenoid 1, a signal transmits from antenna 1 to antenna 2 without attenuation at a frequency f_1 as the reflection coefficient is equal to unite, a signal cannot propagate from antenna 2 to antenna 1 at this frequency as the reflection coefficient is about zero, and a signal transmits from antenna 2 to antenna 1 without attenuation at a frequency f_2; however, inverse propagation is not possible at this frequency. When voltage is applied to solenoid 2, a signal transmits from antenna 2 to antenna 1 substantially without attenuation at a frequency f_1, and it transmits from antenna 1 to antenna 2 without attenuation at a frequency f_2. Circular aperture of the antennas is used because the Faraday effect (rotation of a polarization plane) is taken into account.

THE PARAMETERS OF THE ISOLATOR

A very important characteristic of an isolator is its amplitude response. During the researches, numerous calculations have been carried out and the optimal structure for isolator is chosen. This structure includes the 12 double-layered periods. The first layer of the structure is FeF_2 and the second one is MnO. The antiferromagnetic resonance frequencies in the considered case are 1.1 THz and 1.8 THz for the first and second layers accordingly. The dependence of the element on a frequency is given in Figure 2. Here the operating frequency f=6.35 THz and the angles $\beta = \pm 30°$ and $\psi = 50°$ are also taken. In our problem, the operating frequency is essentially higher than the resonant one. The losses in FeF_2 and MnO can be neglected at this frequency (Figure 2). Here we consider an above resonance frequency and we have $\mu_{xx} = 1.415$ and $\mu_{xx} = -0.42$ for the first layer of the period at this frequency and $\mu_{xx} = 1.2$ and $\mu_{xx} = -0.63$ for the second layer. In the given work we consider a TE incident wave. The dependence of the reflection coefficient on an incidence angle for this structure (the amplitude characteristic) is presented in Figure 5. It is seen that the reflection coefficient is minimal (R=0.05) at $\alpha = 76.4^\circ$ and the one is equal to unit at $\alpha = -76.4^\circ$ for $\beta = 30^\circ$ (the solid line).

Thus, this structure has isolator properties and it passes a signal only in a forward direction and does not pass a signal in an opposite direction. The stopband for an opposite direction is 9.6° at the level 0.707, 15.8 at the level 0.5, and 27.7 at the level 0.1. Now, let us chose $\beta = -30^\circ$ (the dash line). For this case, the minimum of the reflection coefficient (R=0.03) is at $\alpha = -76.4^\circ$.

The frequency dependence of the reflection coefficient is presented in Figure 6. This graph is presented for the incidence angle $\alpha = 76.4°$. It is obvious that this structure has bad frequency selective properties and low squareness coefficient. The bandwidth is 1.88 THz (29.38%) at the 0.707 level (a half power level), and the bandwidth is 3.13 THz (48.9%) at the 0.5 level of the maximum. The squareness is 1.66 at levels 0.5 and 0.707. It also is verified that changing orientation of magnetic field allows us to shift a reflection minimum in a small range. And changing value of this field does not give us any useful effect.

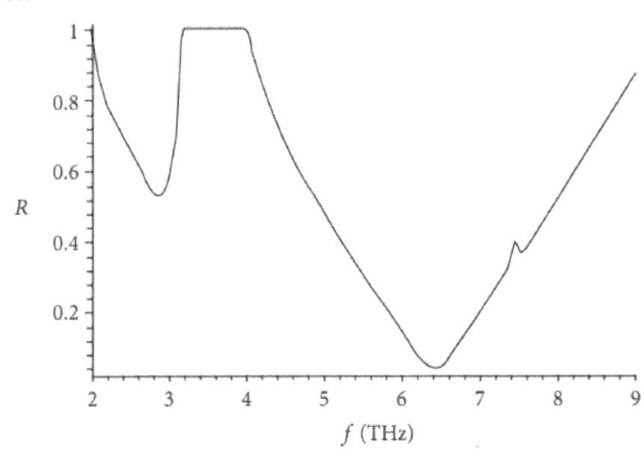

Figure 6: The dependence of reflection coefficient on a frequency at $\alpha = 76.4°$, d_1=0.78 mm, and $d_1/d_2 = 1.3125$.

The results demonstrating the possibility of mechanical tuning of the isolator by changing an angle of an anisotropy axis inclination are shown in Figure 7. The minimum is shifted from $\alpha = 72^\circ$ to $\alpha = 82^\circ$ if an inclination angle is changed from $\beta = 25^\circ$ to $\beta = 40^\circ$. Simultaneously, the angle bandwidth is narrowed and it is equal to 28°, 10°, and 6° correspondingly. Decreasing the inclination angle less than $\beta = 25^\circ$ and increasing it more than $\beta = 50^\circ$ leads to disappearance of the angle selective properties of the structure.

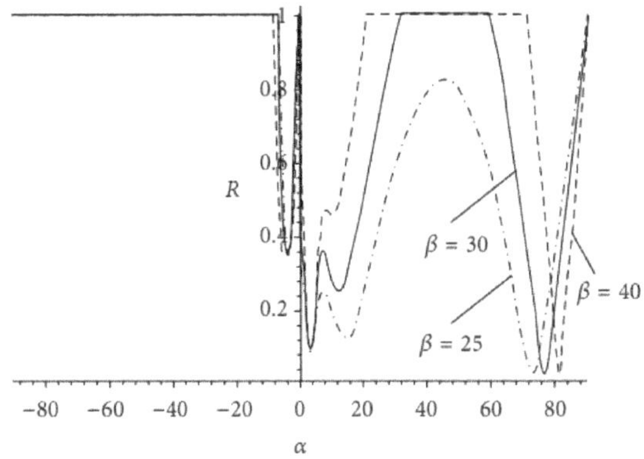

Figure 7: Dependence of the reflection coefficient on an incidence angle for difference inclination angles β, $d_1 = 0.78$ mm, $d_1/d_2 = 1.3125$, and $f = 6.35 \cdot 10^{12}$.

Figure 8 shows the results of calculations, allowing us to improve the described isolator. It is seen in Figure 8that the reflection minimum is observed at $\alpha = 76.4^\circ$ when the frequency is 6.35 GHz and the reflection minimum corresponds to $\alpha = -76.4^\circ$ at the frequency 4.05 GHz. Thus, the structure can be used in duplex terahertz systems as a two-way isolator. Moreover, when the direction of the magnetic field is switched from $\psi = 50^\circ$ to $\psi = -50^\circ$ the above minima are reversed. Therefore, if voltage is applied to solenoid 1 then a signal transmits from antenna 1 to antenna 2 substantially without attenuation at frequency 6.35 GHz, and a signal transmits from antenna 2 to antenna 1 without attenuation at frequency 4.05 GHz. If voltage is applied to solenoid 2 then a signal transmits from antenna 2 to antenna 1 substantially without attenuation at frequency 6.35 GHz, and a signal transmits from antenna 1 to antenna 2 without attenuation at a frequency 4.05 GHz.

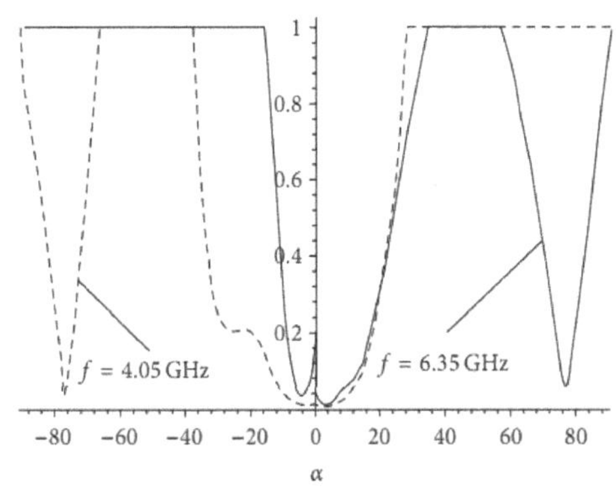

Figure 8: The dependence of the reflection coefficient on an incident angle for $d_1 = 0.78$ mm, $d_1/d_2 = 1.3125$, $\beta = 30^\circ$; (1) $\psi = 50^\circ$ and (2) $\psi = -50^\circ$.

CONCLUSIONS

Electromagnetic properties of a stratified anisotropic slab with an arbitrary orientation of the anisotropy axis are studied analytically and numerically. The translation and reflection matrices methods are used. The analytical expression of the translation matrix is obtained in the convenient compact form for the first time. The dependence of the eigenwave wavenumbers in an anisotropic slab and reflection coefficient on an anisotropy axis orientation and an incidence angle is investigated. It is shown that the structure shows nonreciprocal properties. It is important that these properties are not based on the Faraday effect. Although this effect is still inherently present in an anisotropic structure, the controlled two-way dual-frequency (duplex) isolator is presented here for the first time. Its operation principle is based on the properties of an anisotropic structure described here for the first time. The very complicated geometry of an anisotropic slab is used in this isolator. But this geometry (an arbitrary orientation of magnetic field) allows us to obtain the necessary properties of the isolator and to control a resonant frequency and an angle of a wave transmission mechanically.

Appendix

Finding a solution of Maxwell equations in the form $\exp(j(\omega t - k_x x - k_v y - k_z z))$, we can write

$$\begin{vmatrix} 0 & -jk_z & jk_y & -j\omega\mu_{xx} & -j\omega\mu_{xy} & -j\omega\mu_{xz} \\ jk_z & 0 & -jk_x & -j\omega\mu_{yx} & -j\omega\mu_{yy} & -j\omega\mu_{xz} \\ -jk_y & jk_x & 0 & -j\omega\mu_{zx} & -j\omega\mu_{zy} & -j\omega\mu_{zz} \\ j\omega\varepsilon & 0 & 0 & 0 & -jk_z & jk_y \\ 0 & j\omega\varepsilon & 0 & jk_z & 0 & -jk_x \\ 0 & 0 & j\omega\varepsilon & -jk_y & jk_x & 0 \end{vmatrix} \begin{vmatrix} E_x \\ E_y \\ E_z \\ H_x \\ H_y \\ H_z \end{vmatrix} = 0. \qquad \text{A.1}$$

From (A.1), we have

$$\begin{vmatrix} 0 & -jk_z & jk_y & -j\omega\mu_{xx} & -j\omega\mu_{xy} \\ jk_z & 0 & -jk_x & -j\omega\mu_{yx} & -j\omega\mu_{yy} \\ -jk_y & jk_x & 0 & -j\omega\mu_{zx} & -j\omega\mu_{zy} \\ j\omega\varepsilon & 0 & 0 & 0 & -jk_z \\ 0 & j\omega\varepsilon & 0 & jk_z & 0 \end{vmatrix} \begin{vmatrix} E_x \\ E_y \\ E_z \\ H_x \\ H_y \end{vmatrix} = \begin{vmatrix} j\omega\mu_{xz} \\ j\omega\mu_{yz} \\ j\omega\mu_{zz} \\ -jk_y \\ jk_x \end{vmatrix} H_z. \qquad \text{A.2}$$

The solutions of (A.2) give us the expressions of the matrix elements in (5) in the form

$$\gamma_{mi} = \frac{\Delta_{mi}}{\Delta}, \qquad \text{A.3}$$

where Δ is the determinant of the coefficient matrix in (A.3) and

$$\Delta_{1i} = \begin{vmatrix} j\omega\mu_{xz} & -jk_{zi} & jk_y & -j\omega\mu_{xx} & -j\omega\mu_{xy} \\ j\omega\mu_{yz} & 0 & -jk_x & -j\omega\mu_{yx} & -j\omega\mu_{yy} \\ j\omega\mu_{zz} & jk_x & 0 & -j\omega\mu_{zx} & -j\omega\mu_{zy} \\ -jk_y & 0 & 0 & 0 & -jk_{zi} \\ jk_x & j\omega\varepsilon & 0 & jk_{zi} & 0 \end{vmatrix},$$

$$\Delta_{2i} = \begin{vmatrix} 0 & j\omega\mu_{xz} & jk_y & -j\omega\mu_{xx} & -j\omega\mu_{xy} \\ jk_{zi} & j\omega\mu_{yz} & -jk_x & -j\omega\mu_{yx} & -j\omega\mu_{yy} \\ -jk_y & j\omega\mu_{zz} & 0 & -j\omega\mu_{zx} & -j\omega\mu_{zy} \\ j\omega\varepsilon & -jk_y & 0 & 0 & -jk_{zi} \\ 0 & jk_x & 0 & jk_{zi} & 0 \end{vmatrix},$$

$$\Delta_{3i} = \begin{vmatrix} 0 & -jk_{zi} & jk_y & j\omega\mu_{xz} & -j\omega\mu_{xy} \\ jk_{zi} & 0 & -jk_x & j\omega\mu_{yz} & -j\omega\mu_{yy} \\ -jk_y & jk_x & 0 & j\omega\mu_{zz} & -j\omega\mu_{zy} \\ j\omega\varepsilon & 0 & 0 & -jk_y & -jk_{zi} \\ 0 & j\omega\varepsilon & 0 & jk_x & 0 \end{vmatrix},$$

$$\Delta_{4i} = \begin{vmatrix} 0 & -jk_{zi} & jk_y & -j\omega\mu_{xx} & j\omega\mu_{xz} \\ jk_{zi} & 0 & -jk_x & -j\omega\mu_{yx} & j\omega\mu_{yz} \\ -jk_y & jk_x & 0 & -j\omega\mu_{zx} & j\omega\mu_{zz} \\ j\omega\varepsilon & 0 & 0 & 0 & -jk_y \\ 0 & j\omega\varepsilon & 0 & jk_{zi} & jk_x \end{vmatrix}. \qquad \text{A.4}$$

REFERENCES

1. R. J. Potton, "Reciprocity in optics," Reports on Progress in Physics, vol. 67, no. 5, pp. 717–754, 2004. · ·
2. A. Shuvaev, A. Pimenov, G. V. Astakhov et al., "Room temperature electrically tunable terahertz Faraday effect," Applied Physics Letters, vol. 102, no. 24, Article ID 241902, 2013. · ·
3. M. Shalaby, F. Vidal, M. Peccianti et al., "Terahertz macrospin dynamics in insulating ferrimagnets,"Physical Review B, vol. 88, no. 14, Article ID 140301, 2013. · ·
4. M. Rotter, W. Ruile, and A. Wixforth, "Non-reciprocal SAW devices for RF applications," in Proceedings of the IEEE Ultrasonics Symposium 2000, vol. 1, pp. 35–38, San Juan, Puerto Rico, October 2000. ·
5. J. Koch, A. A. Houck, K. L. Hur, and S. M. Girvin, "Time-reversal-symmetry breaking in circuit-QED-based photon lattices," Physical Review A, vol. 82, no. 4, Article ID 043811, 2010. · ·
6. N. R. Anderson and R. E. Camley, "Attenuated total reflection study of bulk and surface polaritons in antiferromagnets and hexagonal ferrites: propagation at arbitrary angles," Journal of Applied Physics, vol. 113, no. 1, Article ID 013904, 2013. · ·
7. T. J. Fal and R. E. Camley, "Non-reciprocal devices using attenuated total reflection and thin film magnetic layered structures," Journal of Applied Physics, vol. 110, no. 5, Article ID 053912, 2011. · ·
8. M. Shalaby, M. Peccianti, Y. Ozturk, and R. Morandotti, "A magnetic non-reciprocal isolator for broadband terahertz operation," Nature Communications, vol. 4, article 1558, 2013. ·
9. S. Chen, F. Fan, X. Wang, P. Wu, H. Zhang, and S. Chang, "Terahertz isolator based on nonreciprocal

magneto-metasurface," Optics Express, vol. 23, no. 2, pp. 1015–1024, 2015. · ·

10. M. R. F. Jensen, T. J. Parker, K. Abraha, and D. R. Tilley, "Experimental observation of magnetic surface polaritons in FeF_2 by attenuated total reflection," Physical Review Letters, vol. 75, no. 20, pp. 3756–3759, 1995. · ·

11. M. R. F. Jensen, S. A. Feiven, T. Dumelow, et al., "Fourier transform spectroscopy of magnetic materials at terahertz frequencies," International Journal of Terahertz Science and Technology, vol. 2, no. 4, pp. 105–119, 2009. ·

12. B. Lax and K. Button, Microwave Ferrites and Ferrimagnetics, McGRaw-Hill Book, 1962.

13. P. Yeh, A. Yariv, and C. S. Hong, "Electromagnetic propagation in periodic stratified media. I. General theory," Journal of the Optical Society of America, vol. 67, no. 4, pp. 423–436, 1977. ·

14. D. W. Berreman, "Optics in stratified and anisotropic media: 4×4 matrix formulation," Journal of the Optical Society of America, vol. 62, no. 4, pp. 502–510, 1972. ·

15. S. Teitler and B. W. Henvis, "Refraction in stratified anisotropic media," Journal of the Optical Society of America, vol. 60, no. 6, pp. 830–834, 1970. ·

16. F. Fedorov, Optics of Anisotropic Media, Editorial URSS, 2004.

CHAPTER 18

A MODELING METHOD FOR SMALL PACKING PARTICLES IN ELECTROMAGNETIC SIMULATION

Yi Cheng Zhong, Yu Jian Cheng, and Yong Fan

INTRODUCTION

Most natural substances, such as building materials and minerals, present in the mixed state, especially in the state of packing particles. The propagation and reflection properties of the electromagnetic wave in inhomogeneous materials have been widely studied [1, 2]. The model of packing particles plays an important role in the research of the ground-penetrating radar, the microwave heating, the microwave desulfurization, and so forth. Therefore, there is an urgent need for building an accurate model of packing particles to study the propagation characteristics of the electromagnetic wave in nature [3–5].

The particle in this paper is defined as the smallest unit of substances with identical electromagnetic properties. Electromagnetic properties of the individual particle are usually easy to recognize by experiment [6–8]. However, when the material is formed by a number of solid packing particles, it is difficult to investigate electromagnetic properties of this mixed-particle material because of the large number and random distribution of these particles [9].

Citation: Yi Cheng Zhong, Yu Jian Cheng, and Yong Fan, "A Modeling Method for Small Packing Particles in Electromagnetic Simulation," International Journal of Antennas and Propagation, vol. 2016, Article ID 1785751, 7 pages, 2016. doi:10.1155/2016/1785751.

Previous researchers have done rigorous and impressive job on this issue and proposed different methods. Effective complex permittivity of dielectric mixtures can be evaluated by different formulations and rules from theoretical models and experiments [10–14]. However, the theoretical model can only be used in the particles with simple and regular shapes. Besides, the experimental method is affected by the random distribution of particles, and most of the time, the accurate formulations of mixtures cannot be gotten from the experimental results.

Many numerical studies have also been performed to analyze the effective permittivity of dielectric mixtures [15–17]. The dynamical solution method (FDTD) is firstly applied to the two-dimensional dielectric mixture [17]. This method is also used in the three-dimensional analysis in this paper. Besides, there are some other kinds of modeling software, such as PFC 3D, used in fields outside of the electromagnetism [18–23]. They are employed in the model of packing particles in architecture and civil engineering. Their principle is to calculate movements of the particles through boundary conditions and physical properties until a stable state is achieved. The available shapes of particles are only the sphere and the combination of several spheres. Thus, particles with complex shapes are difficult to be modeled. The modeling software also needs a long time as well as a lot of computing resources for the modeling of a large number of particles. This paper aims to establish an accurate model through a simple and quick process to simulate various packing particles, such as sand, soil, and coal powder. Furthermore, the model can be imported into some kinds of commercial electromagnetic simulation software easily. Currently, there are a variety of methods for modeling stacked particles in the region of architecture, mechanics, and geophysics. But very few of them are proper to the electromagnetic research and none of them can be imported into simulation software directly.

This paper will focus on how to model contacted balls and regular polyhedrons to simulate the packing particles. It can also be employed in the modeling of packing particles with complex shapes. The accuracy of the model is validated by experiments of packing density and effective permittivity.

MODELING

Model of Packing Spheres

The completed modeling process is shown in Figure 1. The first step is to distribute balls uniformly within the desired region. Supposing the desired region with a length of L, a width of S, and a height of H, the number of spheres, W , can be calculated by floor[$(L \times S \times H \times 2)/(a + b)$]. Floor is the function of the rounding down. The diameter range of spheres is (a, b). Secondly, centers of spheres are moved and radii of spheres are gradually increased to make them contact each other. Finally, the modeling will be ended if all the spheres contact at least two other spheres or the edge of the analyzed region. When the process is over, the packing density of the spheres becomes relatively stable and within the possible region validated by experiments [24]. Thus, this model is considered to be consistent with the actual distribution of the packing spheres.

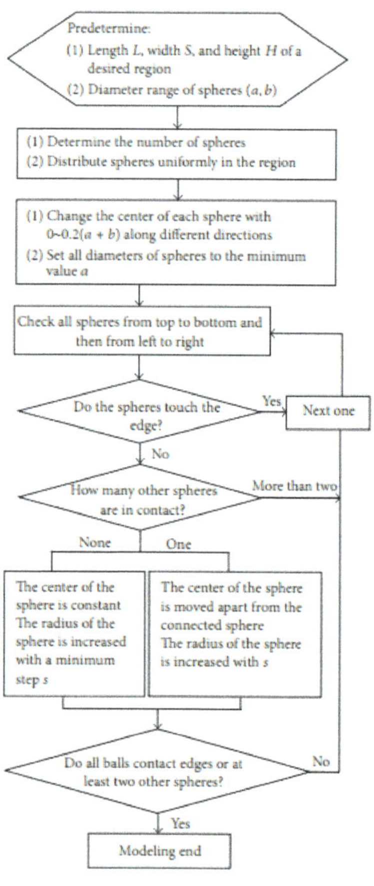

Figure 1: Modeling process.

The modeling method is different from these rigorous modeling and analytical methods in architecture and civil engineering. That is because it does not consider the gravity and friction of the balls. However, the modeling method is particularly used for the electromagnetic analysis of relatively small packing particles; thus it can be accepted without the gravity and the friction in the modeling process. Besides, in the electromagnetic simulation of accumulated particles, the more important influence factors are the shape, size, and packing density of the investigated material [16]. The effects of the gravity and the friction are small, which are usually excluded in the analysis of electromagnetic properties of packing particles.

Figure 2 demonstrates some key steps in the modeling process with different cycle time, P. The present modeling was carried out with balls, which have diameters ranging in 8~10 mm, within a cubic region of 75 × 75 × 105 mm³. The total modeling time is limited in 10 seconds.

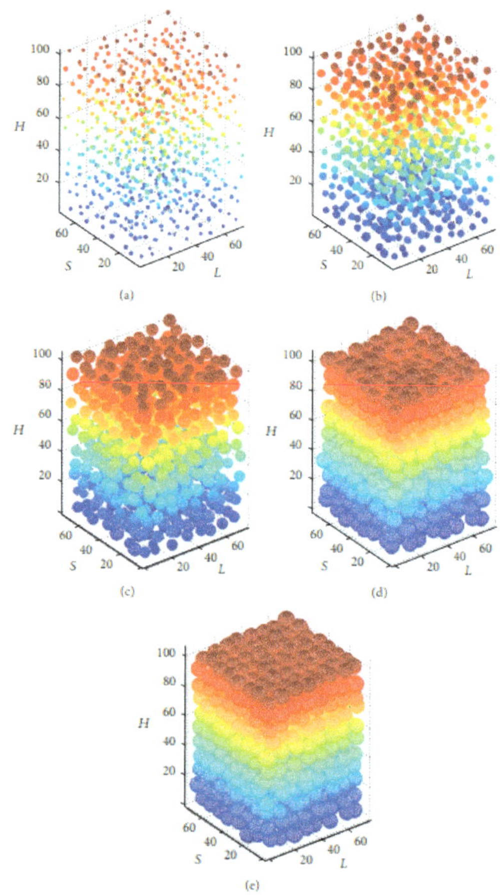

Figure 2: Modeling process with different cycle times (mm). (a) $P = 1$; (b) $P = 20$; (c) $P = 40$; (d) $P = 50$; and (e) $P = 55$.

In the electromagnetic analysis of packing particles, the shape, size, and packing density are the most important parameters. The shape and size of particles in the model should be consistent with the actual situation. The criterion for the reasonable modeling is that the packing density of the model is in good agreement with the experimental result.

The experimental result on the packing density of steel balls in cylindrical containers is published by Scott [24]. The relationship between the packing density and the diameter of the cylindrical container is investigated by experiment.

The packing density of steel balls is from 0.56 to 0.6 in the cylindrical container. The diameter of the cylindrical container is 10.5 times compared with the diameter of the steel ball. When the container becomes bigger, the packing density will rise to 0.58~0.62.

The packing density of the modeling method proposed in this paper should comply with the experimental result. Several models are built in the same condition as described in [24] in order to validate the correctness of our method. The results are shown in Figure 3. Modeling is defined within a cylinder with a diameter of 105 mm and a height of 120 mm. The diameter of balls is limited in the range of 9~11 mm and 4.5~5.5 mm, respectively. This is because the packing density in a limited container is related to the size of balls.

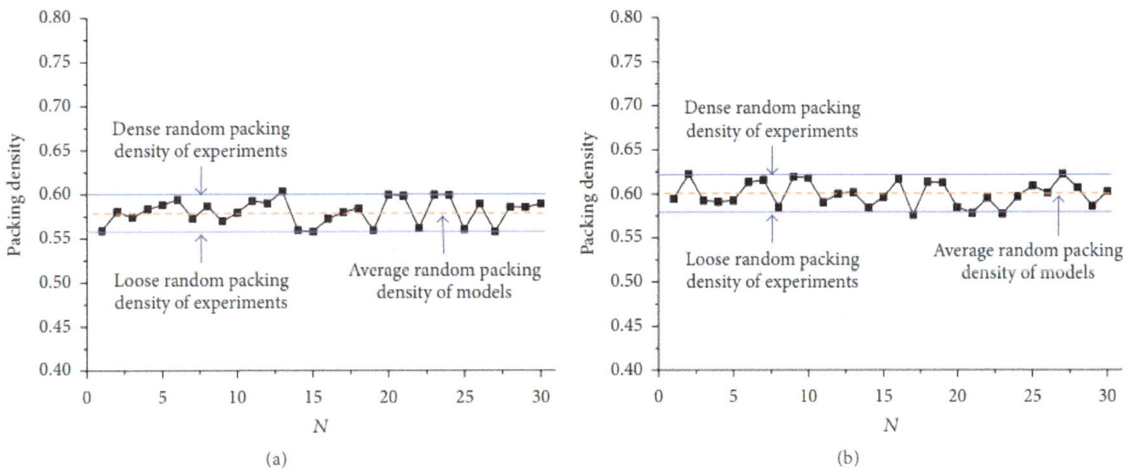

Figure 3: Comparison of the packing density between the model and the experiment. N is the modeling number. (a) Diameter of balls ranging from 9 mm to 11 mm. (b) Diameter of balls ranging from 4.5 mm to 5.5 mm.

Figure 3 shows a good agreement of the packing density between the modeling result and the experimental result in [24]. Modeling has been implemented 30 times in these two cylindrical containers. Most packing density of models agrees with the result described in [24]. The average packing density of 30 models is 0.585 and 0.601, respectively. They are nearly middle values in the packing density range. These indicate the stability and reliability of the modeling method.

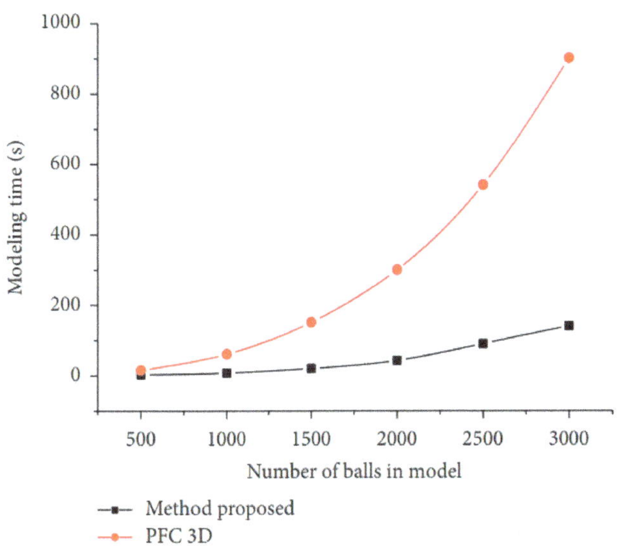

Figure 4: Comparison of the modeling time between PFC 3D and the method proposed in this paper.

There are some other modeling methods applied in other areas [25, 26], such as PFC 3D. Figure 4 shows the time to model packing balls by use of our method and PFC 3D. Both of the modeling methods are carried out with balls having 8~10 mm diameter in the cube having the bottom area of 75 × 75 mm. The density of balls is defined as 2000 kg/m^3 and the stiffness of all balls and boundary is 10^8 N/m. The friction between balls is set as 1. The modeling time is defined to be from the beginning of the modeling to reach a steady state. The comparison shows that the proposed method has much less modeling time, which is more and more obvious with the increase of the number of balls.

Model of Packing Polyhedrons

Generally speaking, the particles are not of the spherical type in most modeling cases. In this modeling method, spheres can be conveniently converted to polyhedrons.

As an example, these spheres are deformed into polyhedrons for the modeling of packing fragmented FR4 substrate. The shape of particles close to reality can increase the accuracy of the following electromagnetic analysis.

As shown in Figure 5, the first step is to choose the edge number, N, of polygons according to the shape of the modeling particle. Then, the balls are divided uniformly by N polygons with N edges. Finally, all the polygons are connected to constitute a polyhedron. This method can transform the spheres into polyhedrons in terms of N, which is good for the modeling of some ball-like particles.

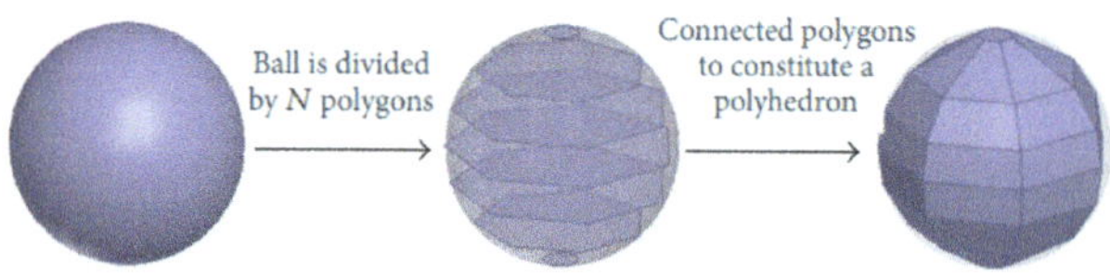

Figure 5: Change spheres into polyhedrons.

This model is built in Matlab software according to the modeling process. The model can be imported into ANSYS HFSS by use of HFSS-MATLAB-Scripting-API directly. Now, the simulation can be started in HFSS.

CORRECTNESS VALIDATION OF MODELING METHOD

To verify the stability and reliability of our modeling method, the model of packing fragmented FR4 as shown in Figure 6 is imported into HFSS and then simulated. The simulated and experimental results will be compared. The investigated frequency band is from 10 GHz to 15 GHz. For a FR4 dielectric substrate, the relative permittivity, ε_r , approximately linearly decreases from 4.29 to 4.21 within the interested frequency band, while the loss tangent, $\tan\delta$, approximately linearly increases from 0.020 to 0.024.

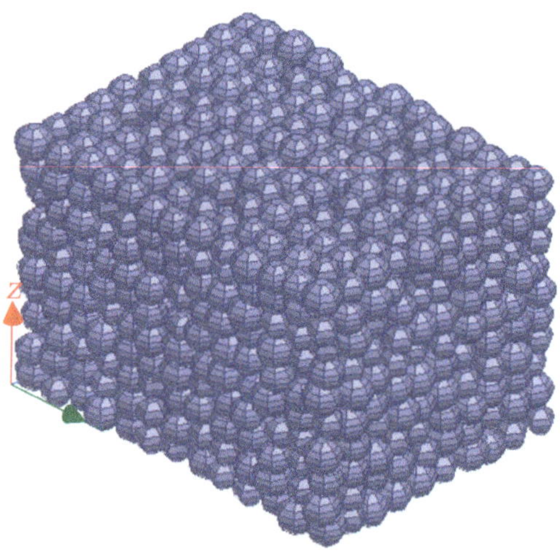

Figure 6: Accumulation of polyhedrons.

The particles of FR4 are filled into a standard rectangular waveguide (19.050 mm × 9.525 mm) to test the parameters [11]. The equivalent ε_r and $\tan\delta$ of the packing particles can be calculated from the S parameters and compared with the results obtained by HFSS. The fragmented FR4 is cut off from a 1 mm thick FR4 substrate with the size of 1 × 1 mm^2. Then, these cube particles are mutually extruded and rubbed to convert them to the ball-like polyhedrons with the diameter of mainly 1.2~1.5 mm.

First, three famous mixing rules about the effective permittivity will be introduced and their results will be compared with simulations and experiments. The Looyenga model [27, 28], which is widely used to calculate of random distributed balls, is

$$\sqrt[3]{\varepsilon_m} = v_1\sqrt[3]{\varepsilon_1} + v_2\sqrt[3]{\varepsilon_2}. \tag{1}$$

The Maxwell-Garnett (MG) mixing rule [29] is

$$\frac{\varepsilon_m - \varepsilon_2}{\varepsilon_m - 2\varepsilon_2} = v_1\frac{\varepsilon_1 - \varepsilon_2}{\varepsilon_1 + 2\varepsilon_2}. \tag{2}$$

The Bruggeman mixing rule [30] is

$$v_1\frac{\varepsilon_1 - \varepsilon_m}{\varepsilon_1 + 2\varepsilon_m} = -v_2\frac{\varepsilon_2 - \varepsilon_m}{\varepsilon_2 + 2\varepsilon_m}, \tag{3}$$

where ε_m is the equivalent dielectric constant of the packing particles and v_1 and v_2 are the volume ratios of the particles and the homogeneous environment, respectively. The volume ratios are calculated by the total volume and the weight of the packing particles. The theoretical ε_m of packing FR4 particles packing in the air ($v_1 = 0.58, \varepsilon_1 = 4.21\sim4.29; v_2 = 0.42, \varepsilon_2 = 1$) of three famous mixing rules is shown in Figure 9.

The measured packing particles in the rectangular waveguide are shown in Figure 7. The vector network analyzer Agilent N5244A is employed to test the S parameters of the waveguide filled with the FR4 particles. Each end of the waveguide is connected to a coaxial converter as shown in Figure 8. A transparent plastic paper with the thickness of 0.08 mm is clipped between the waveguide and the converter in order to confine the particles in the waveguide.

Figure 7: FR4 particles in a rectangular waveguide.

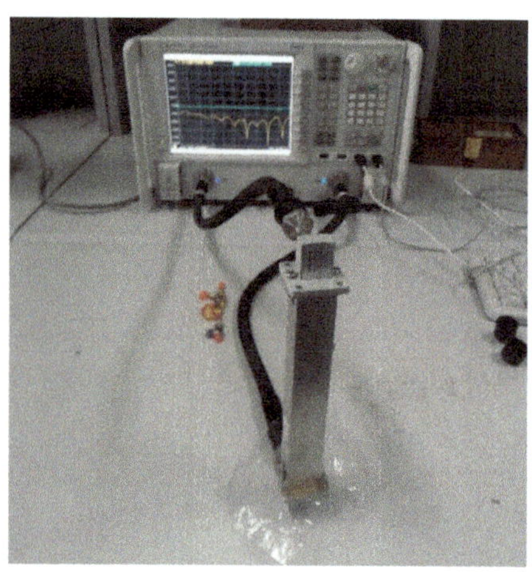

Figure 8: Measurement of electromagnetic properties of packing particles in the rectangular waveguide.

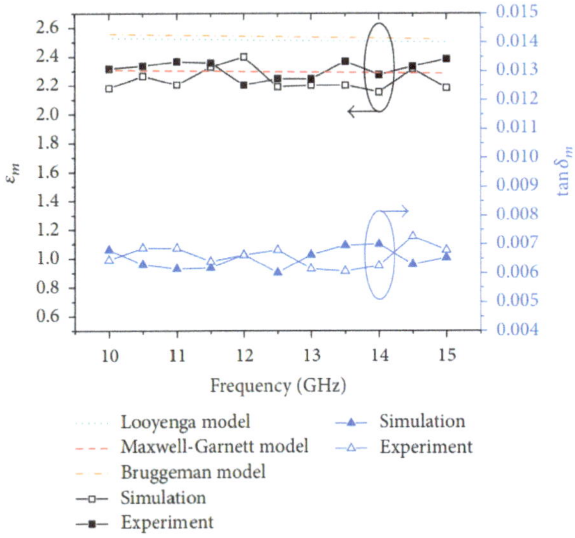

Figure 9: Theoretical, experimental, and simulated results of ε_m and $\tan\delta_m$ in the frequency band with H=3 mm.

The ambient temperature is 25°C and the relative humidity is 65% in the laboratory. The packing depth, H, should be longer than $0.1\lambda_g$ (λ_g is the guide wavelength). Otherwise, the measured result is inaccurate. Besides, H should not be the integer times of $\lambda_g/2$ because the resonant effect in the waveguide will seriously affect the measured result [6, 10]. In the testing frequency band, the range of λ_g is 14.72 mm~30.48 mm. Thus, the effective range of H is selected as 3~10 mm in the measurement.

The theoretical, experimental, and simulated results of ε_m and $\tan\delta_m$ calculated from S parameters [6, 31, 32] are shown in Figures 9 and 10. These results are the average value of 30 repeated simulations and experiments. In Figure 9, the simulated results of the effective permittivity show good agreement with experimental data in the whole band at H=3 mm. The MG mixing rule agrees well with the experimental results of the packing polyhedrons with smaller packing depth H. In Figure 10, the simulated and experimental effective permittivities still agree well especially in the range of large packing depth, and both increase with the packing depth. This is because the random distribution of particles has a significant effect on the result in the case of fewer particles. With the increase of H, the difference between the simulated and experimental results becomes smaller. Both of them get closer to the results of the Looyenga rule and the Bruggeman rule. They still fall between the predictions of the MG rule and the Bruggeman rule, which is consistent with the result in [17] for two-dimensional dielectric mixture by using the FDTD method similar to this paper. It is worth pointing out that the simulated and experimental results satisfy the Wiener bounds and the Hashin-Shtrikman bounds as well, since they are often looser than the bound between the MG rule and the Bruggeman rule. The influence of the random distribution of particles will be weakened with the increasing of number of packing particles.

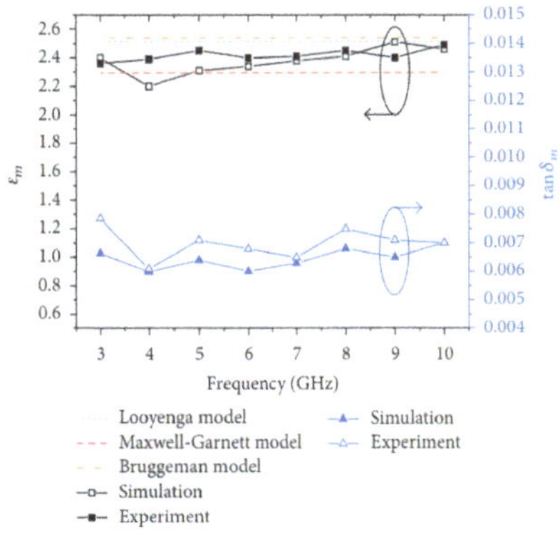

Figure 10: Theoretical, experimental, and simulated results of ε_m and $\tan\delta_m$ at the center frequency, 12.5 GHz, with different packing depth H of the particles.

The advantages of modeling and numerical simulation are embodied in the analysis above. Compared with the empirical formula, it takes more conditions into consideration such as shape, size and distribution of the particles, the thickness of packing, and the frequency. Besides, the modeling and simulation seem to be helpful to select the suitable formulation for the effective permittivity. For example, the MG mixing rule shows better agreement with the simulated and experimental results than the other two rules in the analysis of the packing polyhedrons with smaller packing depth H.

CONCLUSION

A modeling method for the packing particles is proposed in this paper. The packing particles in the shape of sphere or sphere-like polygon can be modeled easily and imported into the electromagnetic simulation software directly. It is much faster than PFC 3D to construct a model with a number of packing balls. It has been verified by experiment and agrees well with formulations and rules of the mixture in literatures. Besides, it is more accurate than the empirical formula to recognize equivalent electromagnetic properties of packing particles due to the fact that more conditions are considered. The modeling method has unique advantages in the field of electromagnetic simulation of packing particles.

Acknowledgments

This work is supported in part by National Basic Research Program of China under Grant no. 2012CB214900, in part by Program for New Century Excellent Talents in University under Grant no. NCET-13-0089, and in part by National Program for Support of Top-Notch Young Professionals.

REFERENCES

1. H. Looyenga, "Dielectric constants of heterogeneous mixtures," Physica, vol. 31, no. 3, pp. 401–406, 1965. · ·
2. J. C. Santamarina, A. Klein, and M. A. Fam, "Soils and waves: particulate materials behavior, characterization and process monitoring," Journal of Soils and Sediments, vol. 1, no. 2, p. 130, 2001. ·
3. M. Meng and F. Wang, "Theoretical analyses and experimental research on a cement concrete dielectric model," Journal of Materials in Civil Engineering, vol.

25, no. 12, pp. 1959–1963, 2013. · ·

4. S. Lambot, I. Van den Bosch, B. Stockbroeckx, P. Druyts, M. Vanclooster, and E. Slob, "Frequency dependence of the soil electromagnetic properties derived from ground-penetrating radar signal inversion," Subsurface Sensing Technologies and Applications, vol. 6, no. 1, pp. 73–87, 2005.
5. M. I. Mishchenko and L. Liu, "Electromagnetic scattering by densely packed particulate ice at radar wavelengths: exact theoretical results and remote-sensing implications," Applied Optics, vol. 48, no. 13, pp. 2421–2426, 2009.
6. A. M. Nicolson and G. F. Ross, "Measurement of the intrinsic properties of materials by time-domain techniques," IEEE Transactions on Instrumentation and Measurement, vol. 19, no. 4, pp. 377–382, 1970. · ·
7. Y.-J. Cheng, Z.-X. Xia, L. Wang, and Y. Fan, "Frequency reconfigurable microwave reaction experiment apparatus for coal desulphurization," Journal of the University of Electronic Science and Technology of China, vol. 43, no. 1, pp. 31–35, 2014.
8. Y. J. Cheng and X. L. Liu, "W-band characterizations of printed circuit board based on substrate integrated waveguide multi-resonator method," IEEE Transactions on Microwave Theory and Techniques, vol. 64, no. 2, pp. 599–606, 2016.
9. D. A. Robinson and S. P. Friedman, "The effective permittivity of dense packings of glass beads, quartz sand and their mixtures immersed in different dielectric backgrounds," Journal of Non-Crystalline Solids, vol. 305, no. 1–3, pp. 261–267, 2002.
10. A. M. Shutko and E. M. Reutov, "Mixture formulas applied in estimation of dielectric and radiative characteristics of soils and grounds at microwave frequencies," IEEE Transactions on Geoscience and Remote Sensing, vol. 20, no. 1, pp. 29–32, 1982.
11. G. Y. Chernyak and H. Heimann, Dielectric Methods for Investigating Moist Soils, Israel Program for Scientific Translations, Jerusalem, Israel, 1967.
12. G. C. Topp, J. L. Davis, and A. P. Annan, "Electromagnetic determination of soil water content: measurements in coaxial transmission lines," Water Resources Research, vol. 16, no. 3, pp. 574–582, 1980. · ·
13. D. A. Robinson and S. P. Friedman, "Electrical conductivity and dielectric permittivity of sphere packings: measurements and modelling of cubic lattices, randomly packed monosize spheres and multi-size mixtures," Physica A: Statistical Mechanics and its Applications, vol. 358, no. 2–4, pp. 447–465, 2005. · ·
14. Z. X. Xia, Y. J. Cheng, and Y. Fan, "Frequency-reconfigurable TM010-mode reentrant cylindrical cavity for microwave material processing," Journal of Electromagnetic Waves and Applications, vol. 27, no. 5, pp. 605–614, 2013.
15. B. Sareni, L. Krähenbühl, A. Beroual, and C. Brosseau, "Effective dielectric constant of periodic composite materials," Journal of Applied Physics, vol. 80, no. 3, pp. 1688–1696, 1996. · ·
16. B. Sareni, L. Krähenbühl, A. Beroual, and C. Brosseau, "Effective dielectric constant of random composite materials," Journal of Applied Physics, vol. 81, no. 5, pp. 2375–2383, 1997. · ·
17. K. K. Karkkainen, A. H. Sihvola, and K. I. Nikoskinen, "Effective permittivity of mixtures: numerical validation by the FDTD method," IEEE Transactions on Geoscience and Remote Sensing, vol. 38, no. 3, pp. 1303–1308, 2000.
18. M. Suzuki and T. Oshima, "Verification of a model for estimating the void fraction in a three-component randomly packed bed," Powder Technology, vol. 43, no. 2, pp. 147–153, 1985.
19. L. Oger, J. P. Troadec, D. Bideau, J. A. Dodds, and M. J. Powell, "Properties of disordered sphere packings I. Geometric structure: statistical model, numerical simulations and experimental results," Powder Technology, vol. 46, no. 2-3, pp. 121–131, 1986.
20. P. R. Rios, "Comparison between a computer simulated and an analytical grain size distribution," Scripta Materialia, vol. 40, no. 6, pp. 665–668, 1999.
21. R. Al-Raoush and M. Alsaleh, "Simulation of random packing of polydisperse particles," Powder Technology, vol. 176, no. 1, pp. 47–55, 2007.
22. M. Elimelech, J. Gregory, and X. Jia, Particle Deposition and Aggregation: Measurement, Modelling and Simulation, Butterworth-Heinemann, 2013.
23. D. He, N. N. Ekere, and L. Cai, "Computer simulation of random packing of unequal particles," Physical Review E, vol. 60, no. 6, pp. 7098–7104, 1999. ·
24. G. D. Scott, "Packing of spheres: packing of equal spheres," Nature, vol. 188, no. 4754, pp. 908–909, 1960. · ·
25. Y. Konakawa and K. Ishizaki, "The particle size distribution for the highest relative density in a compacted body," Powder Technology, vol. 63, no. 3, pp. 241–246, 1990.
26. J. Rodríguez, C. H. Allibert, and J. M. Chaix, "A computer method for random packing of spheres of unequal size," Powder Technology, vol. 47, no. 1, pp. 25–33, 1986.
27. L. D. Landau, J. Bell, M. Kearsley, L. Pitaevskii, E. Lifshitz, and J. Sykes, Electrodynamics of Continuous Media, Elsevier, Philadelphia, Pa, USA, 1984.
28. J. A. Reynolds and J. M. Hough, "Formulae for dielectric constant of mixtures," Proceedings of the Physical Society, Section B, vol. 70, no. 8, pp. 769–775, 1957.
29. J. C. Maxwell-Garnett, "Colours in metal glasses and in metallic films," Philosophical Transactions of the Royal Society of London, vol. 203, no. 359–371, pp. 385–420, 1904. ·
30. Ć. Blanchard, J. A. Portí, J. A. Morente, A. Salinas, and E. A. Navarro, "Determination of the effective permittivity of dielectric mixtures with the transmission line matrix method," Journal of Applied Physics, vol. 102, no. 6, Article ID 064101, 2007.

31. C. J. F. Böttcher, O. C. van Belle, P. Bordewijk, and A. Rip, Theory of Electric Polarization, Elsevier Science, Philadelphia, Pa, USA, 1978.

32. W. B. Weir, "Automatic measurement of complex dielectric constant and permeability at microwave frequencies," Proceedings of the IEEE, vol. 62, no. 1, pp. 33–36, 1974.

CHAPTER 19

RECTIFICATION OF RF FIELDS IN LOAD DEPENDENT COUPLED SYSTEMS: APPLICATION TO NON-INVASIVE ELECTROCEUTICALS

Sree N. Koneru, Charles R. Westgate, Kenneth J. McLeod

Keywords: RF Demodulation in Tissue, Non-Linear Coupling, Complex Load-Line, Electroceuticals, Neuromodulation, Stochastic Resonance

INTRODUCTION

Electroceuticals are bioelectronic devices which rely on electrical fields induced into tissues to influence neural activity and thereby achieve therapeutic effects [1] . While the term "electroceuticals" was coined only recently in light of renewed efforts to expand research in the area, electric fields have been used in medicine for the larger part of the 20th century. Diathermy devices have existed since the 1920's to treat a broad range of health conditions from inflammation to tumors; defibrillators were developed in the late 1930's; and, pacemakers, muscle stimulators, transcutaneous nerve stimulators (TENS), spinal cord stimulators and most recently, deep brain stimulators, have subsequently been introduced into medicine. A critical development in the field of electroceuticals has been the transition from reliance on relatively high intensity (1 - 100 V/m), supra-threshold, electric fields which are sufficient to deterministically excite the neuromuscular system (i.e. neurostimulation) [2] to neuromodulation approaches, wherein the induced electric fields are sub-threshold [3] . Neuromodulation approaches do not ensure initiation of action potentials, but are sufficient to bias neural activity. Nonetheless, most of the current neuromodulatory technologies being developed still rely on skin contact or other invasive means to deliver electrical currents into the tissue. As a result, even when using the lower level, high frequency stimuli, skin irritation or damage can arise from chronic transcutaneous neurostimulation. Further, surgical risks involved in invasive neurostimulation techniques and the need for carefully monitored use have slowed the development and widespread use of electroceuticals in medicine.

The need for much smaller induced electric field intensities in neuromodulation processes, as compared to neurostimulation, creates an opportunity to pursue alternative field induction mechanisms. Specifically, the ability of magnetic fields (EMF) to readily penetrate clothing, bandages, and skin, raises the possibility of developing "contactless" electroceuticals. The magnitude of the electric field induced in tissue is directly proportional to the magnitude and frequency of an applied magnetic field, and so high frequency, that is, radio-frequency (RF) fields would seem to have certain advantages in this application as RF based devices would not require large magnetic field intensities. However, while RF fields readily penetrate into the body, physiologic processes are responsive only to relatively low frequencies, typically in the KHz range or lower. Therefore, in order to influence the electro-physiologic behavior of tissue with RF fields, it is necessary to modulate the RF waveform and then ensure the RF waveform is demodulated within the tissue. Such demodulation requires a

Citation: Sree N. Koneru, Charles R. Westgate, Kenneth J. McLeod (2016) Rectification of RF Fields in Load Dependent Coupled Systems: Application to Non-Invasive Electroceuticals. Journal of Biomedical Science and Engineering,09,112-121. doi: 10.4236/jbise.2016.92007

non-linear or rectification process within the tissue.

At frequencies below 10 MHz electric fields induced into tissues are largely excluded from the interior of cells due to the low conductivity of the cell plasma membrane. The charge build-up on the plasma membraneperturbs the normal transmembrane potential and can influence the gating of the non-linear ion channels present in the cell membrane [4] . That is, for fields below 10 MHz, the cell membrane can, in principle, demodulate alternating electric fields in the tissue. Alternatively, electric fields higher than about 10 MHz pass through the cell as a result of the low capacitive reactance of the membrane at these higher frequencies. Consistent with this biophysical premise, experimentation has confirmed that for frequencies greater than 10 MHz, the non-linear rectifying behavior of the cell membrane is insufficient to initiate any significant biological effects at field intensities below those which create significant heating, and to date, no other non-linear processes in tissues has been identified which could serve to demodulate RF fields over 10 MHz [5] .

However, when low-power RF transmitters are operated in close proximity to the body, a very different exposure condition exists. Specifically, to maximize efficiency, RF amplifiers are typically operated near saturation, but operation near saturation increases system non-linearity [6] . This non-linearity is strongly evident when the load is complex, resulting in the operational "load-line" contorting into an ellipse at high output levels (Figure 1) (Salib, et al., 1987). This behavior is commonly observed in commercially available RF power amplifiers (Figure 2). The net result is that while the load (i.e. tissue) may be linear, the coupled RF system may behave similarly to an ideal electric field generator driving a non-linear load. We hypothesize that this load-amplifier interaction could result in significant RF field demodulation in a living system. We approach this question in two parts: First, we investigate whether "demodulation" can be demonstrated under complex load conditions, using a numerical simulation approach. Second, using a low power RF amplifier operating near saturation, we test the predictions made by the numerical simulations. With this prototype system, we were able to demonstrate a 7.5% demodulation efficiency, confirming the plausibility of such effects occurring in vivo.

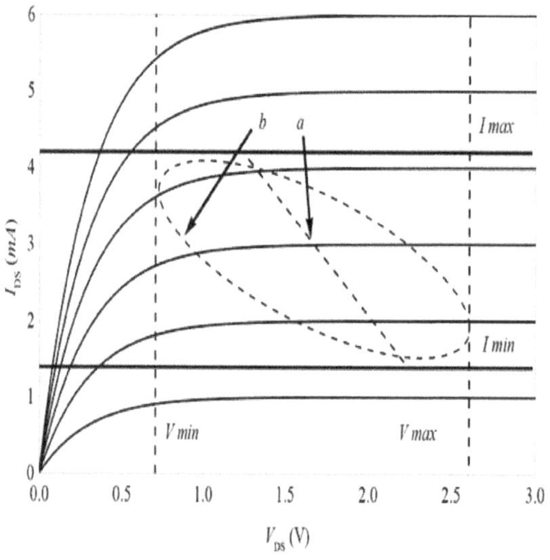

Figure 1. Load-lines, representing the trajectory of all possible instantaneous values of current (I_{DS}) and voltage (V_{DS}), for a typical electronic amplifying device operating under two loading conditions: (a) small signal, resistive loading; (b) large signal (operation in or near saturation) with a complex load resulting in an elliptical load line.

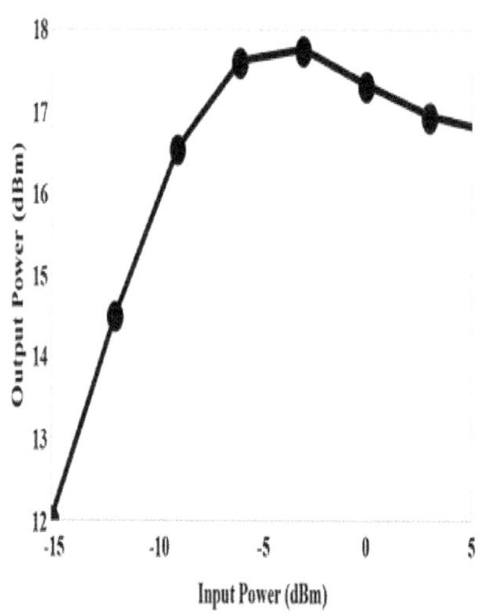

Figure 2. Output power measurements from a 150 mW wideband (10 MHz-3 GHz) power amplifier (RF-Lambda Model RAMP01M03GC), driven at 915 MHz (Wavetek Model 1080) and terminated into a loop antenna. The output response when approaching and reaching saturation is typical of RF power amplifiers where a decrease in output power occurs for increased input power.

METHODS

Part 1—Numerical Simulations

We initially tested whether demodulation can be demonstrated under coupled loading conditions, by employing numerical simulation. More specifically, we investigated if demodulation can occur when a weak RF amplifier operating near saturation is coupled into complex loads. At frequencies above 10 MHz, human tissue appears to be strictly resistive, that is, it presents a linear load. However, human tissues tend to be physically large relative to the wavelengths of interest. For example, at 100 MHz, the dielectric permittivity of tissue is essentially the permittivity of water, or close to 100 F/m. Correspondingly, the wavelength of 100 MHz in body tissue will be around 30 cm. Therefore, while for such frequencies there are no explicit capacitive or inductive reactance elements in human tissue, the time delays inherent to wave transmission in the body gives rise to wave behavior as if physical capacitive or inductive load elements were present.

Analysis of RF amplifier systems, driven to near saturation, with complex loading, generally precludes numerical analysis, and so developers commonly rely on "load-pull" evaluation techniques when developing such systems [7] . Load-pull is an empirical technique wherein the load is altered and the output of the amplifier is directly measured. Here, to permit numerical analysis, we utilized a much simplified one-dimensional model (Fig- ure 3). The model incorporates an RF oscillator/amplifier driving a loop antenna as the field source. The exposed tissue is represented as a pick-up coil with impedance R_p. To introduce the necessary time delays associated with the large physical structure of human tissues, we utilized a transmission line with length on the order of a fraction of the wavelength of interest. The transmission line, of impedance Z_T, terminates with an impedance Z_L, which may be matched or unmatched to that of the transmission line. This terminating impedance may reflect, for example, that of bone, while the bulk of the tissue (transmission line) would have the impedance of muscle, thereby creating an unmatched load condition.

The electric field existing at the exposed tissue site, i.e. across R_p, is evaluated through a time series simulation as the sum of the incident and reflected waves in the system. The amplitude of the reflected wave is calculated as [8] :

$$V_R(\omega) = V_I(\omega) \times \Gamma(\omega) \tag{1}$$

where:

V_R = Intensity of reflected wave

ω = Oscillator frequency

V_I = Intensity of incident wave

Γ = Reflection coefficient = $(Z_L - Z_T)/(Z_L + Z_T)$

We evaluated system behavior under conditions of both small-signal and large-signal amplifier operation into a complex load, the former using linear I-V load line characteristics, and the latter using elliptical load line characteristics. For the latter analyses, we approximated the load line as:

$$V_I = a \times e^{\frac{-(I_P - b)^2}{c}} \tag{2}$$

where I_P = Current in pick-up coil; and parameters a, b & c represent the height, position and width of the elliptical load-line portion respectively.

Numerical Model Algorithm

The modeling algorithm involved an eight step process which tracked each half-cycle of the oscillator output as it traversed the model system of Figure 4:

1) Production of an RF current (and corresponding magnetic field) in the loop antenna.
2) Induction of an RF current in the pick-up coil.
3) The development of an electric potential across the resistive load in the pick-up coil that includes R_p in parallel with the transmission line input impedance.
4) Transmission of an incident traveling half-cycle down the transmission line.
5) Reflection of the travelling half-cycle at the terminating load.
6) Transmission of the reflected half-cycle back towards the pick-up coil.

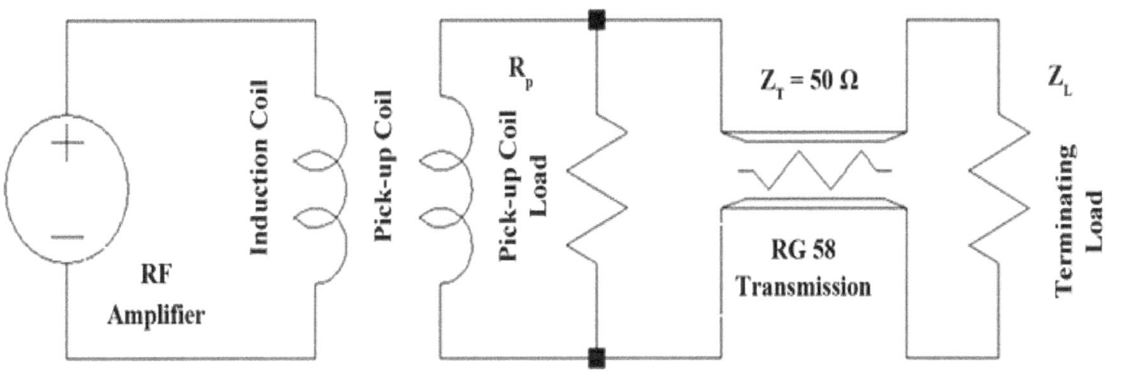

Figure 3. Schematic of the transmitter-tissue model. The transmitter and complex load (representing body tissue) are inductively coupled to each other. The layers of tissue (thickness) are represented by the length of a lossy transmission line. The RF amplifier induces a field in a pick-up coil (first layer of tissue) which is coupled to a terminating load andmeasured via a transmission line, where the transmission line represents layers of tissue. The terminating load is varied, representing, for example, the impedance of bone or air, resulting in a range of reflective coefficients.

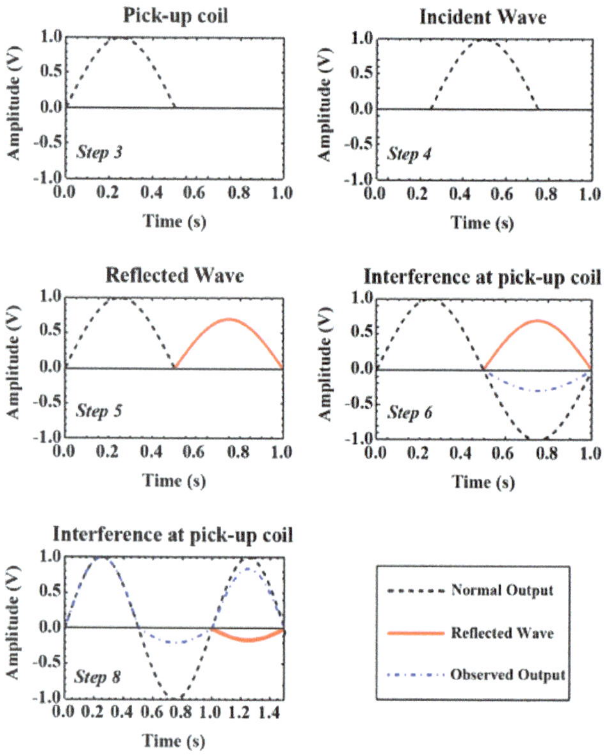

Figure 4. Results at each time step of the simulation model performed for the conditions of Γ = 0.7, transmission line length = 1/4 λ. Steps 3 through 6 follow the sequence of the first 1/2 wave cycle induced into the pick-up coil. Step 8 shows the beginning of the second half wave with its reduced amplitude.

7) Development of a feedback current in the pick-up coil with the corresponding creation of a back-EMF.
8) Determination of the amplifier output current under the combined loading of the pick-up coil load current and the back EMF

100 points per half-cycle were simulated to ensure a smooth representation of the waveforms, and the simulation was continued until a steady state in the output voltage was observed. For all simulations, the D.C. component of the output voltage across the pick-up coil load (R_P) was calculated by averaging over each complete cycle for the duration of simulation period. Simulations for various transmission line lengths (i.e. time delays) and reflection coefficients were performed to investigate the influence of the reflected wave components on the potential recorded across the pick-up coil load (R_P).

Part 2—Physical Measurements

We also tested whether demodulation under complex loading of a saturated RF amplifier could be demonstrated in a physical model. A prototype RF oscillator device was built to verify the outcomes predicted by the numerical model. The physical model consisted of: 1) A battery operated 40 MHz Colpitts-type RF oscillator with a 10 cm circular loop antenna. A timing circuit pulsed the oscillator at 1 KHz utilizing a 10% duty cycle. 2) A 10 cm circular pick-up coil with a 500 Ω resistive load, represented a human tissue load 3) RG 58 co-axialcable (50 Ω; velocity factor 0.67) of a fixed length was used to represent a thick layer of muscle 4) An oscilloscope (Agilent Technologies, model MSO7104B) was used to measure electric potential as well as to present a terminating resistive load (Z_L) of either 50 Ω or 1 MΩ, in parallel with 15 pF.

The two parameters varied in the testing of the physical model were electrical length of the co-axial cable and terminating impedance. Four different lengths of RG-58 co-axial cables, representing electrical lengths (at 40 MHz) of 1/4 λ, 1/2 λ, 3/4 λ and 1 λ were used. Recordings were obtained with both terminating load conditions, resulting in eight possible measurement combinations using the two parameters.

RESULTS

Numerical Results—Small Signal Operation

Simulation results under conditions for which the amplifier output was not load dependent (i.e. small signal, non-saturated amplifier operation), using a 1/4 λ transmission line length with a terminating load condition of Γ = 0.7 illustrates the overall behavior of the system (Figure 5(a)). The initial half-cycle induced in the pick-up coil travels down the transmission line and is reflected back with the same sign as the incident half-cycle, but at reduced amplitude. The reflected wave arrives back at the pick-up coil as the amplifier output is changing polarity. For a load independent oscillator, the output of the oscillator is no different than that expected in the absence of a reflected wave, and so the voltage across R_P due to the reflected wave and that due to the generator's output simply add. This results in reduced amplitude of the subsequent half-wave, which in turn proceeds down the transmission line. Correspondingly, the average value of the potential across the R_P will have a net DC component when averaged over a full cycle of the oscillator output with this DC component decaying over time.

Simulation results were obtained for reflection coefficients ranging from 0.6 - 0.9 (Figure 5(b)). The magnitude of the reflection coefficient is seen to have a significant effect on the duration of transient response, that is, the period of time for which a D.C. offset occurs. However, at the operating frequencies of interest (>10 MHz) this corresponds to transient time of less than 25 microseconds.

Numerical Results—Large Signal Operation

When an RF amplifier is operating near saturation into a complex load, the output of the device becomes load dependent (i.e. large signal) in a non-trivial manner. We modeled this load dependence per Equation 3 using values of a = 1.5, b = 0.5, and c = 1 such that increased current demand on the amplifier results in an initial rise in output voltage. This behavior results in a positive feedback such that when the reflected wave increases the load on the output amplifier, the output of the oscillator increases. The positive feedback allows the system to attain a steady state with a net D.C. offset (Figure 6). We simulated operation of the system under a range of reflection coefficients Γ = 0.6 - 0.9 and for transmission line lengths ranging from 0 - 0.5 λ (Figure 7). Rectifi-

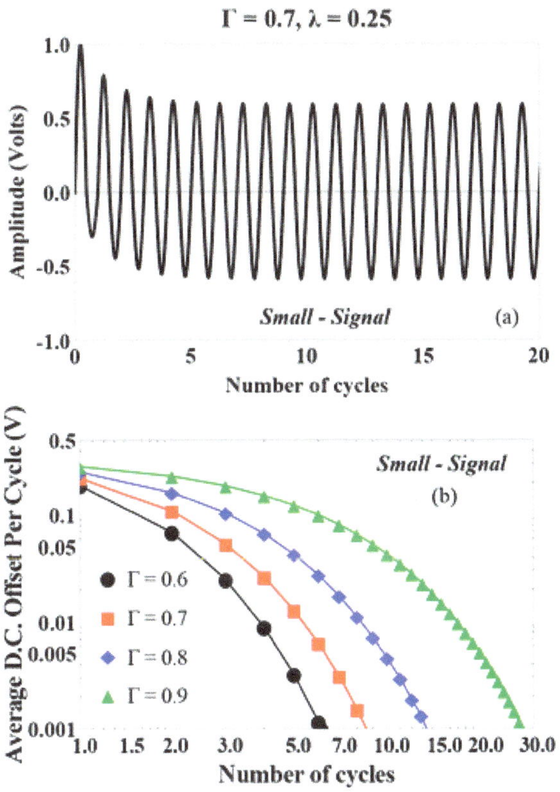

Figure 5. (a) Output at the pick-up coil load (R_p) over 20 cycles. The waveform has the highest magnitude DC component in the first cycle which then decays rapidly over time, approaching zero after about 5 cycles. (b) D.C Offset calculated for various reflection coefficients. Higher reflection coefficients result in a slower decay of the DC component, but even at a reflection coefficient of 0.9, the DC component decays to zero in fewer than 20 cycles.

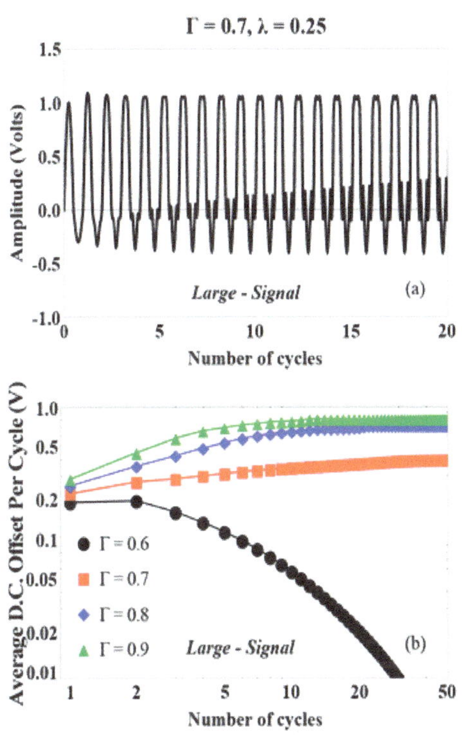

Figure 6. (a) Simulation using an elliptical load line. Output across the pick-up coil load stabilizes after an initial start-up transient with a positive peak that is much higher than the negative peak. (b) The maximum D.C. offset is about 0.6V and a steady state is reached with that amplitude after approximately 20 cycles.

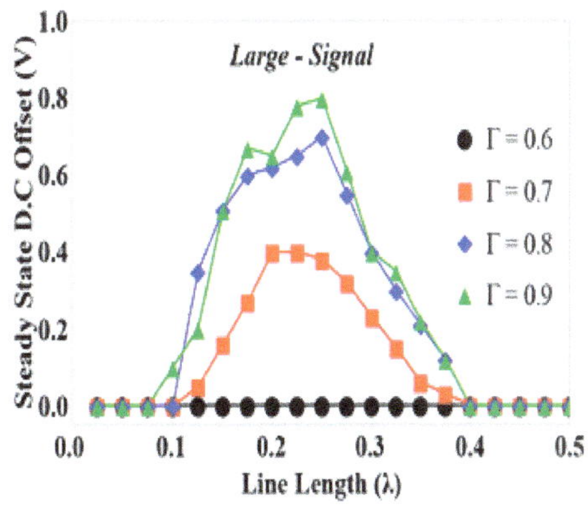

Figure 7. Steady state D.C offset amplitude for various reflection coefficients as a function of the total delay time (in units of wavelength). A reflection coef- fecient of 0.9 results in amaximum DC offset (rectification) of 80%. The optimaltotal delay time with respect to providing the maximum DC offset is 0.5 λ, but a rectification effect is observed over delay times of 0.3 λ to 0.7 λ.

cation factors of up to 80% were obtained for reflection coefficients of 0.9 and time delays of 1/4 wavelengths, but a wide range of parameter values (reflection coefficients and time delays) were found to produce significant rectification factors.

Laboratory Model Observations

Measurements made with the prototype 40 MHz RF oscillator/amplifier circuit, driving a transmission line of fixed length showed distortions in the measured potential across the terminating load, resulting in a distinct D.C. offset. However, this was observed only when there was an impedance mismatch between the transmission line and terminating load. When the transmission lines were terminated with a 50 Ω, 15 pF (Z = 42 Ω, Γ = −0.09) load, the resultant waveform lacked any discernable distortion. As such, distorted waveforms were observed for only half the conditions, i.e. when the transmission lines were terminated into a 1 MΩ resistive load in parallel with 15 pF (Z = 265 Ω; Γ = 0.68). Figure 8 shows a representative waveform when a 1/4 λ transmission line

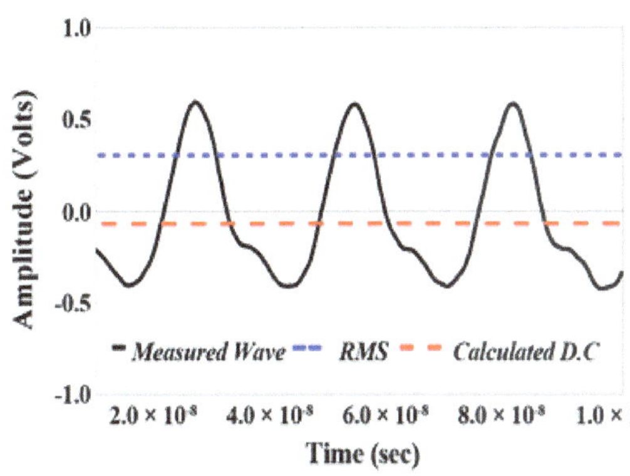

Figure 8. Waveform observed across the pick-up coil load of the prototype device when the coil load (R_p) is linked to a unmatched terminating loadthrough a 1/4 λ transmission line. The waveform has peak positive and peak negative voltages of 0.6 and 0.4 respectively, with an RMS value of 0.33 V. The D.C. component is calculated to be 0.025 V representing about a 7.5% demodulation efficiency.

was mismatched with the terminating load, the condition under which a maximum D.C. offset was observed. Even small deviations from the 1/4 λ delay significantly suppressed the D.C. offset to non-discernable levels.

DISCUSSION

It is well known that demodulation of RF electromagnetic fields in living tissue cannot occur, at frequencies greater than 10 MHz, leading to the suggestion that high frequency stimulation should not be biologically effective. Nonetheless, a remarkably large number of experimental papers have reported biological effects associated with the use of pulsed RF stimulation. In this paper, we hypothesized that a form of demondulation could occur under certain complex loading conditions, and tested this hypothesist using both numerical and physical models.

Numerical simulation indicated that that a D.C. offset can be sustained in tissue when an RF transmitter operating at saturation, is driven into a complex load. The offset in the generated waveforms was evident in its distorted shape and was verified through numerical integration. The interference between the incident and reflected waveforms results in a time varying load being presented to the amplifier. This results in a synchronous time varying output of the amplifier, resulting in a "D.C. shift" in the output. If this RF field were modulated, for example, presented a short burst, which a relaxation time period, the field observed at the load would include a frequency component at the modulation frequency, as if demodulation had occurred. A range of reflection co-efficients (Γ = 0.6 - 0.9) and transmission line lengths (0 - 0.5 λ) were investiated, with the best demodulation efficiency predicted for λ = 1/4 and Γ = 0.9.

Irrespective of the line length used, the physical model presented a Γ of 0.68 when the transmission line terminated into the 1 MΩ, 15 pF load. Consistent with the predictions made by the numerical model, the physical model showed significant distortion in the recorded waveforms and a net D.C. offset. Moreover, the best demodulation efficiency was observed when a line length of ¼ λ was used, as predicted by the numerical modeling. The results from both the numerical and physical model lend support to our hypothesis that under complex loading conditions, an RF amplifier, operating near saturation, and closely coupled to the load, can produce a sustained D.C. offset in biological tissue. Correspondingly, if the RF output is modulated, energy at the modulation frequency will be observed in the tissue, even in the absence of any explicit non-linear (rectifying) component in the tissue.

A system is said to be coupled if some physical process allows components of the sytem to influence each other [9] . Consideration of the amplifier and tissue as closely coupled components as opposed to modeling the system as if the transmitter were independently acting upon the tissue (i.e. a cause-effect type relationship) is the key to our observations. In simple systems, the interactions are generally linear such that an effect is directly proportional to the causal factor. In complex systems, the interactions are, by definition, non-linear. Such systems typically give rise to emergent behaviors, that is, behaviors which cannot be predicted by evaluation of the system components in isolation. Here, the reflecting load, combined with time delays in the tissue model and the load dependent behavior inherent to low power RF amplifiers, introduces sufficient non-linear coupling that a D.C. component in the output emerges. If a more powerful amplifier were utilized such that it did not need to operate near saturation (large signal mode), or the tissue load was more remote from the source such that there was less feedback coupling, this non-linear pheneomenon would not be observed. Of course, the design goal in portable devices is to maximize battery lifetime, and to do so in any RF device requires operation near saturation.

While neurostimulation typically requires DC field intensities on the order 10 V/m to be present for a duration of about 50 - 500 μsec [10] , neuromodulation technologies utilize far lower field intensities. In classical systems theory, random fluctuations in signals are referred to as "noise" and are nearly always considered detrimental, and so great efforts are made to eliminate this noise. However, in physiological systems, such as neural systems, while "noise" is ubiquitous, it is, in fact, widely considered to be essential in facilitating information processing in the body [11] . Stochastic Resonance (SR) is a phenomenon where the presence of "noise" in non-linear systems can be used to enhance the detection of sub-threshold stimuli. For example, in a situation where there is a detection threshold, such as exists for nerves, a subthreshold "signal" in the presence of noise may randomly exceed the necessary threshold to activate the nerve. A characteristic feature of SR in a system is that the system output, when plotted against system noise power, will produce a peak response at a non-zero value [12] .

CONCLUSION

In summary, employing weak RF amplifiers, operating near saturation, in close proximity to the body presents the possibility of creatingnon-deterministic neuromodulation effects resulting from the development of small fields (<10 V/m) at the frequency of RF modulation, which combine with inherent system noise in a stochastic manner (observed as a stochastic resonance). Irrespective of the type of modality, if nerve activity can be altered using modulated RF field exposure, we envision several important applications. Specifically, a popular non-pharmacological neurostimulation therapy for treating chronic pain is neural masking using Transcutaneous Electrical Nerve Stimulation [13] . However, use of this technology is limited given the risk of skin burning, and the inability to use TENS through bandages or clothing. Sub-threshold neuromodulation therapy, whether used transcutaneously or invasively, has already shown its potential in modulating a variety of nerves: the occipital nerve for treating migraines [14] , sacral nerve for treating bowel incontinence [3] and vagal sympathetic complex to treat heart failure [15] . Correspondingly, the ability to modulate nerve activity without invasive procedures/skin contact would have significant advantages for those capable of benefiting from traditional neurostimulation therapies. We believe that RF electroceutical technologies, based on the principles discussed here, hold promise for the development of wearable, non-invasive devices capable of modulating larger nerve fibers and nerve plexi such as the brachial nerve plexus and sacral nerve plexus, among many others.

REFERENCES

1. Famm, K. (2013) A Jump Start for Electroceuticals. Nature, 159-161. http://dx.doi.org/10.1038/496159a

2. Markand, O.N., Kincaid, J.C., Pourmand, R.A., Moorthy, S.S., King, R.D., Mahomed, Y. and Brown, J.W. (1984) Electrophysiologic Evaluation of Diaphragm by Transcutaneous Phrenic Nerve Stimulation. Neurology, 604. http://dx.doi.org/10.1212/WNL.34.5.604
3. Duelund-Jakobsen, J.E.A. (2013) Sacral Nerve Stimulation at Subsensory Threshold Does Not Compromise Treatment Efficacy: Results from a Randomized, Blinded Crossover Study. Annals of Surgery, 219-223. http://dx.doi.org/10.1097/SLA.0b013e318269d493
4. McLeod, K.J., Rubin, C.T. and Donahue, H.J. (1995) Electromagnetic Fields in Bone Repair and Adaptation. Radio Science, 233-244. http://dx.doi.org/10.1029/94RS00832
5. Sheppard, A.R. (2008) Quantitative Evaluations of Mechanisms of Radiofrequency Interactions with Biological Molecules and Processes. Health Physics, 95, 365-396. http://dx.doi.org/10.1097/01.HP.0000319903.20660.37
6. Maas, S. (2003) Non-Linear Microwave and RF Circuits. 2nd Edition, Artech House, Norwood.
7. Ghannouchi, F.M. and Hashmi, M.S. (2013) Load Pull Techniques with Application to Power Amplifier Design. Springer. http://dx.doi.org/10.1007/978-94-007-4461-5
8. Sadiku, M.N. (2010) Electromagnetic Wave Propagation. Elements of Electromagnetics, 500-501.
9. Strogatz, S. (2003) Preface. Sync, Hyperion Books, 3.
10. Mogyoros, I., Kiernan, M.C. and Burke, D. (1996) Strength-Duration Properties of Human Peripheral Nerve. Brain, 439-447. http://dx.doi.org/10.1093/brain/119.2.439
11. McDonnell, M.D. and Ward, L.M. (2011) The Benefits of Noise in Neural Systems: Bridging Theory and Experiment. Nature Reviews Neuroscience, 415-426. http://dx.doi.org/10.1038/nrn3061
12. McDonnell, M. and Abott, D. (2009) What Is Stochastic Resonance? Definitions, Misconceptions, Debates, and Its Relebance to Biology. PLoS Computational Biology, 5, 1-9. http://dx.doi.org/10.1371/journal.pcbi.1000348
13. Nnoaham, K. and Kumbang, J. (2010) Transcutaneous Electrical Nerve Stimulation (TENS) for Chronic Pain. The Cochrane Collaboration, 1-62.
14. Slotty, P.J.E.A. (2015) Occipital Nerve Stimulation for Chronic Migraine: A Randomized Trial on Subthreshold Stimulation. Cephalalgia, 73-78. http://dx.doi.org/10.1177/0333102414534082
15. Kember, G.E.A. (2014) Vagal Nerve Stimulation Therapy: What Is Being Stimulated? Plos One, e114498. http://dx.doi.org/10.1371/journal.pone.0114498

CHAPTER 20 MISSILE-BORNE SAR RAW SIGNAL SIMULATION FOR MANEUVERING TARGET

Weijie Xia, Yuanyuan Qi, Linlin Huang, and Xue Jin

INTRODUCTION

Synthetic aperture radar (SAR) raw signal generation and image simulation [1, 2] play a significant role in the development of SAR system. On the one hand, SAR raw signal generation is an effective and financial tool that enables us to obtain the echo data needed for the validation of the radar imaging algorithms; on the other hand, SAR image simulation provides a feasible way to establish a target feature database, which is the foundation of the target automatic recognition.

Because of the importance of SAR echo and image simulation, a lot of researches focused on the field have been carried out. In the published literature, approaches to simulating SAR echo signal can be divided into three categories: time-domain (TD) methods, two-dimensional (2D) frequency domain methods, and hybrid time/frequency (TF) domain methods [3]. Classical TD method was once employed by Mori to present an interferometric SAR simulator [4]. This algorithm has a high precision but is time consuming, and the scattering characteristic of the whole target is difficult to take into account. To improve the processing efficiency, the 2D method, taking advantage of the efficiency of fast Fourier transform algorithm, has been studied in many papers [5–8]. Unfortunately, this method cannot be directly employed under the situation of nonideal trajectory. As a tradeoff of the aforementioned two methods, hybrid TF simulation, with the ability to consider both the nonideal trajectory and the moderate computational efficiency, has also been an effective way to operate SAR simulation [9, 10].

So far, quite a lot of studies have been addressed to simulate SAR raw data [11–16]. However, unlike the traditional SAR configuration in which the platform travels an ideal trajectory, for realistic missile-borne SAR system, both the platform and the target move along curvilinear trajectories with acceleration and even jerk, which may result in two-order even high-order terms in the range history. With this special range history, the traditional approaches cannot be directly adopted to obtain its 2D frequency spectrum for the overall SAR system transfer function which depends on the azimuth and range coordinates of the target. Consequently, the improved SAR raw signal simulator with the ability to consider the curvilinear trajectory is strongly required. With respect to this, Franceschetti et al. [11] proposed an efficient SAR raw signal simulator based on full 2D Fourier transform, but this simulator is only effective for cases of narrow beam and slow trajectory deviation. Meng et al. developed a fast raw data simulator based on the effectual access of 2DFS [17], but there was still phase space variance

Citation: Weijie Xia, Yuanyuan Qi, Linlin Huang, and Xue Jin, "Missile-Borne SAR Raw Signal Simulation for Maneuvering Target," International Journal of Antennas and Propagation, vol. 2016, Article ID 2598024, 12 pages, 2016. doi:10.1155/2016/2598024.

which could lead to the inaccurateness of the raw signal. Besides, to our knowledge, the most important feature of the complex target, namely, the dynamic Radar Cross Section (RCS), has not been introduced to the echo generation so far in the literature, and available methods often need to know discrete reflectivity distribution map or digital elevation model (DEM) of the imaged scene as an input. Furthermore, there has not been an integrated method for dynamic target echo simulation which can deal with both monostatic SAR and bistatic SAR in the published literatures.

Aiming at the above problems, in this paper, we will fully take the target RCS feature into account to propose a novel RFAT echo generation method for missile-borne SAR. In this method, the high frequency electromagnetic scattering calculation method based on physical optics (PO), geometrical optics (GO), and Incremental Length Diffraction Coefficients (ILDC) is adopted to calculate the scattering characteristic of target; the SaG model and the IPM model are conducted into the real-time motion simulation; the 1DFFT based echo simulation method is used to generate SAR raw data; finally, the RDA (range-Doppler algorithm) is adopted to obtain the simulated SAR image; simulation results show that the method is effective and suitable for different kinds of SAR (monostatic and bistatic) and motion situations.

ANALYSIS OF THE PROPOSED SIMULATION ALGORITHM

Signal Model

In this section, the main employed symbols (nomenclature) are defined as follows.

f_0: carrier frequency.

c: light speed.

λ: carrier wavelength.

T_r: chirp duration.

R: target-to-antenna distance.

x_n: sensor position azimuth coordinate at t_n.

k: chirp rate.

t_n: slow time.

t': fast time.

N_a: number of slow time samples.

N_r: number of fast time samples.

x': continuous form of x_n.

Supposing that the radar is illuminating a point scatterer P with a series of chirp signals, the received raw data for P can be described by the following expression:

$$s(t,t_n,R) = \gamma \cdot \text{rect}\left[\frac{t-t_n-2R/c}{T_r}\right] \cdot \exp\left[-j2\pi f_0\frac{2R}{c} + j\pi k\left(t-t_n-\frac{2R}{c}\right)^2\right], \quad (1)$$

where γ is the reflectivity function. For simplicity, replacing $t-t_n$ in (1) by t', we obtain

$$s(t',x_n,R) = \gamma \cdot \text{rect}\left[\frac{ct'/2-R}{cT_r/2}\right] \cdot \exp\left[-\frac{j4\pi R}{\lambda} + \frac{j4\pi k}{c^2}\left(\frac{ct'}{2}-R\right)^2\right], \quad (2)$$

where $ct'/2$ is the spatial sampling interval in the range direction; let $r' = ct'/2$; the echo equation of the point scatterer in the range-time domain can be written as follows:

$$s(x_n,r',R) = \gamma \cdot \text{rect}\left[\frac{r'-R}{cT_r/2}\right] \exp\left[-j\frac{4\pi R}{\lambda} + j\pi k\frac{4}{c^2}\left(r'-R\right)^2\right]. \quad (3)$$

Actually, the complex target consists of many point scatterers; the echo signal of the target can be calculated by summing up each point scatterer echo signal as follows:

$$s(x',r') = \int \gamma(x',R)\,\text{rect}\left[\frac{r'-R}{cT_r/2}\right] \exp\left(-j\frac{4\pi R}{\lambda}\right) \cdot \exp\left[j\pi k\frac{4}{c^2}\left(r'-R\right)^2\right] dR, \quad (4)$$

where $\gamma(x',R)$ is the overall reflectivity of all points at distance R from the sensor. Through (4), we can simulate the SAR raw data in the time domain directly, but it suffers from the disadvantages of high computation complexity. Furthermore, it is hard to obtain the scattering center points of a complex target. To solve this problem, a Fourier transform is implemented in the range direction to (4):

$$
\begin{aligned}
S(x',\eta) &= \int s_r(x',r')\exp(-j\eta r')\,dr' \\
&= \iint \gamma(x',R)\exp\left(-j\frac{4\pi R}{\lambda}\right)\exp\left[j\pi k\frac{4}{c^2}(r'-R)^2\right]\mathrm{rect}\left[\frac{r'-R}{cT_r/2}\right]\exp(-j\eta r')\,dR\,dr' \\
&= \int \gamma(x',R)\exp\left(-j\frac{4\pi R}{\lambda}\right)\int \exp\left[j\pi k\frac{4}{c^2}(r'-R)^2\right]\mathrm{rect}\left[\frac{r'-R}{cT_r/2}\right]\exp(-j\eta r')\,dr'dR \\
&= \int \gamma(x',R)\exp\left(-j\frac{4\pi R}{\lambda}\right)\exp(-j\eta R)\,\mathrm{rect}\left[\frac{c\eta}{4k\pi T_r}\right]\exp\left(-j\frac{\eta^2c^2}{16k\pi}\right)dR \\
&= \mathrm{rect}\left[\frac{c\eta}{4k\pi T_r}\right]\exp\left(-j\frac{\eta^2c^2}{16k\pi}\right)\int \gamma(x',R)\exp\left[-j\left(\frac{4\pi}{\lambda}+\eta\right)R\right]dR \\
&= \mathrm{rect}\left[\frac{c\eta}{4k\pi T_r}\right]\exp\left(-j\frac{\eta^2c^2}{16k\pi}\right)\Gamma\left[x',\left(\frac{4\pi}{\lambda}+\eta\right)\right],
\end{aligned} \tag{5}
$$

where $\Gamma[x',(4\pi/\lambda+\eta)]$ represents the Fourier transformation of the scattering coefficient of the target which will be explained specifically in the next section and η refers to the corresponding spatial frequency of the range coordinate variable r' with the unit of rad/m. According to the stretchable nature of the Fourier transform, the relationship between range frequency and spatial frequency can be described as

$$
f_r = \frac{c}{4\pi}\eta, \quad \text{or } \eta = 2\pi\frac{2f_r}{c}. \tag{6}
$$

Substituting (6) to (5), we obtain

$$
S(x',f_r) = \mathrm{rect}\left[\frac{f_r}{kT_r}\right]\exp\left[-j\frac{\pi f_r^2}{k}\right]\cdot\Gamma\left\{x',\left[\frac{4\pi(f_0+f_r)}{c}\right]\right\}, \tag{7}
$$

which is the echo equation in the range frequency domain. Then, the original time-domain echo signal can be obtained after a 1D inverse Fourier transform.

Dynamic Modeling for Electromagnetic Scattering

From the final expression described in (7), we can find that the key procedure of generating the echo data is the computation of $\Gamma\{x',f_0+f_r\}$. As mentioned in the previous section, $\Gamma\{x',f_0+f_r\}$ is the Fourier transformation of the scattering coefficient, namely, the Radar Cross Section (RCS) of the target. Thus, in this section, the PO, GO, and ILDC based RCS calculation method is first introduced and then the Doppler correction method for dynamic target is given.

The PO/GO and ILDC Method

The GO is a ray-based method intended for the consideration of electrically large structures, which employs ray-launching and transmission, reflection, and refraction theory to model the interaction between the dielectric regions [18]. The PO utilizes ray optics to estimate the field on a surface and then calculate the transmitted or scattered field through Stratton-Chu formulation [19]. The backscatter contributions are evaluated by both GO and PO, which can be used to compute multiple reflections effectively. Furthermore, the edge diffraction effect of electrically large scaled complex objects is also considered and calculated with the theory of Incremental Length Diffraction Coefficients (ILDC) [20].

Doppler Correction Method

For dynamic missile target, the distance change between target and radar over time could lead to the Doppler effects in the echo [21]. Thus, it is needed for modified results of RCS calculation to take Doppler effects into account.

The modified RCS can be expressed as

$$
\Gamma_1(f,t) = \exp\left[\frac{-j4\pi R^c(t)f}{c}\right]\Gamma_0(f,t) \tag{8}
$$

In this formula, $R^c(t)$ represents the distance from the geometric center of the complex target to the sensor, $\exp(-j4\pi R^c(t)f/c)$ is the Doppler phrase term, denotes the electromagnetic wave frequency, and $\Gamma_0(f,t)$ and $\Gamma_1(f,t)$ are the RCS before phrase correction and the modified RCS, respectively.

Dynamic Target Range History Analysis

In order to realize the phase correction described in (8), dynamic target range history analysis is presented in this section.

Monostatic Radar

Consider the missile-borne SAR illuminating an aerial maneuvering target as shown in Figure 1. The missile-borne SAR platform denoted by $\vec{Q}_0$ moves along the nonideal trajectory $\vec{Q}(t)$ which obeys a high-order polynomial-type representation as a function of the continuous time:

$$
\vec{Q}(t) = \vec{Q}_0 + \vec{v}_q t + \vec{a}_q t^2 + \vec{g}_q t^3 + \cdots, \tag{9}
$$

where $\vec{v}_q$ is the initial velocity vector, $\vec{a}_q$ is the initial acceleration vector, and $\vec{g}_q$ is the jerk vector. Similarly, the target denoted by $\vec{P}_0$ moves along the nonideal trajectory $\vec{P}(t)$ which has the following expression:

$$
\vec{P}(t) = \vec{P}_0 + \vec{v}_p t + \vec{a}_p t^2 + \vec{g}_p t^3 + \cdots, \tag{10}
$$

where $\vec{v}_p$, $\vec{a}_p$, and $\vec{g}_p$ are the initial velocity vector, the acceleration vector, and the jerk vector, respectively.

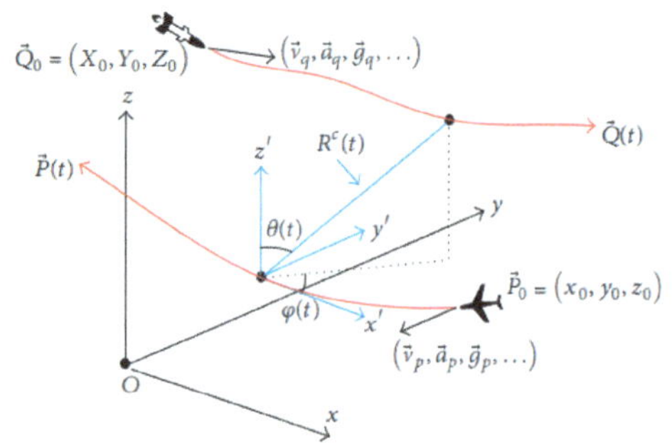

Figure 1: Monostatic missile-borne SAR geometry.

Then, the instantaneous slant range between target and radar can be expressed as

$$R^c(t) = \left|\vec{Q}(t) - \vec{P}(t)\right| = \left|\vec{r}_0 + \vec{v}_{pq}t + \vec{a}_{pq}t^2 + \vec{g}_{pq}t^3 + \cdots\right|, \quad (11)$$

where $\vec{r}_0 = \vec{Q}_0 - \vec{P}_0$ represent the initial vector between radar $\vec{Q}$ and target $\vec{P}$ and $\vec{v}_{pq}$ $\vec{a}_{pq}$, and $\vec{g}_{pq}$ refer to the relative velocity, acceleration, and jerk vector between radar and the target, respectively.

For the situation with a low relative speed, "Stop and Go" (SaG) models are generally employed to describe the motion state. As a quasistatic approximate method, this model supposes that the platform remains stationary between transmitting and receiving pulse. Figure 2 shows the simple graph of the instantaneous distance between radar and target, where the blue line represents the real-time distance and the red line is the SaG distance which can be written as

$$R^c_{\text{sa}}(t) = R^c(t_n) = \left|\vec{r}_0 + \vec{v}_{pq}t_n + \vec{a}_{pq}t_n^{\,2} + \vec{g}_{pq}t_n^{\,3} + \cdots\right|, \quad t \in \left[t_n - \frac{T_r}{2}, t_n + \frac{T_r}{2}\right]. \quad (12)$$

Then, the SaG echo signal for a simple point target can be obtained by substituting (12) into (2):

$$s_{\text{sa}}(t_n, R^c) = \gamma \cdot \text{rect}\left[\frac{t - t_n - 2R^c_{\text{sa}}(t)/c}{T_r}\right] \cdot \exp\left[-j2\pi f_0 \frac{2R^c_{\text{sa}}(t)}{c} + j\pi k\left(t - t_n - \frac{2R^c_{\text{sa}}(t)}{c}\right)^2\right]. \quad (13)$$

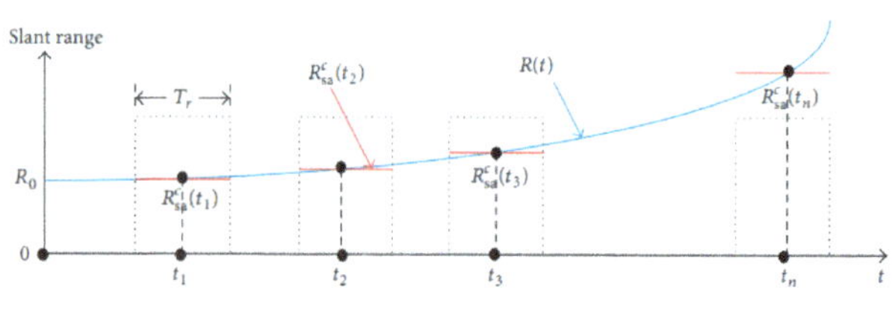

Figure 2: Range history of the SaG model.

It can be seen from the type above that the SaG method has a simple form of echo signal which costs low computational expense. However, this hypothesis ignores the distance error caused by the motion of radar and target within the pulse time. This error can be negligible at the case of low speed but could have a significant impact in the high-speed situation which will be analyzed in the following.

Focusing on the error, we can obviously draw the following conclusions.

(1) The error is minimum (0) when taking the value of SaG distance as that of the real-time distance in the middle of echo pulse signal, that is, when $t' = 0$.

(2) The error is the maximum when the sample is located at the endpoints of echo pulse signal; namely, $t' = \pm T_r/2$. The maximum range error can be expressed as

$$\Delta R_{\max} = \left|\frac{\vec{v}_{pq}T_r}{2} + \vec{a}_{pq}\left(\frac{T_r}{2}\right)^2 + \vec{g}_{pq}\left(\frac{T_r}{2}\right)^3 + \cdots\right|. \quad (14)$$

When the maximum range error is smaller than a distance resolution cell, the impact it caused can be neglected, namely,

$$\left|\frac{\vec{v}_{pq}T_r}{2} + \vec{a}_{pq}\left(\frac{T_r}{2}\right)^2 + \vec{g}_{pq}\left(\frac{T_r}{2}\right)^3 + \cdots\right| < \frac{2B}{c}, \quad (15)$$

that is, also the restricted condition of the SaG model; namely, the displacement between the target and the platform in half of the pulse time must be less than one range resolution cell.

High-Speed Cases

As mentioned above, the conventional SaG model is generally suitable under low speed cases, but there may be extreme cases for which nonnegligible error can be caused by such an approximation, for example, those involving very high relative speed between the platform and the target, or for radars exploiting long time pulse waveforms. Therefore, a new motion model with low complexity and small error should be derived to calculate the range history in extreme cases.

Based on (14), it can be concluded that $\Delta R_{\max}$ is determined by the motion parameters $(\vec{v}_{pq}, \vec{a}_{pq}, \vec{g}_{pq})$ and the time parameter T_r. Since the motion parameters are out of control, the only way to reduce the maximum range error lies in the pulse during time T_r. According to the fast time sampling, we can divide T_r into N_r parts as shown in Figure 3.

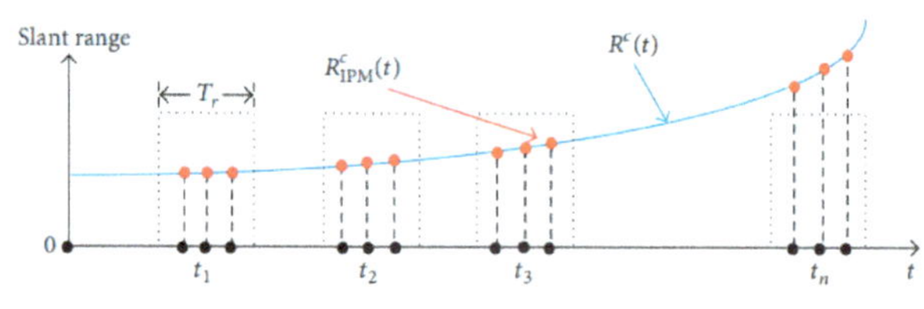

Figure 3: Range history of the IPM model.

So that the red points described new range history can be written as

$$R^c_{\text{IPM}}(t) = \left|\vec{r}_0 + \vec{v}_{pq}t(t_n, m) + \vec{a}_{pq}t(t_n, m)^2 + \vec{g}_{pq}t(t_n, m)^3\right|,$$

$$t(t_n, m) = t_n - \frac{T_r}{2} + \frac{mT_r}{N_r}, \quad m = 0, 1, 2, \ldots, N_r, \quad (16)$$

where N_r is the number of fast time sample points, m is the order number, and $R^c_{\text{IPM}}(t)$ is the simulation slant range of the new model. From the expression, it can be found that the new model takes the motion in the duration time into consideration, so it is called the Inner Pulse Motion (IPM) model.

Bistatic Radar

Now let us consider the cases of bistatic missile-borne SAR. The geometric configuration is shown in Figure 4.

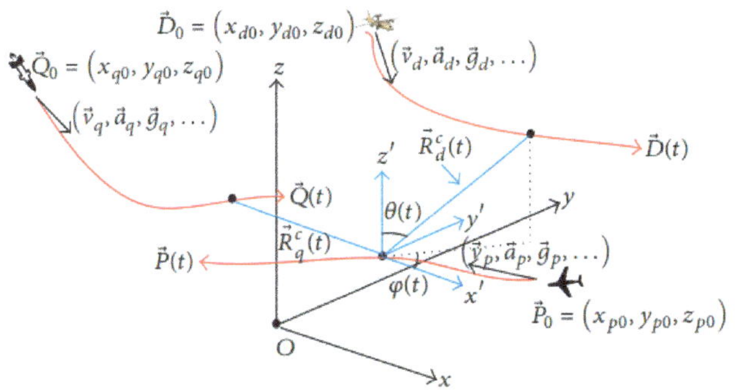

Figure 4: Bistatic missile-borne SAR geometry.

In this figure, the transmitter denoted by $\vec{Q}$ moves along the nonideal trajectory $\vec{Q}(t)$, the receiver denoted by $\vec{D}$ moves along the trajectory $\vec{D}(t)$, and the target denoted by $\vec{P}$ moves along the trajectory $\vec{P}(t)$; they are all considered to obey a high degree polynomial-type representation as a function of time:

$$\vec{Q}(t) = \vec{Q}_0 + \vec{v}_q t + \vec{a}_q t^2 + \vec{g}_q t^3 + \cdots,$$
$$\vec{P}(t) = \vec{P}_0 + \vec{v}_p t + \vec{a}_p t^2 + \vec{g}_p t^3 + \cdots,$$
$$\vec{D}(t) = \vec{D}_0 + \vec{v}_d t + \vec{a}_d t^2 + \vec{g}_d t^3 + \cdots. \quad (17)$$

Then, the equivalent instantaneous slant range can be calculated by summing up the transmitter distance and the receiver distance as follows:

$$R^c_q(t) = \left|\vec{Q}(t) - \vec{P}(t)\right| = \left|\vec{R}^c_{q0} + \vec{v}_{pq} t + \vec{a}_{pq} t^2 + \vec{g}_{pq} t^3 + \cdots\right|,$$
$$R^c_d(t) = \left|\vec{P}(t) - \vec{T}(t)\right| = \left|\vec{R}^c_{d0} + \vec{v}_{dp} t + \vec{a}_{dp} t^2 + \vec{g}_{dp} t^3 + \cdots\right|,$$
$$R^c_{\text{bis}}(t) = \frac{\left[R^c_d(t) + R^c_q(t)\right]}{2}, \quad (18)$$

where $R^c_q(t)$ is the distance between transmitter and target, $R^c_d(t)$ is the distance between receiver and target, and $\vec{v}_{pq}$, $\vec{a}_{pq}$, and $\vec{g}_{pq}$ are the relative velocity vector, acceleration vector, and jerk vector between the transmitter and the target, respectively, while $\vec{v}_{dp}$, $\vec{a}_{dp}$, and $\vec{g}_{dp}$ are those between the target and the receiver.

Range-Doppler Algorithm

So far, we have completed the echo generation for complicated moving target; next, the typical RDA (range-Doppler algorithm) is applied in this section for data processing. The specific flow diagram is shown as seen in Figure 5.

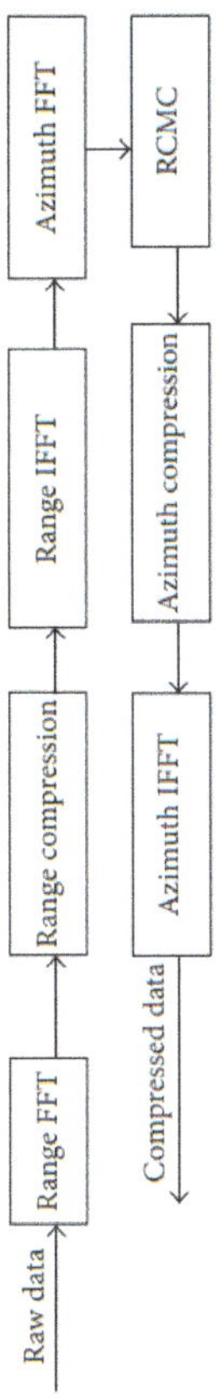

Figure 5: Flowchart of RDA.

The SAR image of one point target is simulated to verify the algorithm efficiency. Then, we get Figure 6(a)showing real part of the echo signal; Figure 6(b) shows the range compression results of echo signal; there is one object simulated on the scene, and the range mitigation can be seen clearly. Figure 6(c) shows the time-domain signal after range migration correction, and Figure 6(d) shows the final SAR image after range migration correction and azimuth compression in which the point target can be clearly distinguished.

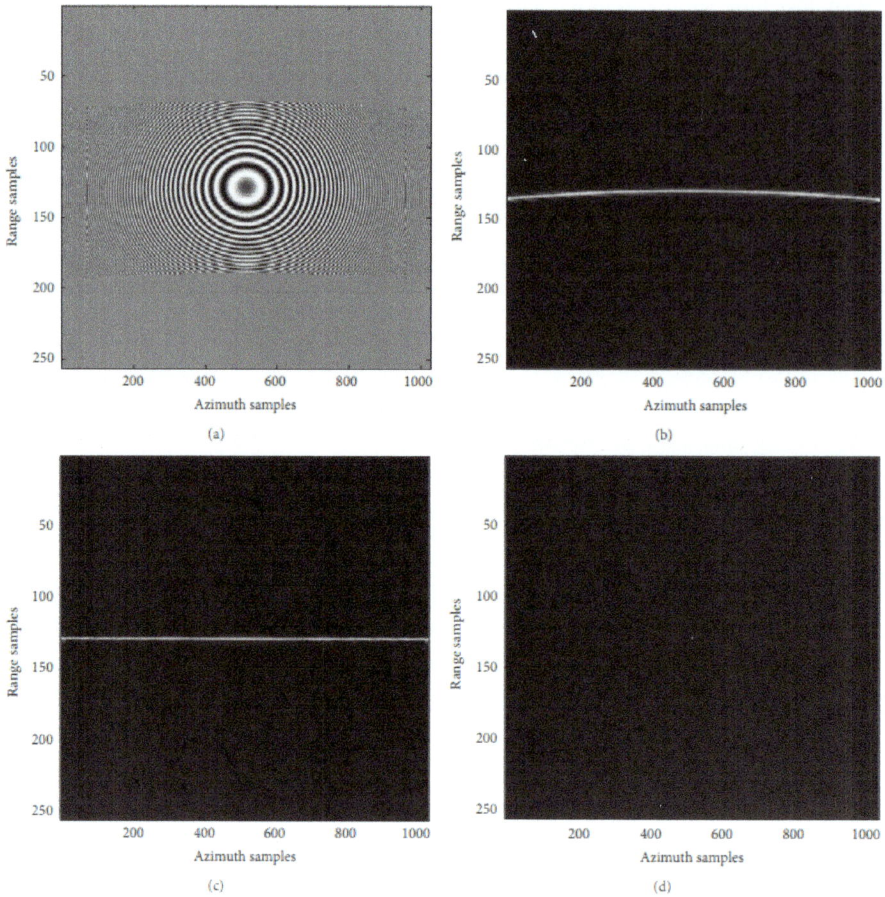

Figure 6: RDA validation for the point target. ((a) The real part of the raw signal; (b) the range compression results of echo signal; (c) the range-Doppler image; (d) the SAR image.)

It must be noted that in this paper we focus our research on the SAR echo and image simulation, not with the SAR imaging algorithm. The RDA is employed here to further verify the results of echo simulation with given parameters of the simulation, while the Doppler parameters estimation and the range cell migration correction (RCMC) it includes are beyond the scope of our research.

Implementation of the Proposed Method

Based on the theoretical analysis in the previous chapter, the block scheme of the proposed echo generation and image simulation method for SAR is shown in Figure 7.

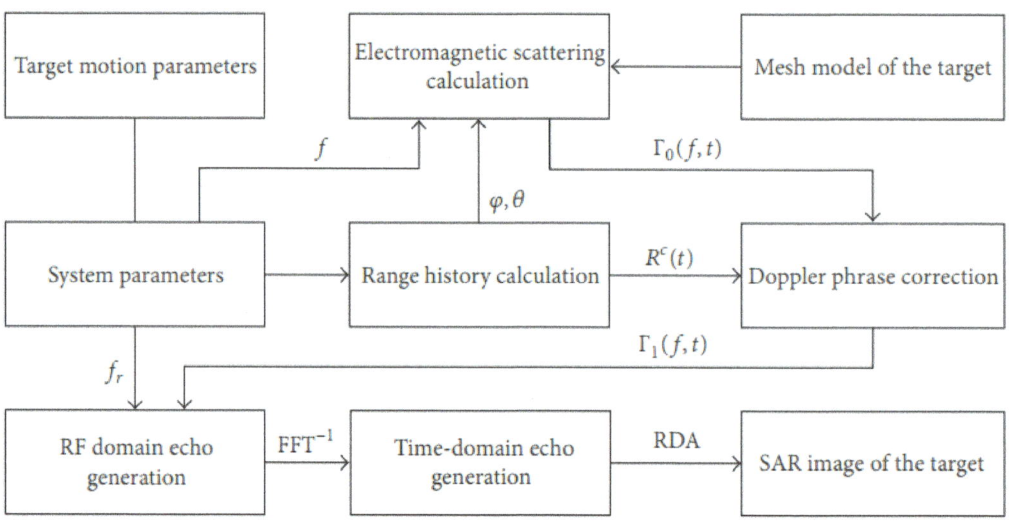

Figure 7: Block scheme of the simulation method.

Consequently, the basic steps are outlined as follows:(1) Set the SAR system parameters and the motion parameters, including bandwidth, sampling rate, and pulse repetition frequency.(2)Get the 3D model and the mesh model of the target. To ensure electrical connectivity, triangles must therefore share an edge. Similarly, segments must connect to other segments at nodes or to mesh triangles at vertices.(3)For range history analysis according to the systems parameters and motion parameters, calculate the range history along with the azimuth and the pitching angle of the target relative to the SAR platform.(4)Calculate the corresponding RCS matrix using the method described in the second part of the Section2.2.(5) Calculate the echo data with the scattering characteristic of the target in the frequency domain and then get the original time-domain echo data through inverse FFT.(6)Finally, the output raw signal is subsequently processed by RD algorithm to obtain the SAR images, in order to prove the validity of the simulator.

SIMULATIONS RESULTS

Simulations and Results

Raw signals simulation, relevant first to single scattering point and then to complex target, is now presented in this section to test the effectiveness of the simulator we proposed.

Point Target Verification

Firstly, in order to show the precision of our algorithm, we consider a fixed single point scatterer placed at the center of the irradiation belt. For such a scatterer, the SAR raw signal can be exactly computed directly in time domain. Accordingly, it is possible to compare the raw signal simulated via the approach we proposed to the exact one. The main parameters are listed in Table 1.

Table 1: Simulation parameters.

Initial distance	10 km	Pitching angle	90 deg
Pulse duration	1 us	Squint angle	0 deg
Carrier frequency	1 GHz	Pulse bandwidth	100 MHz
Number of the azimuth samples	1024	Number of the range samples	256
Pulse repetition frequency	200 Hz	Fast time sampling rate	120 MHz

First of all, the simulated SAR raw data using TD method and our RFAT method are depicted in Figures 8(a)and 8(b). Two types of raw data look quite like each other, which confirm the effectiveness of our proposed algorithm.

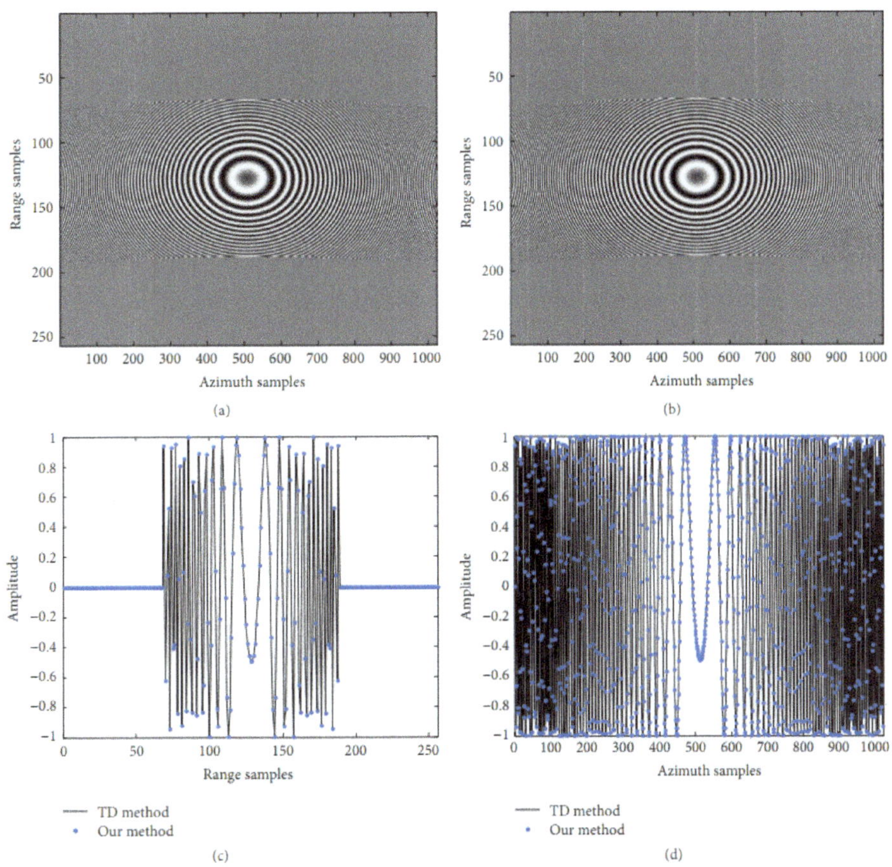

Figure 8: Simulated SAR raw data using different methods: (a) TD method; (b) our method; (c) range cut; (d) azimuth cut.

To compare the raw data results clearly, the central signals of both the range direction and the azimuth direction are shown, respectively, in Figures 8(c) and 8(d), which are extracted from Figures 8(a) and 8(b)signal and intuitively reflect the linear frequency modulation (LFM) characteristic. By contrast, we can observe that the two lines almost superposed each other, which proves the accuracy of the proposed method.

Complex Target Simulation

A plane model, with the size of 7.4 m × 8.8 m × 1 m shown in Figure 9, is utilized as a complex target to generate its raw data in this section. Besides, the monostatic case, high-speed case, and bistatic case are all implemented.

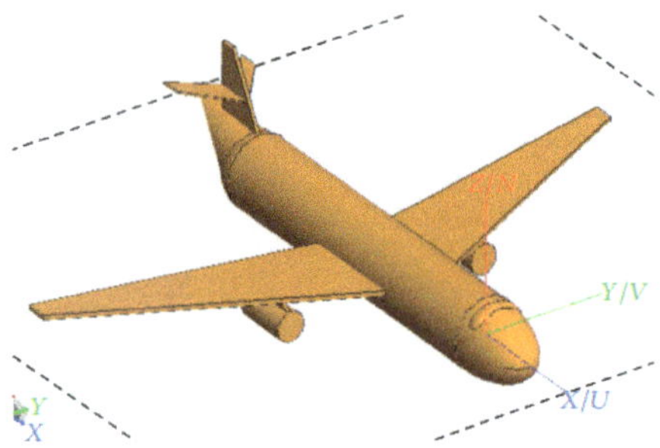

Figure 9: The 3D model of the plane.

(a) Monostatic SAR Cases. First of all, in the case of rectilinear motion, consider the same typical X-band missile-borne SAR system of Figure 1, whose main parameters are listed in Table 2. It should be noted that, in this part of simulation, the x-axis is defined as the azimuth direction, the y-axis represents the range direction, and the -axis is the vertical direction.

According to the basic theory of SAR, the distance resolution and the azimuth resolution after pulse compression are, respectively, 0.25 m and 0.5 m.

The simulation results are shown in Figure 10. Between them, (a) is the real part of the generated raw data and (b) is the obtained SAR image, which clearly reflects the structure information of the aircraft including the fuselage and wings, and the plane in the image tilts up to 10 degrees which matches well with the actual posture.

Table 2: Simulation parameters.

Initial distance	2.0 km	Pitching angle	75 deg
Pulse duration	0.4 us	Squint angle	10 deg
Carrier frequency	10 GHz	Pulse bandwidth	600 MHz
Synthetic aperture time	0.3584 s	Antenna length	0.98 m
Azimuth samples number	163	Range samples number	252
Relative velocity vector	(150, 0, 0) m/s	Relative acceleration vector	(0, 0, 0) m/s^2

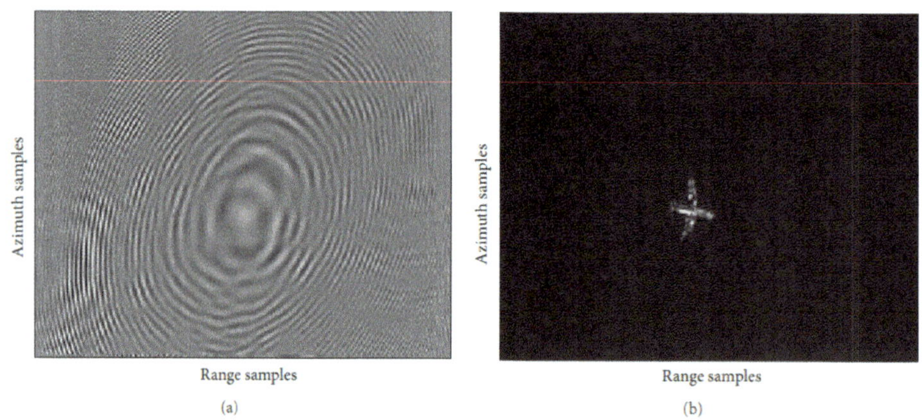

Figure 10: The simulated echo (a) and image (b) results of the plane model in the case of uniform linear motion.

Then, in the case of curvilinear motion, the curve track is shown in Figure 11.

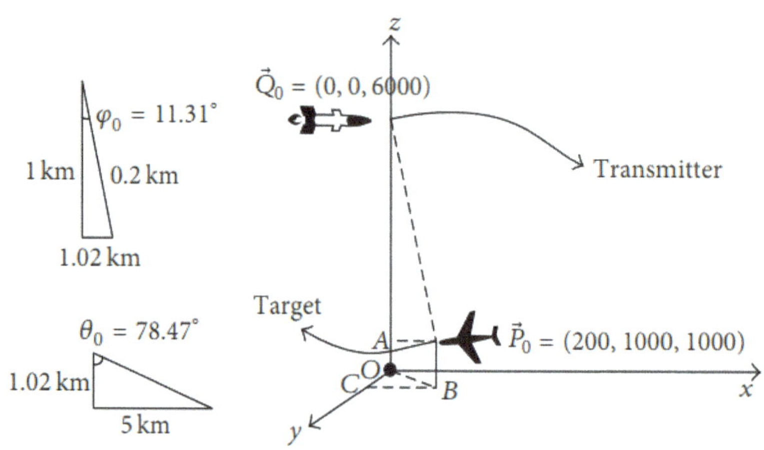

Figure 11: The trajectory diagram in the case of nonlinear motion.

It can be seen from Figure 11 that the missile platform locates at (0, 0, 6000) initially, and the plane target is at (200, 1000, 1000). By solving the triangle AP_0Q_0 and OCB, we obtain the initial azimuth angle and pitching angle which, respectively, are $\varphi_0 = 11.31^\circ$ and $\theta_0 = 78.47^\circ$. Besides, the relative velocity is set as $\vec{v}_{pq} = (100, 20, 10)$ m/s, and the relative acceleration is set as $\vec{a}_{pq} = (10, 10, 10)$ m^2/s.

The results of echo signal and SAR image are, respectively, shown in Figure 12. It is easy to distinguish a plane shape in Figure 12(b). By comparing Figures 10(b) and 12(b), we can see that, in the case of curve track, the simulated SAR image is fuzzier; this is due to the nonuniform sampling caused by the nonideal movement.

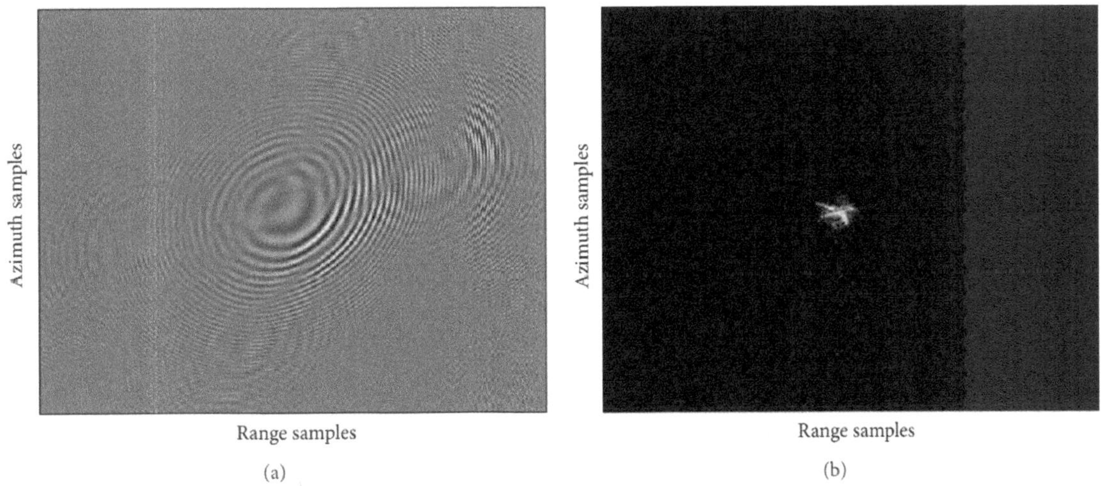

Figure 12: The simulated echo (a) and image (b) results of the plane model in the case of nonlinear motion.

(b) High-Speed Cases. We also performed a series of high-speed missile-borne SAR simulation experiments. The system also works in X band. The bandwidth of the transmitted chirp signal is 1 GHz; PRF is 2000 Hz. In the vertical direction, the relative height is initially 6000 m and the relative velocity is 0 m/s, while the acceleration is controlled to be 20m/s^2. In the azimuth direction, the relative velocity is set to be 1000 m/s, and both the initial distance and the acceleration are zero. Finally, in the range direction, the relative velocity and acceleration are also set to be zero. At the imaging moment, the pitching angle and the azimuth angle are about 50° and 60°, respectively.

Figure 13 shows the simulated SAR echo result and image result. Between them, (a) is the real part of the raw data and (b) is the SAR image obtained by processing the raw data with the RD algorithm. It can be seen from Figure 15(b) that a plane was flying in the direction of the 30 degrees.

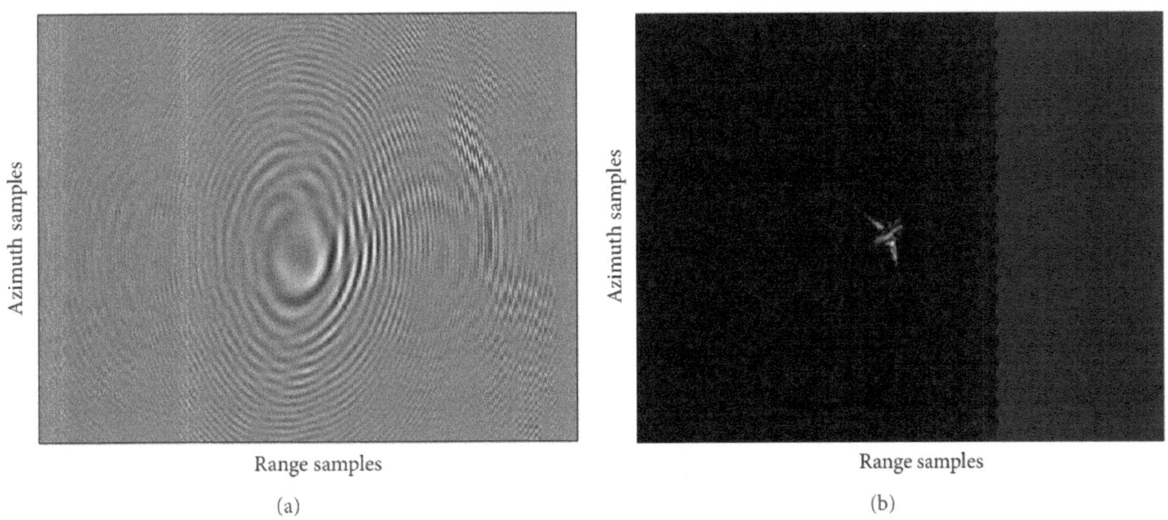

Figure 13: The simulated echo (a) and image (b) results of the plane model in the case of high- speed.

(c) Bistatic SAR Cases. Finally, we implemented a missile-borne bistatic SAR simulation. In this simulation, the receiver and transmitter travel along different but parallel tracks with constant and equal velocities which are controlled to be 150 m/s, and they share the same squint angle of 0°. The bistatic angle achieves 53.14°, which means that the system has quite a high bistatic degree as well. The heights of the transmitter and receiver both are 3000 m, respectively. The geometry configuration is shown in Figure 14.

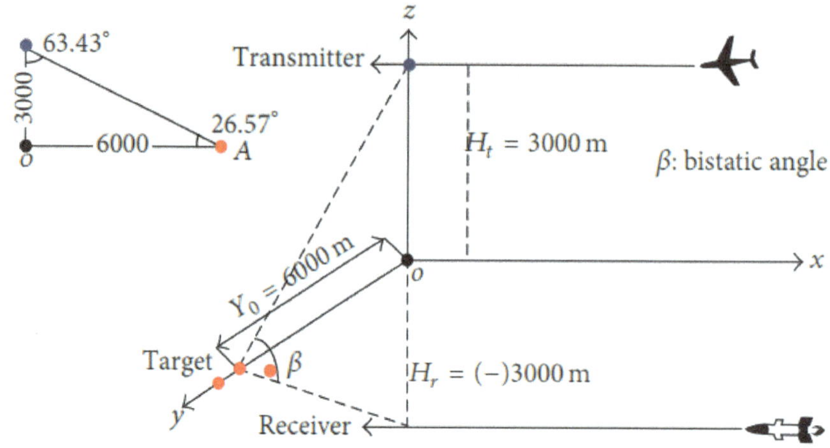

Figure 14: The geometry diagram in the simulation of bistatic case.

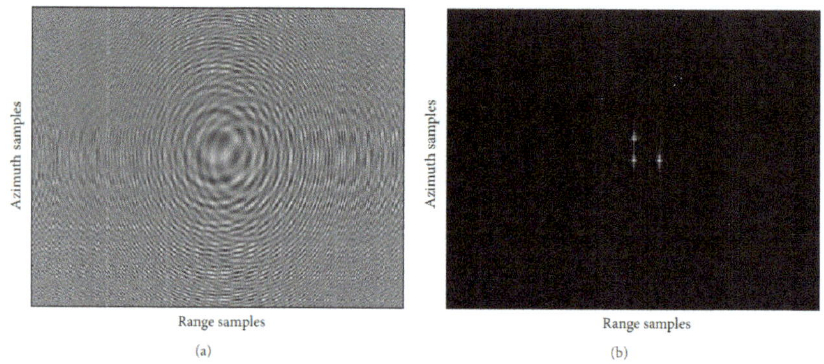

Figure 15: The simulated echo (a) and image (b) results of the plane model in the case of bistatic SAR.

The simulated scene consists of three point scatterers, whose locations and velocity are shown in Table 3. Scatterer A locates in the center of the scene; A and B locate in the same instantaneous Doppler frequency but different range cells, while A and C have the same range cell but different instantaneous Doppler frequency.

Table 3: The velocity and location of the three point scatterers.

Point scatterer	Location	Velocity along the x-axis direction
A	$(0, 6000, 0)$	50 m/s
B	$(0, 6010, 0)$	50 m/s
C	$(10, 6000, 0)$	50 m/s

Figure 15(a) shows the real part of the simulated raw data matrix; Figure 15(b) shows the imaging result where we can see that all the point scatterers are well focused at the expected position.

Computational Efficiency Analysis

Compared with the TD method, the proposed simulator has the advantage of computational efficiency which can be expressed as

$$C_{\mathrm{Tol}} = N_r N_a C_R + \frac{1}{2N_r N_a \log_2(N_r)} + N_r N_a, \tag{19}$$

where $1/2N_r N_a \log_2(N_r)$ originates the IFFT operation, $N_r N_a$ originates the dot product of the matrix, and C_r represents the complexity for computing single frequency RCS of the whole target which determines the whole processing time to a large extent. For the above-mentioned plane model in Figure 9 (with the size of 7.4 m × 8.8 m × 1 m), the single frequency RCS calculating time reaches to 2.54 s in the case of a computer with 8 cores. In fact, we conduct our simulation along with 10 programs in a computer with 16 cores; in this case, to obtain the simulation results of Figure 10, the total processing time is about

$$2.54 \times 252 \times 163/10/2/3600 \approx 1.45\,\text{h}. \quad (20)$$

It is clear that the proposed method is much more computationally efficient than the TD method; however, as a tradeoff, it is more time consuming than the 2D method. In particular, the improving calculation condition like the cluster technology could make the simulation of larger targets possible in a more reasonable time.

CONCLUSION

To support SAR system design for target recognition, a novel RFAT SAR simulation method which realized the transformation from geometric model to SAR image has been presented. The method not only is faster compared with the TD simulator but also can be useful in applications where a 2D simulator cannot be used. The simulation results prove that the wide radar scattering characteristic of target has a great impact on the SAR imaging. Different target postures and different SAR system can cause a huge difference in the SAR image, which further demonstrates the difficulty of high-resolution SAR image interpretation. Thus, a lot more research is needed in the SAR imaging mechanism.

Acknowledgments

The work was supported by "The Fundamental Research Funds for the Central Universities" (no. NS2016040).

REFERENCES

1. I. G. Cumming and F. H. Wong, Digital Processing of Synthetic Aperture Radar Data: Algorithms and Implementation, Artech House, Norwood, Mass, USA, 2005.
2. G.-O. Glentis, K. Zhao, A. Jakobsson, H. Abeida, and J. Li, "SAR imaging via efficient implementations of sparse ML approaches," Signal Processing, vol. 95, no. 2, pp. 15–26, 2014.
3. B. Deng, X. Li, H. Wang, Y. Qin, and J. Wang, "Fast raw-signal simulation of extended scenes for missile-borne SAR with constant acceleration," IEEE Geoscience and Remote Sensing Letters, vol. 8, no. 1, pp. 44–48, 2011.
4. A. Mori and F. De Vita, "A time-domain raw signal simulator for interferometric SAR," IEEE Transactions on Geoscience and Remote Sensing, vol. 42, no. 9, pp. 1811–1817, 2004.
5. S. Cimmino, G. Franceschetti, A. Iodice, D. Riccio, and G. Ruello, "Efficient spotlight SAR raw signal simulation of extended scenes," IEEE Transactions on Geoscience and Remote Sensing, vol. 41, no. 10, pp. 2329–2337, 2003.
6. L. Yang, W. Yu, S. Zheng, and L. Zhang, "Efficient bistatic SAR raw signal simulator of extended scenes,"International Journal of Antennas and Propagation, vol. 2014, Article ID 130784, 9 pages, 2014.
7. Y. Wang, Z. Zhang, and Y. Deng, "Squint spotlight SAR raw signal simulation in the frequency domain using optical principles," IEEE Transactions on Geoscience and Remote Sensing, vol. 46, no. 8, pp. 2208–2215, 2008.
8. Y. Tian, S. Guo, Y. Wang, et al., "A novel geo-SAR echo simulation method," in Proceedings of the 10th European Conference on Synthetic Aperture Radar (EUSAR '14), pp. 1–4, VDE, Berlin, Germany, June 2014.
9. B.-N. Wang, F. Zhang, and M.-S. Xiang, "SAR raw signal fast algorithm in mixed domain," Journal of Electronics and Information Technology, vol. 33, no. 3, pp. 690–695, 2011.
10. G. Franceschetti, R. Guida, A. Iodice, D. Riccio, and G. Ruello, "Efficient simulation of hybrid stripmap/spotlight SAR raw signals from extended scenes," IEEE Transactions on Geoscience and Remote Sensing, vol. 42, no. 11, pp. 2385–2396, 2004.
11. G. Franceschetti, A. Iodice, S. Perna, and D. Riccio, "SAR sensor trajectory deviations: fourier domain formulation and extended scene simulation of raw signal," IEEE Transactions on Geoscience and Remote Sensing, vol. 44, no. 9, pp. 2323–2333, 2006.
12. X. Liu, L.-R. Zhang, and Y. Zhou, "2-D Fourier domain fast algorithm for UWB ground penetrating SAR raw signal simulation," Journal of Chang'an University (Natural Science Edition), vol. 34, no. 4, pp. 109–114, 2014. ·
13. X. Liu, C. Zeng, J. Wang, and W. Zhao, "Improved method for SAR echo signal simulation using two-dimensional frequency domain transformation," Journal of Xidian University, vol. 40, no. 3, pp. 42–49, 2013.
14. R.-J. Li, K.-F. Ji, H.-X. Zou, et al., "Simulation of SAR imagery of target based on electromagnetic scattering characteristic computation," Radar Science and Technology, no. 5, pp. 395–405, 2010.
15. J. Qin and Z. M. Zhang, "Implementation and optimization of SAR echo simulation based on GPU,"Science Technology and Engineering, vol. 14, no. 3, pp. 85–89, 2014.
16. G. Franceschetti, A. Iodice, D. Riccio, and S. Perna, "Efficient simulation of airborne SAR raw data of extended scenes," IEEE Transactions on Geoscience and Remote Sensing, vol. 44, no. 10, pp. 2851–2860, 2006.
17. Z. Meng, Y. Li, C. Li, M. Xing, and Z. Bao, "A raw data simulator for Bistatic Forward-looking High-speed Maneuvering-platform SAR," Signal Processing, vol. 117, pp. 151–164, 2015. · ·
18. R. G. Kouyoumjian, L. Peters, and D. T. Thomas, "A modified geometrical optics method for scattering by dielectric bodies," IEEE Transactions on Antennas and Propagation, vol. 11, no. 6, pp. 690–703, 1963. · ·
19. X. Weijie and W. Xiaojie, "SAR image simulation and verification for urban structures," International Journal of Electronics, vol. 103, no. 2, pp. 247–260, 2016. ·
20. K. M. Mitzner, "Incremental length diffraction coefficients," Tech. Rep. AFAL-TR-73-296, Aircraft Division, Northrop Corporation, 1974.
21. C. Ning, C.-Z. Dong, J. Huang, and X.-Y. Zhang, "Simulation of dynamic RCS data on space flight target,"Computer Engineering and Design, vol. 35, no. 4, pp. 1367–1371, 2014.

CHAPTER 21

ELECTROMAGNETIC ACOUSTIC TRANSDUCERS

Maria Kogia, Liang Cheng, Abbas Mohimi, Vassilios Kappatos, Tat-Hean Gan, Wamadeva Balachandranand Cem Selcuk

Keywords: Electromagnetic Acoustic Transducer (EMAT); guided waves; solar thermal plants; non-destructive testing; high temperatures

INTRODUCTION

Solar thermal industry is an environmentally friendly means of power generation, converting solar energy into electricity. Concentrated Solar Plants (CSPs) are employed in the solar thermal industry for collecting sunlight and converting it firstly into heat and later into electricity. There are several types of CSPs such as solar towers, dish concentrators, and parabolic trough CSPs; the latter is the most widely used [1] and is shown in Figure 1a.

The two essential components of parabolic trough CSPs are the long, parabolic trough shaped reflectors and the absorber tubes. The reflector collects and focuses the sunlight to the absorber tube, which is located at the focal point of the reflector and runs its whole length. The absorber tubes are composed of long, thin, stainless steel tubes covered by a glass envelope under vacuum, as shown in Figure 1b,c. Their entire surface should be exposed to sunlight and therefore there are few access points for any extra hardware to be attached to them. Inside the tubes, working fluid, either water/molten salt or synthetic oil, is flowing, absorbing the heat of the sunlight and transmitting it into heat/steam engines for the final stages of the power generation procedure.

The high temperatures (400–500 °C), the contraction/expansion endured due to the fast cooling of the tube at high temperatures, and the vibrations can result in the appearance of defects either on the surface or inside the absorber tube. Problems with absorber tubes have been reported [5,6,7], such as creep, thermo-mechanical fatigue, and hot corrosion. The turbulent mixing of hot and cold flow streams of the working fluid may result in temperature variations along the pipe and consequently in thermomechanical fatigue [8,9,10,11]. Corrosion is another common problem absorber tubes may suffer from [12,13,14]; local pitting corrosion can cause the initiation of stress corrosion cracking or result in small-scale leaks. Most stainless steel pipes are vulnerable to pitting corrosion and stress corrosion cracking as well [15]. The low flow of the working fluid may also lead the absorber tube to overheat; consequently, the absorber tube may be subject to creep damage, thermal oxidation, softening, and/or stress rupture [16].

Non Destructive Testing (NDT) techniques are suggested to be applied for the structural health monitoring or inspection

Citation: Maria Kogia, Liang Cheng, Abbas Mohimi, Vassilios Kappatos, Tat-Hean Gan, Wamadeva Balachandran and Cem Selcuk, Electromagnetic Acoustic Transducers, doi:10.3390/app5041715.

of the absorber tubes. Several NDT techniques can be used for either the inspection or the monitoring of high temperature structures; Acoustic Emission (AE), Eddy Current (EC), Holographic Interferometry, Laser Ultrasonic, Guided Wave Testing (GWT), and Infrared Thermography (IR) are some of those. These techniques have been reported for operating at temperatures up to 300 °C, though with shortcomings. Some of the drawbacks are qualitative results, sensitivity to noise, laboratorial utilization, and the need for coupling media [17,18,19,20]. AE is a passive NDT technique that has been widely deployed in Structural Health Monitoring (SHM); it monitors the elastic waves generated after the initiation or propagation of a crack [21]. Nevertheless, AE is sensitive to noise and gives qualitative results. EC is a non-contact technique that can be employed for the inspection of any electrically conductive material; however, it is subject to the skin effect, leading it to be mainly used for the detection of surface and subsurface defects [22]. Holographic Interferometry can give detailed results and be used for the detection of small defects, but it is mainly used for laboratorial tests as its setup is complicating and it is sensitive to vibrations [23]. Laser Ultrasonic has a small and adjustable footprint; therefore, it can be used for the inspection of irregular surfaces and samples of small and complex geometry. It induces high frequency ultrasound and thus very small defects can be detected as well. However, its setup is complicated and it is mainly used for laboratorial tests. GWT is used for the inspection or monitoring of large structures; mainly piezoelectric transducers are employed upon the structure being tested exciting/receiving guided waves [24]. Piezoelectric transducers require direct access to the specimen and a coupling medium (usually a water-based gel), and their response cannot travel through a vacuum. IR can be used in this application mainly for overheating identification; however, the length of the pipes makes this technique practically inefficient. The camera needs to scan the whole length of the pipe, which is time-consuming and may not be possible while the absorber tube is operating [25].

The monitoring of absorber tubes requires a non-contact NDT technique that can be applied at high temperatures without the need for a couplant and can inspect the whole length of the tube from a single point. Therefore, GWT can be applied with the use of non-contact transducers that can withstand high temperatures and excite/receive guided waves and more particularly a T(0,1) wave mode (or SH_0) [26,27]. The EMAT can be used for this application, since it is a non-contact technology that has been used in GWT and can be applied for the inspection of structures under hostile conditions such as moving specimen and elevated temperatures.

Figure 1. (**a**) Parabolic Trough Concentrated Plants [2]; (**b**) reflector and absorber tube [3]; (**c**) absorber Tube [4].

In the following section the main operating principles of EMATs are described and emphasis is also placed on the potentials and limitations of EMATs on guided waves and high temperatures. A brief description of this study follows. The results obtained from the theoretical study are shown in Section 4, where the dispersion curves of the absorber tubes were calculated and processed and a PPM EMAT with racetrack coil has been evaluated mainly regarding its guided wave purity characteristics at both room and high temperatures. In Section 5 and Section 6 the experimental setup and the results of the experimental evaluation of a pair of PPM EMATs are presented. The EMATs were tested as far as their efficiency to excite SH_0 is concerned and the parameters that may affect their performance at both room and high temperatures. The EMATs were validated for their wave purity, their sensitivity to noise, their lift-off limitations, their power requirements, and their performance at high temperatures while they were exciting/receiving guided waves. Hence, a comparison between the simulated and the experimental results is finally presented. A high temperature EMAT exciting SH_0 wave and operating up to 400–550 °C for either monitoring or long-term inspection is also to be accomplished in the near future.

EMAT TECHNOLOGY

EMAT is a non-contact technology of transducers that can be used for the inspection of moving structures or a specimen that operates at elevated temperatures. Their response can travel through a vacuum as well, making them even more suitable for this application compared to piezoelectric transducers. EMATs can excite/receive guided waves and thus they can be employed in GWT for either the inspection or the monitoring of large structures [27,28,29,30]. Hence, high-temperature EMATs may be used for the inspection or monitoring of absorber tubes. Nevertheless, the high-temperature EMATs that have been reported so far have been designed for thickness measurements [31,32,33]. Therefore, a high-temperature EMAT that excites/receives guided waves for the long-term inspection of high-temperature structures is still required.

A typical EMAT transducer is composed of either a permanent magnet or an electromagnet for the generation of a static magnetic field, and a coil. The coil is driven by an alternating current that generates a dynamic magnetic field. This dynamic magnetic field induces an eddy current in the specimen placed below the coil. If the specimen being tested is an electrical conductor and non-ferromagnetic, mainly Lorentz force is exerted upon the material particles; Equation (1) shows that Lorentz force is equal to the product of eddy current density and the overall magnetic field:

$$F_L = Je \times (B_{st} + B_{dyn}) \quad (1)$$

where F_L is Lorentz force, J_e is the eddy current density, B_{st} stands for the static magnetic field, and B_{dyn} refers to the dynamic magnetic field. In this application, the absorber tubes are made of 316 L stainless steel, which is paramagnetic; therefore the main force generated in them would be Lorentz force.

The EMAT configuration and the material properties of its main components affect EMAT performance at both room and high temperatures. The material the coil is made of influences the EMAT performance. The impedance of the coil affects the energy transmitted to the specimen and the SNR/quality of the signal received; Equation (2) shows how the resistance increases with temperature rise and Equation (3) demonstrates the relationship between the temperature and the noise level of the signal received:

$$R(T) = R_0[1 + \alpha(T - T_0)] \quad (2)$$

where R_0 is the resistance of a single turn coil, α is the temperature coefficient of resistance, and T_0 is the room temperature.

$$V_{Noise} = (4K\beta TR_{EMAT})^2 \quad (3)$$

where K is the Boltzmann constant, T is the temperature measured in Kelvin, β is the bandwidth, and R_{EMAT} is the resistance of the EMAT coil per turn. Consequently, the coil should be preferably made of a low-resistance material such as copper. However, copper is known to oxidize at high temperatures and, therefore, other materials could be used for this application such as silver, platinum, nickel, or constantan. Hernandez-Valle (2011) has already made an electromagnet EMAT that can operate at up to 600 °C without any cooling system for thickness measurements and apart from designing a high temperature electromagnet he has also tested several coil designs at high temperatures [32].

The permanent magnets also have limitations regarding their maximum operating temperature. The Maximum Operating Temperature (MOT) of a magnet is equal to half of its Curie Point; beyond Curie Temperature, the strength of the magnet decreases rapidly. Consequently, high Curie Point magnets are preferable for high-temperature applications. Nevertheless, the magnetic strength of the main two types of high-temperature magnets, Alnico (with a maximum operating temperature of 500 °C) and SmCo (with a maximum operating temperature of 300 °C), is smaller than the magnetic strength of Neo (NdFeB) magnets. However, NdFeB magnets cannot be used at temperatures higher than 200 °C.

A cooling system may also be required so that both the magnets and the coil will operate efficiently. Idris *et al.* have designed and tested a water-cooled EMAT that can obtain signals up to 1000 °C for thickness measurements. The EMAT was exposed to the heat source for as much time as it needed for the signal to be recorded and then it was removed [33]. Oil- and air-cooled EMATs have also been designed. Generally, the reported high-temperature EMATs seem to operate at high temperatures for short periods of time, which are suitable for inspection, but they cannot be used in long-term condition monitoring. Consequently, an EMAT operating at high temperatures (500 °C) for long periods of time is required and could be used for the structural health monitoring of high-temperature structures such as the absorber tubes.

Our METHODOLOGY

This study is divided into two main parts, the theoretical validation of the PPM EMAT design and the experimental evaluation of a pair of PPM EMATs. Figure 2 shows the schematic of the methodology followed in this study. In the theoretical part, the dispersion curves of a 3 mm thick, 316 L stainless steel plate were calculated for both room and high temperatures. The resonant frequency of EMAT and the wave velocity of SH_0 at the resonant frequency of EMAT for room temperature, 60 °C, 100 °C and 180 °C are also calculated. Electromagnetic simulations in COMSOL were also carried out for the theoretical validation of this EMAT design regarding its wave purity characteristics at both room and high temperatures. Experiments were conducted with a pair of PPM EMATs with racetrack coil on a stainless steel plate. During the experiments the EMATs were tested regarding their wave purity potentials, their lift-off limitations, their power requirements, and their performance at high temperatures (up to 180 °C). The results obtained from the experimental procedure were compared with the results from the theoretical validation of the EMAT and the conclusions of this study are summarized and presented in the last part of this paper.

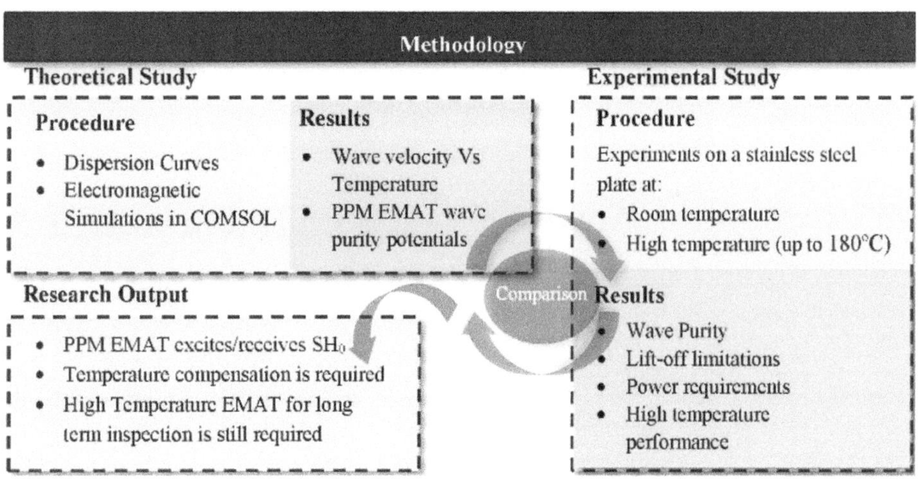

Figure 2. Methodology schematic.

THEORETICAL STUDY

Dispersion Curves

GWT is an NDT technique that can be employed in this applicaton. Low-frequency waves can travel big distances without being significantly attenuated; however, their relatively large wavelength limits the size of defect that can be detected. A 10 mm length of defect was set as the minimum size defect that should be detected; considering that the wavelength should be smaller than the double of the minimum size defect, in this case the wavelength should be smaller than 20 mm. A SH_0 wave of 12 mm wavelength was introduced to this structure.

As the wave propagates inside the specimen, it strikes at the boundaries of the medium, resulting in the change of its waveform. Consequently, there is an infinite number of possible wave modes that may appear in the material; their velocity changes in respect not only to the material properties of the specimen but to its geometry as well. Hence, the same mode at a different frequency may propagate with a different mode shape and velocity. This phenomenon is called dispersion; most of the wave modes in guided waves are dispersive Nevertheless, T(0,1) (or, alternatively, SH_0 on plates) is not dispersive, making the interpretation of the signal received less complicated. Its displacement is also in-plane, making it more suitable for this application, since the wave will propagate all along the pipe without it being affected by the working fluid that flows inside the pipe; T(0,1) cannot propagate in liquids. The number of wave modes that may appear in a plate is also smaller compared to a pipe and therefore the interpretation of the signal received from a plate may be less complicated as well.

In this preliminary study, the EMAT is simulated and designed for exciting SH_0 waves, which will propagate axially on a 316 L stainless steel plate. The dispersion curves of a 3 mm thick, 316 L stainless steel plate of 8000 kg/m^3 density, 195 GPa Young's modulus, and 0.285 Poisson ratio have been calculated and demonstrated in Figure 3. This figure shows that the wavelength curve crosses the SH_0 curve at 256 kHz frequency and thus this should be the resonant frequency of the EMAT. The wavelength curve also crosses the S_0, A_0, and A_1 curves; Table 1 shows at which frequency the wavelength curve crosses each wave mode curve and their wave velocity at these frequencies. Hence, an EMAT designed to excite SH_0 of 12 mm wavelength should be driven with an AC current of 256 kHz frequency. Otherwise if the EMAT is tuned to any other frequency, it will be likely for it to excite Lamp waves instead of SH_0. This is more possible if the coil design is meander instead of racetrack, for meander coils are used for Lamp waves as well. At 256 kHz, the SH_0 and the A_0 have the same group velocity. As a result, both wave modes can be excited/ received from the EMAT, simultaneously resulting in a more complicated signal, since A_0 is dispersive at this frequency. Nevertheless, the orientation of the displacement of each wave mode is different; the SH_0 has an in-plane displacement while the A_0 has an out-of-plane displacement. Thus, a further study should be conducted regarding the wave mode, and more particularly the displacement to which the PPM EMAT is sensitive. Consequently, an electromagnetic simulation calculating and showing the amplitude and the direction of the wave modes generated by a PPM EMAT is needed.

Material properties such as the Young's Modulus, Poisson ratio, and density of a specimen change with temperature rise as well, resulting in variations in the velocity of the propagating wave mode. The Young's modulus and density of the specimen decrease with an increase in temperature, while the Poisson ratio increases. These changes result in a decrease in the velocity of the SH_0. Table 1 summarizes the wave velocity of SH_0, S_0, A_0, and A_1 at room temperature, 60 °C, 100 °C, and 180 °C. Hence, as the temperature rises and the wave velocity decreases, any reflections received from the EMAT should shift in time. Hence, the temperature should be kept stable during the inspection or temperature compensation should take place during the signal interpretation.

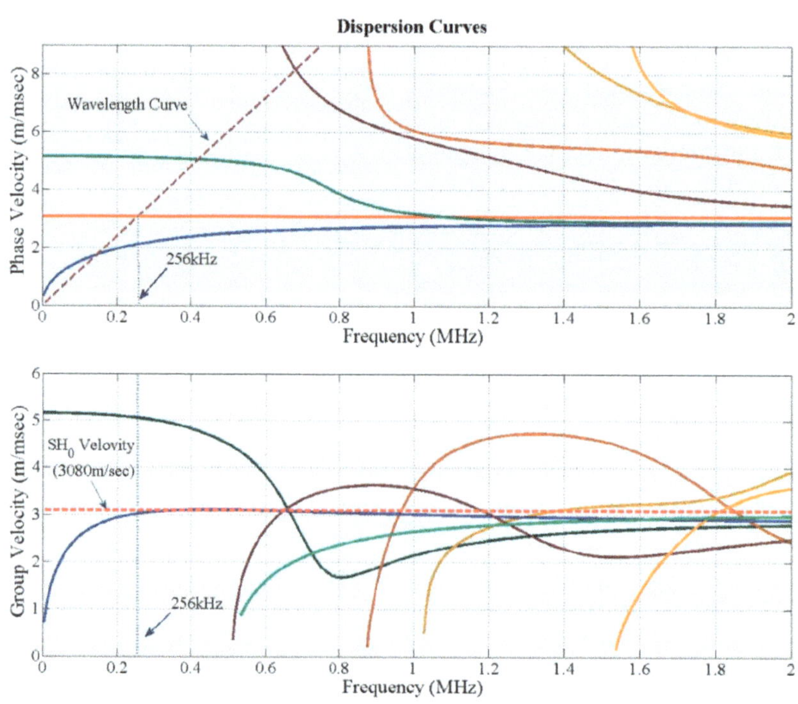

Figure 3. Dispersion curves of a 3 mm thick 316 L stainless steel plate.

Table 1. Dispersion Curves, wave velocity, and frequency.

Dispersion Curves—$SH_0/S_0/A_0/A_1$ Frequency & Wave Velocity								
Value	Frequency (kHz)				Velocity (m/s)			
Temperature (°C)	22	60	100	180	22	60	100	180
SH_0	256	255	254.2	250.6	3080	3067	3051	3008
S_0	420	419	415.5	405	5044	5022	5003	4944
A_0	143	140	139	136	1736	1720	1711	1685
A_1	676	673	668	657	8243	8217	8203	8129

Electromagnetic Simulations

Both the coil design and the magnet configuration affect the distribution of Lorentz force in space, its direction, and its amplitude. As a result, the wave mode the EMAT is to generate and/or receive gets affected by the configuration of its two main components. An EMAT configuration that can be used for the SH_0 is the PPM EMAT with racetrack coil [26,27,28,29 ,30,31,32,33,34,35,36,37,38,39,40].

In a PPM EMAT, the distance between two adjacent magnets whose magnetic field has the same direction (pitch) is equal to the wavelength; the arrangement of the magnets is illustrated in Figure 4. Racetrack coil is also used in this EMAT configuration. In this coil design there are no gaps within its turns, resulting in its broadband response in frequency. Hence, only the magnets' arrangement and the frequency of the AC current can affect the frequency response of EMAT.

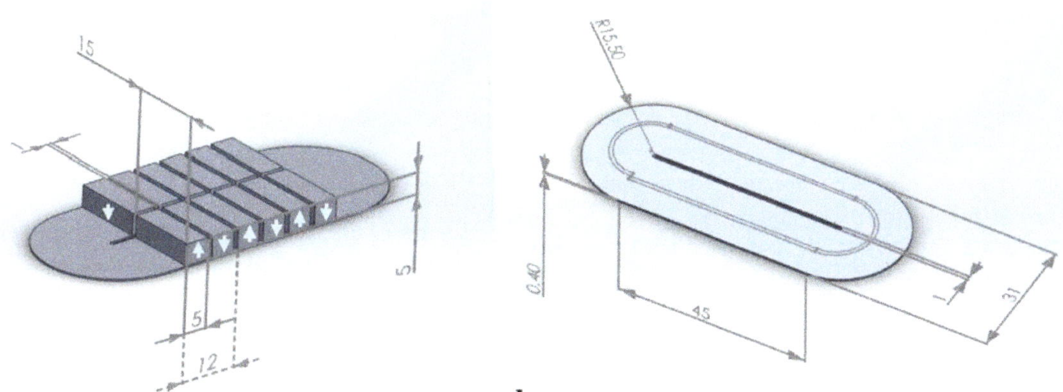

a b

Figure 4. (**a**) Magnet arrangement in a PPM EMAT; (**b**) racetrack coil.

Finite Element Analysis (FEA) was used for the theoretical validation of this EMAT design regarding its guided wave purity characteristics at room temperature and at 180 °C. A coupled electromagnetic and mechanical analysis was carried out in COMSOL. A 3D model was created, in which a 12-magnet PPM EMAT was evaluated regarding eddy current, magnetic flux, and Lorentz force distribution. In this model two arrays of six Nd-Fe-B magnets each was simulated; each magnet has a 15 mm width, 5 mm depth, 5 mm height, and magnetization of 750 kA/m, and the direction of their magnetic flux is on the *z* axis. Their arrangement in space is the same as Figure 4 and, thus, the distance between one another is 1 mm. Due to the high requirements in time and computational power these complicated models have, the copper coil has been simplified and designed as two rectangular blocks of 35 mm width, 15 mm depth, and 0.4 mm height. The AC current driven to the coil is a two-cycle, Hanning windowed sinusoidal wave of 20 A amplitude and 256 kHz central frequency with direction on the *x* axis. The orientation of the excitation current flowing inside one rectangular coil is opposite to the orientation of the current inside the other coil. The EMAT has a lift-off of 0.6 mm from the specimen, which is a 316 L stainless steel plate of 3 mm thickness, 750 mm width, and 750 mm depth and of the same material properties as the plate simulated in the previous section at room temperature and at 180 °C. The material, magnetic, and electrical properties of the EMAT did not change with temperature.

Figure 5a,b show the excitation/eddy current at 4 μs and the static magnetic flux distribution, respectively. Both the electric and the magnetic field are uniformly distributed. The orientation of the eddy current alters between the two sides of the coil and it is on the *x*–*y* plane. The orientation of the magnetic field alters as it is depicted in Figure 5b; its orientation is mainly on the *z* axis and its maximum strength is observed at the center of each magnet separately, as is expected. Thus, the Lorentz force generated should mainly result in an in-plane displacement. A probe was also placed 30 cm away from the EMAT for obtaining the in-plane (*x*–*y* plane) and out-of-plane (*x*–*z* plane) displacement. Figure 5c shows that the in-plane displacement maximizes at 98 μs while the out-of-plane displacement maximizes at 100 μs; the Time of Flight (ToF) of the in-plane displacement matches with the wave velocity of the SH_0, as was calculated from the dispersion curves. while the out-of-plane displacement can be the A_0. Hence, this EMAT configuration may excite the A_0 wave mode as well. However, the maximum value of the out-of-plane displacement is significantly smaller than the in-plane displacement and therefore experimental validation of this EMAT design regarding its guided wave purity characteristics is still required.

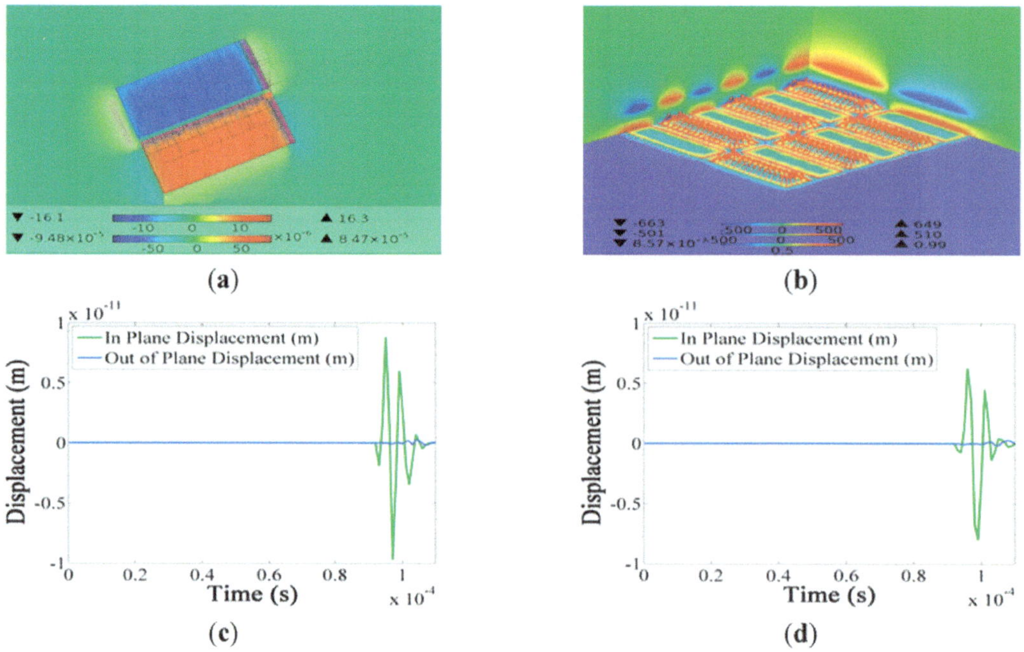

Figure 5. (**a**) Excitation/eddy current distribution at 12 μs; (**b**) magnetic flux distribution of the 12 Nd-Fe-B; (**c**) in-plane/out-of-plane displacement 30 cm away from the EMAT at room temperature; (**d**) in-plane/out-of-plane displacement 30 cm away from the EMAT at 180 °C.

Figure 5d shows the in-plane and out-of-plane displacement 30 cm away from the EMAT at 180 °C; both displacements are shifted in time, as was expected since the wave velocity changes with temperature rise and the ToF matches with the wave velocity as it was calculated from the dispersion curves. The amplitude of the in-plane displacement decreased; however, the amplitude of the out-of-plane displacement did not decrease with the temperature rise. As was mentioned, in the FEA model only the material properties of the specimen changed so that the effect of the temperature rise would be simulated. Nevertheless, the temperature rise affects the components of the EMAT as well, as was mentioned in the introduction, and thus a further decrease in the amplitude of both displacements may be observed in the experimental results, which the current model does not take into account.

Consequently, this EMAT configuration can mainly generate in-plane displacement (SH_0); however, it may also generate a small out-of-plane displacement (A_0) at this specific frequency. This can result in complicating signal analysis, since the out-of-plane displacement can be dispersive and mode conversion can also occur, which can lead to incorrect conclusions regarding the structural integrity of the specimen. In GWT, EMATs should excite a single wave mode, so that the signal received from the specimen can provide valid information regarding the structural integrity of the specimen. Also, the temperature rise affected the ultrasonic response of the EMAT, as was expected; amplitude attenuation and shifting in time were the main two changes in the signal received as the temperature increased. Hence, the wave purity characteristics of the PPM EMAT for GWT are of great importance for this application as well as the effect of temperature on the ultrasonic response of EMATs and thus an experimental validation of this EMAT design regarding its guided wave characteristics and its high-temperature performance is still required. The simulation results will be compared with the experimental results.

EXPERIMENTAL SETUP

A pair of PPM EMATs with racetrack coil, manufactured by Sonemat Limited (Warwick, UK), was experimentally evaluated at both room and high temperatures. The pitch of the magnets is 12 mm and equal to the wavelength of the SH_0[41]. During these experiments, the EMATs were tested regarding defect detection using guided waves at both room and high temperatures. Their sensitivity to noise was also experimentally evaluated; the influence of common mode noise on their performance and the effect of common ground connection between the EMAT and the specimen on the noise reduction were tested. Their lift-off limitations were validated; the thickness of the glass envelope, under which is the stainless steel pipe of the absorber tubes, will attenuate the signal generated by the EMAT and will perform as a fixed lift-off between the EMAT and the stainless steel pipe. Hence, the maximum lift-off EMATs can reach when they are used for GWT needs to be known. Their power requirements were also investigated, since only Lorentz force is generated in stainless steel, making the EMATs less efficient and more power demanding.

The experimental setup used is shown in Figure 6. The specimen used is a 316Ti stainless steel square plate of 1.25 m length and 3 mm thickness. The EMAT coil is made of lacquered copper wires of 0.315 mm diameter and the maximum operating temperature of the EMATs is 250 °C. Ritec RAM 5000 SNAP pulser/receiver (RITEC Inc., Warwick, RI, USA) was used for driving the EMAT transmitter with a six-cycle, Hanning-windowed pulse of 256 kHz frequency. Ritec was also used for amplifying the signal received with a gain of 80 dB and filtering it within the bandwidth of 10 kHz and 20 MHz. The signal is finally collected, averaged, and recorded in a 2-channel Agilent oscilloscope (Keysight Technologies Inc., Santa Rosa, CA, USA).

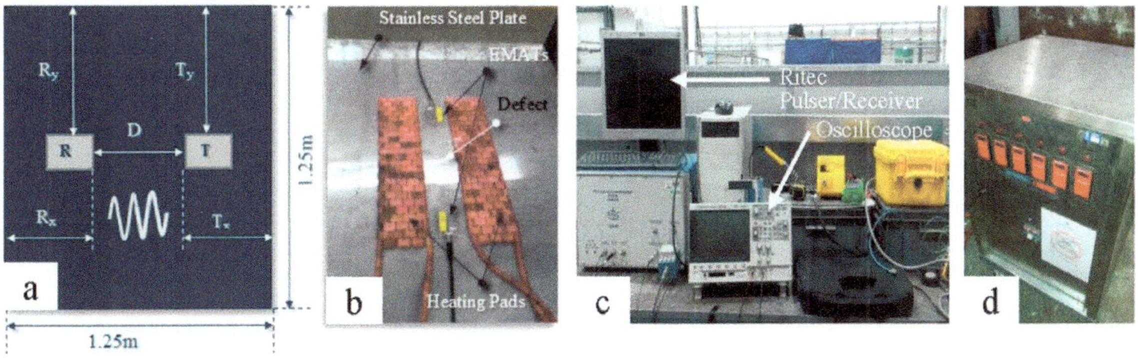

Figure 6. (**a**) Schematic of experimental setup; (**b**) experimental setup (EMATs, specimen, heating pads); (**c**) pulser/receiver—oscilloscope; (**d**) heating unit.

During the room-temperature experiments both defect-free and defective areas were tested. The distance between the transmitter and the receiver was equal to 30 cm. Five defects were created with different length and mass loss each and located 10 cm away one from the other. The defect tested was 20 mm long and with 66.6% mass loss; the transmitter was 15 cm away from one edge of the defect and the receiver was 15 cm away from the other edge of the defect. The effect of the voltage difference between the EMAT and the specimen on the quality of the signal received has been also investigated; in fact, an additional thin, stainless steel cover was also placed all around the transducers, touching both the EMATs and the specimen, for establishing a common ground connection. The influence of lift-off on EMAT response was investigated from zero to 1 mm lift-off with a step of 0.1 mm. A study regarding the power supply requirements of these EMATs was also accomplished by gradually decreasing the power output of Ritec with a step of 5% starting from its maximum power level (5000 W) and stopping at 20% of its maximum power, where no useful information could be retrieved anymore from the signal received. For the high-temperature experiments, the distance between the transmitter and the receiver was equal to 30 cm with the defect located 15 cm away from the transmitter and 15 cm away from the receiver as well (similar to the room temperature setup). The temperature was increased from ambient to 180 °C with a step of 10 °C, using a three-phase heating unit. During the high-temperature experiments the EMATs were continuously in contact with the specimen.

EXPERIMENTAL RESULTS AND DISCUSSION

Room Temperature Experiments

During this set of experiments, the EMATs were tested for their sensitivity to common mode noise. Four case studies were investigated; the EMATs were employed in both a defect-free and a defective area, with and without common ground connection with the specimen (shielding). More particularly, the connection/interaction between the EMAT and the specimen is differential, since the specimen induces to the EMAT coil an alternating, differential mode current. Parasitic capacitance exists between the specimen and the EMAT as a result of their physical spacing and the presence of dielectric between them [35]. Tranformers perform in a similar way and they also suffer from common mode noise; in real transformers, a small capacitance links the primary to the secondary winding and also serves as a path for the common mode current across the transformer. As a result, in both cases the common current flows to the ground via the parasitic capacitance and thus no current flows to the EMAT coil/specimen or secondary winding. Nevertheless, an autotransformer acts as a high-value parallel impedance that does not attenuate the differential current significantly but presents zero impedance to the common mode signals by shorting them to ground potential [42]. Autotransformers have smaller resistance and leakage reactance compared to conventional two winding transformers; therefore, the former is more efficient than the latter [43]. Hence, if we presume that the interaction between the EMAT and the specimen is equivalent to a transformer and we connect them so as to perform as an autotransformer, then the common mode noise should be cancelled and the EMAT receiver should work more efficiently. In this case, the alternating, differential mode current will be induced to the EMAT coil and the signal received will have an enhanced SNR and valid information would be retrieved from it. The autotransformer connection is also more robust, reasulting in an increase of the amplitude of the "wanted" signal. If an extra layer of stainless steel is attached to EMATs, touching both the EMAT housing and the specimen, then a common ground connection (autotransformer) is established. Figure 7 shows the equivalent electrical circuit of the EMAT/specimen connection.

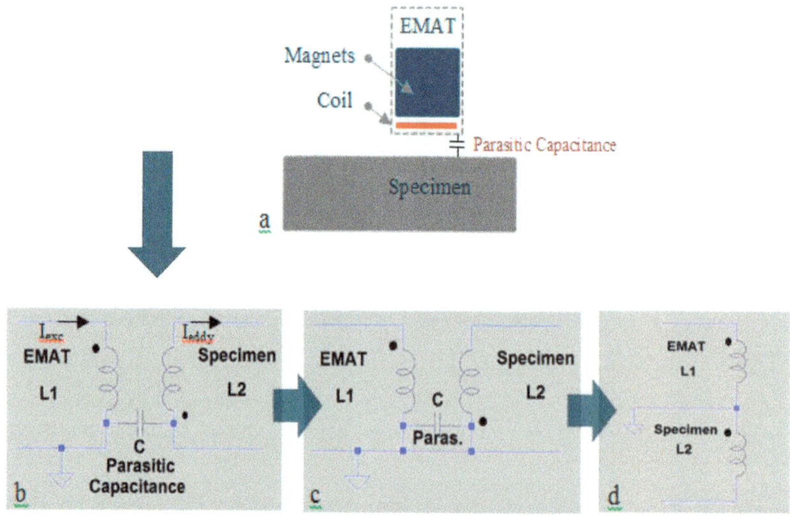

Figure 7. (**a**) Schematic of EMAT/specimen connection; (**b**) equivalent electrical circuit of EMAT/specimen without common ground connection; (**c**) common ground connection; (**d**) EMAT/specimen—autotransformer schematic.

Figure 8a shows the signal received from a defect-free area when the EMAT and the specimen do not have any common ground connection. In this figure the first reflection is the signal transmitted from the transmitter to the receiver and the other three are coming from the edges of the plate. In this case the noise level is high and the amplitude of all the reflections is low, leading to a low SNR. Figure 8b illustrates the signal received when the defective area was tested without the common ground connection. Similar to Figure 8a, the signal transmitted and the three reflections from the edges of the plate are clearly obvious in the signal received. However, no reflections from the defect are obvious. Figure 9a demonstrates the signal received when the EMATs test the defect-free area while a common ground connection between the EMATs and the specimen has been established. In this case, the amplitude of the reflections increased, the noise level decreased, and as a result the SNR increased by five times. Figure 9b shows the signal received from the defective area when the EMATs and the specimen had a common ground connection. Similar to Figure 9a, the SNR of the signal increased four times more compared to the signal shown in Figure 8b. Actually, in Figure 9b both reflections from the crack are clearly obvious.

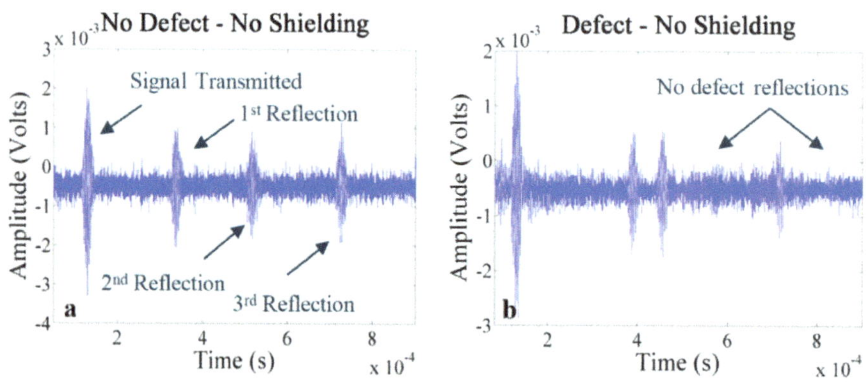

Figure 8. (**a**) Signal received from the defect-free area; (**b**) signal received from the defect.

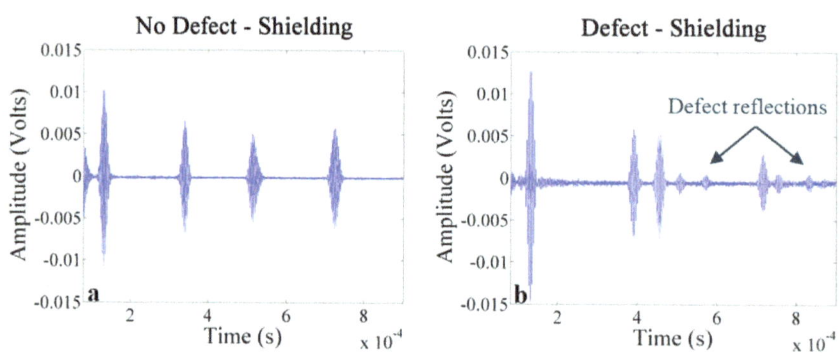

Figure 9. (**a**) Signal received from the defect-free area with shielding; (**b**) signal received from the defect with shielding.

Hence, the shielding has resulted in an enhanced SNR due to the noise cancellation. The electromagnetic coupling between the specimen and the EMAT receiver is weak when they are connected as a two-winding transformer (no shielding); in this case, the electromagnetic losses between the EMAT and the specimen are greater than in a conventional two-winding transformer, since no ferrite connects the EMAT and the specimen. The air between them increases the noise level and attenuates the electromagnetic coupling/"wanted" signal (there is no ferrite to drive the electromagnetic wave; on the contrary, the electromagnetic wave scatters when air is between the EMAT and the specimen). When the specimen/EMAT connection performs as an autotransformer, the noise level decreases and the amplitude of the "wanted" signal (ultrasonic response of the specimen) increases; no noise interferes with the EMAT receiver and thus more current is induced to the coil. The autotransformer is more efficient than a two-winding transformer and as a result more current is induced in the coil when the autotransformer connection is established between the EMAT and the specimen. The unshielded EMAT has greater losses than a conventional transformer would have due to the weak connection between the EMAT and the specimen, while an autotransformer has smaller losses compared to both the unshielded EMAT and a conventional transformer. Thus, a significant increase in the amplitude of the "wanted" signal is observed when the EMAT is shielded. Consequently, the voltage difference between the EMAT and the specimen significantly affects the quality of the signal received and when there is no voltage difference and both components are connected to the ground, the probability of defect detection increases as well. Also, the frequency selected based on the size defect and the dispersion curves is proven to be correct for the detection of the 20 mm long defect.

The attenuation of the signal and the ToF of the main four reflections from the edges of the plate remain features that enable us to distinguish the defective from the defect-free areas when there was no common ground connection. Although the velocity of SH_0 was not to change during the experiments, for no temperature rise occurred, the ToF of both the first and the second reflection from the edges of the plate change from one case study to the other. It is likely that the defect causes this time delay by trapping a portion of the energy/spectrum of the wave propagating inside the plate.

From all the above figures and more especially from Figure 9a,b, it is obvious that only the SH_0 wave mode had been received from the EMAT; the time of arrival of the reflections matches with the SH_0 velocity, as it was calculated from the dispersion curves and the electromagnetic model. Hence, the experimental results match with the theoretical. However, further experimental investigation should be conducted regarding the wave purity characteristics of this EMAT design. In this set of experiments the EMAT receiver was mainly evaluated regarding its ultrasonic potentials; laser interferometry tests, during which the actual displacement generated from the EMAT transmitter can be observed, may also be required for further experimental evaluation of the EMAT transmitter regarding its wave purity characteristics.

Figure 10a illustrates how the amplitude of the signal transmitted changes (%) with respect to the lift-off. According to the literature, EMATs are sensitive to lift-off [28] and therefore their efficiency decreases with the increase of lift-off. The amplitude of the signal transmitted decreases almost linearly with the lift-off increase; when the lift-off is equal to 1 mm only the signal transmitted can be clearly observed. This confirms the high sensitivity of EMATs to lift-off; more particularly, the lift-off limitations of EMATs differ depending on the application the EMAT is designed for. EMATs for thickness measurements can still operate efficiently when the lift-off exceeds 1 mm; however, EMATs for guided waves are more sensitive to lift-off. A parameter that influences the performance of EMAT regarding lift-off is the impedance of the coil. The impedance changes with lift-off as well as with the material properties of the specimen. The inductance due to the magnetization of the specimen is smaller when the EMAT is employed on paramagnetic materials compared to ferromagnetic materials. Consequently, an alternative coil design should be used so that EMATs can be efficiently used in the inspection of absorber tubes, since the thickness of the glass envelope is 7 mm; stacked coils may be more efficient.

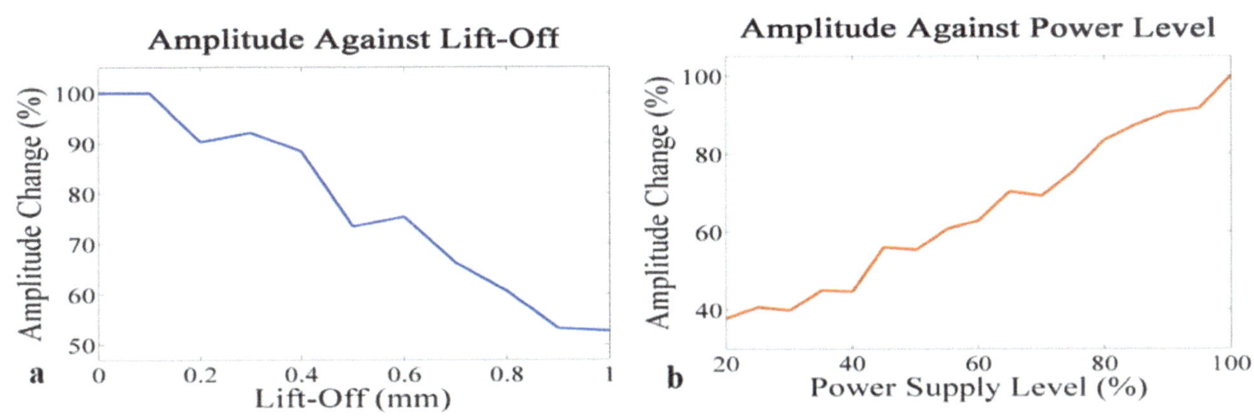

Figure 10. Amplitude change of signal transmitted against (**a**) lift-off increase; (**b**) power increase.

An experimental evaluation of these EMATs regarding their power requirements was also conducted. The power level decreased gradually from 100% to 20% with a step of 5%; we stopped there as no useful information could be retrieved from the signal received when the power level was smaller than 20%. Figure 10b shows how the amplitude of the signal transmitted increases with power supply increase; it can be observed that the amplitude increases almost linearly with the power increase. Similar to lift-off, the impedance of the coil affects the power requirements of EMATs. If the pulser/receiver unit drives the EMAT transmitter through an output resistor, the magnitude of the impedance of the coil should be equal to the output resistor of the pulser unit, so that the voltage drop in the coil will be minimized. Therefore, impedance matching is always required, which means that either a unit with zero output impedance should be chosen or an impedance matching network should be added between the pulser and the EMAT transmitter. Impedance matching should be used for the coil to be driven with the maximum power possible so that strong signals will be obtained. Nevertheless, the power supply level of EMATs remains high, leading to the conclusion that EMATs are considered to be more efficient as receivers rather than transmitters.

High-Temperature Experiments

As was mentioned in the theoretical section, high-temperature EMATs have been designed so far only for thickness measurements and thus an EMAT that can withstand high temperatures and excite guided waves is still required. A first approach for that would be the selection of the suitable high Curie magnets and high-temperature coil or the design of a cooling system so that the EMAT would be as efficient as possible at high temperatures. However, all of the above may result in a more complicated design. Hence, a further study about room temperature, guided wave EMATs, and their performance at high temperatures should be conducted prior to the design of a new EMAT.

Hence, the EMATs were tested from ambient temperature to 180 °C with a step of 10 °C; the maximum operating temperature of these EMATs is 250 °C and therefore they were tested up to 180 °C only, so that any serious and irreversible damage will be avoided. The EMATs were continuously exposed to the heat source with zero lift-off during the rise in temperature, while the overall time they were exposed to the heat was equal to 15 min. This set of experiments was conducted three times. Figure 11a–d show the signal received at room temperature, 60 °C, 100 °C, and 180 °C, respectively. Firstly, in the signal obtained at room temperature, both the reflections from the plate edges and the first two reflections from the defect are clear. However, it is obvious that the amplitude of the second reflection from the plate diminishes greatly after 60 °C, while at 100 °C and 180 °C it can hardly be noticed. Similarly, the amplitute of the first reflection from the defect and the forth reflection from the plate decreases with the increase in temperature.

Figure 12 shows how the amplitude of the signal transmitted decreases with temperature rise; it is clear that the amplitute dwindles almost linearly with the rise in temperature. However, the amplitude of the signal transmitted in 30 °C and 40 °C was slightly larger than the amplitude at room temperature in the areas marked in red in Figure 12. Also, the amplitude error alters with temperature rise. A reason for that may be the ground connection between the EMAT and the specimen. The thermal conductivity of stainless steel is low and therefore the specimen was not heated up uniformly; as a result, the plate bended and the mechanical connection between the EMAT ground and the specimen altered with temperature rise due to the gradient of the bend. This mechanical/electrical connection significantly influences the amplitude of the signal transmitted and thus may be the reason for the amplitude increase at 30 °C and 40 °C.

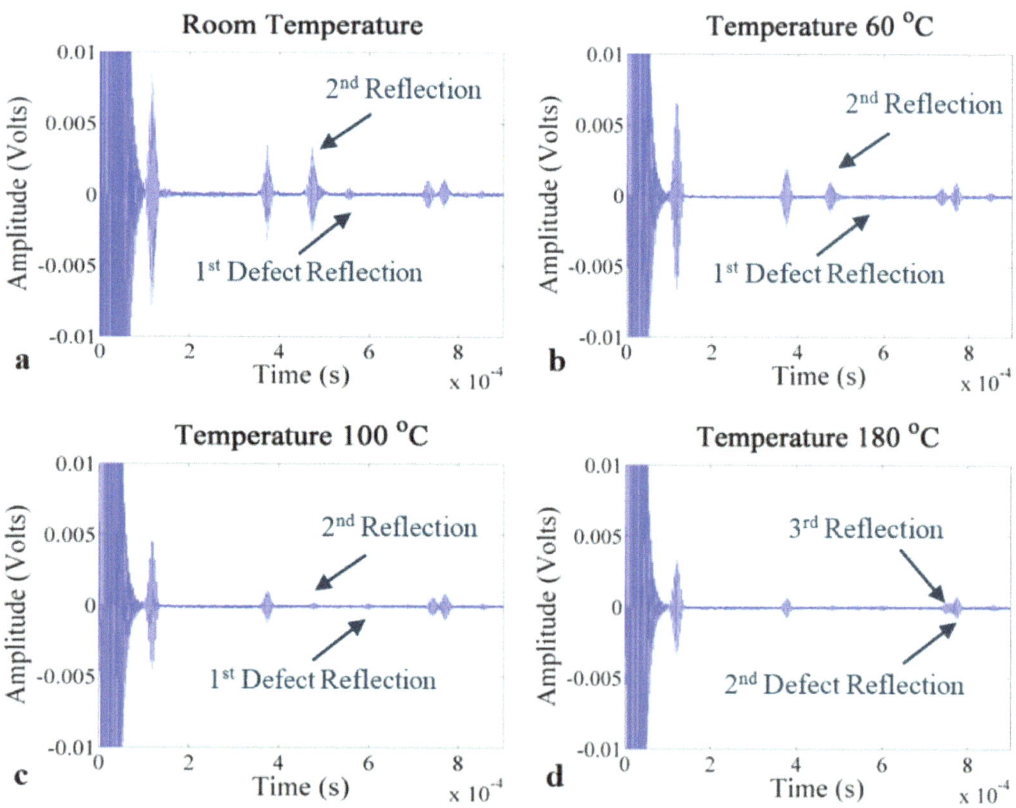

Figure 11. Signal received at (**a**) room temperature; (**b**) 60 °C; (**c**) 100 °C; (**d**) 180 °C.

Time shifting is also observed, as was expected due to the change in the wave velocity at high temperatures, presented in the theoretical study. The third reflection from the plate shifts in time and starts coming closer to the second reflection from the crack, leading to an increase of the magnitude of the latter. Consequently, the experimental results match the theoretical. Temperature compensation should take place and both the ToF of the reflections as well as the drop of their amplitude may be two features that can be further processed and used for the identification of the temperature of the structure being tested. Room-temperature EMATs cannot be used for the monitoring of high-temperature structures (>100 °C); however, a high-temperature EMAT specifically designed for the monitoring of high-temperature structures (>200 °C) should be compared with the performance of this EMAT up to 100 °C.

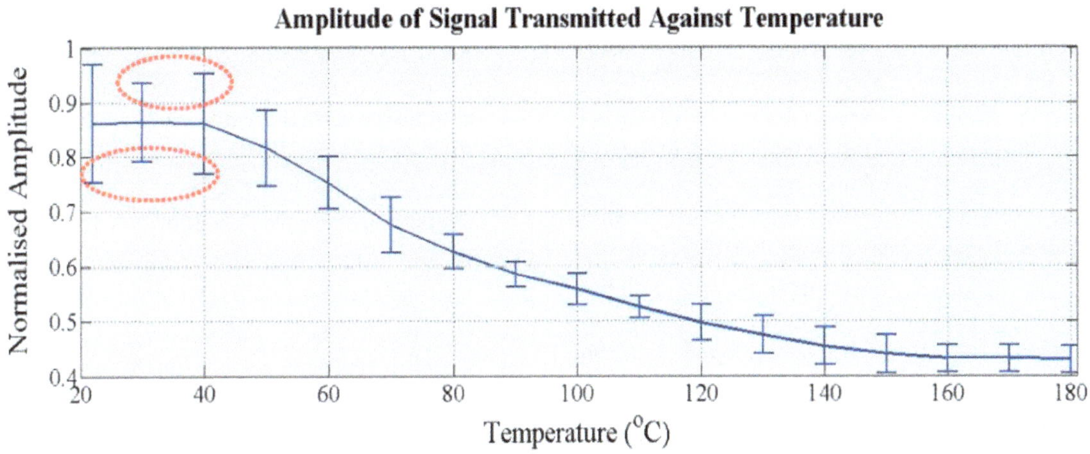

Figure 12. Amplitude of the signal transmitted against temperature.

CONCLUSIONS AND FUTURE WORK

The absorber tube is an essential part of Parabolic Trough CSPs and is very likely to get damaged due to its hostile operating conditions. Hence, NDT techniques are required for their monitoring and/or inspection. A promising technique for this application is the use of high-temperature EMAT transducers for the excitation of guided waves and more particularly of the T(0,1) wave mode. A theoretical study about the GWT of absorber tubes, their dispersion curves, and the wave mode that should be applied for their inspection were calculated. The ultrasonic response of a PPM EMAT and its wave purity characteristics were also presented. A pair of PPM EMATs was experimentally evaluated regarding its wave purity, its sensitivity to noise, its lift-off limitations, its power requirements, and its performance at high temperatures while it was exciting/receiving guided waves. It was found that a PPM EMAT receiver can mainly detect SH_0 and thus it can be used for the GWT of the absorber tubes in terms of wave purity characteristics, as the theoretical study also showed. However, laser interferometry tests are also required for the transmitter to be validated. Also, a common ground connection between the EMAT and the specimen can significantly enhance the SNR of the signal received. The current EMAT design is not efficient enough to inspect a stainless steel pipe with a lift-off larger than 1 mm and therefore it cannot be employed for the inspection of absorber tubes. A room-temperature EMAT cannot be applied at high temperatures (<200 °C) and thus a high-temperature EMAT is still required for the guided wave monitoring of high-temperature structures.

The design and manufacturing of a high-temperature PPM EMAT with racetrack coil operating efficiently at temperatures higher than 300 °C for long-term inspection or monitoring is our next step. Several thermal, electromagnetic, and mechanical simulations have already been carried out and have given encouraging results. Moreover, a further experimental investigation regarding the impedance of the EMAT and its relationship with the lift-off and the material properties of the specimen being tested is also another part of our future research.

ACKNOWLEDGMENTS

The authors are indebted to the European Commission for the provision of funding through the INTERSOLAR FP7 project. The INTERSOLAR project is coordinated and managed by Computerized Information Technology Limited and is funded by the European Commission through the FP7 Research for the benefit of SMEs program under Grant Agreement Number: GA-SME-2013-1-605028. The INTERSOLAR project is a collaborative research project between the following organizations: Computerized Information Technology Limited, PSP S.A., Technology Assistance BCNA 2010 S.L., Applied Inspection Limited, INGETEAM Service S.A., Brunel University, Universidad De Castilla—La Mancha (UCLM), and ENGITEC Limited.

Author Contributions

M.K. performed the literature review and the finite element simulations, conducted all the experiments and wrote the paper; A.M., V.K. and C.S. reviewed the manuscript; T.-H.G., W.B. and L.C. supervised the research.

REFERENCES

1. Poullikkas, A. Economic analysis of power generation from parabolic trough solar thermal plants for the Mediterranean region—A case study for the island of

Cyprus. *Renew. Sustain. Energy Rev.* **2009**, *13*, 2474–2484.

2. Staff, C.W. Industry Experts Trumpet Untold Solar Potential. 2013. Available online: http://www.constructionweekonline.com/article-23701-industry-experts-trumpet-untold-solar-potential/ (accessed on 12 March 2015).
3. Taggart, S. Parabolic troughs: Concentrating Solar Power (CSP)'s Quiet Achiever. 2008. Available online: http://www.renewableenergyfocus.com/view/3390/parabolic-troughs-concentrating-solar-power-csp-s-quiet-achiever (accessed on 12 March 2015).
4. Steinfeld, A. Topics for Master/Bachelor Thesis and Semester Projects: Spectral Optical Properties of Glass Envelopes of Parabolic trough Concentrators. Available online: http://www.pre.ethz.ch/teaching/topics/?id=87 (accessed on 18 August 2015).
5. Guillot, S.; Faika, A.; Rakhmatullina, A.; Lambert, V.; Verona, E.; Echegut, P.; Bessadaa, C.; Calvet, N.; Py, X. Corrosion effects between molten salts and thermal storage material for concentrated solar power plants. *Appl. Energy*2012, *94*, 174–181.
6. Herrmann, U.; Kearney, D.W. Survey of Thermal Energy Storage for Parabolic Trough Power Plants. *J. Sol. Energy Eng.* **2002**, *124*, 145–152.
7. Papaelias, M.; Cheng, L.; Kogia, M.; Mohimi, A.; Kappatos, V.; Selcuk, C.; Constantinou, L.; Muñoz, C.Q.G.; Marquez, F.P.G.; Gan, T.H. Inspection and structural health monitoring techniques for concentrated solar power plants. *Renew. Energy* **2015**, *85*, 1178–1191.
8. Lee, J.L.; Hu, L.W.; Saha, P.; Kazimi, M.S. Numerical analysis of thermal striping induced high cycle thermal fatigue in a mixing tee. *Nucl. Eng. Des.* **2009**, *239*, 833–839.
9. Noguchi, Y.; Okada, H.; Semba, H.; Yoshizawa, M. Isothermal, thermo-mechanical and bithermal fatigue life of Ni base alloy HR6W for piping in 700 °C USC power plants. *Procedia Eng.* **2011**, *11*, 1127–1132.
10. Jinu, G.R.; Sathiya, P.; Ravichandran, G.; Rathinam, A. Comparison of thermal fatigue behaviour of ASTM A 213 grade T-92 base and weld tubes. *J. Mech. Sci. Technol.* **2010**, *24*, 1067–1076.
11. Sunny, S.; Patil, R.; Singh, K. Assessment of thermal fatigue failure for BS 3059 boiler tube experimental procedure using smithy furnace. *Int. J. Emerg. Technol. Adv. Eng.* **2012**, *2*, 391–398.
12. Posteraro, K. Thwart Corrosion under Industrial Insulation. *Chem. Eng. Prog.* **1999**, *95*, 43–47.
13. Halliday, M. Preventing corrosion under insulation-new generation solutions for an age old problem. *J. Prot. Coat. Linings* **2007**, *24*, 24–36.
14. De Vogelaere, F. Corrosion under insulation. *Process. Saf. Prog.* **2009**, *28*, 30–35.
15. Kumar, M.S.; Sujata, M.; Venkataswamy, M.A.; Bhaumik, S.K. Failure analysis of a stainless steel pipeline. *Eng. Fail. Anal.* **2008**, *15*, 497–504.
16. Stine, W.B.; Harrigan, R.W. *Solar Energy Systems Design*; John Wiley and Sons, Inc.: New York, NY, USA, 1986.
17. Hutchins, D.A.; Saleh, C.; Moles, M.; Farahbahkhsh, B. Ultrasonic NDE Using a Concentric Laser/EMAT System. *J. Nondestruct. Eval.* **1990**, *9*, 247–261.
18. Kirk, K.J.; Lee, C.K.; Cochran, S. Ultrasonic thin film transducers for high-temperature NDT. *Insight* **2005**, *47*, 85–87.
19. Momona, S.; Moevus, M.; Godina, N.; R'Mili, M.; Reynauda, P.; Fayolle, G. Acoustic emission and lifetime prediction during static fatigue tests onceramic-matrix-composite at high temperature under air. *Compos. A Appl. Sci. Manuf.***2010**, *41*, 913–918.
20. Shen, G.; Li, T. Infrared thermography for high temperature pipe. *Insight* **2007**, *49*, 151–153.
21. Beattie, G. Acoustic emission principles and instrumentation. *J. Acoust. Emiss.* **1983**, *2*, 95–128.
22. Hagemaier, D.J. *Fundamentals of Eddy Current Testing*; American Society for Nondestructive, Testing: Columbus, OH, USA, 1990.
23. Malmo, J.T.; Jøkberg, O.J.; Slettemoen, G.A. Interferometric testing at very high temperatures by TV holography (ESPI). *Exp. Mech.* **1998**, *28*, 315–321.
24. Silk, M.G.; Bainton, K.F. The propagation in metal tubing of ultrasonic wave modes equivalent to Lamb waves.*Ultrasonics* **1979**, *17*, 11–19.
25. Pfander, M.; Lupfert, E.; Pistor, P. Infrared temperature measurements on solar trough absorber tubes. *Sol. Energy***2007**, *81*, 629–635.
26. Cawley, P.; Alleyne, D. The use of Lamb waves for the long range inspection of large structures. *Ultrasonics* **1996**, *34*, 287–290.
27. Hirao, M.; Ogi, H. A SH-wave EMAT technique for gas pipeline inspection. *NDT E Int.* **1999**, *32*, 127–132.
28. Bottger, W.; Schneider, H.; Weingarten, W. Prototype EMAT system for tube inspection with guided ultrasonic waves.*Nucl. Eng. Des.* **1986**, *102*, 369–376.
29. Wilcox, P.; Lowe, M.; Cawley, P. Omnidirectional guided wave inspection of large metallic plate structures using an EMAT array. *IEEE Trans. Ultrason. Ferroelectr. Freq. Control* **2005**, *52*, 653–665.
30. Andruschak, N.; Saletes, I.; Filleter, T.; Sinclair, A. An NDT guided wave technique for the identification of corrosion defects at support locations. *NDT E Int.* **2015**, *75*, 72–79.
31. Burrows, S.E.; Fan, Y.; Dixon, S. High temperature thickness measurements of stainless steel and low carbon steel using electromagnetic acoustic transducers. *NDT E Int.* **2014**, *68*, 73–77.
32. Hernandez, V.F. Pulsed-Electromagnet EMAT for High Temperature Applications. Ph.D. Thesis, University of Warwick, Warwick, UK, 2011.
33. Idris, A.; Edwards, C.; Palmer, S.B. Acoustic wave measurement at elevated temperature using a pulsed laser generator and an Electromagnetic Acoustic Transducer detector. *Nondestruct. Test. Eval.* **1994**, *11*, 195–213.

34. Dixon, S.; Edwards, C.; Palmer, S.B. High-accuracy non-contact ultrasonic thickness gauging of aluminium sheet using electromagnetic acoustic transducer. *Ultrasonics* **2001**, *39*, 445–453.

35. Jian, X.; Dixon, S.; Edwards, R.S.; Morrison, J. Coupling mechanism of an EMAT. *Ultrasonics* **2006**, *44*, 653–656.

36. Ribichini, R.; Cegla, F.; Nagy, P.B.; Cawley, P. Study and Comparison of Different EMAT Configurations for SH Wave Inspection. *IEEE Trans. Ultrason. Ferroelectr. Freq. Control* **2011**, *58*, 2571–2581.

37. Gaultier, J.; Mustafa, V.; Chahbaz, A. EMAT Generation of Polarized Shear Waves for Pipe Inspection. In Proceeding of the 4th PACNDT, Toronto, ON, Canada, 14–18 September 1998.

38. Rose, J.L.; Lee, C.M.; Hay, T.R.; Cho, Y.; Park, I.K. Rail Inspection with Guided Waves. In Processing of the 12th A-PCNDT—Asia-Pacific Conference on NDT, Auckland, New Zealand, 5–10 November 2006.

39. Jackel, P.; Niese, F. EMAT Application: Corrosion Detection with Guided Waves in Rod, Pipes and Plates. In Proceedings of the 11th European Conference on Non-Destructive Testing (ECNDT 2014), Prague, Czech, 6–10 October 2014.

40. Hübschen, G. Generation of Horizontally Polarized Shear Waves with EMAT Transducers. *e-J. Nondestruct. Test.* **1998**,*3*, 1–7.

41. Kogia, M.; Mohimi, A.; Liang, C.; Kappatos, V.; Selcuk, C.; Gan, T.H. High temperature Electromagnetic Acoustic Transducer for the inspections of jointed solar thermal tubes. In Proceedings of the First Young Professionals International Conference, Budapest, Hungary, 17–20 September 2014.

42. Pulse Electronics, Understanding Common Mode Noise. Available online: www.pulseelectronics.com/download/3124/g204/pdf (accessed on 10 November 2015).

43. Winders, J. Autotransformers and Three-Winding Transformers. In *Power Transformers: Principles and Applications*, 1st ed.; Marcel Dekker: New York, NY, USA, 2002.

CHAPTER 22

A STUDY OF ANTENNA SYSTEM FOR HIGH ORDER MIMO DEVICE

Zhaoliang Chen, Wen Geyi, Ming Zhang, and Jun Wang

INTRODUCTION

Nowadays, wireless communication technology is developing rapidly with the goals of achieving large channel capacity, high transmission rate, and high reliability. Multiple-input multiple-output (MIMO) arrays have been recognized as a key technology to reach the goals. In a multiple channel environment, the MIMO system can increase the transmission rate and channel capacity without sacrificing additional frequency spectrum and transmitted power [1], and a MIMO antenna system can improve the mobile communication quality when it has good isolation and envelope correlation coefficient (ECC) among its antenna elements simultaneously. Due to the limited space of the mobile terminals, the volume occupied by the antenna elements and the distance between them must be small, which propose major challenges for a MIMO antenna system to realize wideband in frequency, miniaturization in size, and high isolation among the antenna elements [1–3]. In the existing research articles, the number of antenna elements deployed in the MIMO systems investigated ranges from two to eight with different goals in mind: some of them are primarily for achieving the wide bandwidth [4, 5]; some for the miniaturization of antenna element size [6, 7]; and some for the high isolation among antenna elements through a variety of methods [8, 9]. It is rare to see a report that solves all these problems simultaneously.

In MIMO antenna designs, reducing mutual coupling and correlation coefficient between closely packed antenna elements is one of the key challenges. For antenna elements in close proximity, the radiated energy from one element can be delivered to other elements through near field coupling or conductive coupling (via the common ground plane). As the number of antenna elements increases, the radiation efficiency of some antenna elements may drop quickly due to the mutual coupling [10]. For this reason, a number of methods have been developed for improving the isolation and correlation coefficient between closely packed MIMO antenna elements in small portable handsets. These include etching slots on the ground plane, using electromagnetic band gap (EBG) structures, deploying decoupling or matching networks, introducing parasitic structures, properly placing the antenna elements, changing radiation patterns, and introducing neutralization line [11–29]. Most of these methods would increase the complexity of the system and require considerable design effort.

In this paper, MIMO antenna systems for handset applications with four and six elements for a mobile handheld

Citation: Zhaoliang Chen, Wen Geyi, Ming Zhang, and Jun Wang, "A Study of Antenna System for High Order MIMO Device," International Journal of Antennas and Propagation, vol. 2016, Article ID 1936797, 14 pages, 2016. doi:10.1155/2016/1936797.

device of small form factor have been investigated. An attempt is made to achieve good isolation, wide bandwidth and miniaturization simultaneously on a very compact platform, and a FR4 substrate of the dimensions 136 mm × 68.8 mm × 1 mm. The antenna element adopted is the three-dimensional inverted-F antenna with added resonating branches wrapped around a very small dielectric cube of dimensions 10 × 10 × 5 mm^3, which is the smallest volume so far that covers the required frequency bands for GSM1900, LTE2300, 2.4-GHz WLAN, and LTE2500. The MIMO antenna systems do not involve any decoupling circuits and have achieved a good isolation throughout the whole frequency band by properly designing the antenna elements and using the pattern diversity. The envelope correlation coefficients, the mean effective gains (MEGs), and the multiplexing efficiency are calculated based on the measured data. The results indicate that the envelope correlation coefficient between any two elements of the MIMO antenna is less than 0.5; the maximum gain of the antenna is more than 1 dB; the difference between the mean effective gains of the antenna elements is less than 3 dB.

DESIGN OF FOUR-ELEMENT MIMO ANTENNA SYSTEM

Specifications of Antenna System

The unfolded view of the antenna element with detailed dimensions is shown in Figure 1. The antenna is folded along the dashed lines and is then wrapped around a dielectric cube of relative permittivity 4.4 and loss tangent 0.02. The design introduces several radiating branches, independent of each other to some degree, to control different frequency bands [30, 31].

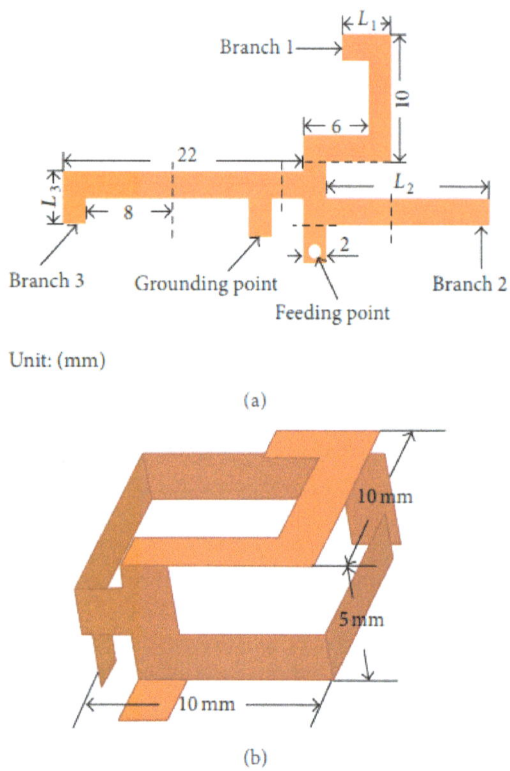

Figure 1: (a) Geometry of the antenna element; (b) 3D graph of the element.

The branches 1 and 3 are mainly used to control the band 1880–1920 MHz; the slot of the ground (see Figure 3) is part of the antenna and is used to achieve the rest of the frequency bands mentioned above; branch 2 affects both low and high frequencies to some extent. The antenna elements are placed at the four corners of the FR4PCB of the size 136 mm × 68.8 mm × 1 mm with relative permittivity 4.4 and loss tangent 0.02. Figures 2 and 3 show the entire structure of the MIMO antenna system and the ground of the PCB, respectively. Figure 4 is the photograph of the fabricated MIMO antenna system.

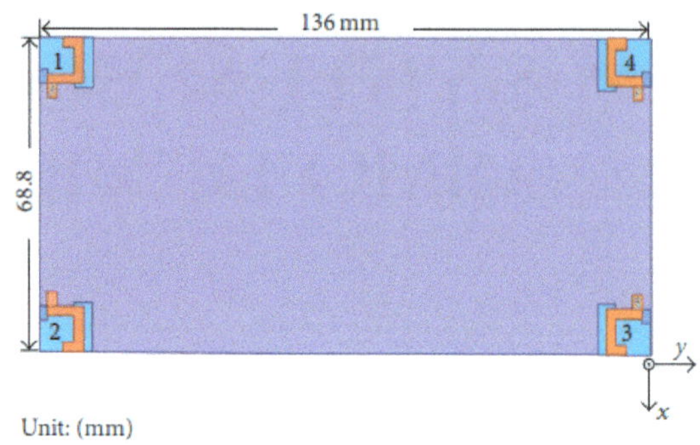

Figure 2: The MIMO antenna system.

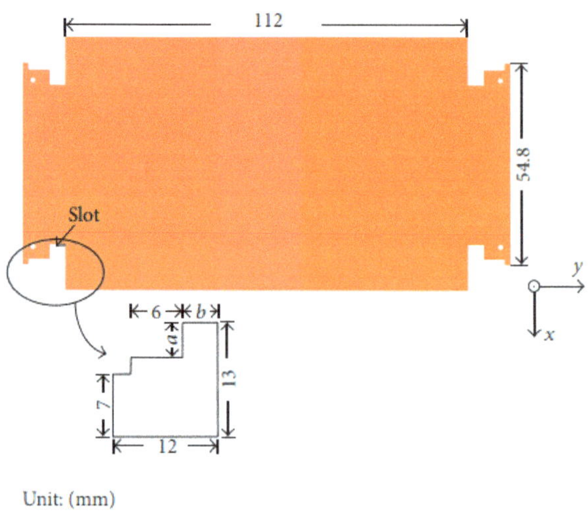

Figure 3: The ground of PCB.

Figure 4: Photograph of the fabricated four-element MIMO system.

Parameters Analyses Based On Simulation and Measurement

To reveal the functions of the resonating branches and the slot (see Figures 1(a) and 3), Figures 5(a)–5(d) show the simulated reflection coefficients of the antenna element for different values of L_1, L_2, L_3, a, and b. The antenna has two resonant frequencies, and after optimization with Ansys HFSS, the optimized values are obtained as follows: L_1=4.1mm, L_2=15 mm, L_3=4mm, a=4mm, andb=4 mm. The measured S-parameters of the four-element MIMO system are shown in

Figure 5(e). Note that the measured −10 dB impedance bandwidth covers the required frequency band, and the mutual coupling is lower than −10 dB across the entire frequency band. Due to the symmetry of the arrangement of the antenna elements, only the S_{11}(1,2,3,4) are plotted.

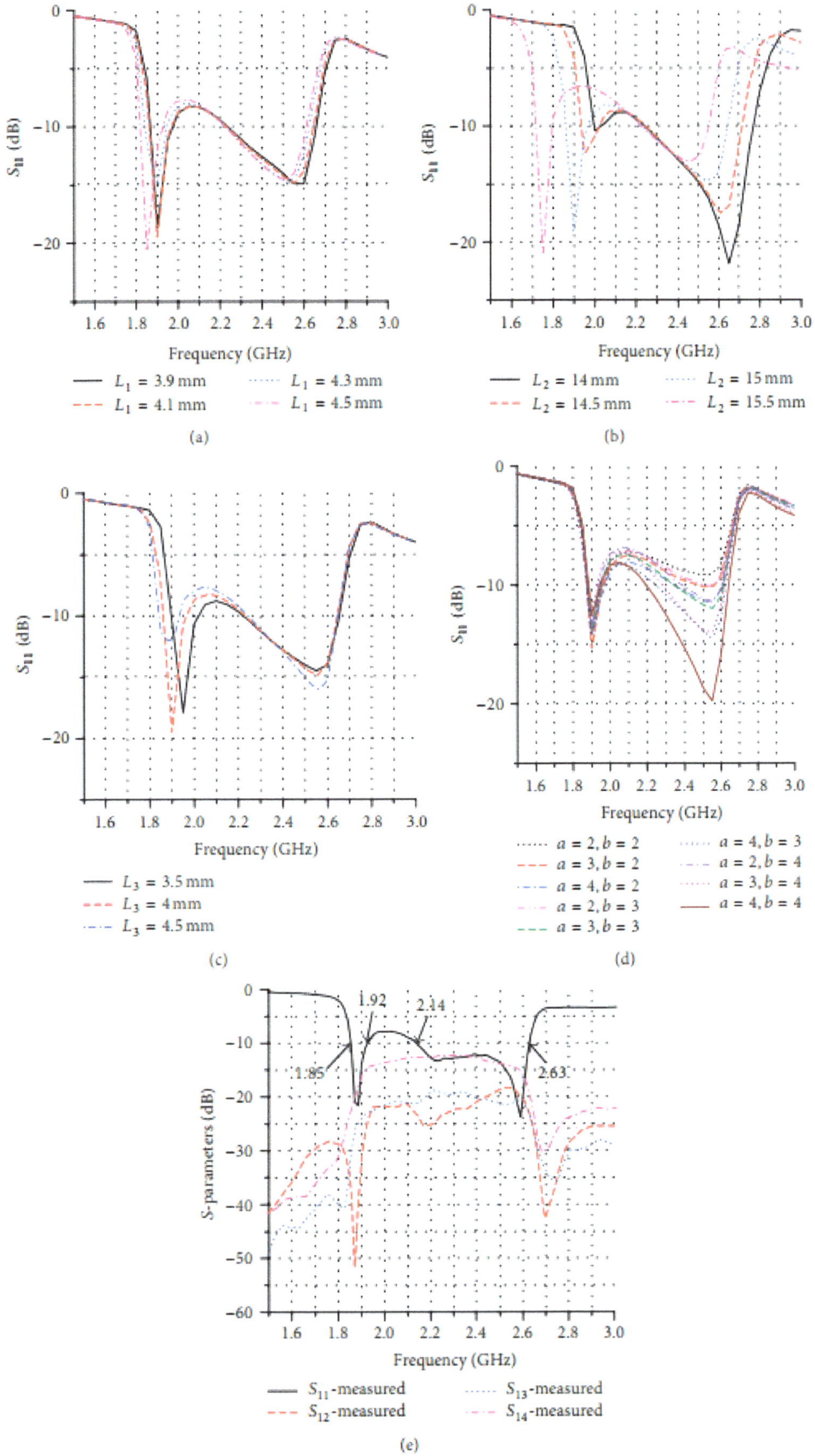

Figure 5: (a)–(d) S-parameters with different L_1, L_2, L_3, a, and b; (e) measured S-parameters.

Figures 6(a) and 6(b) are the current distributions along the controlling branches for 1.9 GHz and 2.35 GHz, respectively.

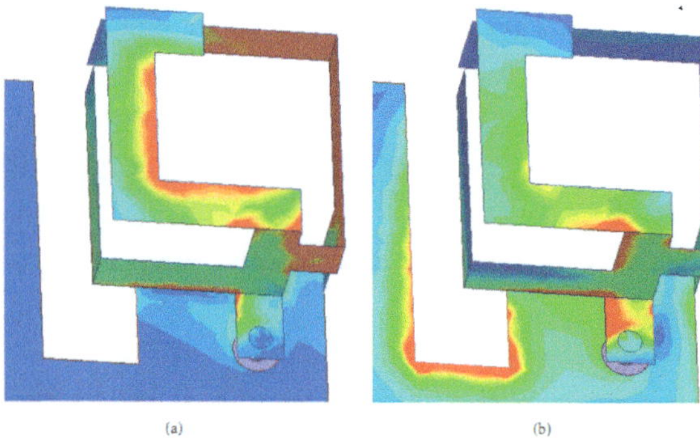

Figure 6: Simulated surface current distributions of the antenna element. (a) At 1.9 GHz; (b) at 2.35 GHz.

The ECC and MEGs are important performance criteria to evaluate the MIMO system [32–35]. The envelop correlation coefficients between antennas i and j can be calculated by [34]

$$\rho_e = \frac{\left|\iint_{4\pi}\left[\mathbf{F}_1(\theta,\varphi)\cdot\mathbf{F}_2(\theta,\varphi)\right]d\Omega\right|^2}{\iint_{4\pi}\left|\mathbf{F}_1(\theta,\varphi)\right|^2 d\Omega \iint_{4\pi}\left|\mathbf{F}_2(\theta,\varphi)\right|^2 d\Omega}, \quad (1)$$

where $\mathrm{F}_i(\theta,\varphi)$ is the field radiation pattern of the antenna system when port is excited and the dot · denotes the Hermitian product.

The field radiation patterns are obtained from the SATIMO Star Lab chamber. The envelope correlation coefficients of the MIMO antenna system are calculated from the measured patterns in two different scenarios, the uniform 3D and the Verizon LTE test plan. The uniform 3D test plan is based on the uniform environment, and the Verizon LTE test plan is based on LTE-A MIMO channel model, as illustrated in Figure 7 [36–38]. The measured ECCs are shown in Figure 8. It can be seen that the ECCs are lower than 0.1 in both scenarios.

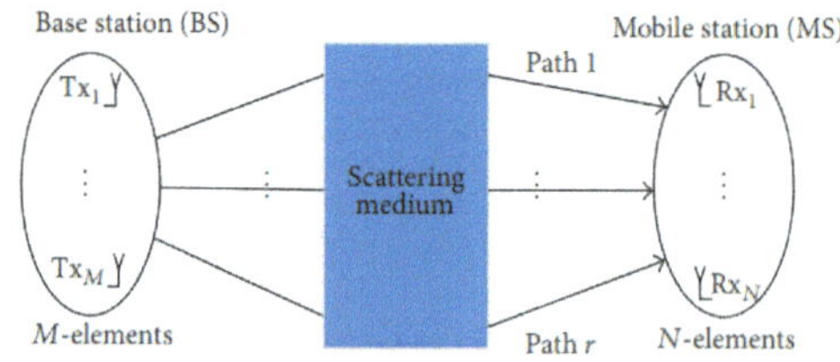

Figure 7: The MIMO channel model.

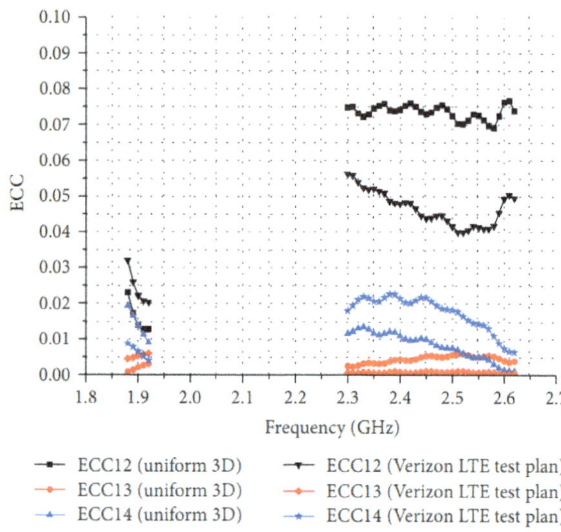

Figure 8: Envelope correlation coefficient between the antennas.

The simulated and measured radiation patterns in the xz-plane and yz-plane for the antenna element are shown in Figure 9. The radiation patterns are obtained under the condition that one element is excited while the others are terminated in a matching load. The solid and dashed lines denote the simulated and measured results, respectively. The black and red lines, respectively, represent the θ and φ polarized components of the antenna power gain.

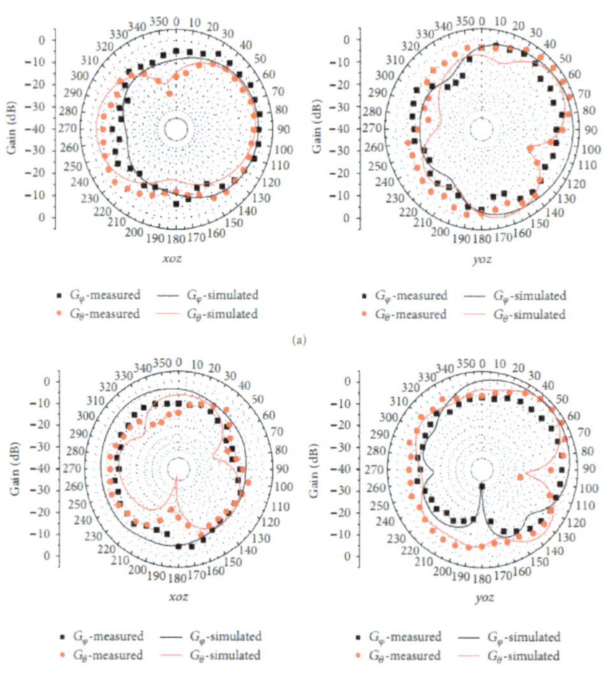

Figure 9: Measured and simulated radiation patterns. (a) At 1.9 GHz; (b) at 2.6 GHz.

The MEG is defined as the ratio of the mean received power to the mean incident power of the antenna. A series of reasonable assumptions may be adopted to simplify the calculation [39, 40]. Based on these approximations, the MEG can be expressed as [41]

$$= \frac{1}{2\pi}\int_0^{2\pi}\left[\frac{\Gamma}{\Gamma+1}G_\theta\left(\frac{\pi}{2},\varphi\right)+\frac{1}{\Gamma+1}G_\varphi\left(\frac{\pi}{2},\varphi\right)\right]d\varphi, \tag{2}$$

where G_θ and G_φ are the θ and φ polarized components of the antenna power gain pattern; Γ is the cross-polarization discrimination (XPD). In this paper, Γ is chosen as 0 dB and 6 dB [42, 43], which are the average values, respectively, for an indoor (0 dB) and an urban (6 dB) fading environment. The MEGs are listed in Table1 and are obtained from the measured data. Due to the symmetry, the MEGs of four elements are almost the same at the same frequency.

Table 1: Mean effective gains of the antennas.

Frequency (GHz)	Γ (dB)	MEG1 (dBi)	MEG2 (dBi)	MEG3 (dBi)	MEG4 (dBi)
1.9	0	−6.4	−6.2	−6	−6.4
	6	−6.3	−6.1	−6.3	−6.2
2.6	0	−8.1	−8	−8.2	−8
	6	−7.8	−7.9	−8	−8.1

The efficiencies of the antenna elements measured at two typical frequencies are shown in Table 2. The efficiencies are obtained from the SATIMO Star Lab chamber. Note that the efficiency is lower at the GSM1900 than other bands due to the absorption of the adjacent elements. The multiplexing efficiency is defined as [44–46]

$$\eta_{\text{mux}} = \frac{P_{T,0}}{P_T}, \tag{3}$$

where $P_{T,0}$ and P_T denote the required power for achieving the same capacity performance with the reference antenna and the MIMO AUT [44]. For high SNRs, η_{mux} can be obtained as [45]

$$\tilde{\eta}_{\text{mux}} = \lim_{P_T\to\infty}\eta_{\text{mux}} = \frac{P_{T,0}}{P_T} = \frac{\det(R)^{1/M}}{\det(R_0)^{1/M}}, \tag{4}$$

where R and R_0 are the receive correlation matrices when using the MIMO AUT and the reference antenna in the same propagation channel; M denote the number of the antenna elements. The multiplexing efficiencies with the uniform 3D scenario are evaluated as shown in Table 2.

Table 2: Antenna efficiency.

Frequency (GHz)	Efficiency (%)	Multiplexing efficiency (dB)
1.9	52	−6.13
2.6	70	−5.12

DESIGN OF SIX-ELEMENT MIMO ANTENNA SYSTEM

In a MIMO system, increasing the number of antenna elements can enhance the channel capacity and the data transmission rate, and the performance of the system can be improved accordingly. Now the number of antenna elements is increased from four to six and two different arrangements will be examined. The first one is shown in Figure 10(a), where two antenna elements are introduced in the middle of the long sides of the substrate and are placed symmetrically about y-axis. From the measured S-parameters (see Figure 12(a)), the isolations between antennas 3, 6 and 4, 5 are higher than −10 dB and cannot meet the practical requirement. From simulation, we find that the radiation intensity of antennas 3 and 4 is very strong along the negative direction of the y-axis. As a result, the radiated energy from antennas 3 and 4 is coupled to antenna 6 and 5, respectively. For this reason, we rearrange the antenna elements so that they are rotationally symmetric about z-axis to reduce the mutual coupling due to the radiation, as illustrated in Figure 10(b). Figure 11 is the photograph of the fabricated system of Figure 10(b). In this case, all S-parameters meet the practical requirements as indicated by Figure 12(b).

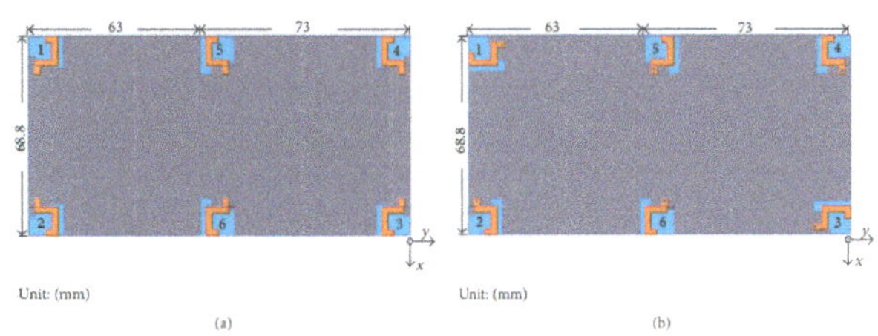

Figure 10: The six-element MIMO system.

Figure 11: Photograph of the fabricated six-element MIMO system.

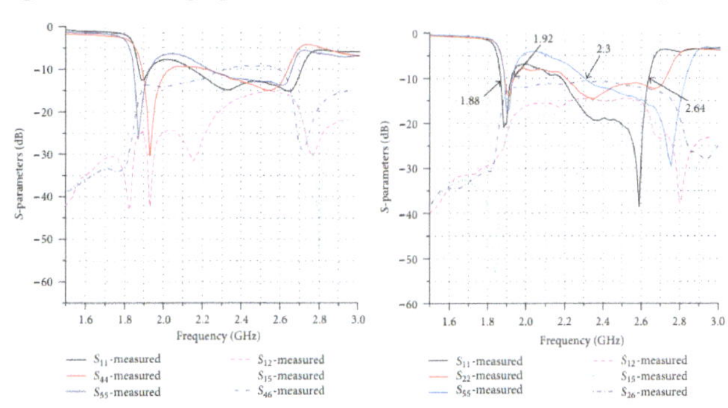

Figure 12: The S-parameters of six-element MIMO system.

Figure 13 shows the envelope correlation coefficients for the MIMO system rotationally symmetric about z-axis. Two scenarios have also been considered.

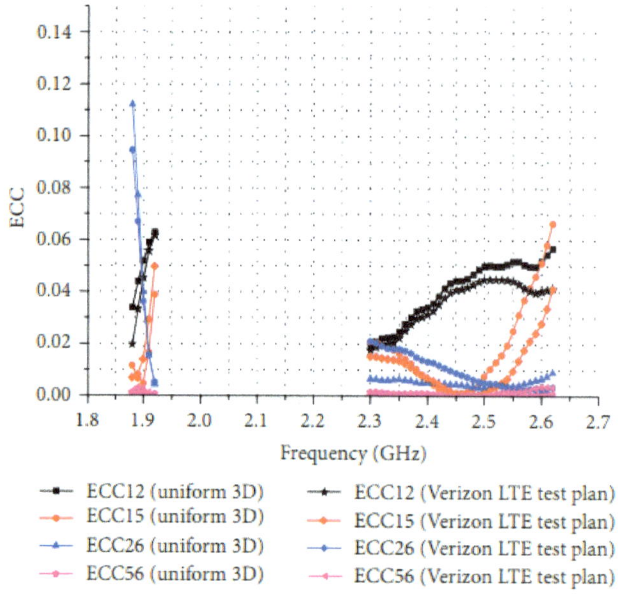

Figure 13: The ECC of six-element MIMO system.

Figure 14 shows the simulated and measured radiation patterns of antennas 1, 2, and 5 at 1.9 GHz and 2.6 GHz. The mean effective gain of each antenna element can be calculated according to the measured radiation pattern and (2) (see Table 3). Table 4 is the measured efficiencies of the antenna elements and the multiplexing efficiencies.

Table 3: Mean effective gains of the antennas.

Frequency (GHz)	Γ (dB)	MEG1 (dBi)	MEG2 (dBi)	MEG5 (dBi)
1.9	0	−6.5	−4.6	−6.8
	6	−6.1	−4.8	−7.2
2.6	0	−5.5	−6.7	−8.1
	6	−5.4	−6.6	−7.6

Table 4: Antenna efficiency.

Frequency (GHz)		Efficiency (%)	Multiplexing efficiency (dB)
1.9	Antenna 1	37	−6.71
	Antenna 2	50	
	Antenna 5	44	
2.6	Antenna 1	68	−5.74
	Antenna 2	65	
	Antenna 5	79	

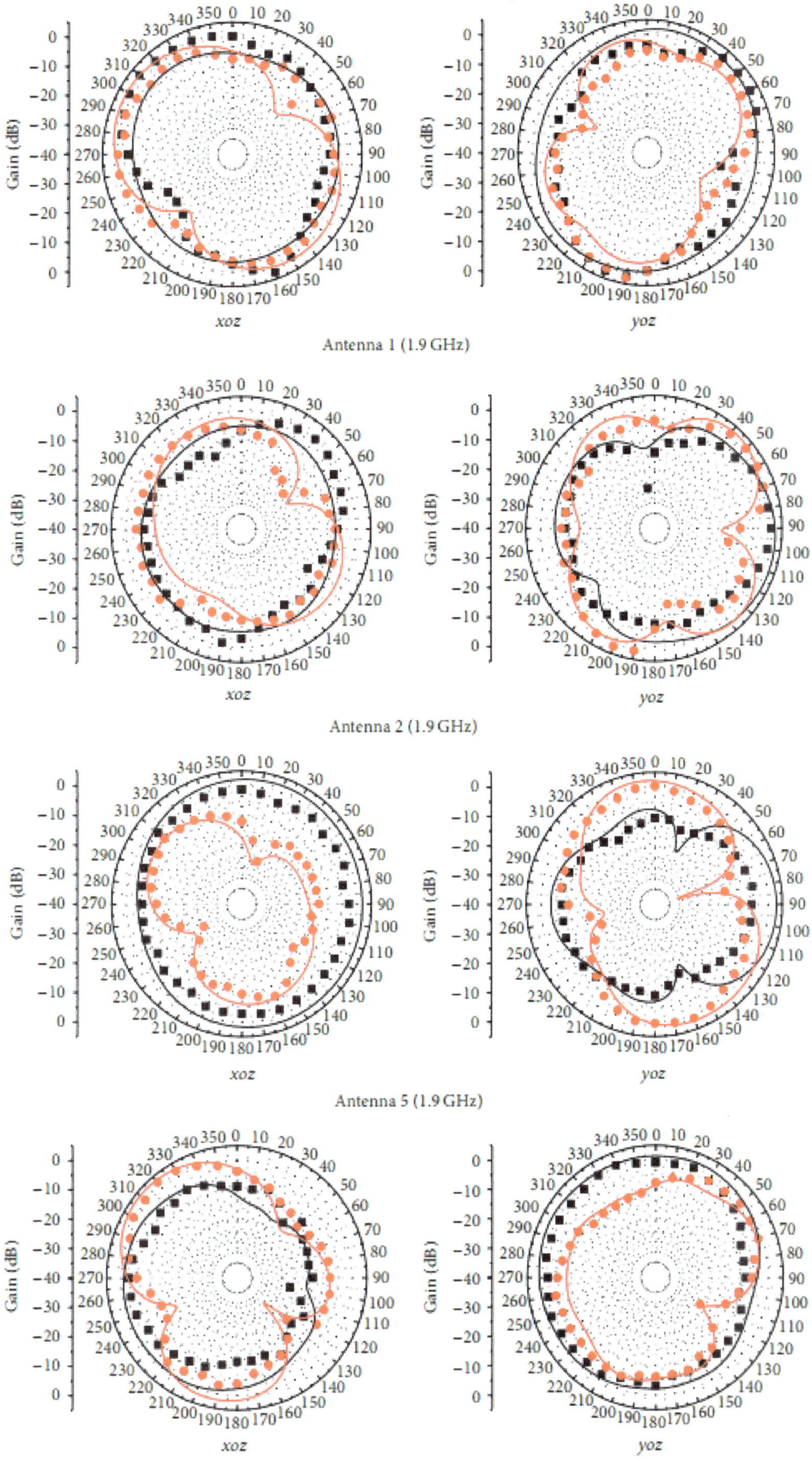
Gain (dB)
xoz
yoz
Antenna 1 (1.9 GHz)
Antenna 2 (1.9 GHz)
Antenna 5 (1.9 GHz)
Antenna 1 (2.6 GHz)

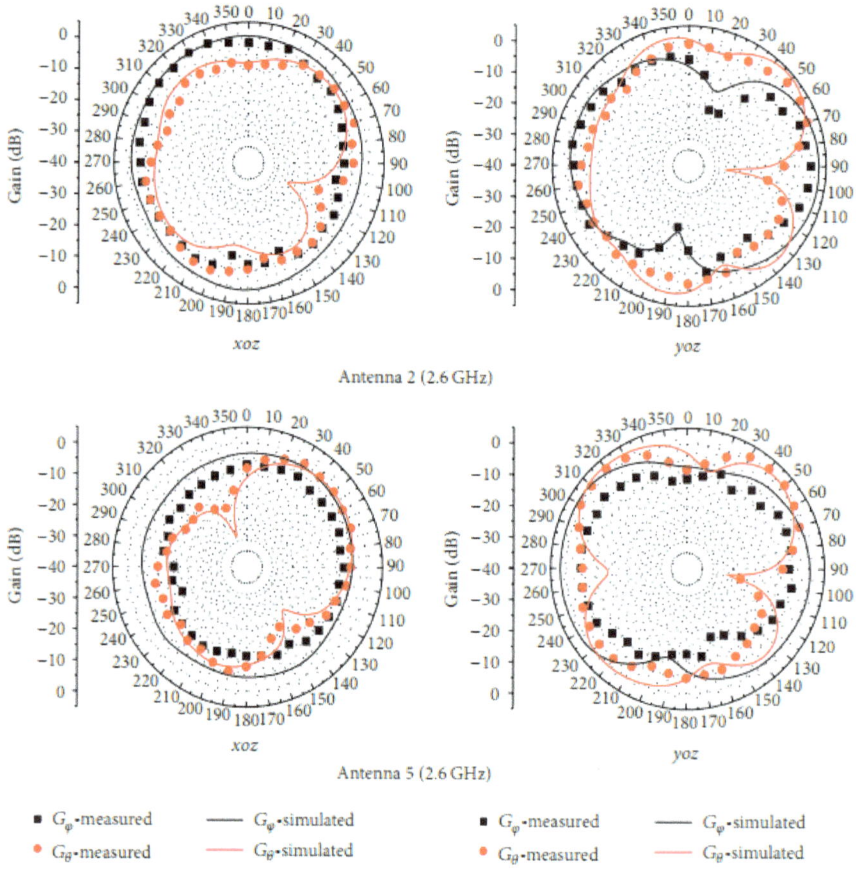

Figure 14: Measured and simulated radiation patterns of antennas 1, 2, and 5 at 1.9 GHz and 2.6 GHz.

As the second arrangement, we move the two middle antenna elements along the long sides to middle position of the short sides, as shown in Figure 15. This arrangement may reduce the influence of human hands on the antennas. Figure 16 shows the measured S-parameters. It is noted that all the isolations are less than −10 dB. Figure 17 is the photograph of the fabricated MIMO system.

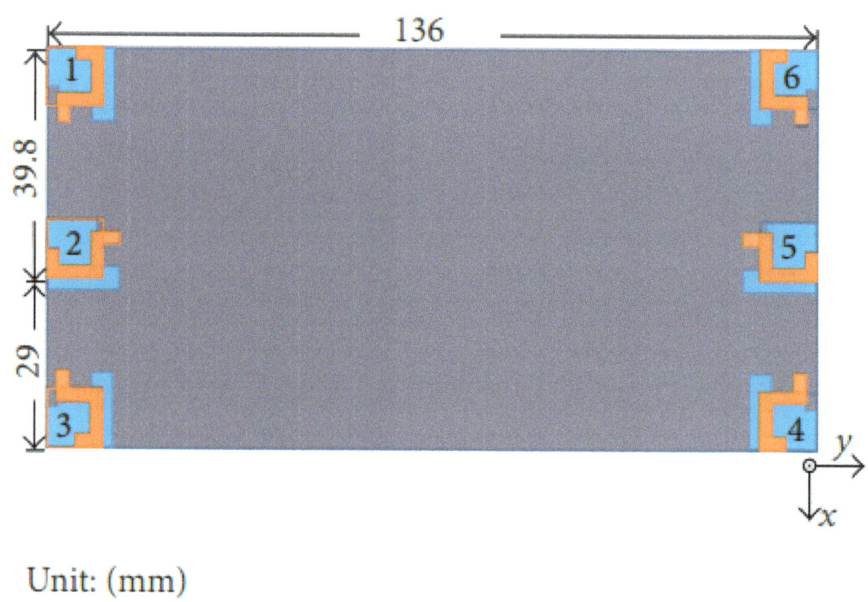

Figure 15: The structure of antennas.

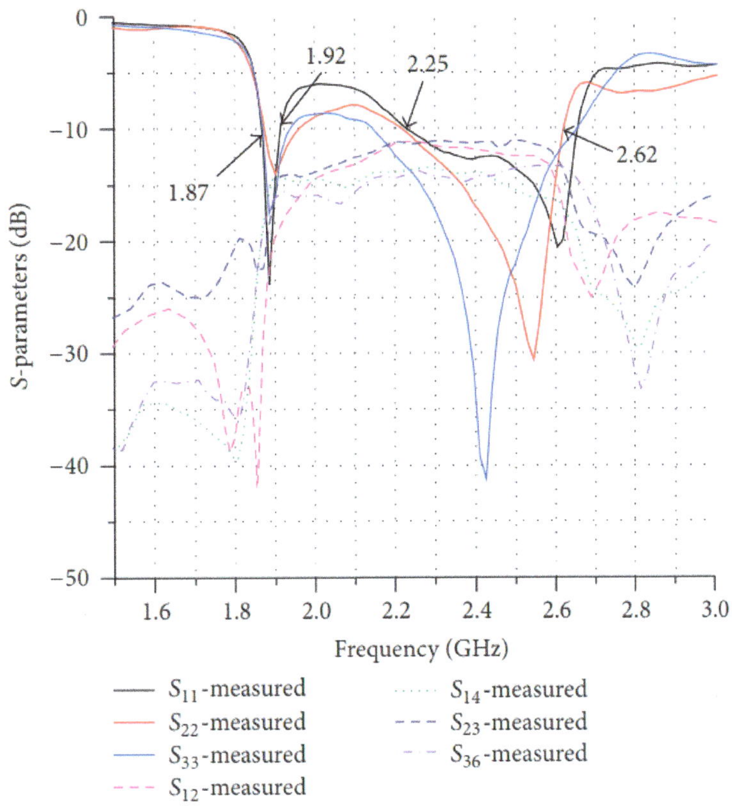

Figure 16: The S-parameters of antennas 1, 2, and 3.

Figure 17: Photograph of the fabricated MIMO antenna system.

Figure 18 shows the envelope correlation coefficients of antennas 1, 2, and 3 in two scenarios, which are less than 0.16 across the frequency bands. The radiation patterns of the antenna elements at 1.9 GHz and 2.6 GHz are shown in Figure 19, from which the MEGs of the antenna elements can be obtained and are listed in Table5. The efficiencies are measured at two frequencies shown in Table 6. Note that the lowest efficiency is about the same with [47], which has only two elements. Also note that the efficiencies for the second arrangement are generally lower than those for the first arrangement due to the fact that the antenna elements are closer to each other and the power absorption by adjacent elements increases. The multiplexing efficiencies are also shown in Table 6.

Table 5: Mean effective gains of the antennas.

Frequency (GHz)	Γ (dB)	MEG1 (dBi)	MEG2 (dBi)	MEG3 (dBi)
1.9	0	-7.1	-8	-7.9
	6	-6.6	-7.8	-7.4
2.6	0	-7.1	-6	-7.8
	6	-7	-7.3	-7.7

Table 6: Antenna efficiency.

Frequency (GHz)		Efficiency (%)	Multiplexing efficiency (dB)
1.9	Antenna 1	39	-7.71
	Antenna 2	53	
	Antenna 3	44	
2.6	Antenna 1	66	-6.22
	Antenna 2	59	
	Antenna 3	72	

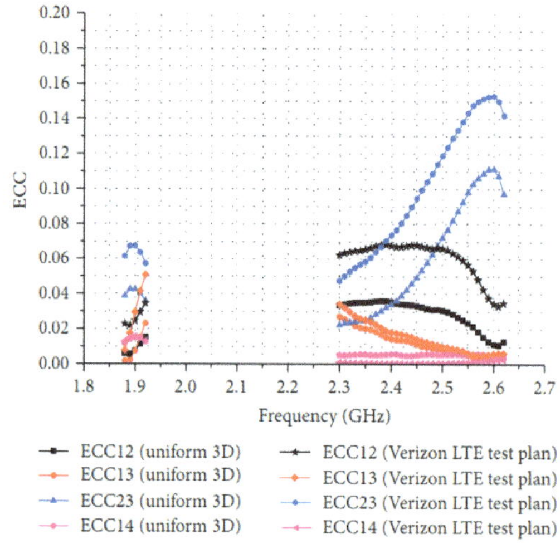

Figure 18: The ECC of antennas 1, 2, and 3.

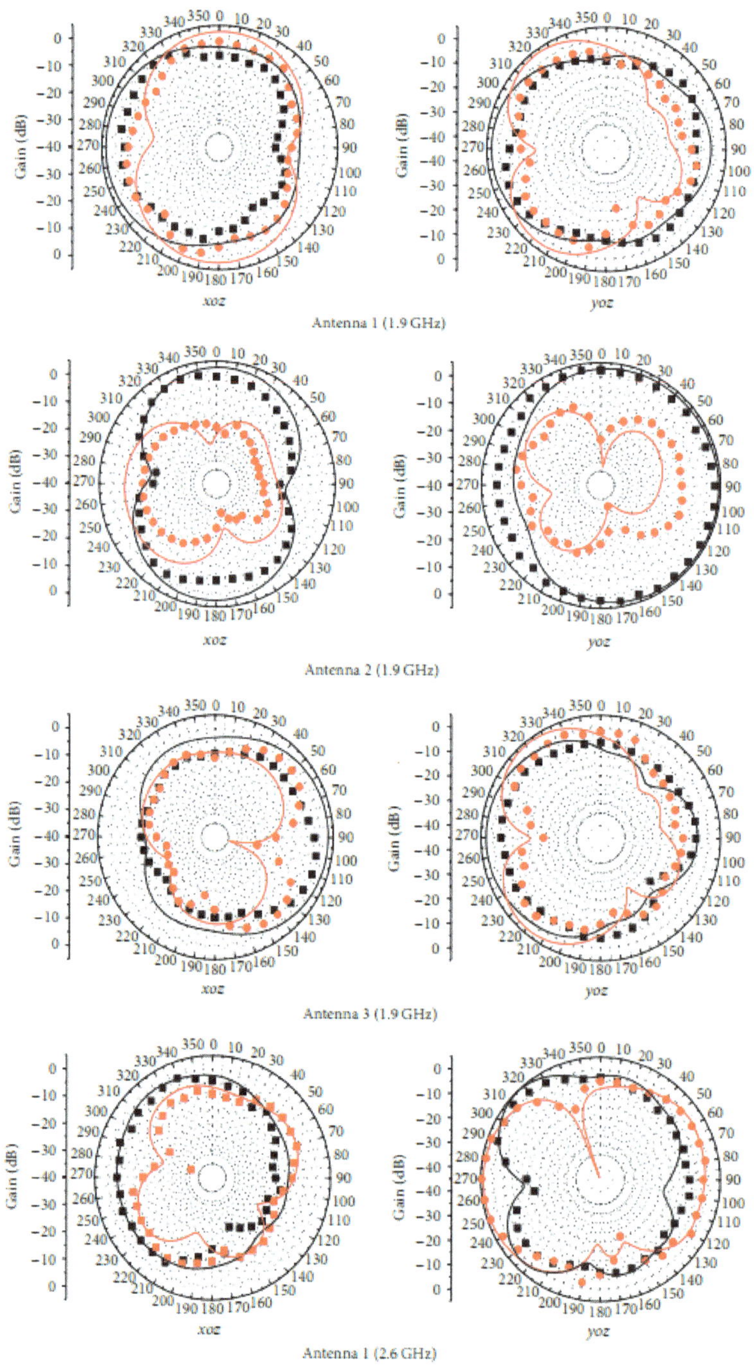

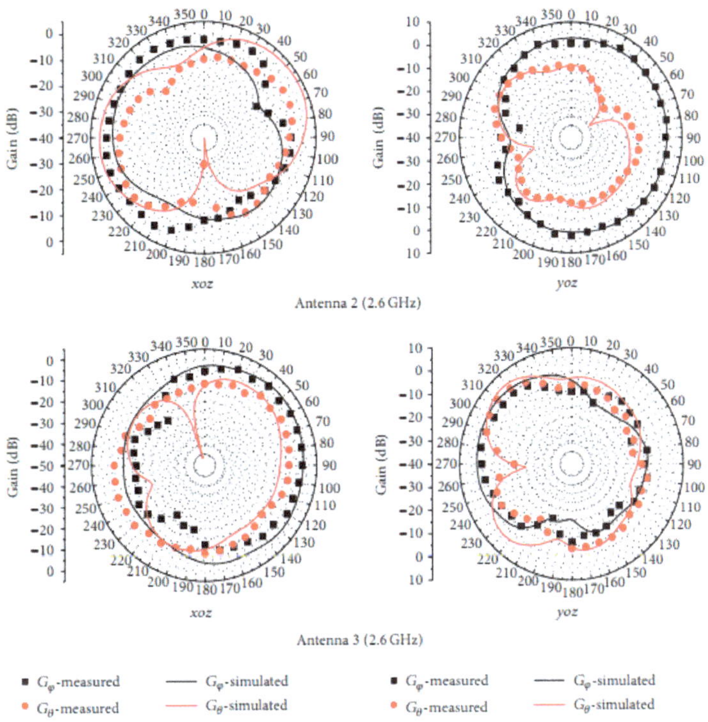

Figure 19: Measured and simulated radiation patterns of antennas 1, 2, and 3 at 1.9 GHz and 2.6 GHz.

Compared with the four-element MIMO arrays, the performances of the six-element MIMO arrays are also acceptable, which implies that a higher capacity can be achieved with more antenna elements. The impact of human body is also an important factor to be considered in the design of handheld devices and have been discussed by many authors (e.g., [48]). In general, the human body will affect various aspects of the performances of the MIMO system, especially degrading the efficiencies of the system.

CONCLUSION

Three kinds of MIMO antenna systems for mobile handheld devices have been investigated in this paper, which are compact in size and cover GSM1900, LTE2300, LTE2500, and WLAN 2.4-GHz for the 5G handset application. The three MIMO systems, all built on a FR4 PCB of small form factor with the size 136 mm × 68.8 mm × 1 mm, consist of four or six three-dimensional planar inverted-F antenna elements wound on a dielectric cube of the size 10 mm × 10 mm × 5 mm (the smallest volume occupied by the antenna that has been reported so far) with relative permittivity 4.4. Different configurations of antenna elements have been examined. The pattern diversity has been used in all our designs and the isolations are higher than 10 dB without using any decoupling circuits. The ECCs are very small according to the measured data, and the MEGs meet the design requirements that the difference of MEGs between elements must be less than 3 dB. The effects of different element number on the performance are discussed. It reveals that the antenna efficiencies depend on the number of antenna elements as well as the separation between them. The antenna efficiency will degrade if the number of elements increases or if the separation between them decreases. These seem to be the big challenges for the design of MIMO system packed in a small space.

ACKNOWLEDGMENTS

This work was supported in part by the Jiangsu Innovation & Entrepreneurship Group Talents Plan and in part by a grant from Huawei Technologies Co. Ltd.

REFERENCES

1. X. Chen, S. Shoaib, I. Shoaib, N. Shoaib, and C. G. Parini, "MIMO antennas for mobile handsets," IEEE Antennas and Wireless Propagation Letters, vol. 14, pp. 799–802, 2015. ·
2. X. Zhao and J. Choi, "Multiband MIMO antenna for 4G mobile terminal," in Proceedings of the IEEE Asia-Pacific Microwave Conference (APMC '13), pp. 41–59, Seoul, South Korea, November 2013. ·
3. K. Zhao, S. Zhang, S. He, K. Ishimiya, and Z. Ying, "Body-insensitive multimode MIMO terminal antenna of double-ring structure," IEEE Transactions on Antennas and Propagation, vol. 63, no. 5, pp. 1925–1936, 2015.
4. A. Toktas and A. Akdagli, "Wideband MIMO antenna with enhanced isolation for LTE, WiMAX and WLAN mobile handsets," Electronics Letters, vol. 50, no. 10, pp. 723–724, 2014.
5. Y. Ding, Z. Du, K. Gong, and Z. Feng, "A novel dual-band printed diversity antenna for mobile terminals," IEEE Transactions on Antennas and Propagation, vol. 55, no. 7, pp. 2088–2096, 2007.
6. K.-S. Min, D.-J. Kim, and M.-S. Kim, "Multi-channel MIMO antenna design for WiBro/PCS band," inProceedings of the IEEE Antennas and Propagation Society International Symposium (APS '07), pp. 1225–1228, IEEE, Honolulu, Hawaii, USA, June 2007.
7. I. P. Kovalyov and D. M. Ponomarev, "Small-size 6-port antenna for three-dimensional multipath wireless

channels," IEEE Transactions on Antennas and Propagation, vol. 54, no. 12, pp. 3746–3754, 2006. · ·

8. W. Li, "A compact MIMO antenna used for LTE terminals with high isolation," in Proceedings of the IEEE 5th International Symposium on Microwave, Antenna, Propagation and EMC Technologies for Wireless Communications (MAPE '13), pp. 308–310, Chengdu, China, October 2013.
9. S. Shoaib, I. Shoaib, N. Shoaib, X. Chen, and C. G. Parini, "A 4x4 MIMO antenna system for mobile tablets," in Proceedings of the 8th European Conference on Antennas and Propagation (EuCAP '14), pp. 2813–2816, The Hague, The Netherlands, April 2014.
10. B. Lee, F. J. Harackiewicz, and H. Wi, "Closely mounted mobile handset MIMO antenna for LTE 13 band application," IEEE Antennas and Wireless Propagation Letters, vol. 13, pp. 411–414, 2014.
11. S. Zhang, Z. Ying, and S. He, "Diagonal chassis mode for mobile handset LTE MIMO antennas and its application to correlation reduction," in Proceedings of the 3rd IEEE International Workshop on Electromagnetics Applications and Student Innovation (IWEM '12), pp. 1–2, IEEE, Chengdu, China, August 2012.
12. Q. Rao and D. Wang, "A compact dual-port diversity antenna for long-term evolution handheld devices,"IEEE Transactions on Vehicular Technology, vol. 59, no. 3, pp. 1319–1329, 2010.
13. A.-D. Capobianco, F. M. Pigozzo, A. Assalini, M. Midrio, S. Boscolo, and F. Sacchetto, "A compact MIMO array of planar end-fire antennas for WLAN applications," IEEE Transactions on Antennas and Propagation, vol. 59, no. 9, pp. 3462–3465, 2011.
14. A. C. K. Mak, C. R. Rowell, and R. D. Murch, "Isolation enhancement between two closely packed antennas," IEEE Transactions on Antennas and Propagation, vol. 56, no. 11, pp. 3411–3419, 2008.
15. C.-C. Hsu, K.-H. Lin, and H.-L. Su, "Implementation of broadband isolator using metamaterial-inspired resonators and a T-shaped branch for MIMO antennas," IEEE Transactions on Antennas and Propagation, vol. 59, no. 10, pp. 3936–3939, 2011.
16. P. J. Ferrer, J. M. González-Arbesú, and J. Romeu, "Decorrelation of two closely spaced antennas with a metamaterial AMC surface," Microwave and Optical Technology Letters, vol. 50, no. 5, pp. 1414–1417, 2008.
17. S.-C. Chen, Y.-S. Wang, and S.-J. Chung, "A decoupling technique for increasing the port isolation between two strongly coupled antennas," IEEE Transactions on Antennas and Propagation, vol. 56, no. 12, pp. 3650–3658, 2008.
18. C.-Y. Chiu, C.-H. Cheng, R. D. Murch, and C. R. Rowell, "Reduction of mutual coupling between closely-packed antenna elements," IEEE Transactions on Antennas and Propagation, vol. 55, no. 6, pp. 1732–1738, 2007.
19. S. Kahng, J. Jeon, J. Anguera, and T. Park, "A compact MIMO antenna using CRLH configuration double-layered folded ring radiations with planar mushroom decoupling structure," IEEE Antennas and Propagation Magazine, vol. 57, no. 2, pp. 123–130, 2015. ·
20. I. Szini, A. Tatomirescu, and G. F. Pedersen, "On small terminal MIMO antennas, harmonizing characteristic modes with ground plane geometry," IEEE Transactions on Antennas and Propagation, vol. 63, no. 4, pp. 1487–1497, 2015.
21. J.-F. Li, Q.-X. Chu, and T.-G. Huang, "A compact wideband MIMO antenna with two novel bent slits,"IEEE Transactions on Antennas and Propagation, vol. 60, no. 2, pp. 482–489, 2012.
22. F. Ahmed, Y. Feng, and R. Li, "Dual wide-band four-unit MIMO antenna system for 4G/LTE and WLAN mobile phone applications," in Proceedings of the Loughborough Antennas and Propagation Conference (LAPC '13), pp. 202–207, Loughborough, UK, November 2013.
23. S. Shoaib, I. Shoaib, N. Shoaib, X. Chen, and C. G. Parini, "Design and performance study of a dual-element multiband printed monopole antenna array for MIMO terminals," IEEE Antennas and Wireless Propagation Letters, vol. 13, pp. 329–332, 2014.
24. M.-S. Han and J. Choi, "MIMO antenna using a decoupling network for next generation mobile application," in Proceedings of the 9th International Symposium on Communications and Information Technology (ISCIT '09), pp. 568–571, IEEE, Icheon, The Republic of Korea, September 2009.
25. J.-M. Lee, K.-B. Kim, H.-K. Ryu, and J.-M. Woo, "A compact ultrawideband MIMO antenna with WLAN band-rejected operation for mobile devices," IEEE Antennas and Wireless Propagation Letters, vol. 11, pp. 990–993, 2012.
26. H. Wang, D. G. Fang, and X. L. Wang, "Mutual coupling reduction between two microstrip patch antennas by using the parasitic elements," in Proceedings of the Asia Pacific Microwave Conference (APMC '08), pp. 1–4, Macau, China, December 2008.
27. G. Li, H. Zhai, Z. Ma, C. Liang, R. Yu, and S. Liu, "Isolation-improved dual-band MIMO antenna array for LTE/WiMAX mobile terminals," IEEE Antennas and Wireless Propagation Letters, vol. 13, pp. 1128–1131, 2014.
28. D. F. Kelley and W. L. Stutzman, "Array antenna pattern modeling methods that include mutual coupling effects," IEEE Transactions on Antennas and Propagation, vol. 41, no. 12, pp. 1625–1632, 1993.
29. V. Ssorin, A. Artemenko, A. Sevastyanov, and R. Maslennikov, "Compact bandwidth-optimized two element MIMO antenna system for 2.5–2.7 GHz band," in Proceedings of the 5th European Conference on Antennas and Propagation (EUCAP '1), pp. 319–323, Rome, Italy, April 2011.
30. Q. Rao and W. Geyi, "Compact multiband antenna for handheld devices," IEEE Transactions on Antennas and Propagation, vol. 57, no. 10, pp. 3337–3339, 2009.
31. W. Geyi, Q. Rao, S. Ali, and D. Wang, "Handset antenna design: practice and theory," Progress in Electromagnetics Research, vol. 80, pp. 123–160, 2008.
32. S. Stein, "On cross coupling in multiple-beam antennas," IRE Transactions on Antennas and Propagation, vol. 10, no. 5, pp. 548–557, 1962. ·

33. I. Salonen and P. Vainikainen, "Estimation of signal correlation in antenna arrays," in Proceedings of the 12th International Symposium on Antenna (JINA '02), vol. 2, pp. 383–386, Nice, France, November 2002.

34. S. Blanch, J. Romeu, and I. Corbella, "Exact representation of antenna system diversity performance from input parameter description," Electronics Letters, vol. 39, no. 9, pp. 705–707, 2003.

35. R. G. Vaughan and J. B. Andersen, "Antenna diversity in mobile communications," IEEE Transactions on Vehicular Technology, vol. 36, no. 4, pp. 149–172, 1987. ·

36. J. P. Kermoal, L. Schumacher, K. I. Pedersen, P. E. Mogensen, and F. Frederiksen, "A stochastic MIMO radio channel model with experimental validation," IEEE Journal on Selected Areas in Communications, vol. 20, no. 6, pp. 1211–1226, 2002.

37. Q. Yao, Y. Yuan, A. Ghazal, C.-X. Wang, L. Luan, and X. Lu, "Comparison of the statistical properties of the LTE-A and IMT-A channel models," in Proceedings of the IEEE Wireless Communications and Networking Conference (WCNC '12), pp. 393–398, IEEE, Shanghai, China, April 2012.

38. A. Jassal, H. Khanfir, and S. Martinez Lopez, "Preliminary system-level simulation results for the 3GPP 3D MIMO channel model," in Proceedings of the IEEE 80th Vehicular Technology Conference (VTC-Fall '14), pp. 1–5, Vancouver, Canada, September 2014.

39. G. F. Pedersen and J. B. Andersen, "Handset antennas for mobile communications: integration, diversity and performance," in Review of Radio Science, pp. 119–138, Oxford University Press, Oxford, UK, 1999.

40. G. Kang, Z. Du, and K. Gong, "Novel compact quasi-dipole diversity antenna for mobile terminals," IEEE Antennas and Wireless Propagation Letters, vol. 10, pp. 1166–1169, 2011.

41. T. Taga, "Analysis for mean effective gain of mobile antennas in land mobile radio environments," IEEE Transactions on Vehicular Technology, vol. 39, no. 2, pp. 117–131, 1990.

42. S. C. K. Ko and R. D. Murch, "Compact integrated diversity antenna for wireless communications," IEEE Transactions on Antennas and Propagation, vol. 49, no. 6, pp. 954–960, 2001.

43. Y. Wang and Z. Du, "A printed dual-antenna system operating in the GSM1800/GSM1900/UMTS/LTE2300/LTE2500/2.4GHz WLAN bands for mobile terminals," IEEE Antennas and Wireless Propagation Letters, vol. 13, pp. 233–236, 2014.

44. R. Tian, B. K. Lau, and Z. Ying, "Multiplexing efficiency of MIMO antennas," IEEE Antennas and Wireless Propagation Letters, vol. 10, pp. 183–186, 2011.

45. R. Tian, B. K. Lau, and Z. Ying, "Multiplexing efficiency of MIMO antennas in arbitrary propagation scenarios," in Proceedings of the 6th European Conference on Antennas and Propagation (EuCAP '12), pp. 373–377, IEEE, Prague, Czech Republic, March 2012.

46. R. Vaughan and J. B. Andersen, Channels, Propagation and Antennas for Mobile Communications, Institution of Electrical Engineers (IEE), London, UK, 2003.

47. S. Wang and Z. Du, "Decoupled dual-antenna system using crossed neutralization lines for LTE/WWAN smartphone applications," IEEE Antennas and Wireless Propagation Letters, vol. 14, pp. 523–526, 2015. · ·

48. K. Zhao, S. Zhang, Z. Ying, and S. He, "MIMO performance study of different antennas for LTE mobile phones in CTIA test mode," in Proceedings of the 7th European Conference on Antennas and Propagation (EuCAP '13), pp. 727–731, Gothenburg, Sweden, April 2013.

CHAPTER 23

DESIGN OPTIONS FOR THE UPGRADE OF THE CMS ELECTROMAGNETIC CALORIMETER

R. Paramatti

Keywords: electromagnetic calorimeter; scintillation; crystal; lead-tungstate; radiation; upgrade;HL-LHC

THE ELECTROMAGNETIC CALORIMETER OF CMS

The CMS ECAL is a compact and homogeneous calorimeter made of 75848 lead-tungstate ($PbWO_4$) crystals, arranged in a quasi-projective geometry [1]. The barrel region (EB) covers a pseudorapidity range up to $|\eta| = 1.48$ while the two endcaps (EE) extend up to $|\eta| < 3.0$. The scintillation light is readout with avalanche photodiodes (APDs) in EB and with vacuum phototriodes (VPTs) in EE. The transverse size of the crystals is very close to the PbWO4 Moliere radius of ` 2.2 cm while the crystal depth is about 26 and 25 radiation lengths ($X_0 = 0.89$ cm) respectively in EB and EE. A preshower detector (ES), based on lead absorbers equipped with silicon strip sensors, is placed in front of the endcap crystals.

PERFORMANCE OF CMS ECAL DURING HL-LHC

The HL-LHC conditions are a significant challenge to both detector longevity and performance. The challenge is particularly difficult in the endcap region, where the radiation levels have a strong gradient along η and typically change by a factor 100 between $|\eta| = 1.48$ and $|\eta| = 3.0$. For the full 3000 fb^{-1} planned for HL-LHC, the highenergy charged hadron fluence at $|\eta| = 2.6$ in the ECAL will be $2 \cdot 10^{14}$ particle/cm^2 and the absorbed dose will be 300 kGy with an expected dose rate of 30 Gy/h. As discussed below, these dose and fluence levels would result in significant loss to the crystal light transmission and the photo-detector (VPT) performance and motivate the replacement of the endcap calorimeter of CMS for HL-LHC.

In the barrel ECAL, radiation damage is not problematic even at the highest η. However, the EB detector has to cope with higher pile-up and increased photodetector (APD) noise resulting from the foreseen instantaneous luminosity of $5 \cdot 10^{34} cm^{-2} s^{-1}$. For these reasons, keeping a high trigger efficiency is a challenge. Finally the electronics must be upgraded to meet the Level 1 readout rate and latency requirements (see Section 4). Lead tungstate crystals are subject to two types of damage: electromagnetic and hadron irradiation damage. The

Citation: R. Paramatti, Design options for the upgrade of the CMS electromagnetic calorimeter, Nuclear and Particle Physics Proceedings, Volumes 273–275, April–June 2016, Pages 995-1001, ISSN 2405-6014, http://dx.doi.org/10.1016/j.nuclphysbps.2015.09.156.

effects of ionizing electromagnetic radiation on lead tungstate were studied in depth using γ radiation from ^{60}Co sources. The crystal light transmission is reduced through the formation of colour centres, while the scintillation mechanism is unchanged. The crystal transparency largely recovers through spontaneous annealing at room temperature [2]. The light output tends towards an equilibrium level, which depends on the dose rate. A precise monitoring system using injected laser light at 447 nm is used to track and correct for response changes during LHC operation. The dominant concern for HL-LHC is the loss of response due to hadron irradiation. In CMS the flux of fast hadrons is dominated by charged pions with energies of order 1 GeV. The effect of charged hadrons has been studied on crystals irradiated with pions and protons from the CERN Proton Synchrotron [3, 4]. Figure 1 shows that the band edge of the transmission spectrum is shifted to higher wavelengths, the transmission is attenuated and therefore the light output is seriously affected. Contrary to the electromagnetic radiation, this damage is cumulative and does not anneal out at room temperature.

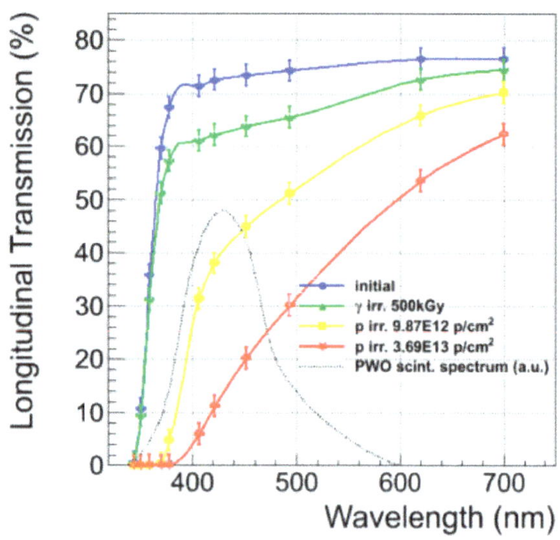

Figure 1: Light transmission of lead tungstate crystals as a function of the wavelength after proton and Co60 photon irradiation; the scintillation spectrum of the crystal is superimposed.

An extensive test-beam campaign on hadron irradiated matrices of 3×3 crystals has been performed to measure the degradation of the energy resolution for several hadron fluences [5]. The results are shown in Figure 2 where the irradiation dose is expressed in terms of the radiation induced absorption coefficient $\mu_{ind}(\lambda)$:

$$\mu_{\text{ind}}(\lambda) = -\frac{1}{L} \cdot ln\left(\frac{LTO(\lambda)}{LTO_0(\lambda)}\right) \tag{1}$$

where LTO is the attenuation of the optical transmission along the crystal length L, with respect to its initial value LTO_0. The $\mu_{ind}(\lambda)$ is found to be linearly dependent on the hadron fluence in the range of fluence under study ($5 \cdot 10^{11}$ to $5 \cdot 10^{13}$protons/cm^2) [3, 4].

The worsening of the energy resolution with the increasing dose, from a non-irradiated crystal matrix (μ_{ind} = 0 m−1) up to the equivalent dose at the end of HL-LHC in the region $|\eta|$ = 2.6 (μ_{ind} = 20 m^{-1}), is clearly visible in the figure. The beam test has also shown that the linearity and timing response of the calorimeter are also degraded at high values of μ_{ind}. Based on the above measurements, a model of crystal ageing has been developed. The radiation damage is modelled using the MARS simulation [6] for the hadron fluence and γ dose estimate and the experimental results for the attenuation of the optical transmission. A full simulation of the electromagnetic shower development in the crystals is performed with Geant4 [7]. The ray-tracing program SLitrani [8] is used to model the light output as a function of μ_{ind}. The γ damage is simulated as a function of the instantaneous luminosity, by assuming the equilibrium value of the light transmission is reached. The hadron damage is simulated by modelling the value of μ_{ind} measured in [3] as a function of the hadron fluence.

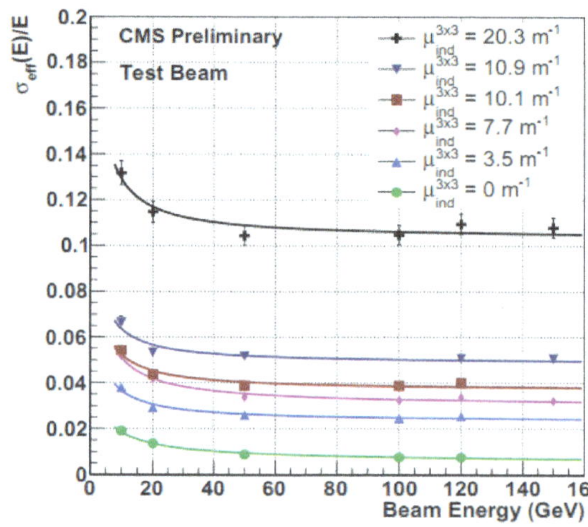

Figure 2: Energy resolution measured with electron beam on hadron irradiated crystals with different dose from non irradiated crystal (μ_{ind} = 0 m^{-1}) up to the equivalent dose at the end of HL-LHC in the region $|\eta|$ = 2.6 (μ_{ind} = 20 m^{-1}).

Figure 3 shows the predicted response reduction, S/S_0, for an electron shower of 50 GeV as a function of η for various integrated luminosities. The uncertainty in this prediction is about 30%, and it comes from uncertainty in the hadron fluence and γ dose estimates, spread in the crystal response to γ radiation, uncertainty in the effects of the hadron radiation on the crystals, and extrapolation of these effects from low to high fluence. Based on the aging model, the barrel ($|\eta| < 1.48$) will suffer less than a factor of 2 light output loss up to 3000 fb^{-1}, while in the endcaps there will be a factor of more than 10 reduction after 3000 fb^{-1} for $\eta > 2$. The large response reduction causes an unacceptable worsening of the energy resolution of electrons and photons, requiring the replacement of the endcap part of the calorimeter. More in detail, the reduction of scintillation light output with respect to undamaged crystals affects the energy resolution causing:

- a degradation of the stochastic term due to the reduced light output, proportional to $\sqrt{S_0/S}$
- an amplification of the noise term due to the increasing energy equivalent noise, proportional to S_0/S ,
- a worsening of the constant term due to the increased light collection non-uniformity along the crystal depth z, proportional to $\partial(S/S_0)/\partial z$.

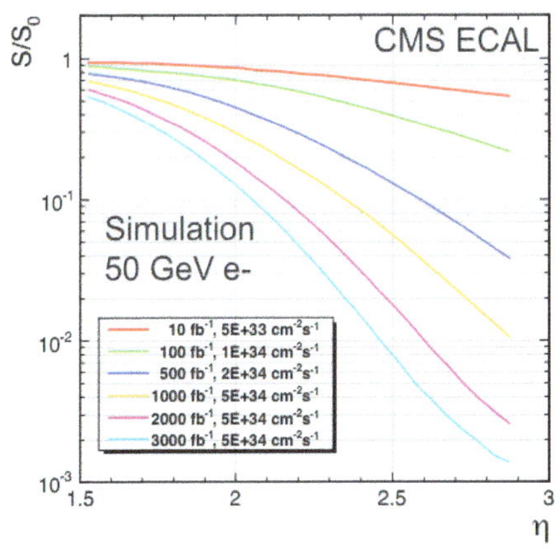

Figure 3: Relative light output S/S_0 for an electron shower of 50 GeV as a function of crystal pseudorapidity, estimated with the model described in the text, for different ageing conditions.

Ageing of Avalanche Photodiodes

The ECAL barrel APDs are silicon photodiodes with internal gain, developed and produced by Hamamatsu Photonics in collaboration with CMS. They are operated at gain G = 50 and at a temperature of 18°C. As with all silicon devices, they are sensitive to γ and hadron damage [9]. The γ radiation causes an increase in the surface current and decreases the quantum efficiency. However the principal concern for the HL-LHC is hadron damage. Neutrons dominate the hadron spectrum behind the ECAL crystals. Hadrons create defects in the silicon bulk which induce an increase in dark current of the detector and consequently in the levels of electronic noise. The evolution of the dark current of APDs in channels located at different pseudorapidity during LHC Run1 is shown in Figure 4 together with the LHC delivered luminosity as a function of time. These data confirm the linearity in the rate of creation of defects in the APD with neutron fluence. During technical stops and winter shutdowns a part of the APD defects anneals and the current reduces.

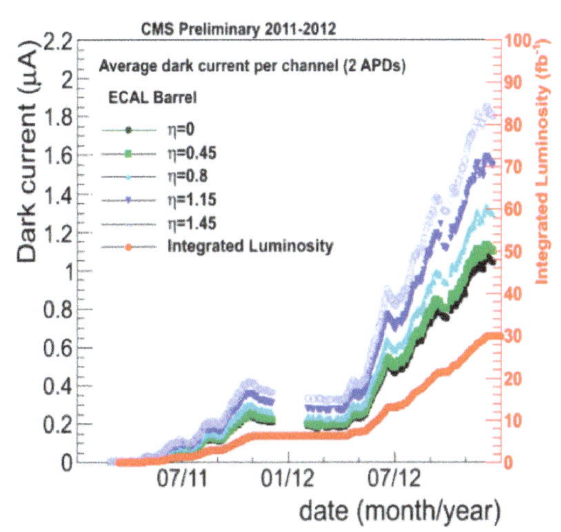

Figure 4: Evolution of the dark current in Avalanche Photodiodes installed in the ECAL Barrel for different pseudorapidity during LHC Run1. The red points show the LHC delivered luminosity as a function of time. During technical stops and winter shutdowns a partial reduction of the current is observed.

The neutron fluence predicted with the FLUKA simulation [10] is about Φ_n = 1.2–2.4 · 10^{14} n/cm^2 (1 MeV neutron equivalent) at $\eta = 0 - 1.45$ for a delivered luminosity of 3000 fb^{-1}. The expected dark current of a pair of APDs at gain 50 at $\eta = 1.45$, considering only the permanent bulk damage, is 200-250 μA. Other effects due to neutron damage at high fluences are a shift of the bias voltage required to obtain a gain of 50 and a reduction of quantum efficiency

The preamplifier noise depends on the square root of the APD bulk current times the shaping time τ, which is currently 43 ns. The evolution of noise as a function of the dark current has been measured with APDs irradiated up to the HL-LHC expected fluence. The noise is expected to increase up to 9-10 ADC counts for a dark current of 200-250 μA (one ADC count corresponds to about 40 MeV). Figure 5 shows the predicted increase of the noise as a function of the integrated luminosity with the present ECAL electronics. The noise levels in EB will reach almost 350 MeV per channel at $\eta = 1.45$ after 3000 fb^{-1}, neglecting changes in crystal light output and quantum efficiency. Considering that electromagnetic showers in ECAL are typically reconstructed summing the energy of at least a 5×5 crystal matrix, the level of electronics noise will be the limiting factor of the EB energy resolution for photons after 1500 fb^{-1}.

The APD dark current is strongly dependent on the temperature and a reduction of a factor 2.5 in dark current is expected if the ECAL operating temperature is reduced from 18 to 8 °C, which translates into a reduction of the noise by 35%. Figure 5 also presents the expected noise levels in the detector at lower temperature and it shows that operating the APDs at 8°C after 3000 fb^{-1} corresponds in terms of noise to operating at 18°C after about 1000-1200 fb^{-1}.

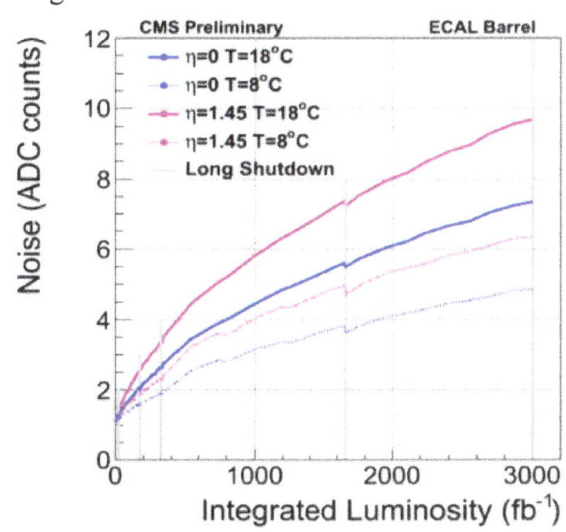

Figure 5: Expected noise with current ECAL electronics as a function of LHC integrated luminosity in the centre of the Barrel ($|\eta| = 0$) and at the edge ($|\eta| = 1.45$). The expected noise is shown also in the case of the decreasing of the ECAL operating temperature to 8 °C.

NEW BARREL ELECTRONICS

In the present calorimeter, the channel readout is formed by two APDs mounted on the rear face of the crystal and connected in parallel. The APDs are connected to the passive motherboard (MB) which distributes high voltage and low voltage and connects the APDs to the very-front-end (VFE) cards. Each VFE card has five readout channels consisting of a multi-gain pre-amplifier (MPGA) with a 43 ns shaping time and a 12-bit analog-to-digital converter (ADC). The digitized signals from five VFE cards are passed to an FE card. The FE card forms the trigger primitive for the 5×5 crystal array and contains a digital latency buffer and the primary event buffer. All the boards were tested to resist the radiation environment of the ECAL endcaps and for ten-year operation. The trigger primitive consists of the transverse energy deposited in a trigger tower (the 5×5 crystal array) and a single bit to qualify the energy deposit along η. The per-crystal information is buffered in the FE for transmission to the data concentrator card, with a maximum latency in the event buffer of 6.4 μs and a maximum Level 1 accept rate of about 100 kHz. The ECAL barrel electronics will be replaced to meet the requirements of the Phase 2 Level 1 trigger system, which foresees an increase of the latency up to 20 μs and an increase of the bandwidth up to 1 MHz. This upgrade will also allow the provision of individual crystal information to the Level 1 calorimeter trigger. By including the VFE in the upgrade, the pulse shaping and sampling can be optimised to mitigate the increase in APD noise and pile-up and improve timing resolution. Finally it will reduce the risk of using electronics for 30 years and 3000 fb^{-1} that were designed for 10 years running and 500 fb^{-1}. The performance goals of the EB upgrade can be achieved through the replacement of the VFE and FE cards, together with the associated low-voltage distribution system and optical links, without changing the APDs and the motherboards as there is no identified concern with these items. Figure 6 shows a possible new architecture. The individual boards will follow the same configuration and form factor as the present electronics in order to fit into the same physical space and use the existing services whenever possible.

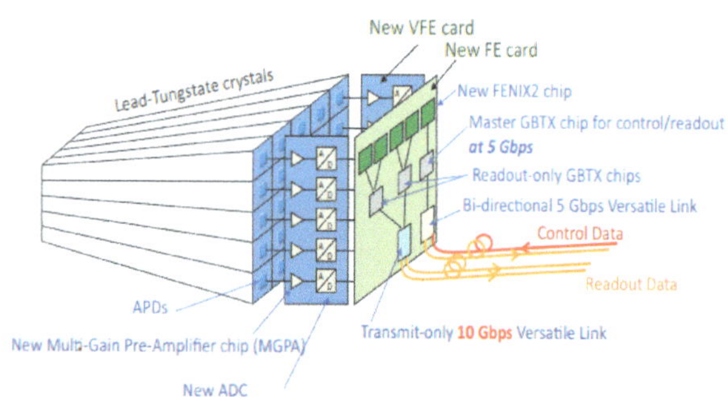

Figure 6: A possible new architecture for the upgraded ECAL Barrel Electronics.

THE ENDCAP CALORIMETRY UPGRADE

As discussed above, the physics performance of CMS ECAL in the HL-LHC phase will be significantly compromised without the replacement of the endcap part. Two conceptual designs for the replacement of endcap calorimetry systems are presented here. One design proposes a standalone tungsten/LYSO(CeF_3) sampling electromagnetic calorimeter optimised for a small Moliere radius and high radiation tolerance, followed by a full-depth hadron calorimeter based on scintillators. In the second conceptual design the endcaps of both the electromagnetic and hadronic calorimeters are replaced by a silicon-based calorimeter with a very high transverse and longitudinal granularity to manage pile-up by tracking shower development in detail.

New sampling ECAL Endcap

The EE sampling structure proposed here uses a very dense design to minimize the effects of pile-up. The structure consists of alternating layers of dense absorber and scintillating crystal tiles. Various materials have been simulated using standalone GEANT4 and fast simulation techniques and the sampling combination of tungsten absorbers with LYSO(Ce) or CeF_3 scintillators provides the best energy resolution, dimensional compactness, and the highest light yield for electromagnetic calorimetry in the endcap/forward region, comparable with the light yield of NaI. These combinations can provide an energy resolution of about $10\%/\sqrt{E}$ and excellent potential for minimizing the effects of pile-up due to the small Moliere radius (14 mm for the W ` /LYSO combination to be compared with 21 mm of the lead tungstate) with a very similar emission wavelength and decay time. The new calorimeter will also be more compact with respect to the current ECAL endcap thanks to the shorter radiation length: 5.1 mm and 8.9 mm respectively for the W/LYSO and the $PbWO_4$.

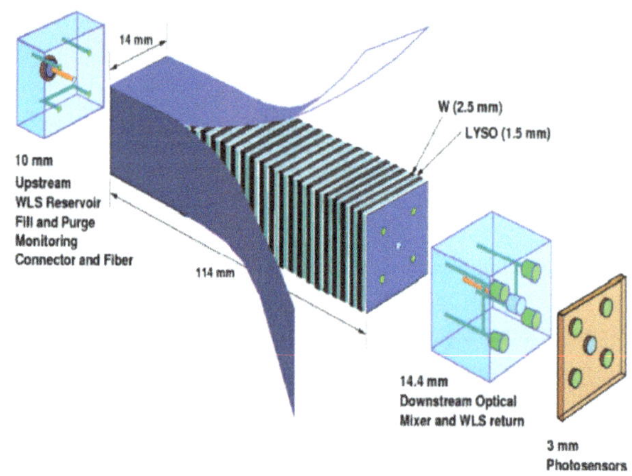

Figure 7: The EE Shashlik configuration based upon interleaved W and LYSO layers. Twenty-nine LYSO crystals and twenty-eight W plates comprise the module. Four WLS quartz capillaries are used to read out the scintillation light from the tiles. A central calibration fiber is used for monitoring. (An alternative technology uses the same sampling structure and thicknesses, but has CeF_3 replacing the LYSO.)

Two different options for the light extraction have been proposed: the shashlik configuration (Fig. 7) with the wavelength shifting (WLS) quartz capillaries running inside the crystals and a more standard configuration with the fibers located along the depolished chamfers of the channel. Beam tests of Shashlik detectors have been performed at Fermilab. A supermodule containing 16 shashlik modules in a 4x4 array has been tested with muon, electron and pion beams with energy up to 120 GeV at the MTest facility at Fermilab. Analysis of the data is currently underway. Beam tests of a W/CeF_3 module read out with WLS fibers in a beam of 491.4 MeV electrons at Frascati have also been completed few months ago. Figure 8 shows the energy spectrum of the CeF_3 where multiple electrons interacting in the structure can be clearly resolved and spectra obtained with single electron selection using external beam line devices are also shown [11]. Further data analysis is ongoing. Both configuration will be tested soon at CERN in H4 area.

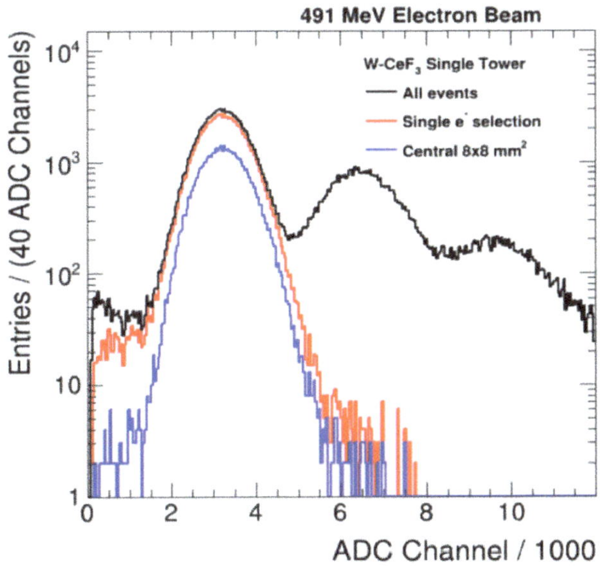

Figure 8: Energy spectrum of the W/CeF_3 channel tested with 491 MeV electron beam in Frascati. Single, double and triple electron peaks are clearly visible (black distribution). Spectra obtained with single electron selection criteria are also shown.

High Granularity Endcap Calorimeter

A dense high-granularity calorimeter running in the High Luminosity LHC offers the opportunity of high performance in the presence of high pile-up environment of the forward (endcap) regions providing very high resolving power for single particles and jets and a good jet energy resolution also exploiting the particle flow technique. The overall design of the high-granularity calorimeter, HGC, is shown in Fig. 9. Moving outwards from the interaction region, the volume begins near the location of the front face of the current electromagnetic endcap calorimeter with a tungsten/lead-silicon sampling electromagnetic calorimeter, the E-HG, with a depth of 25 X_0 and about 1 λ. This is followed by a brasssilicon hadron calorimeter, 3.5 λ deep, the H-HG, which is followed by a 5.5 λ brass-scintillator sampling backing calorimeter, the B-HG, to bring the total calorimeter depth to 10 λ. The E-HG and H-HG will use planes of silicon as the active medium, while the B-HG, situated where the radiation levels are low, can be constructed with either plastic scintillator, as is the current HE, or with GEM or Micromegas gas chambers.

In this structure there will be only a small gap between the electromagnetic and hadron calorimeters, which is expected to improve the jet energy measurements. Additionally, by bringing the hadron calorimeter into the space currently used for the EE electronics, there will be space behind the B-HG where the radiation levels will be sufficiently low so that not only electronics stations and other services, but also additional muon chambers could be situated. For E-HG a variable longitudinal sampling is proposed, determined by the absorber thickness:

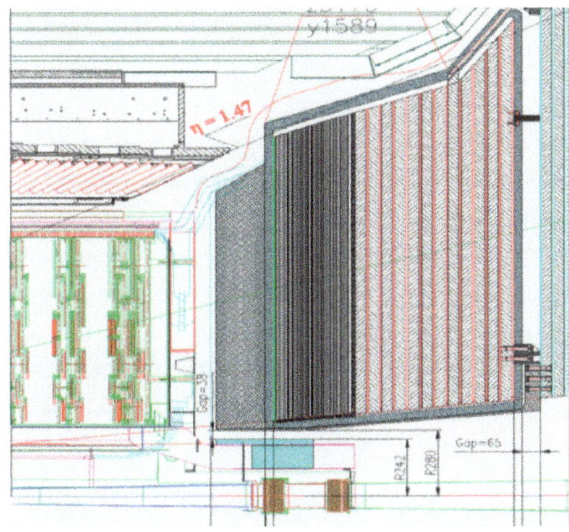

Figure 9: Sketch of a possible structure of HGC. The E-HG has its front face at the same location as the front face of the current EE. Directly behind it there is the H-HG, which is a 3.5 λ silicon-brass calorimeter. Behind that is a 5.5 λ backing hadron calorimeter, which, since the radiation levels are low, could use the same technology as the current HE.

- 10 layers: 0.5 X_0 thickness absorber followed by a plane of silicon,
- 10 layers: 0.8 X_0 thickness absorber followed by a plane of silicon,
- 10 layers: 1.2 X_0 thickness absorber followed by a plane of silicon.

Most of the calorimeter uses cells of 0.9 cm^2. There will be six million cells in the E-HG for a surface of 420 m^2 of silicon. The absorber layers are composed of tungsten in an alveolar mechanical structure, and copper/lead in the units which slide into the alveolar structure. A similar configuration, but with tungsten as the absorbing material, has been studied in test beams by the CALICE collaboration for future ILC/CLIC detectors and, when instrumented with 525 micron-thick silicon sensors, gave an energy resolution of around 16.5%/ $\sqrt{E}$ with a linear energy response [12].

PILE-UP MITIGATION WITH PRECISE TIMING

Although precise time measurement has not been a primary goal of calorimeters at hadron colliders, during LHC Run1 the CMS ECAL measured the timing of electrons from Z decays with an excellent resolution, about 200 ps [13]. During HL-LHC a more precise timing may allow the subtraction of spurious energy deposits coming from pile-up, help identify pile-up jets, and improve the estimation of the position of the production vertex for di-photon events. A time measurement with precision of a few tenths of ps would allow the association of energy deposits to the primary vertex from which they originate within 1 cm, thus reducing significantly the probability of misreconstructing photons and electrons. R&D is ongoing to determine if such a goal can be obtained by adding a dedicated timing detector or by introducing more sophisticated timing capabilities to the new electronics in EB and to the new EE detector. For a dedicated detector, micro-channel plates for direct ionization measurement, gas electron multipliers, and avalanche photodiode detectors are being investigated. If a pre-radiator were included, such detectors could provide timing information for both neutral and charged particles. The current ECAL geometry has been used in a simplified study using Geant where crystals are divided in different longitudinal sub-cells. In each sub-cell the timing information is extracted by averaging the time of simulated hits within that volume, weighted by the energy of each deposit, restricting to deposits of energy above 5 keV. The optimal thickness and position of the subcell has been studied in terms of intrinsic spread of the timing information. The smallest spread is obtained for 1 cm thick layer when there are about 8-10 radiation lengths in front of it, i.e. at the maximum of the shower. The time spread as a function of depth for photons of transverse momentum of 25 GeV is shown in the Figure 10 and is less than 10 ps already after 2 X_0.

CONCLUSIONS

The High Luminosity LHC environment poses severe requirements to detectors in terms of performance and radiation hardness. We have shown that the ECAL barrel will survive up to 3000 fb^{-1} providing good performance during HL-LHC while the ECAL endcaps should be replaced at the end of the LHC Phase 1 (after 500 fb^{-1}). Two new endcap calorimeter options are being studied: a sampling calorimeter composed of tungsten absorber and LYSO or CeF_3 scintillators and a high granularity calorimeter (HGC). Finally we discussed the possibility to provide precise timing resolution, which may add important information for pile-up mitigation.

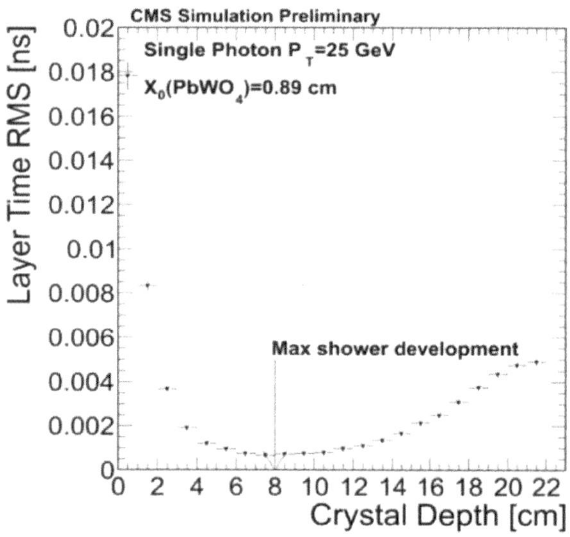

Figure 10: Intrinsic spread of the Geant hits time for 25 GeV p_T photons as a function of the crystal depth with 1 cm thick layers.

REFERENCES

1. CMS Collaboration, "The CMS experiment at the CERN LHC", JINST 3 (2008) S08004
2. P. Adzic et al., "Radiation hardness qualification of PbWO4 scintillation crystals for the CMS Electromagnetic Calorimeter", JINST 5 (2010) P03010
3. M. Huhtinen et al., "High-energy proton induced damage in PbWO4 calorimeter crystals", Nucl. Instrum. Meth. A545 (2005), 63-87
4. E. Auffray et al., "Experimental study of the lead tungstate scintillator proton-induced damage and recovery", IEEE Trans. Nucl. Sci. 59 n.5 (2012) 2219-2223
5. M.T. Lucchini et al., "Test beam evaluation of CMS ECAL Endcap Supercrystals constructed from hadron-damaged crystals", CMS DN-2013/007
6. I. Azhgirey and V.Talanov, "Proc. of XVIII Workshop on the Charged Particles Accelerators", Technical Report CMS-2005- 010, (2000)
7. GEANT4 Collaboration, "GEANT4 - a simulation toolkit", Nucl. Instrum. Meth. A506 (2003) 250
8. F. Gentit, "Litrani: A General purpose Monte Carlo program simulating light propagation in isotropic or anisotropic media", Nucl.Instrum.Meth. A486 (2002) 35-39
9. S. Baccaro et al., "Radiation damage effect on avalanche photodiodes", Nucl.Instrum.Meth. A426 (1999) 206-211
10. G. Battistoni, F. Cerutti, A. Fasso' et al., "The FLUKA code: Description and benchmarking", AIP Conf. Proc. 896 (2007)
11. P. Meridiani et al., to be published in Proc. IEEE NSS 2014
12. C. Adloff, Y. Karyotakis, J. Repond et al., "Response of the CALICE Si-W electromagnetic calorimeter physics prototype to electrons", Nucl.Instrum.Meth. A608 (2009) 372-383
13. CMS Collaboration, "Energy calibration and resolution of the CMS electromagnetic calorimeter in pp collisions at sqrt(s) = 7 TeV", JINST 8 (2013) P09009

CHAPTER 24

ELECTROMAGNETIC CONTRIBUTION TO CHARGE SYMMETRY VIOLATION IN PARTON DISTRIBUTIONS

X.G. Wang, A.W. Thomas, R.D. Young

Keywords: Charge symmetry; Parton distributions; Electromagnetic correction

INTRODUCTION

Charge symmetry (CS) refers to the invariance of the QCD Hamiltonian under the operator $e^{i\pi I_2}$, a rotation by 180 deg about the 2-axis in iso-space. Under this operationu-quarks rotate into d -quarks and vice-versa, while protons and neutrons are also interchanged. As a result, if QCD were to respect this symmetry, the up-quark distribution in the proton, u^p, and the down quark distribution in the neutron, d^n, would be identical. Similarly one would have $d^p = u^n$. Precisely these relations have been almost universallyassumed for the past 40 years, as without such an assumption it would have been impossible to separate the flavor dependence of the parton distribution functions (PDFs).

Studies of strongly interacting systems have established that CS is typically respected at the level of a fraction of a percent [1], much better than isospin symmetry, which requires the invariance of the Hamiltonian under all rotations in iso-space. Nevertheless, as one uses tests of symmetries to probe for physics beyond the Standard Model, or aims for higher precision at the LHC, it is vital to know just how well the PDFs satisfy CS. Furthermore, subtle tests of such a symmetry can also yield information on how QCD itself works. Thus the study of charge symmetry violation (CSV) in PDFs may also lead to a deeper understanding of the structure of the nucleon itself. For reviews of CSV in PDFs we refer to Londergan et al. [2] and [3].

There are two dominant sources of CSV in the nucleon, the electromagnetic interaction and the mass differences of the u and d quarks, $\delta m = m_d - m_u$. The first investigations of CSV in the PDFs were based on the effect of δm within the MIT bag model [4], [5] and [6]. These calculations showed CSV violating effects as large as 5% at large-x, while they were at the percent level in the momentum fractions:

$$\delta U = \int_0^1 dx\, x\, \delta u(x)\ ;\ \delta D = \int_0^1 dx\, x\, \delta d(x)\ , \qquad (1)$$

where the CSV PDFs are $\delta u = u^p - d^n$ and $\delta d = d^p - u^n$. The major effect, which could be understood in terms of the dominant role played by di-quark correlations [7], arose from the mass difference between the dd and uu spectator pairs to the struck u-quark in a neutron and d-quark in a proton. It was found that δu and δd had a similar magnitude and opposite sign.

Citation: X.G. Wang, A.W. Thomas, R.D. Young, Electromagnetic contribution to charge symmetry violation in parton distributions, Physics Letters B, Volume 753, 10 February 2016, Pages 595-599, ISSN 0370-2693, http://dx.doi.org/10.1016/j.physletb.2015.12.062.

In the context of the NuTeV experiment [8], where these published effects were sufficiently large to reduce the anomaly to 2σ or less [9] and [10] considerable work was carried out to establish the extent to which these results were model independent. Recently, lattice QCD studies of these moments [11] and [12] (although necessarily the charge conjugation positive combination, rather than the valence combination calculated in the bag model) confirmed the sign and magnitude of the pioneering calculations.

The importance of QED radiation on DIS processes was recognised more than 40 years ago in the context of photon radiation from quarks in charged current neutrino interactions [13]. In the context of DGLAP evolution of PDFs, Spiesberger [14]summarised the potentially large effects associated with mass singularities involving $\ln(Q_0^2/m_q^2)$, where m_q is a light quark current mass. To avoid such problems he proposed to redefine the PDFs at the starting scale, Q_0^2, to include the effects of these singularities. The residual effects of photon radiation are then relatively small and at any given scale,Q^2, could be shown to be equivalent to a shift of scale of the PDFs by a charge-dependent factor (analogous to "dynamical rescaling").

In the modern era, Martin et al. [MRST] [15] and Glück et al. [16] sought to improve on the work of Spiesberger by including a photon distribution at the starting scale. In both cases this meant that CS was violated at that scale and both the initial photon distribution and the CSV PDFs were estimated in terms of the large logarithms associated with the mass singularities. Again in the context of the NuTeV experiment, it is important that the sign of the CSV associated with QED radiation was the same as that arising from δm [16], even though they enter with opposite signs in the neutron–proton mass difference. It is therefore vital [17] to have a consistent treatment of both effects and this is the aim of the present work.

The appearance of current masses in the QED logarithm is at odds with the modern understanding of non-perturbative QCD. At low scales the phenomenon of spontaneous chiral symmetry breaking [18] means that what naturally appears is a constituent quark mass, rather than the current quark mass. For example, a naive evaluation of the electromagnetic self-energy of a quark including the non-perturbative quark propagator, naturally yields a result proportional to $e_q^2\alpha M(0)$, where $M(0)\sim 0.4$ GeV[19] and e_q is the charge of the quark in units of the positron charge. Since the quark–photon splitting function is derived by cutting this self-energy diagram, one is rather led to a correction to the PDFs at the starting scale of order $\ln(Q_0^2/M(0)^2)$, which is necessarily much smaller than proposed in Refs. [15] and [16].

Very recently Martin and Ryskin [20] re-examined the issue of the initial photon distribution. They observed that at the low scale Q_0^2 the major part of the input photon distribution "comes from the coherent emission of the photon from the 'elastic' proton". In the present work we use this insight to make a new and more consistent calculation of the electromagnetic contributions to CSV. Since the scale at which typical, valence dominated quark models like NJL are matched to QCD is somewhat lower that used in Refs. [15] and [16], we modify the initial quark distribution and generate the initial photon distribution for the coherent radiation from the elastic proton. The evolution from that scale to a typical scale at which one might compare with experiment, say 5 GeV², then includes incoherent radiation of both gluons and photons from the quarks. In comparison with simply adding the effect of the quark mass difference to the original estimates of QED radiation by MRST and Glück et al., the extent of CSV on the u-quarks increases a little, while that for the d -quarks decreases in magnitude. Overall, at a scale of 5 GeV², the total CSV effect on the difference between the d and u-quark distributions, which is the combination relevant to the NuTeV experiment, is some 20% larger than the value based on quark mass differences alone.

QUARK DISTRIBUTION FUNCTIONS

The dynamical parton distributions, generated radiatively from valence-like inputs at low scales, are determined from global fit by Glück, Reya and Vogt (GRV) [21], taking into account small-x data on deep inelastic and other hard scattering processes. The leading order (LO) input distributions of proton at $Q_0^2=\mu_{LO}^2=0.26$ GeV² are then given by

$$\begin{aligned}
&\times(1-1.8\sqrt{x}+9.5x)\\
xd_v(x,\mu_{LO}^2)&=0.614(1-x)^{0.9}xu_v(x,\mu_{LO}^2)\\
x\Delta(x,\mu_{LO}^2)&=0.23x^{0.48}(1-x)^{11.3}\\
&\times(1-12.0\sqrt{x}+50.9x)\\
x(\bar{u}+\bar{d})(x,\mu_{LO}^2)&=1.52x^{0.15}(1-x)^{9.1}\\
&\times(1-3.6\sqrt{x}+7.8x)\\
xg(x,\mu_{LO}^2)&=17.47x^{1.6}(1-x)^{3.8}\\
xs(x,\mu_{LO}^2)&=x\bar{s}(x,\mu_{LO}^2)=0\,,
\end{aligned} \tag{2}$$

where $\Delta\equiv\bar{d}-\bar{u}$. The corresponding next-to-leading order (NLO) input at $Q_0^2=\mu_{NLO}^2=0.40$ GeV² is

$$\begin{aligned}
xu_v(x,\mu_{NLO}^2)&=0.632x^{0.43}(1-x)^{3.09}(1+18.2x)\\
xd_v(x,\mu_{NLO}^2)&=0.624(1-x)^{1.0}xu_v(x,\mu_{NLO}^2)\\
x\Delta(x,\mu_{NLO}^2)&=0.20x^{0.43}(1-x)^{12.4}\\
&\times(1-13.3\sqrt{x}+60.0x)\\
x(\bar{u}+\bar{d})(x,\mu_{NLO}^2)&=1.24x^{0.20}(1-x)^{8.5}\\
&\times(1-2.3\sqrt{x}+5.7x)\\
xg(x,\mu_{NLO}^2)&=20.80x^{1.6}(1-x)^{4.1}\\
xs(x,\mu_{NLO}^2)&=x\bar{s}(x,\mu_{NLO}^2)=0.
\end{aligned} \tag{3}$$

PHOTON DISTRIBUTION FUNCTIONS

The additional contribution to the valence quark charge asymmetries arises from radiative QED effects. The so-called DGLAP evolution equations are modified by introducing the photon PDF, $\gamma^N(x,Q^2)$. Following Martin and Ryskin, as shown in Ref.[20], the major part of the input photon PDF, $\gamma^p(x,Q_0^2)$, comes from the coherent emission of the photon from the elastic proton. Below the model scale, Q_0^2, we assume that the contribution from incoherent emission of photons from quarks within the nucleon is negligible. Above the model scale we utilise the APFEL program [22] to perform combined LO/

NLO QCD and LO QED evolution in the variable–flavor–number scheme (VFNS). That is, in that region both QCD and QED radiation is treated as incoherent radiation from the quarks.

The coherent emission from the proton is given by [20]

$$\gamma^p_{coh}(x, Q_0^2) = \frac{\alpha}{2\pi}\frac{1+(1-x)^2}{x}\int_0^{|t|<Q_0^2} dq_t^2 \frac{q_t^2}{(q_t^2+x^2m_p^2)^2}F_1^2(t)\,, \tag{4}$$

where q_t is the transverse momentum of the emitted photon and

$$t = -\frac{q_t^2+x^2m_p^2}{1-x}\,. \tag{5}$$

F_1 is the Dirac form factor of proton. Letting Q^2=–t, $F_1(Q^2)$ is given by

$$F_1(Q^2) = \frac{4M_p^2G_E^p(Q^2)+Q^2G_M^p(Q^2)}{Q^2+4M_p^2}\,, \tag{6}$$

with a dipole parametrisation for the electric and magnetic form factors,

$$G_E^p(Q^2) = \frac{G_M^p(Q^2)}{\mu_p} = \frac{1}{(1+\frac{Q^2}{\Lambda^2})^2}\,, \tag{7}$$

where μ_p=2.793 and $\Lambda^2 = 0.71$ GeV^2.

For the neutron, we neglect the small F_1 form factor of the neutron and set

$$\gamma^n_{coh}(x, Q_0^2) = 0\,. \tag{8}$$

The proton momentum fraction carried by the photon is

$$p_\gamma(Q_0^2) = \int_0^1 x\gamma^p_{coh}(x, Q_0^2)dx\,. \tag{9}$$

Correspondingly, the initial distribution functions of the valence quarks in proton should be modified as

$$\begin{aligned} u_v^p(x, Q_0^2) &= \left[u_v^p(x, Q_0^2)\right]_{GRV} - \beta^u f(x, Q_0^2)\,,\\ d_v^p(x, Q_0^2) &= \left[d_v^p(x, Q_0^2)\right]_{GRV} - \beta^d f(x, Q_0^2)\,, \end{aligned} \tag{10}$$

where f (x, Q^2) at Q^2 = 1 GeV^2 is taken from Ref. [23],

$$f(x, Q^2 = 1\,\text{GeV}^2) = x^{-0.5}(x-0.0909)(1-x)^4 \tag{11}$$

The quantity f(x) is chosen since its x dependence has roughly the same form as the MRST initial valence quark parton distribution functions in both the limit x→0 andx→1 at $Q^2 = 1$ GeV^2. The first moment of f(x) is fixed to be zero, in agreement with the valence quark normalisation. $f(x, Q_0^2)$ at $Q_0^2 = 0.26$ GeV^2 and $Q_0^2 = 0.40$ GeV^2 are obtained by LO and NLO QCD evolution from $Q^2 = 1$ GeV^2, respectively.

The coefficients β^u and β^d are determined by assuming that the momentum loss of the valence u and d quarks in the proton are $\frac{2}{3}p_\gamma$ and $\frac{1}{3}p_\gamma$, respectively,

$$\beta^u \int_0^1 dx\,xf(x, Q_0^2) = 2\beta^d \int_0^1 dx\,xf(x, Q_0^2) = \frac{2}{3}p_\gamma(Q_0^2)\,. \tag{12}$$

The momentum fraction carried by the photon and the corresponding β parameters are shown in Table 1.

Table 1. The momentum fraction carried by the photon and the corresponding β parameters.

Q_0^2 (GeV^2)	$p_\gamma(Q_0^2)$	β^u	β^d
0.26	0.00105	0.0584	0.0292
0.40	0.00113	0.0614	0.0307

RESULTS

In each case, we evolve to the final scales $Q^2 = 4\ \text{GeV}^2$, $10\ \text{GeV}^2$ and $20\ \text{GeV}^2$. The pure QED contributions to the isospin-violating majority and minority valence distributions, $x\delta u_v$ and $x\delta d_v$, respectively, are shown in Fig. 1. The corresponding contributions to the second moments of δu_v and δd_v are given in Table 2.

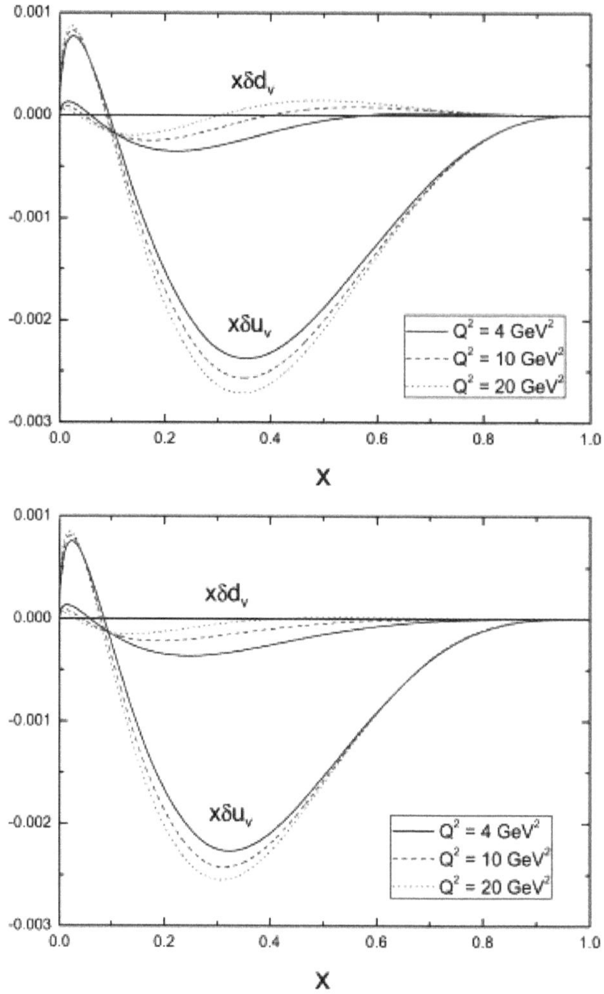

Figure 1. (Colour online.) The pure QED contributions to isospin-violating majority xδuv and minority xδdv valence parton distributions at $Q^2 = 4$ GeV², 10 GeV² and 20 GeV². (Upper) LO QCD evolution from LO initial distributions at $Q_0^2 = 0.26\ \text{GeV}^2$; (Lower) NLO QCD evolution from NLO initial distributions at $Q_0^2 = 0.40\ \text{GeV}^2$. In both cases, the QED evolution is of LO in α.

Table 2. The pure QED contributions to the second moments of δuv and δdv at $Q^2 = 4$ GeV², 10 GeV² and 20 GeV². The photon and valence quark distribution functions at the initial scale are given by Eqs. (4), (8) and (10), respectively.

Q_0^2 (GeV²)	$Q^2 = 4$ GeV²		$Q^2 = 10$ GeV²		$Q^2 = 20$ GeV²	
	δU_v	δD_v	δU_v	δD_v	δU_v	δD_v
0.26	−0.00099	−0.00009	−0.00107	−0.00003	−0.00113	0.00001
0.40	−0.00089	−0.00013	−0.00095	−0.00007	−0.00099	−0.00003

At the initial scale, both δU_v and δD_v are negative in the valence region. At higher scales, δU_v will decrease and therefore always remain negative. The sign of δD_v depends on the final scale Q^2. There exists a critical scale, Q_c^2, above which δD_v will be positive. At $Q^2 = 10\ \text{GeV}^2$, which is appropriate for the NuTeV experiment, the QED contributions to the second moments for both the u and d quarks are significantly smaller than the predictions of Refs. [15] and [16].

At $Q^2 = 4\ \text{GeV}^2$, the QCD contributions to the second moments are derived in Ref. [24] by extrapolating the first lattice simulations [11] to the physical point,

$$\delta U_V = -0.0023(7)\ ,\quad \delta D_V = 0.0017(4)\ , \quad (13)$$

where the number in brackets indicates the error in the last significant figure. These values are in good agreement with previous phenomenological estimates of CSV, both those calculated within the MIT bag model [5] and [10] and those found in the MRST analysis [23]. Using the simplest phenomenological parametrisation

$$\delta q_v(x, Q^2) = \kappa_q f(x, Q^2) \tag{14}$$

where $f(x,Q^2)$ is obtained by NLO QCD evolution from $Q_0^2 = 1\ \mathrm{GeV}^2$, and the normalisation factors are determined by taking the constraint, Eq. (13),

$$\kappa_u = -0.26(8)\ , \quad \kappa_d = 0.19(4)\ . \tag{15}$$

Combining with the pure QED contributions, given by the lower plot of Fig. 1, we show the total isospin violating distributions at $Q^2 = 4\ \mathrm{GeV}^2$ and $10\ \mathrm{GeV}^2$ in Fig. 2. Here we see that the influence of QED has only a small effect on the down-quark (or minority quark) CSV. For the up (or majority) quark, we see that the QED effects enhance the overall magnitude of the quark-mass induced CSV.

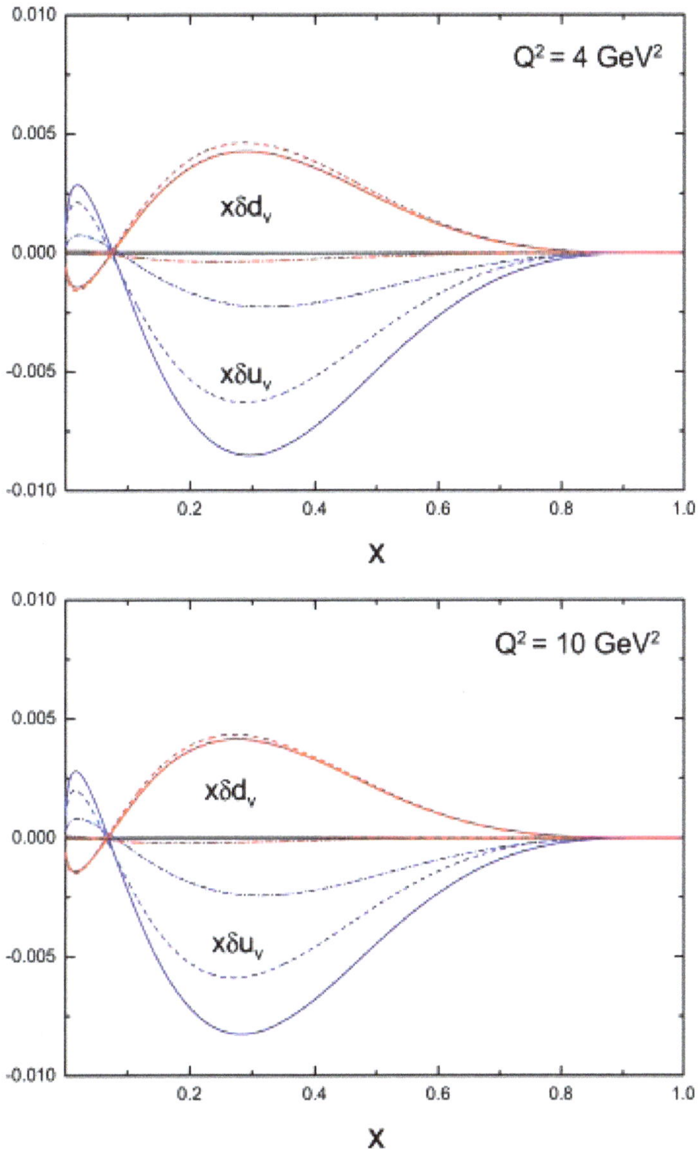

Figure 2. (Colour online.) The isospin-violating majority xδuv and minority xδdv valence parton distributions at $Q^2 = 4\ \mathrm{GeV}^2$ and $10\ \mathrm{GeV}^2$. Dash-dotted, dashed and solid curves represent pure QED, pure QCD and the total contributions, respectively.

Incorporating this new determination of partonic CSV, including the effect of QED, we revisit the total effect of CSV in the extraction of $\sin^2\theta_W$ by the NuTeV Collaboration. The total correction, Δs_W^2, to s_W^2 arising from charge symmetry violation may be calculated using the very convenient functional, $F[s_W^2, \delta q; x]$, provided by the collaboration[25]

$$\Delta s_W^2 = \int_0^1 F[s_W^2, \delta q; x] x \delta q(x, Q^2) dx \tag{16}$$

at the central value $Q^2 = 10\ \mathrm{GeV}^2$. The individual contributions to Δs_W^2 are listed in Table 3. Therefore, the total correction arising from valence quark charge symmetry violation becomes

$$\Delta s_W^2|_{\mathrm{total}} = \Delta s_W^2|_{\mathrm{QED}} + \Delta s_W^2|_{\mathrm{QCD}} = -0.0022 \pm 0.0004\ , \tag{17}$$

where the error is calculated by combining the errors on the individual contributions in quadrature. For the electromagnetic contribution the errors are taken as the differences between matching at μ_{LO}^2 and μ_{NLO}^2, while for the quark mass contribution the errors arise from Eq. (15). This value is consistent with that reported by Bentz et al.[17], namely $\Delta s_W^2 = -0.0026 \pm 0.0011$, but now with a significantly improved estimate of the uncertainty associated with the QED contribution.

Table 3. The QED and QCD corrections to Δs_W^2 arising from valence quark charge symmetry violation.

Δs_W^2	δu_v	δd_v	Total
QED	−0.00043(6)	0.00004(2)	−0.00039(6)
QCD	−0.00102(31)	−0.00074(17)	−0.00176(35)

CONCLUSION

In summary, we have revisited the electromagnetic contribution to charge symmetry violation (CSV) in the parton distribution functions of the nucleon, which contributes the largest uncertainty associated with the CSV correction to the NuTeV anomaly. At very lowQ^2 we treat the radiation of photons from the nucleon coherently, following the suggestion of Martin and Ryskin [20], while above the scale typically associated with valence dominated quark models the photon emission is associated with the individual quarks, through QED evolution [22]. The resulting electromagnetic contribution to the combination of second moments relevant to the NuTeV anomaly, namely $\delta D_v - \delta U_v$ is of order 0.0010 (at 10 GeV^2). When used with the NuTeV functional this yields a correction of less than 10% of the NuTeV anomaly. Adding the latest lattice QCD estimate of this moment [24], which is consistent with the older model dependent calculations [4],[5] and [6], results in a total CSV correction to Δs_W^2 of −0.0022±0.0004, which constitutes a reduction in the anomaly of more than 40%. If one were to add the isovector EMC from Ref. [26], the total correction would be −0.0041±0.0007 and comparing with the quoted anomaly, −0.0050±0.0016, the discrepancy with the Standard Model appears to be resolved. The major remaining issue is the potential asymmetry between the s and $\bar{s}$ distributions [27], [28], [29] and [30] and resolving that issue is now of even greater importance.

ACKNOWLEDGEMENTS

We wish to acknowledge important discussions with Ian Cloët and Lei Chang. This work has been supported by the University of Adelaide and by the Australian Research Council through the ARC Centre of Excellence for Particle Physics at the Tera-Scale and through grants FL0992247 and DP151103101 (A.W.T.) and DP110101265 andFT120100821 (R.D.Y.).

REFERENCES

1. G.A. Miller, B.M.K. Nefkens, I. Slaus, Phys. Rep. 194 (1990) 1.
2. J.T. Londergan, J.C. Peng, A.W. Thomas, Rev. Mod. Phys. 82 (2010) 2009.
3. J.T. Londergan, A.W. Thomas, Prog. Part. Nucl. Phys. 41 (1998) 49.
4. E. Sather, Phys. Lett. B 274 (1992) 433.
5. E.N. Rodionov, A.W. Thomas, J.T. Londergan, Mod. Phys. Lett. A 9 (1994) 1799.
6. J.T. Londergan, G.T. Carvey, G.Q. Liu, E.N. Rodionov, A.W. Thomas, Phys. Lett. B 340 (1994) 115.
7. F.E. Close, A.W. Thomas, Phys. Lett. B 212 (1988) 227.
8. G.P. Zeller, et al., NuTeV Collaboration, Phys. Rev. Lett. 88 (2002) 091802; G.P. Zeller, et al., Phys. Rev. Lett. 90 (2003) 239902 (Erratum).
9. J.T. Londergan, A.W. Thomas, Phys. Rev. D 67 (2003) 111901.
10. J.T. Londergan, A.W. Thomas, Phys. Lett. B 558 (2003) 132.
11. R. Horsley, Y. Nakamura, D. Pleiter, P.E.L. Rakow, G. Schierholz, H. Stuben, A.W. Thomas, F. Winter, et al., Phys. Rev. D 83 (2011) 051501.
12. I.C. Cloët, R. Horsley, J.T. Londergan, Y. Nakamura, D. Pleiter, P.E.L. Rakow, G. Schierholz, H. Stuben, et al., Phys. Lett. B 714 (2012) 97.
13. J.E. Kiskis, Phys. Rev. D 8 (1973) 2129.
14. H. Spiesberger, Phys. Rev. D 52 (1995) 4936.
15. A.D. Martin, R.G. Roberts, W.J. Stirling, R.S. Thorne, Eur. Phys. J. C 39 (2005) 155.
16. M. Glück, P. Jimenez-Delgado, E. Reya, Phys. Rev. Lett. 95 (2005) 022002.
17. W. Bentz, I.C. Cloët, J.T. Londergan, A.W. Thomas, Phys. Lett. B 693 (2010) 462.
18. A.W. Thomas, W. Weise, The Structure of the Nucleon, Wiley–VCH, Berlin, Germany, 2001, 389 pp.
19. A. Bashir, L. Chang, I.C. Cloët, B. El-Bennich, Y.X. Liu, C.D. Roberts, P.C. Tandy, Commun. Theor. Phys. 58 (2012) 79.
20. A.D. Martin, M.G. Ryskin, Eur. Phys. J. C 74 (2014) 3040.
21. M. Glück, E. Reya, A. Vogt, Eur. Phys. J. C 5 (1998) 461.
22. V. Bertone, S. Carrazza, J. Rojo, Comput. Phys. Commun. 185 (2014) 1647.
23. A.D. Martin, R.G. Roberts, W.J. Stirling, R.S. Thorne, Eur. Phys. J. C 35 (2003) 325.
24. P.E. Shanahan, A.W. Thomas, R.D. Young, Phys. Rev. D 87 (2013) 094515.
25. G.P. Zeller, et al., Phys. Rev. D 65 (2002) 111103(R); G.P. Zeller, et al., Phys. Rev. D 67 (2003) 119902 (Erratum).
26. I.C. Cloet, W. Bentz, A.W. Thomas, Phys. Rev. Lett. 102 (2009) 252301.
27. A.I. Signal, A.W. Thomas, Phys. Lett. B 191 (1987) 205.
28. W. Melnitchouk, M. Malheiro, Phys. Lett. B 451 (1999) 224.
29. M. Traini, Phys. Lett. B 707 (2012) 523.
30. D. Mason, et al., NuTeV Collaboration, Phys. Rev. Lett. 99 (2007) 192001.

INDEX

J

L

M

N

O

P

R

S

T

U

V